Lecture Notes in Physics

Springer-Verlag Berlin Heidelberg GmbH

Physics and Astronomy ONLINE LIBRARY
http://www.springer.de/phys/

The Editorial Policy for Edited Volumes

The series *Lecture Notes in Physics* (LNP), founded in 1969, reports new developments in physics research and teaching - quickly, informally but with a high degree of quality. Manuscripts to be considered for publication are topical volumes consisting of a limited number of contributions, carefully edited and closely related to each other. Each contribution should contain at least partly original and previously unpublished material, be written in a clear, pedagogical style and aimed at a broader readership, especially graduate students and nonspecialist researchers wishing to familiarize themselves with the topic concerned. For this reason, traditional proceedings cannot be considered for this series though volumes to appear in this series are often based on material presented at conferences, workshops and schools.

Acceptance

A project can only be accepted tentatively for publication, by both the editorial board and the publisher, following thorough examination of the material submitted. The book proposal sent to the publisher should consist at least of a preliminary table of contents outlining the structure of the book together with abstracts of all contributions to be included. Final acceptance is issued by the series editor in charge, in consultation with the publisher, only after receiving the complete manuscript. Final acceptance, possibly requiring minor corrections, usually follows the tentative acceptance unless the final manuscript differs significantly from expectations (project outline). In particular, the series editors are entitled to reject individual contributions if they do not meet the high quality standards of this series. The final manuscript must be ready to print, and should include both an informative introduction and a sufficiently detailed subject index.

Contractual Aspects

Publication in LNP is free of charge. There is no formal contract, no royalties are paid, and no bulk orders are required, although special discounts are offered in this case. The volume editors receive jointly 30 free copies for their personal use and are entitled, as are the contributing authors, to purchase Springer books at a reduced rate. The publisher secures the copyright for each volume. As a rule, no reprints of individual contributions can be supplied.

Manuscript Submission

The manuscript in its final and approved version must be submitted in ready to print form. The corresponding electronic source files are also required for the production process, in particular the online version. Technical assistance in compiling the final manuscript can be provided by the publisher‘s production editor(s), especially with regard to the publisher's own LaTeX macro package which has been specially designed for this series.

LNP Homepage (springerlink.com)

On the LNP homepage you will find:
–The LNP online archive. It contains the full texts (PDF) of all volumes published since 2000. Abstracts, table of contents and prefaces are accessible free of charge to everyone. Information about the availability of printed volumes can be obtained.
–The subscription information. The online archive is free of charge to all subscribers of the printed volumes.
–The editorial contacts, with respect to both scientific and technical matters.
–The author's / editor's instructions.

T. Pöschel N. Brilliantov (Eds.)

Granular Gas Dynamics

Springer

Editors

Thorsten Pöschel and Nikolai Brilliantov
Humboldt-Universität
Charité, Institut für Biochemie
Monbijoustr. 2
10117 Berlin, Germany

Cataloging-in-Publication Data applied for

A catalog record for this book is available from the Library of Congress.

Bibliographic information published by Die Deutsche Bibliothek
Die Deutsche Bibliothek lists this publication in the Deutsche Nationalbibliografie; detailed bibliographic data is available in the Internet at http://dnb.ddb.de

ISSN 0075-8450
ISBN 978-3-642-05757-1 ISBN 978-3-540-39843-1 (eBook)
DOI 10.1007/978-3-540-39843-1

springer.de

Originally published by Springer-Verlag Berlin Heidelberg New York in 2003
Softcover reprint of the hardcover 1st edition 2003

Typesetting: Camera-ready by the authors/editor
Cover design: *design & production*, Heidelberg

Printed on acid-free paper
54/3141/du - 5 4 3 2 1 0

Preface

While the dynamics and statics of dense Granular Materials still lack a general theory, in the past few years significant progress has been achieved in understanding the properties of Granular Gases, i.e., of dilute Granular Systems where the mean free path of inelastically colliding particles is much larger than the average particle size. Although conceptually rather simple, the theoretical description of Granular Gas dynamics is far from being trivial. It is the dissipative nature of particle collisions which causes a variety of exciting phenomena, such as structure formation, a non-Maxwellian velocity distribution function, spontaneous shock wave formation and many others, but on the other hand represents the main reasons for severe complications in the Granular Gas theory.

A necessary condition for the application of the Boltzmann equation is the validity of molecular chaos. Dissipative particle collisions in Granular Gases are source of correlations of different type which lead to significant deviations from molecular chaos, much more pronounced than for molecular gases. To investigate this difficult problem simplified systems can be considered, which may be treated analytically but nevertheless exhibit the most salient properties of Granular Gases. One important example of such systems is the inelastic Maxwell Model which has been introduced into the Granular Gas research only recently.

Much progress has been achieved in the last few years in understanding the mathematical properties of the Boltzmann equation for dissipative gases and new methods for its analysis have been developed. The hydrodynamic description of Granular Gases is very useful when the length and time scales significantly exceed the molecular scales. There are still many open questions concerning the justification of granular hydrodynamics and the range of its validity. Recent studies shed some light on these severe problems.

Many fascinating phenomena occur in driven Granular Gases. Some of them resemble processes in molecular fluids, however, the nature of these processes is different.

The present book collects contributions in which these and further problems are discussed. Many of the authors contributed important parts to the present body of knowledge in the field of dissipative gases. Insofar the contributions represent the present state of the art.

The volume is structured into three parts. Part I addresses general problems of the Kinetic Theory of Granular Gases. Part II deals with granular hydrodynamics and, finally, Part III is devoted to driven Granular Gases and structure formation.

The book is aimed not only at experts, but also at students and newcomers to the field. Therefore, the authors have been asked to present their material in a self-contained and pedagogical way. All articles went through a careful refereeing process.

The editors are grateful to all authors for their contributions and to the referees whose comments have improved the quality of the volume significantly. The Centre Européen de Calcul Atomique et Moléculaire, Lyon (CECAM) is appreciated for financial support to perform the workshop "Granular Gases" from September 2 to 5, 2002, where the idea to edit this book was born.

Berlin, June 2003

Thorsten Pöschel
Nikolai Brilliantov

Contents

Part I. Kinetic Theory

Asymptotic Solutions of the Nonlinear Boltzmann Equation for Dissipative Systems
Matthieu H. Ernst, Ricardo Brito 3
1 Introduction 3
2 Inelastic BGK Model 6
3 Basics of Inelastic Scattering Models 9
4 Analysis of Inelastic Scattering Models 18
5 Inelastic Maxwell Models 25
6 Conclusions and Perspectives 31
References 34

The Homogeneous Cooling State Revisited
Isaac Goldhirsch, S. Henri Noskowicz, Oded Bar-Lev 37
1 Introduction 37
2 Setting Up the Problem 39
3 Heuristic Analysis 40
4 The Near-Maxwellian Range of Speeds 45
5 Reduction of the Boltzmann Equation for the HCS 47
6 Concluding Remarks 53
References 61

The Inelastic Maxwell Model
Eli Ben-Naim, Paul L. Krapivsky 65
1 Introduction 65
2 Uniform Gases: One Dimension 67
3 Uniform Gases: Arbitrary Dimension 73
4 Impurities 80
5 Mixtures 86
6 Lattice Gases 88
7 Conclusions 90
References 92

Velocity Fluctuations in Cooling Granular Gases
Andrea Baldassarri, Umberto Marini Bettolo Marconi, Andrea Puglisi . 95
1 Introduction . . . 95
2 Instabilities of the Homogeneous Cooling State . . . 96
3 A Starting Point: The Homogeneous Inelastic Maxwell Model . . . 100
4 The One-Dimensional Gas . . . 103
5 The Two-Dimensional Gas . . . 107
6 Conclusions . . . 115
References . . . 116

Self-Similar Asymptotics for the Boltzmann Equation with Inelastic Interactions
Alexander V. Bobylev, Carlo Cercignani . . . 119
1 Introduction . . . 119
2 Isotropic Equation and Preliminary Result . . . 121
3 A Recent Extension: Preliminaries . . . 123
4 Complete Proof of the Conjecture . . . 127
References . . . 129

Kinetic Integrals in the Kinetic Theory of Dissipative Gases
Thorsten Pöschel, Nikolai V. Brilliantov . . . 131
1 Introduction . . . 131
2 A Simple Example . . . 133
3 Granular Gases of Viscoelastic Particles . . . 136
4 Evaluation of Kinetic Integrals . . . 137
5 Computational Formula Manipulation to Evaluate Kinetic Integrals . . . 145
6 Kinetic Integrals in the Kinetic Theory of Granular Gases . . . 152
7 Conclusion . . . 161
References . . . 161

Kinetics of Fragmenting Freely Evolving Granular Gases
Ignacio Pagonabarraga, Emmanuel Trizac . . . 163
1 Introduction . . . 163
2 Model . . . 164
3 Kinetics . . . 166
4 Numerical Simulations . . . 172
5 Discussion . . . 181
References . . . 183

Part II. Granular Hydrodynamics

Shock Waves in Granular Gases
Alexander Goldshtein, Alexander Alexeev, Michael Shapiro 187
1 Introduction 187
2 One-Dimensional Waves 190
3 Two-Dimensional Waves 202
4 Waves in Vibrated Granular Media 207
5 Summary 220
References 222

Linearized Boltzmann Equation and Hydrodynamics for Granular Gases
J. Javier Brey, James W. Dufty, María J. Ruiz-Montero 227
1 Introduction 227
2 Nonlinear Boltzmann Equation and the Homogeneous Cooling State 229
3 Linearized Boltzmann Equation 233
4 Eigenvalue Problem 236
5 Navier-Stokes and Green-Kubo Expressions 241
6 Discussion 246
References 248

Development of a Density Inversion in Driven Granular Gases
Yaron Bromberg, Eli Livne, Baruch Meerson 251
1 Introduction 251
2 The Model Problem and Hydrodynamic Equations 253
3 Steady State Profiles and Density Inversion 257
4 Formation of a Density Inversion: Low-Mach-Number Flow 259
5 Formation of a Density Inversion: Early Times 262
6 Discussion 264
References 265

Kinetic Theory for Inertia Flows of Dilute Turbulent Gas-Solids Mixtures
Cliff K.K. Lun, Stuart B. Savage 267
1 Introduction 267
2 Gas-Solids Interactions 269
3 Granular Transport Theory 271
4 Moment Method 275
5 Mixture Theory 278
6 Turbulence Modulation 280
7 Application 282

8 Comparisons between Predictions and Experiments 283
9 Conclusion 285
References 286

Part III. Driven Gases and Structure Formation

Driven Granular Gases
Stefan Luding, Raffaele Cafiero, Hans J. Herrmann 293
1 Introduction 293
2 The Model 295
3 Mean Field Evolution Equations 298
4 Numerical Simulations 301
5 Simulations with Rotation 305
6 Analytical Study of the Velocity Distribution 307
7 Summary and Conclusions 311
References 314

Van der Waals-Like Transition in Fluidized Granular Matter
Rodrigo Soto, Médéric Argentina, Marcel G. Clerc 317
References 332

Birth and Sudden Death of a Granular Cluster
Ko van der Weele, Devaraj van der Meer, Detlef Lohse 335
1 Introduction 335
2 Flux Model 337
3 More Than Two Compartments: Hysteresis 338
4 Coarsening and Sudden Death 339
5 The Limit for $N \to \infty$ Compartments: Anti-diffusion 343
6 Extensions and Applications 345
References 346

Vibrated Granular Media as Experimentally Realizable Granular Gases
Sean McNamara, Eric Falcon 347
1 Introduction 347
2 Description of the Simulations 348
3 Comparison of Simulation and Experiment 350
4 Effects of Clustering ($n > 2$ or 3) 356
5 Conclusions 364
References 365

List of Contributors

Alexander Alexeev
Laboratory of Aerosol Research
Faculty of Mechanical Engineering
Technion
Haifa 32000
Israel
alik@ttd.tu-darmstadt.de

Médéric Argentina
Departamento de Física
Facultad de Ciencias Físicas
y Matemáticas
Universidad de Chile
Casilla 487-3
Santiago
Chile
mederic@dfi.uchile.cl

Andrea Baldassarri
INFM Udr Roma 1
University of Rome "La Sapienza"
piazzale Aldo Moro 2
00185 Roma
Italy
andrea.baldassarri
@roma1.infn.it

Oded Bar-Lev
School of Mathematics
Faculty of Exact Sciences
Tel-Aviv University
Ramat-Aviv
Tel-Aviv 69978
Israel
fnbarlev@post.tau.ac.il

Eli Ben-Naim
Theoretical Division and
Center for Nonlinear Studies
Los Alamos National Laboratory
Los Alamos NM 87545
USA
ebn@lanl.gov

Alexander V. Bobylev
Department of Mathematics
Karlstad University
Karlstad
Sweden
Alexander.Bobylev@kau.se

J. Javier Brey
Física Teórica
Facultad de Física
Universidad de Sevilla
41080 Sevilla
Spain
brey@us.es

Nikolai V. Brilliantov
Institut für Biochemie, Charité
Humboldt-Universität zu Berlin
Monbijoustraße 2
10117 Berlin
Germany
nbrillia@polly.phys.msu.ru

Ricardo Brito
GISC
and Depto de Física Aplicada I
Universidad Complutense
28040 Madrid
Spain
brito@seneca.fis.ucm.es

Yaron Bromberg
School of Physics and Astronomy
Tel Aviv University
Tel Aviv 69978
Israel
yaronbr@pob.huji.ac.il

Raffaele Cafiero
P.M.M.H.
École Supérieure de Physique
et de Chimie Industrielles (ESPCI)
10, rue Vauquelin
75251 Paris CEDEX 05
France
cafiero@pmmh.espci.fr

Carlo Cercignani
Dipartimento di Matematica
Politecnico di Milano
Milano
Italy
carcer@mate.polimi.it

Marcel G. Clerc
Departamento de Física
Facultad de Ciencias Físicas y Matemáticas
Universidad de Chile
Casilla 487-3, Santiago
Chile
marcel@dfi.uchile.cl

James W. Dufty
Department of Physics
University of Florida
Gainesville
Florida 32611
USA
dufty@phys.efl.edu

Matthieu H. Ernst
Instituut voor Theoretische Fysica
Universiteit Utrecht
Postbus 80.195
3508 TD Utrecht
The Netherlands
ernst@phys.uu.nl

Eric Falcon
Laboratoire de Physique
École Normale Supérieure de Lyon
46, allée d'Italie
69007 Lyon
France
Eric.Falcon@ens-lyon.fr

Isaac Goldhirsch
Department of Fluid Mechanics
and Heat Transfer
Faculty of Engineering
Tel-Aviv University
Ramat-Aviv
Tel-Aviv 69978
Israel
isaac@eng.tau.ac.il

Alexander Goldshtein
MediGuide Ltd.
P.O.B. 15003
Haifa 31053
Israel
alex_g@mediguide.co.il

Hans J. Herrmann
Institute for Computer Applications 1
University of Stuttgart
Pfaffenwaldring 27
70569 Stuttgart
Germany
hans@ica1.uni-stuttgart.de

Paul L. Krapivsky
Center for Polymer Studies
and Department of Physics
Boston University
Boston, MA 02215
USA
paulk@sid3.bu.edu

Eli Livne
Racah Institute of Physics
Hebrew University of Jerusalem
Jerusalem 91904
Israel
eli@frodo.fiz.huji.ac.il

Detlef Lohse
Department of Applied Physics
and J.M. Burgers Centre for
Fluid Dynamics
University of Twente
P.O. Box 217
7500 AE Enschede
The Netherlands
D.Lohse@tn.utwente.nl

Stefan Luding
Particle Technology
DelftChemTech
TU Delft
Julianalaan 136
2628 BL Delft
The Netherlands
and
Institute for Computer Applications 1
University of Stuttgart
Pfaffenwaldring 27
70569 Stuttgart
Germany
s.luding@tnw.tudelft.nl

Cliff K. K. Lun
Center for Multiphase
Flows Research
55 Cairnside Crescent
Toronto
Canada
Cliff.Lun@utoronto.ca

Umberto Marini Bettolo Marconi
Dipartimento di Matematica e Fisica
and INFM Udr Camerino
University of Camerino
Via Madonna delle Carceri
62032 Camerino
Italy
umberto.marini.bettolo
@roma1.infn.it

Sean McNamara
Centre Européen de Calcul
Atomique et Moléculaire
Ecole Normale Supérieure
46, Allée d'Italie
69007 Lyon
France

Devaraj van der Meer
Department of Applied Physics
and J.M. Burgers Centre
for Fluid Dynamics
University of Twente
P.O. Box 217
7500 AE Enschede
The Netherlands
R.M.vanderMeer@tn.utwente.nl

Baruch Meerson
Racah Institute of Physics
Hebrew University of Jerusalem
Jerusalem 91904
Israel
meerson@cc.huji.ac.il
and
Institute for Computer Applications 1
Universität Stuttgart
70569 Stuttgart
Germany
sean@ica1.uni-stuttgart.de

S. Henri Noskowicz
Department of Fluid Mechanics
and Heat Transfer
Faculty of Engineering
Tel-Aviv University
Ramat-Aviv
Tel-Aviv 69978
Israel
henri@eng.tau.ac.il

Ignacio Pagonabarraga
Departament de Física Fonamental
Universitat de Barcelona
Carrer Martí i Franqués, 1
08028-Barcelona
Spain
ignacio@hermes.ffn.ub.es

Thorsten Pöschel
Institut für Biochemie, Charité
Humboldt-Universität zu Berlin
Monbijoustraße 2
10117 Berlin
Germany
thorsten.poeschel@charite.de

Andrea Puglisi
INFM Center for Statistical
Mechanics and Complexity
University "La Sapienza"
piazzale Aldo Moro 2
00185 Roma
Italy
Andrea.Puglisi@roma1.infn.it

María J. Ruiz-Montero
Física Teórica
Facultad de Física
Universidad de Sevilla
41080 Sevilla
Spain
majose@us.es

Stuart B. Savage
McGill University
Department of Civil Engineering
and Applied Mechanics
Montreal
Canada
stuart.savage@mcgill.ca

Michael Shapiro
Laboratory of Transport Processes
in Porous Materials
Faculty of Mechanical Engineering
Technion
Haifa 32000
Israel
mersm01@tx.technion.ac.il

Rodrigo Soto
Departamento de Física
Facultad de Ciencias Físicas y
Matemáticas
Universidad de Chile
Casilla 487-3
Santiago
Chile
rsoto@dfi.uchile.cl

Emmanuel Trizac
Laboratoire de Physique Théorique
(UMR 8627 du CNRS)
Bâtiment 210
Université de Paris-Sud
91405 Orsay Cedex
France
Emmanuel.Trizac@th.u-psud.fr

Ko van der Weele
Department of Applied Physics
and J.M. Burgers Centre for
Fluid Dynamics
University of Twente
P.O. Box 217
7500 AE Enschede
The Netherlands
J.P.vanderWeele@tn.utwente.nl

Part I

Kinetic Theory

Asymptotic Solutions of the Nonlinear Boltzmann Equation for Dissipative Systems

Matthieu H. Ernst[1] and Ricardo Brito[2]

[1] Instituut voor Theoretische Fysica, Universiteit Utrecht, Postbus 80.195, 3508 TD Utrecht, The Netherlands. ernst@phys.uu.nl
[2] GISC and Depto de Física Aplicada I, Universidad Complutense, 28040 Madrid, Spain. brito@seneca.fis.ucm.es

Abstract. Analytic solutions $F(v,t)$ of the nonlinear Boltzmann equation in d dimensions are studied for a new class of dissipative models, called inelastic soft spheres, interacting through pseudo-power law repulsions, and characterized by a collision frequency $\propto g^{\nu}$, where g is the relative speed of the colliding particles. These models embed inelastic hard spheres ($\nu = 1$) and inelastic Maxwell models ($\nu = 0$). The systems are either freely cooling without energy input or driven by thermostats, e.g. white noise, and approach stable nonequilibrium steady states, or marginally stable homogeneous cooling states, where the data, $v_0^d(t)F(v,t)$ plotted versus $c = v/v_0(t)$, collapse on a scaling or similarity solution $f(c)$, where $v_0(t)$ is the r.m.s. velocity. The dissipative interactions generate overpopulated high energy tails, described generically by stretched Gaussians, $f(c) \sim \exp[-\beta c^b]$ with $0 < b < 2$, where $b = \nu$ with $\nu > 0$ in free cooling , and $b = 1 + \frac{1}{2}\nu$ with $\nu \geq 0$ when driven by white noise. Power law tails, $f(c) \sim 1/c^{a+d}$, are only found in marginal cases, where the exponent a is the root of a transcendental equation. The stability threshold depend on the type of thermostat, and is for the case of free cooling located at $\nu = 0$.
Moreover we analyze an inelastic BGK-type kinetic equation with an energy dependent collision frequency coupled to a thermostat, that captures all qualitative properties of the velocity distribution function in Maxwell models, as predicted by the full nonlinear Boltzmann equation, but fails for harder interactions with $\nu > 0$.
This review was completed early February 2003 and covers most of the relevant literature available at that time.

1 Introduction

Classic kinetic theory [1–5] deals with elastic particles with energy conserving dynamics. The system is described by the single particle distribution function, whose time evolution is governed by the nonlinear Boltzmann equation. The asymptotic states of such systems follow the universal laws of thermodynamics, and the distribution function is the Maxwell Boltzmann distribution. This scenario does not apply to dissipative systems, where energy is lost in inelastic interactions.

In elastic systems the approach to asymptotic states is characterized by a *kinetic* stage of rapid relaxation in velocity space to a locally homogeneous equilibrium state, followed by a *hydrodynamic* stage of slow approach to a globally homogeneous equilibrium state. The time scale in the kinetic stage is the mean free time t_{mf} between collisions. In the kinetic theory of inelastic systems [6–12] the type of decay depends on the energy supply to the dissipative system. Without energy supply there is first a kinetic stage of rapid relaxation on the time scale t_{mf} to a locally homogeneous adiabatic state, the *homogeneous cooling* state, described by *scaling* or *similarity* solutions with a slowly changing parameter, at least for weakly inelastic systems. With energy supply the evolution is more similar to the elastic case with, however, equilibrium states replaced by non-equilibrium steady states. The velocity distributions in these adiabatic or steady states are very different from a Maxwell Boltzmann distribution. The subsequent stage of evolution involves transport phenomena and complex hydrodynamic phenomena of clustering and pattern formation [12, 13].

The interest in granular matter in general has strongly stimulated new developments in the kinetic theory of granular fluids and gases, which show surprising new physics. A granular fluid is a collection of small or large macroscopic particles, with short range repulsive hard core interactions, in which energy is lost in inelastic collisions, and the system cools when not driven. When rapidly driven, gravity can be neglected. The dynamics is based on binary collisions and ballistic motion between collisions, which *conserve total momentum*. So these systems can be considered to be a granular *fluid or gas*. The prototypical model for these so-called rapid granular flows is a fluid or gas of perfectly smooth mono-disperse inelastic hard spheres, and its non-equilibrium behavior can be described by the nonlinear Boltzmann equation [6–12]. The inelastic collisions are modeled by a *coefficient of restitution* α $(0 < \alpha < 1)$, where $\left(1 - \alpha^2\right)$ measures the degree of inelasticity.

This review focuses on the first stage of evolution, and studies the velocity distribution $F(v, t)$ in spatially homogeneous states of inelastic systems. For that reason most of the citations, given in this article, only refer to kinetic theory studies of $F(v, t)$, based on the nonlinear Boltzmann equation or on simplified versions of it, like the BGK- kinetic equation. We refer to the literature for further interesting developments in the theory of inelastic fluids of more complicated systems with internal degrees of freedom, unlike particles, lack of equipartition (see reviews in this volume by Luding et al. [14], and by Baldassarri et al. [15]), about Green-Kubo formulas in granular fluids (see the review by Brey et al. [16], and Goldhirsch and van Noije [17]), and the article by Cáceres [18] about anomalous velocity distributions in non-equilibrium steady states of general systems out of equilibrium, based on the Langevin and Fokker-Planck equations. The revival [19–31] in kinetic theory of inelastic systems has been strongly stimulated by the increasing sophistication of experimental techniques [32, 33], which make direct measurements of velocity distributions feasible in non-equilibrium steady states. In this review we also include inelastic generalizations [34] of the classical repulsive power law interactions [2–4], which embed both the inelastic hard spheres ($\nu = 1$), as well as the recently much studied [35–45] inelastic Maxwell models ($\nu = 0$) in a single class of models, which are characterized by a collision frequency $\propto g^\nu$, where g is the relative speed of the colliding particles.

In fact, the kinetic theory for such models is of interest in its own right, as the majority of inter-particle interactions in macroscopic systems involve some effects of inelasticity. Our goal is to expose the generic and universal features of the velocity distributions in dissipative fluids, and to compare them with conservative fluids to highlight the differences.

A classical problem in kinetic theory is the possibility of overpopulated high energy tails in velocity distributions [46, 47], as many physical and chemical processes only occur above a certain energy threshold. Consequently, this old problem has received a new stimulus through a large amount of recent theoretical and experimental studies on tail distributions in many particle systems with inelastic interactions. From the point of view of kinetic theory the intriguing question is, what is the generic feature causing overpopulated tails, possibly even power law tails, in systems of *inelastic* particles, how does the overpopulation depend on the scattering cross sections, and on the different forms of energy input. The generic feature is the mechanism for overpopulation, and not the specific shape of the tails.

Finally, from the point of view of nonequilibrium steady states, the structure of velocity distributions in elastic and dissipative systems, including the high energy tail, is a subject of continuing research, as the universality of the Gibbs' state of thermal equilibrium is lacking outside thermal equilibrium, and a possible classification of generic structures would be of great interest in many fields of non-equilibrium statistical mechanics.

The *plan of the paper* is as follows: in Sect. 2 we discuss a simple inelastic BGK- or single-relaxation-time model [48] to illustrate the phenomenon of power law tails. The exponent in the algebraic tail depends qualitatively in the same manner on the degree of inelasticity as in 2- and 3-dimensional Maxwell models. In Sect. 3 the nonlinear Boltzmann equation is constructed for inelastic generalizations of the classical repulsive power law potentials, to which we refer as Inelastic Soft Spheres or ISS-models. This is done for freely cooling as well as for systems driven by thermostats or heat sources. Section 4 gives a systematic analysis for the energy balance equation, it derives the nonlinear integral equation for the scaling or similarity solution, denoted by $f(c)$, and presents an asymptotic analysis of the high energy tails in the form of stretched Gaussians, $f(c) \sim \exp\left[-\beta c^b\right]$ where $b < 2$. The method used can only be applied to ISS-models with $\nu > 0$, where the exponent $b = b(\nu)$ is found as a simple function of ν. The case of freely cooling Maxwell models ($\nu = 0$) forms an exceptional borderline case, discussed in Sect. 5. Here algebraic tails, $f(c) \sim 1/c^{d+a}$, are found, where the exponent a is determined by a transcendental equation. It yields $a = a(\alpha)$ as a function of the degree of inelasticity. We end with some perspectives and conclusions.

2 Inelastic BGK Model

2.1 Kinetic Equations

The goal of this section is to present in the nutshell of a simple Bhatnagar-Gross-Krook (BGK) model [48] a preview of many of the qualitative features of velocity relaxation in homogeneous systems.

A crude scenario for the relaxation without energy input suggests that the system will cool down due to inelastic collisions, and the velocity distribution $F(v,t)$ will approach a Dirac delta function $\delta(\vec{v})$ as $t \to \infty$, while the width or r.m.s. velocity $v_0(t)$ of this distribution, defined as $\langle v^2 \rangle = \frac{1}{2} d v_0^2$, is shrinking. With a constant supply of energy, the system can reach a non-equilibrium steady state (NESS).

To model this evolution we use a simple BGK-type kinetic equation, introduced in 1996 by Brey et al. [49],

$$\partial_t F(v,t) - D\nabla_{\vec{v}}^2 F(v,t) = -\omega_\nu(t)\left[F(v,t) - F_0(v,t)\right]. \tag{1}$$

We have added a heating term, $-D\nabla_{\vec{v}}^2 F(v,t)$ to the usual BGK equation which represents the heating by a white noise of strength D [1]. Here the mean collision frequency, $\omega_\nu(t) = 1/t_{mf}$, is a function of the r.m.s. velocity $v_0(t)$, chosen as $\omega_\nu = v_0^\nu$, in preparation of Sect. 3.1. The kinetic equation describes the relaxation of $F(v,t)$ with a time-dependent rate $\omega_\nu(t)$ towards a Maxwellian with a width proportional to $\alpha v_0(t)$, defined by

$$F_0(v,t) = \left(\sqrt{\pi}\alpha v_0\right)^{-d} \exp\left[-\left(v/\alpha v_0\right)^2\right] \equiv \left(\alpha v_0\right)^{-d} \phi(c/\alpha), \tag{2}$$

where $c = v/v_0$. The constant α $(0 < \alpha < 1)$ is related to the inelasticity, $\gamma = \frac{1}{2}\left(1-\alpha^2\right)$, of the model, and the totally inelastic limit $(\alpha \to 0)$ is ill-defined in this model, as the mean energy is divergent for $\alpha = 0$. Note that the 'loss term', $-\omega_\nu F$, and the 'gain term', $+\omega_\nu F_0$, contribute respectively $-\omega_\nu v_0^2$ and $+\omega_\nu \alpha^2 v_0^2$ to the net energy loss rate in the energy balance equation.

2.2 Free Cooling ($D = 0$)

The cooling law of the mean square velocity $\langle v^2 \rangle = \frac{1}{2} d v_0^2$, or the granular temperature $T \propto v_0^2$, is obtained by applying $\int d\vec{v} v^2 (\ldots)$ to (1) with $\omega_\nu = v_0^\nu$. The result is $\dot{v}_0 = -\gamma v_0^{\nu+1}$, yielding

$$v_0(t) = v_0(0) / \left[1 + \nu\gamma t v_0^\nu(0)\right]^{1/\nu}. \tag{3}$$

The result is a homogeneous cooling law, $T \sim t^{-2/\nu}$, which agrees with Haff's law [9] for $\nu = 1$, corresponding to inelastic hard spheres. Note that for negative ν the homogeneous cooling law takes the form,

$$v_0(t) = v_0(0)\left[1 - |\nu|\gamma t / v_0^{|\nu|}(0)\right]^{1/|\nu|}, \tag{4}$$

[1] For a systematic discussion of driven systems, see Sect. 3.3

i.e. at $t = t_s \equiv v_0^{|\nu|}/|\nu|\gamma$ the r.m.s. velocity and the mean kinetic energy are vanishing, which is *unphysical*, and so are the BGK-models with $\nu < 0$.

As indicated in the introduction, the long time behavior of $F(v,t)$ in free cooling is determined by a scaling or similarity solution of the form $F(v,t) = v_0^{-d}(t)\, f(v/v_0(t))$. We insert this ansatz in (1), eliminate $\dot{v}_0 = -\gamma v_0^{\nu+1}$, and obtain the scaling equation,

$$c\frac{d}{dc}f + (d+a)\,f = \frac{a}{\alpha^d}\phi\left(\frac{c}{\alpha}\right), \tag{5}$$

where a is defined as,

$$a = 1/\gamma = 2/\left(1-\alpha^2\right). \tag{6}$$

We also note that the scaling equation is independent of ν. The exact solution of this equation is:

$$f(c) = \frac{A}{c^{d+a}} + \frac{a}{\alpha^d \pi^{d/2}}\left(\frac{1}{c^{d+a}}\right)\int_0^c du\, u^{d+a-1}\exp\left[-u^2/\alpha^2\right]. \tag{7}$$

The integration constant A is fixed by the normalizations,

$$\int d\vec{c}\{1, c^2\} f(c) = \{1, \tfrac{1}{2}d\}, \tag{8}$$

where $d\vec{c} = \Omega_d c^{d-1} dc$ with $\Omega_d = 2\pi^{d/2}/\Gamma(d/2)$ being the surface area of a $d-$dimensional hyper-sphere. The normalization integral converges only near $c \simeq 0$ if $A = 0$. Then the solution (7) with $A = 0$ is identical to the scaling form, obtained in [49] for $\nu = 1$.

For velocities far above thermal, i.e. $c \gg 1$, the solution has a power law tail,

$$f(c) \sim \frac{a\alpha^a}{\pi^{d/2}}\left(\frac{1}{c^{d+a}}\right)\int_0^\infty du\, u^{d+a-1}e^{-u^2} = \frac{a\alpha^a \Gamma\left(\frac{d+a}{2}\right)}{2\pi^{d/2}}\left(\frac{1}{c^{d+a}}\right), \tag{9}$$

with exponent $a = 1/\gamma = 2/\left(1-\alpha^2\right)$, such that $\langle c^2\rangle$ is bounded for $\alpha > 0$. The exact solution (7), including its high energy tail, is independent of the exponent ν, that determines the energy dependence of the mean collision frequency $\omega_\nu = v_0^\nu$ in the BGK model.

As we shall see in Sect. 5, a similar heavily overpopulated tail, $f(c) \sim 1/c^{d+a}$ with $d > 1$, will also be found in freely cooling Maxwell model with $\omega_0 = 1$ $(\nu = 0)$. There the exponent $a(\alpha)$ takes in the elastic limit $(\alpha \to 1)$ the form $a \simeq 1/\gamma_0 = 4d/\left(1-\alpha^2\right)$. However, in the *general class* of ISS-models with collision frequency $\omega_\nu \sim v_0^\nu$ with $\nu > 0$ the tails are *not* given by *power laws*, but by *stretched Gaussians*, $f(c) \sim \exp\left[-\beta c^b\right]$ with $0 < b = b(\nu) < 2$. These models will be introduced and discussed in Sect. 3.2.

2.3 NESS ($D \neq 0$)

Next we extend the result of [49] to an inelastic BGK-model with $\nu \geq 0$, driven by white noise. By applying $\int d\vec{v}\, v^2$ to (1) we obtain the temperature balance equation at *stationarity*,

$$\frac{dv_0^2}{dt} = 4D - 2\gamma v_0^{\nu+2} = 0, \tag{10}$$

where the collisional dissipation, $2\gamma v_0^{\nu+2}$, is compensated by energy input from the external white noise. Also note that for $\nu < -2$ the fixed point solution $v_0(\infty)$ in (10) still exists, but it is unstable [34]. If $v_0(0) < v_0(\infty)$, then $v_0(t)$ vanishes as $t \to \infty$, and if $v_0(0) > v_0(\infty)$ then $v_0(t)$ diverges [2].

To obtain the solution $F(v, \infty)$ of (1) in the NESS we rescale $F(v, \infty) = v_0^{-d}(\infty) f(v/v_0(\infty))$ to the standard width $\langle c^2 \rangle = \frac{1}{2}d$, substitute the rescaled form in the kinetic equation, and eliminate D, using the stationarity condition (10) as well as (6). This yields the rescaled equation in *universal* form,

$$\frac{1}{c^{d-1}}\frac{d}{dc}c^{d-1}\frac{d}{dc}f(c) - 2af(c) = -\frac{2a}{\alpha^d}\phi\left(\frac{c}{\alpha}\right), \tag{11}$$

where the normalizations (8) are imposed. The O.D.E. shows that $f(c)$ is independent of the noise strength, D, and does not contain any dependence on the initial data. This equation can be solved exactly, and more details will be published in [34]. However, for the purpose of this section, we only want to extract from the differential equation (11) the asymptotic form of $f(c)$. In that case we may neglect in (11) the inhomogeneity $\phi(c/\alpha) \sim \exp\left[-(c/\alpha)^2\right]$ for $c \gg \alpha$, and find the asymptotic solution for the BGK-model driven by white noise, i.e.

$$\begin{aligned} &f(c) \sim \exp\left[-\beta c^b\right] \\ &b = 1;\ \beta = \sqrt{2a} = 2/\sqrt{1-\alpha^2}. \end{aligned} \tag{12}$$

The constant β is independent of the parameter ν. The exponentially decaying high energy tail is also a 'stretched' Gaussian, which is overpopulated when compared to a Maxwellian, but the overpopulation is much less heavy than is the freely cooling case (9) with an algebraic tail.

As we shall see in Sect. 4.3, a similar exponential high energy tail will be found in the white noise driven Maxwell model ($\nu = 0$; $b = 1$; $\beta = \sqrt{8/(1-\alpha^2)}$), but not in the general class of ISS-models, where the stretching exponent b takes a value in the interval $0 < b = \overline{b}(\nu) \leq 2$.

---

[2] Thanks are due to E. Trizac and A. Barrat for pointing out to us that the energy balance equations in this article have stability thresholds that are ν−dependent.

3 Basics of Inelastic Scattering Models

3.1 Boltzmann Equation as a Stochastic Process

The nonlinear Boltzmann equation for dissipative interactions in the *homogeneous cooling state* can be put in a broader perspective, that covers both elastic and inelastic collisions, as well as interactions where the scattering of particles is described either by conservative (deterministic) forces, or by stochastic ones. To do so it is convenient to interpret the Boltzmann equation as a stochastic process, similar to the presentations in the classical articles of Waldmann [1], and Uhlenbeck and Ford [5], or in Ulam's stochastic model [50] showing the basics of the approach of a one-dimensional gas of elastic particles towards a Maxwellian distribution.

Consider a spatially homogeneous fluid of elastic or inelastic particles in $d-$dimensions, specified by their velocities $(\vec{v}, \vec{w}, \cdots)$, and interacting through binary collisions, $(\vec{v}, \vec{w}) \to (\vec{v}\,', \vec{w}')$, that are described by transition probabilities. To describe fluids out of equilibrium the total momentum $\vec{G} = \frac{1}{2}(\vec{v} + \vec{w})$ needs to be conserved in a binary collision. The outgoing or post-collision velocities $(\vec{v}\,', \vec{w}')$ can be parametrized in terms of the incoming velocities $(\vec{v}, \vec{w})$, and an impact (unit) vector $\vec{n}$, that is chosen on the surface of a unit sphere with a certain probability, proportional to the collision frequency $\text{A}(\vec{g}, \vec{n})$, that in general depends on the relative speed $g = |\vec{v} - \vec{w}|$ of the colliding particles, and the angle between $\hat{\vec{g}}$ and $\vec{n}$, where $\hat{\vec{a}}$ denotes a unit vector.

In this article we consider the simplest case of modeling the inelastic collisions through a velocity-independent coefficient of restitution α $(0 < \alpha < 1)$, where the component $\mathrm{g}_{\|} \equiv \vec{g} \cdot \vec{n}$ is not only reflected, as in elastic collisions, but also reduced in size by a factor α, i.e.

$$\mathrm{g}_{\|}{}' = -\alpha \mathrm{g}_{\|}. \tag{13}$$

The components $\vec{g}_{\perp} = \vec{g} - \mathrm{g}_{\|}\vec{n}$, orthogonal to $\vec{n}$, remain unchanged. More explicitly, with the help of momentum conservation we find for the post-collision velocities resulting from the direct collisions $(\vec{v}, \vec{w}) \to (\vec{v}\,', \vec{w}') = (\vec{v}^*, \vec{w}^*)$,

$$\begin{aligned} \vec{v}^* &= \vec{v} - \tfrac{1}{2}(1+\alpha)\,\vec{g}\cdot\vec{n}\vec{n} \\ \vec{w}^* &= \vec{w} + \tfrac{1}{2}(1+\alpha)\,\vec{g}\cdot\vec{n}\vec{n}. \end{aligned} \tag{14}$$

The corresponding energy loss in such a collision is,

$$\Delta E = \tfrac{1}{2}\left(v^{*2} + w^{*2} - v^2 - w^2\right) = -\tfrac{1}{4}\left(1-\alpha^2\right)\left(\vec{g}\cdot\vec{n}\right)^2, \tag{15}$$

where $\left(1-\alpha^2\right)$ measures the degree of inelasticity. The value $\alpha = 1$ describes elastic collisions. In the special case of hard spheres the impact vector $\vec{n}$ is the unit vector along the line of centers of the colliding spheres at contact, as illustrated in Fig. 1.

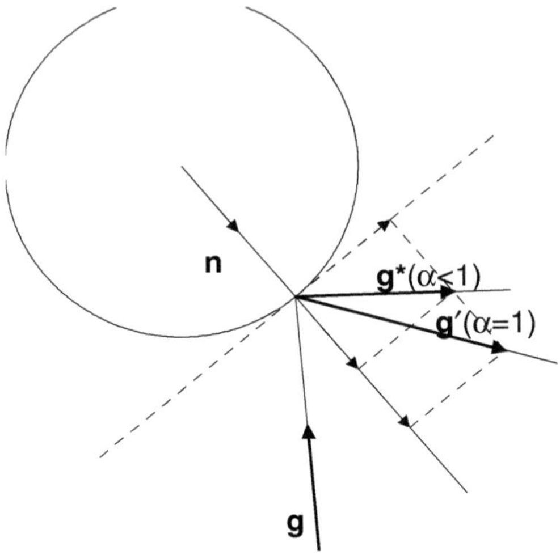

Fig. 1. Elastic ($\alpha = 1$) and inelastic ($\alpha < 1$) scattering of hard spheres, where α is the coefficient of restitution, and the parallel component $g_{\|}$ is reflected as in (13). In elastic collisions the scattering angle $\chi = \cos^{-1}\left(\hat{\vec{g}} \cdot \hat{\vec{g}}'\right) = \pi - 2\phi$ with $\phi = \cos^{-1}\left(\hat{\vec{g}} \cdot \vec{n}\right)$ and in inelastic ones $\chi = \cos^{-1}\left(\hat{\vec{g}} \cdot \hat{\vec{g}}^*\right) = \pi - \phi - \phi^*$ with $\phi^* = \cos^{-1}\left(\hat{\vec{g}}^* \cdot \vec{n}\right)$

The system above can be described by an isotropic velocity distribution, $F(v,t) = F(|\vec{v}|, t)$, as long as one only considers isotropic initial distributions. Its time evolution is given by the Boltzmann equation,

$$\partial_t F(v,t) = I(v|F) \equiv \int d\vec{w} d\vec{v}' d\vec{w}' \int d\vec{n} \left[W(\vec{v}, \vec{w}|\vec{v}', \vec{w}'; \vec{n}) \right.$$
$$\left. \times F(v',t) F(w',t) - W(\vec{v}', \vec{w}'|\vec{v}, \vec{w}; \vec{n}) F(v,t) F(w,t) \right]. \quad (16)$$

Here $W(\vec{v}', \vec{w}'|\vec{v}, \vec{w}; \vec{n})$ is the transition probability per unit time that the incoming pair state $(\vec{v}, \vec{w})$ at impact vector $\vec{n}$ is scattered into the outgoing pair state $(\vec{v}', \vec{w}')$. The loss term is the sum over all parameters of the direct collisions $(\vec{v}, \vec{w}) \to (\vec{v}', \vec{w}')$ at fixed $\vec{v}$, including a sum over impact vectors $\vec{n}$. Similarly the gain term is the sum over all parameters of the restituting collisions, $(\vec{v}', \vec{w}') \to (\vec{v}, \vec{w})$ at fixed $\vec{v}$. The transition probability for the scattering event $(\vec{v}, \vec{w}) \to (\vec{v}', \vec{w}')$, obeying the inelastic reflection law (13) and momentum conservation, is in general proportional to the collision frequency $\mathrm{A}(\vec{g}, \vec{n})$, and contains delta functions, selecting the allowed collisions,

$$W(\vec{v}', \vec{w}'|\vec{v}, \vec{w}; \vec{n}) = \mathrm{A}(\vec{g}, \vec{n})\, \delta^{(d)}\left(\vec{G}' - \vec{G}\right) \delta^{(d-1)}\left(\vec{g}_\perp' - \vec{g}_\perp\right) \delta\left(g_{\|}' + \alpha g_{\|}\right). \quad (17)$$

For conservative interactions the total energy is conserved, as well as total momentum and total number of particles. Moreover (17) shows that the transition probability is symmetric for $\alpha = 1$,

$$W\left(\vec{v}\,',\vec{w}'|\vec{v},\vec{w};\vec{n}\right) = W\left(\vec{v},\vec{w}|\vec{v}\,',\vec{w}';\vec{n}\right). \tag{18}$$

This means that the transition probabilities for elastic binary collisions satisfy the condition of *detailed balance.* Energy conservation and the detailed balance relation in combination with the $H-$theorem guarantee that the Maxwellian velocity distribution is approached at large times.

Once the detailed balance relation is obeyed, it is trivial to prove the $H-$theorem. Defining the $H-$function or entropy $S(t) = -H(t)$ as,

$$H(t) = \int d\vec{v} F(\vec{v},t) \ln F(\vec{v},t), \tag{19}$$

we obtain its decay rate by applying $\int d\vec{v} \ln F(\vec{v},t)$ to (16), then symmetrizing over $\vec{v} \leftrightarrow \vec{w}$ and $\vec{v}\,' \leftrightarrow \vec{w}'$, and subsequently over $(\vec{v},\vec{w}) \leftrightarrow (\vec{v}\,',\vec{w}')$. The result is,

$$\partial_t H(t) = \tfrac{1}{4} \int d\vec{v} d\vec{w} d\vec{v}\,' d\vec{w}' \int d\vec{n} W \left[Y - X\right] \ln\left(X/Y\right) \leq 0, \tag{20}$$

where $X = F(v) F(w)$ and $Y = F(v') F(w')$ and the inequality follows from $[Y - X] \ln(X/Y) \leq 0$. The equality sign holds if and only if $X = Y$. This implies that H decreases monotonically and becomes stationary only when F approaches the Maxwellian.

The reason for reviewing these 'obvious' properties in elastic systems is that several of them, such as energy conservation, the detailed balance relation, the $H-$theorem, and the approach to a stationary Maxwellian velocity distribution no longer hold in dissipative systems ($\alpha < 1$), as we shall see.

3.2 Boltzmann Equation in Standard Form

The standard form of the collision term $I(v|F)$ in the Boltzmann equation for *elastic* interactions ($\alpha = 1$) contains the differential scattering cross-section $\sigma(g,\chi)$. It can be calculated from the pair potential $V(r)$ [4], and depends on the relative speed $g = |\vec{g}|$ and the scattering angle $\chi = \pi - 2\varphi$, where χ and φ are defined in Fig. 1. The collision frequency is then given by $\mathrm{A}(\vec{g},\vec{n}) = g\sigma(g,\chi)$. For repulsive power law potentials $V(r) \sim r^{-s}$ the collision frequency is $\mathrm{A}(\vec{g},\vec{n}) = g^{\nu} \mathrm{A}\left(\hat{\vec{g}} \cdot \vec{n}\right)$ with $\nu = 1 - 2(d-1)/s$ in $d-$dimensions [51]. By definition Maxwell molecules have a collision frequency, which is independent of g, corresponding to $\nu = 0$ or $s = 2(d-1)$. For hard spheres, $s \to \infty$ or $\nu = 1$ with $\mathrm{A}(\vec{g},\vec{n}) = |\vec{g}\cdot\vec{n}|\,\sigma^{d-1} = \left|\mathrm{g}_{\|}\right|\sigma^{d-1}$, and σ is the hard sphere diameter. Using these relations the collision term in (16) can be reduced to the classical Boltzmann equation for elastic energy-conserving collisions. For *inelastic* hard spheres the collision rate is again given by $\mathrm{A}(\vec{g},\vec{n}) = \left|\mathrm{g}_{\|}\right|\sigma^{d-1}$.

In general, a given positive function $\text{A}(\vec{g},\vec{n})$ defines a stochastic scattering model, and in particular the choice, $\text{A}(\vec{g},\vec{n}) = \text{A}\left(\hat{\vec{g}}\cdot\vec{n}\right)$, defines inelastic Maxwell models [29–31]. As a simple realization of a dissipative scattering model we consider the ISS-models with a collision frequency $\text{A}(\vec{g},\vec{n}) = \text{A}_0\,|\vec{g}\cdot\vec{n}|^{\nu}$ with $\nu \geq 0$. This class includes inelastic hard spheres ($\nu = 1$) and inelastic Maxwell models ($\nu = 0$). In the remainder of this article we restrict ourselves to this class of models.

To reduce the Boltzmann equation to its standard form we use the transition probabilities for dissipative interactions (17) with $\text{A}(\vec{g},\vec{n}) = \text{A}_0\left|\text{g}_{\|}\right|^{\nu}$. Consider first the loss term in (16), insert the relation $d\vec{v}'d\vec{w}' = d\vec{g}_{\perp}'\,d\text{g}_{\|}'d\vec{G}'$, and carry out the integrations over the delta functions. The result is,

$$(\partial_t F(v))_{\text{loss}} = -\int d\vec{w}\int d\vec{n}\text{A}_0\left|\text{g}_{\|}\right|^{\nu} F(w)\,F(v)\,. \tag{21}$$

The gain term involves the transition probability for the restituting collisions $(\vec{v}',\vec{w}') \to (\vec{v},\vec{w})$, obtained from (17) by interchanging primed and unprimed velocities,

$$\begin{aligned}(\partial_t F(v))_{\text{gain}} &= \int d\vec{w}d\vec{g}_{\perp}'\,d\text{g}_{\|}'d\vec{G}'\int d\vec{n}\text{A}_0\left|\text{g}_{\|}'\right|^{\nu}\delta^{(d)}\left(\vec{G}-\vec{G}'\right)\\ &\quad\times\,\delta^{(d-1)}\left(\vec{g}_{\perp}-\vec{g}_{\perp}'\right)\frac{1}{\alpha}\delta\left(\text{g}_{\|}'+\frac{1}{\alpha}\text{g}_{\|}\right)F(w')\,F(v')\\ &= \int d\vec{w}\int d\vec{n}\,(1/\alpha)\,\text{A}_0\left|\text{g}_{\|}/\alpha\right|^{\nu} F(w^{**})\,F(v^{**})\,.\end{aligned} \tag{22}$$

In the second integral we have carried out the integrations over the primed velocities, and used the following relations for the restituting velocities,

$$\begin{aligned}\vec{v}' &= \vec{G}' + \tfrac{1}{2}\vec{g}_{\perp}' + \tfrac{1}{2}\text{g}_{\|}'\vec{n} = \vec{G} + \tfrac{1}{2}\vec{g}_{\perp} - \tfrac{1}{2\alpha}\text{g}_{\|}\vec{n}\\ &= \vec{v} - \tfrac{1}{2}\left(1+\tfrac{1}{\alpha}\right)\vec{g}\cdot\vec{n}\vec{n} \equiv \vec{v}^{**}\\ \vec{w}' &= \vec{w} + \tfrac{1}{2}\left(1+\tfrac{1}{\alpha}\right)\vec{g}\cdot\vec{n}\vec{n} \equiv \vec{w}^{**}.\end{aligned} \tag{23}$$

In the first equality $\vec{v}'$ has been expressed in center of mass and relative velocities. In the second equality we have used the inelastic collision law (13) and conservation of total momentum, and the very last equality defines the restituting velocities, $(\vec{v}^{**},\vec{w}^{**})$. They are the incoming velocities that result in the scattered velocities $(\vec{v},\vec{w})$, described by the inverse of the transformation (14).

The space-homogeneous Boltzmann equation in its standard form is then obtained by combining (21) and (22) with $\text{A}(\vec{g},\vec{n}) = |\vec{g}\cdot\vec{n}|^{\nu}$ to yield,

$$\begin{aligned}\partial_t F(v) &= I(v|F)\\ I(v|F) &= \int_{\vec{n}}\int d\vec{w}\,|\vec{g}\cdot\vec{n}|^{\nu}\left[\frac{1}{\alpha^{\nu+1}}F(v^{**})\,F(w^{**}) - F(v)\,F(w)\right],\end{aligned} \tag{24}$$

where $\int_{\vec{n}}(\cdots) = (1/\Omega_d)\int d\vec{n}\,(\cdots)$ is an average over a $d-$dimensional unit sphere, and we have absorbed constant factors in the time scale. Here $\nu = 1$ corresponds to inelastic hard spheres and $\nu = 0$ to inelastic Maxwell models. Velocities and time have been dimensionalized in terms of the width and the mean free time of the initial distribution. Moreover, the Boltzmann equation obeys conservation of particle number and total momentum, but the average kinetic energy or granular temperature, $T \propto \langle v^2 \rangle$, decreases in time on account of the dissipative collisions, i.e.

$$\int d\vec{v}\,(1,\,\vec{v},\,v^2)\,F(v,t) = \left(1,\,0,\,\tfrac{1}{2}dv_0^2(t)\right), \tag{25}$$

where $v_0(t)$ is the r.m.s. velocity. The inelastic scattering models with collision frequency $\mathrm{A} \sim g^\nu$ $(\nu > 0)$, are the inelastic analogs of the deterministic models with repulsive power law potentials, $V(r) \sim r^{-s}$ with $2(d-1) < s < \infty$, introduced at the start of this section. The inelastic case with $\nu = 2$ corresponds to an exactly solvable stochastic scattering model with energy conservation, known as Very Hard Particle model [51].

The most basic and most frequently used model for dissipative systems with short range hard core repulsion is the Enskog-Boltzmann equation for inelastic hard spheres in $d-$dimensions [6], which simplifies in the spatially homogeneous case to (24) with $\nu = 1$. Recently, inelastic Maxwell models have been studied extensively. Ben-Naim and Krapivsky [29] introduced the one-dimensional version of (24) with $(d = 1, \nu = 0)$, and Bobylev et al [30] have introduced a three-dimensional Maxwell model with $\mathrm{A}(\vec{g},\vec{n}) = \mathrm{A}_0 \left|\hat{\vec{g}}\cdot\vec{n}\right|$. The Maxwell models in (24) for general d were first considered in [37,39]. We frequently use the terms 'harder' (larger ν) and 'softer' interactions (smaller ν), which only refers to the energy dependence of the collision frequency, $\mathrm{A} \propto g^\nu$, at large relative speed g. We only use a relative criterion for 'harder' and 'softer', and not an absolute one. Moreover the faster the collision frequency increases at large impact velocities, the more rapidly the high energy tail of $F(v,t)$ relaxes relative to the bulk values of $F(v,t)$ with v in the thermal range.

3.3 Cooling and Driven Systems

Homogeneous Cooling and Scaling:
An inelastic fluid without energy input will cool down due to the collisional dissipation in (15). In experimental studies of granular fluids energy has to be supplied at a constant rate to keep the system in a non-equilibrium steady state, while in analytic and simulation studies freely cooling systems can be studied directly. Without energy input the velocity distribution $F(v,t)$ will approach a Dirac delta function $\delta(\vec{v})$ as $t \to \infty$, and all moments approach zero, including the width $v_0(t)$.

However, an interesting structure is revealed when velocities, $\vec{c} = \vec{v}/v_0(t)$, are measured in units of the instantaneous width $v_0(t)$, and the long time limit is taken while keeping $\vec{c}$ constant, the so-called *scaling limit.* Monte Carlo simulations [35] of the Boltzmann equation suggest that in this limit the rescaled velocity distribution of the homogeneous cooling state can be collapsed on a scaling form or similarity solution $f(c)$. These observations seem to indicate that the long time behavior of $F(v,t)$ in freely cooling systems approaches a simple, and to some extent *universal*, form $f(c)$, which is the same for different initial distributions. Such *scaling* or *similarity* solutions have the structure,

$$F(v,t) = (v_0(t))^{-d} f(v/v_0(t)), \tag{26}$$

where $\vec{c} = \vec{v}/v_0(t)$ is the scaling argument. Then $f(c)$ satisfies the normalizations,

$$\int d\vec{c} f(c) = 1; \quad \int d\vec{c} c^2 f(c) = \tfrac{1}{2} d, \tag{27}$$

on account of (25). Substitution of the scaling ansatz (26) in the Boltzmann equation (24) then leads to a *separation of variables*, i.e.

$$\begin{aligned} I(c|f) =& \gamma (df + c df/dc) = \gamma \vec{\nabla}_{\vec{c}} \cdot \vec{c} f \\ \dot{v}_0 =& -\gamma v_0^{\nu+1}, \end{aligned} \tag{28}$$

where γ is a separation constant.

The question is then: Can one determine physically acceptable solutions? Here a short history. The first kinetic model with dissipative interactions, that has been solved exactly for the case of free cooling, *and* exhibits a heavily overpopulated high energy tail, $f(c) \sim 1/c^{d+a}$, is the inelastic BGK model, discussed in Sect. 2. Asymptotic solutions ($c \gg 1$) of the scaling equation for inelastic hard spheres [21, 22] have predicted the existence of exponential *high energy tails*, $f(c) \sim \exp[-\beta c]$. We talk about *tails*, *over-populations* or stretched Gaussians in $f(c)$, when the ratio of $f(c)$ and a Gaussian is an increasing function of c at $c \gg 1$. The predictions about high energy tails were later confirmed in great detail by Monte Carlo simulations of the long time solutions of the nonlinear Boltzmann equation for inelastic hard spheres [25, 26], as well as in further analytic work [23].

Similarity solutions for freely cooling inelastic Maxwell models [29, 30] where first studied in terms of the scaling variable, $(1-\alpha^2)t$, relating large times and small inelasticities [30], but the solutions obtained turned out to be unphysical, i.e. *non-positive*. The first exact positive similarity solution was found by Baldassarri et al. [35] for the one-dimensional Maxwell model. It shows a surprisingly strong high energy tail of algebraic type, $\sim 1/c^4$. Using Monte Carlo simulations of the Boltzmann equation, these authors also showed that rather general initial distributions approach towards this scaling form for long times. Algebraic tails, $f(c) \sim 1/c^{d+a}$, for Maxwell models in higher dimensions have been obtained analytically in [37–41], and the actual approach in time towards such scaling forms has been studied analytically in [41, 42] for general classes of initial distributions.

Driving and Steady States:
Heating may be described by applying an external stochastic force to the particles in the system, or by connecting the system to a thermostat, which may be modeled by a frictional force. For example, the friction force $\gamma\vec{v}$ – here with a negative friction coefficient $(-\gamma)$ – is called a Gaussian thermostat. Complex fluids (e.g. granular) subject to such forces can be described by the microscopic equations of motion for the particles, $\dot{\vec{r}}_i = \vec{v}_i$, and $\dot{\vec{v}}_i = \vec{a}_i + \vec{\tilde{\xi}}_i$ $(i = 1, 2, \cdots)$, where $\vec{a}_i$ and $\vec{\tilde{\xi}}_i$ are respectively the systematic and random forces per unit mass. Here $\vec{a}_i$ contains frictional forces, which may depend on velocity, and in the present case $\vec{\tilde{\xi}}_i$ represents external noise (modeling energy input), which is taken to be Gaussian white noise with zero mean, and variance,

$$\overline{\tilde{\xi}_{i,\alpha}(t)\,\tilde{\xi}_{j,\beta}(t')} = 2D\delta_{ij}\delta_{\alpha\beta}\delta(t-t'), \tag{29}$$

where α, β denote Cartesian components, and D is the noise strength. The Boltzmann equation for system driven in this manner takes the form,

$$\begin{aligned} &\partial_t F(\vec{v}) + \mathcal{F} F(\vec{v}) = I(v|F) \\ &\mathcal{F} = \vec{\nabla}_{\vec{v}} \cdot \vec{a} - D\nabla^2_{\vec{v}}. \end{aligned} \tag{30}$$

A detailed derivation on how to include frictional and stochastic forces in kinetic equations can be found, for instance, in [19, 22, 52].

When energy is supplied at a constant rate, driven systems can reach a NESS. Again there is the question about universality of these asymptotic states as $t \to \infty$. Does the NESS depend on the inelasticity, on the type of thermostat, and on the initial distribution? The basic idea to show universality is always essentially the same. Rescale the velocity distribution $F(v, \infty)$ by measuring velocities in terms of their typical size, i.e. the width $v_0(\infty)$,

$$F(v,\infty) = (v_0(\infty))^{-d} f(v/v_0(\infty)), \tag{31}$$

and analyze the scaling form $f(c)$. The scaling equation for the NESS function $f(c)$ will be analyzed in Sects. 4.2 and 4.3. Here we make some comments about what is known. The scaling solutions for the NESS show again overpopulated tails in the form of stretched Gaussians, albeit with a larger stretching exponent b than in free cooling. The scaling solution for inelastic hard spheres driven by white noise was first analyzed in [22], predicting $f(c) \sim \exp\left[-\beta c^b\right]$ with $b = 3/2$. The experiments of Rouyer and Menon [32, 33], and of Aranson and Olafsen [33] seem to confirm the stretched Gaussian behavior with $b = 3/2$. Maxwell models, driven by white noise, were first studied in [29, 31], but did not give any predictions about overpopulated high energy tails. The high energy tails for Maxwell models, driven by white noise, were predicted in [40, 42] to have exponential tails, $f(c) \sim \exp[-\beta c]$. A number of more detailed analytical and numerical studies about this driven Maxwell model in one dimension [43–45, 53] have appeared as well. Moreover, as will be shown in Sect. 4.2, the integral equation for the scaling form $f(c)$ in free cooling is identical to the integral equation for the NESS distribution for a special thermostat, provided $F(v,\infty)$ is also rescaled to the same width as in (27). How to determine the scaling form $f(c)$ in free cooling and NESS will be described in subsequent sections.

3.4 Qualitative Analysis

In order to illustrate the rich behavior of the inelastic systems, we start with the Boltzmann equation for the one-dimensional inelastic Maxwell model (24) $(d=1, \nu=0)$, where $F(v,t)$ satisfies,

$$\partial_t F(v) - D\nabla_v^2 F(v) = I(v|F), \tag{32}$$

and the collision term has the form,

$$\begin{aligned} I(v|F) &= \int dw \left[\frac{1}{\alpha} F(v^{**}) F(w^{**}) - F(v) F(w)\right] \\ &= -F(v) + \frac{1}{p}\int du F(u) F\left(\frac{v-qu}{p}\right). \end{aligned} \tag{33}$$

All velocity integrations extend over the interval $(-\infty,+\infty)$. Here the outgoing velocities (v,w), and the incoming ones (v^{**},w^{**}) are according to (23) related by,

$$v = qv^{**} + pw^{**}; \qquad w = pv^{**} + qw^{**} \tag{34}$$

with $p = 1-q = \frac{1}{2}(1+\alpha)$. By changing integration variables $w \to v^{**} = u$ with $dw = (\alpha/p)\,du$, and using the relation $w^{**} = (v-qu)/p$ one obtains the second equality in (33). The normalization of mass and mean square velocity $\langle v^2 \rangle = \frac{1}{2}v_0^2$ are given by (25) for $d=1$. The temperature balance equation is obtained from (32) as,

$$\frac{dv_0^2}{dt} = 4D - 2pqv_0^2, \tag{35}$$

and describes the approach to the non-equilibrium steady state (NESS) with a width $v_0^2(\infty) = 2D/pq$, where the heating through random forces, $\sim D$, is balanced by the collisional losses.

To understand the physical processes involved we first discuss in a qualitative way the relevant limiting cases. Without the heating term $(D=0)$, equation (32) reduces to the freely cooling inelastic Maxwell model.

If one takes in addition the elastic limit ($\alpha \to 1$ or $q \to 0$), the collision laws reduce in the *one-dimensional* case to $v^{**} = w, w^{**} = v$, i.e. an exchange of particle labels; the collision term vanishes identically; every $F(v,t) = F(v)$ is a solution; there is no randomization or relaxation of the velocity distribution through collisions, and the model becomes trivial at the Boltzmann level of description, whereas the distribution function in the presence of *infinitesimal* dissipation ($\alpha \to 1$) approaches a Maxwellian.

If we turn on the noise ($D \neq 0$) at vanishing dissipation ($q = 0$), the collision term in (32) vanishes, and the granular temperature follows from (35) as $v_0^2(t) = v_0^2(0) + 4Dt$, and increases linearly with time. With stochastic heating *and* dissipation (even in infinitesimal amounts) the system reaches a NESS.

3.5 Comments

Spatial Dependence:
When *spatial* dependence of the distribution function $F(\vec{r}, \vec{v}; t)$ is relevant, the collision term in the Enskog-Boltzmann equation must be slightly modified, namely the angular integration $\int_{\vec{n}}$ over the full solid angle should be replaced by $2\int_{\vec{n}} \theta(-\vec{g} \cdot \vec{n})$, where the unit step function $\theta(x)$ restricts the $\vec{n}-$integration to the pre-collision hemisphere with $\vec{g} \cdot \vec{n} < 0$ (see Fig. 1). In the spatially uniform case both representations are identical because the restituting velocities (23) are even functions of $\vec{n}$.

Origin of Maxwell Models:
In several papers [30, 37, 41, 43] inelastic Maxwell models have been introduced as more or less *ad hoc mathematical simplifications* of the nonlinear collision term for inelastic hard spheres. This has been done by replacing the relative velocity g in the hard sphere collision frequency $\text{A}(\vec{g}, \vec{n}) = |\vec{g} \cdot \vec{n}|$ by its mean value, $\langle g \rangle \text{A}(\vec{n}) \sim v_0(t) \text{A}(\vec{n})$, where $v_0(t)$ is the root mean square velocity.

This procedure guarantees that the homogeneous cooling law for inelastic Maxwell model constructed in this way, is identical to the one for inelastic hard spheres, and given by Haff's law [9], $T(t) \sim t^{-2}$. The construction of inelastic Maxwell and ISS models, followed in the present article, is more in the spirit of [29], i.e. by defining the collision term through transition probabilities for the scattering process $(\vec{v}, \vec{w}) \leftrightarrow (\vec{v}\,', \vec{w}\,')$ with the proper constraints.

Violation of H-Theorem:
As the detailed balance symmetry and the $H-$theorem are lacking in dissipative interaction models, there is no guarantee that the entropy $S(t) = -H(t)$ is *non-decreasing.* In fact the solutions of the Boltzmann equation for such dissipative interaction models approach [41] for long times to a scaling form, defined in (26). By inserting such solutions into (19) and anticipating the

cooling law (3) in the next section, one easily verifies that the entropy in the *scaling* state keeps *decreasing* as t becomes large, i.e.

$$S(t) = -H(t) \sim -(d/\nu)\ln t. \tag{36}$$

This result is typical for pattern forming mechanisms in configuration space, where spatial order or correlations are building up, as well as in chaos theory, where the rate of irreversible entropy production is negative on an attractor [54]. Moreover, there is no fundamental objection against decreasing entropies in an open system in contact with a reservoir, which is here the energy sink formed by the dissipative collisions. The dynamics in $N-$particle velocity space corresponds to a contracting flow $d\vec{v}^{*}d\vec{w}^{*} = \alpha d\vec{v}d\vec{w}$ where $\alpha < 1$, where the probability is contracting onto an attractor. This is a well known phenomenon in chaos theory.

Particles with Pseudo-power Law Repulsion:
A system of N inelastic hard spheres is a model with microscopic particles, characterized by positions and velocities, and interacting via well-defined force laws, that can be studied by means of MD simulations. However microscopic particles with dissipative inter-particle forces, such as inelastic Maxwell molecules and ISS-models, are stochastic models, only defined in $N-$particle velocity space. Molecular Dynamics simulations can not be performed for such models.

Modeling Dissipation:
There are many ways to model inelastic collisions that dissipate the relative kinetic energy of colliding particles; e.g. by using Hertz' contact law [55, 56], visco-elastic media [57], or coefficients of normal and tangential restitution [6, 11]. The coefficients of restitution may also depend on the relative speed of the colliding particles [55].

For the ISS-models, studied in this article, the scattering laws (scattering angle, collisional energy loss) are independent of ν, i.e. they are the same for inelastic hard spheres, inelastic Maxwell models and for general ISS-models. Only the collision frequency $\mathrm{A}(\vec{g},\vec{n})$ of the ISS-models has the same energy dependence as the collision frequency of elastic particles interacting through repulsive power law potentials.

4 Analysis of Inelastic Scattering Models

4.1 Homogeneous Cooling Laws

We start with the freely evolving case and compute the cooling rate, defined through, $dv_0^2/dt = -\zeta_\nu(t)\, v_0^2$. This can be done by applying $\left(\int d\vec{v}v^2\right)$ to the Boltzmann equation (24), changing integration variables $(\vec{v},\vec{w}) \to (\vec{v}^{**},\vec{w}^{**})$ in the gain term, and using the relations $d\vec{v}d\vec{w} = \alpha d\vec{v}^{**}d\vec{w}^{**}$ together with the energy loss ΔE per collision in (15). In general the above cooling equation is a formal identity, and does not provide a closed equation for $v_0(t)$. However, if $F(v,t)$ rapidly relaxes to a scaling form, then the *subsequent* time evolution of $v_0^2 = (2/d)\langle v^2\rangle$ is described by a closed equation for $v_0(t)$, as we will see, i.e.

$$\begin{aligned}\frac{dv_0^2}{dt} &= \tfrac{2}{d}\int d\vec{v}v^2 I\left(v|F\right) = \tfrac{2}{d}\int_{\vec{n}}\int d\vec{v}d\vec{w}\left|\vec{g}\cdot\vec{n}\right|^{\nu}\Delta E F\left(v,t\right)F\left(w,t\right) \\ &= -\frac{1}{2d}\left(1-\alpha^2\right)v_0^{\nu+2}\int_{\vec{n}}\int d\vec{c}d\vec{c}_1\left|(\vec{c}-\vec{c}_1)\cdot\vec{n}\right|^{\nu+2} f\left(c\right)f\left(c_1\right), \quad (37)\end{aligned}$$

from which the cooling rate can be identified as,

$$\zeta_\nu\left(t\right) \equiv -\frac{dv_0^2}{dt}/v_0^2 = 2\gamma_0\kappa_{\nu+2}\, v_0^{\nu}\left(t\right). \quad (38)$$

Here the coefficient $\gamma_0 = \frac{1}{4d}\left(1-\alpha^2\right)$ measures the inelasticity. Moreover, the constant κ_ν is defined as,

$$\kappa_\nu = \int_{\vec{n}}\int d\vec{c}d\vec{c}_1\left|(\vec{c}-\vec{c}_1)\cdot\vec{n}\right|^{\nu} f\left(c\right)f\left(c_1\right) = \beta_\nu\int d\vec{c}d\vec{c}_1\left|\vec{c}-\vec{c}_1\right|^{\nu} f\left(c\right)f\left(c_1\right), \quad (39)$$

and the $(d-1)$ $-$dimensional angular integral β_ν is evaluated as,

$$\beta_\nu = \int_{\vec{n}}\left|\hat{\vec{a}}\cdot\vec{n}\right|^{\nu} = \frac{\int_0^{\pi/2} d\theta\left(\sin\theta\right)^{d-2}\left(\cos\theta\right)^{\nu}}{\int_0^{\pi/2} d\theta\left(\sin\theta\right)^{d-2}} = \frac{\Gamma\left(\frac{\nu+1}{2}\right)\Gamma\left(\frac{d}{2}\right)}{\Gamma\left(\frac{\nu+d}{2}\right)\Gamma\left(\frac{1}{2}\right)}. \quad (40)$$

In one-dimensional systems $\beta_\nu = 1$ for all ν, and in higher dimensions for the ν-values of interest here β_ν is convergent ($\nu > 0$). As a consequence of the scaling ansatz the coefficient κ_ν is a time independent constant, that does depend on the unknown scaling form. For Maxwell models ($\nu = 0$) the cooling rate ζ_0 can be calculated explicitly from (40), and the constant κ_2 is $\kappa_2 = (1/d)\left\langle\left|\vec{c}-\vec{c}_1\right|^2\right\rangle = 1$ on account of (27). The result is,

$$\zeta_0 \equiv -\frac{dv_0^2}{dt}/v_0^2 = 2\gamma_0. \quad (41)$$

The r.m.s. velocity decays then as $v_0\left(t\right) = v_0\left(0\right)\exp\left[-\gamma_0 t\right]$.

The most important time scale in kinetic theory is the mean free time t_{mf}, which equals the inverse of the *mean collision rate* $\omega_\nu\left(t\right)$, defined as the average of the collision frequency $\mathrm{A}\left(\vec{g},\vec{n}\right) = \left|(\vec{c}-\vec{c}_1)\cdot\vec{n}\right|^{\nu}$ over the velocities of the colliding pair in the scaling state, i.e.

$$\omega_\nu\left(t\right) \equiv -\int d\vec{v}I_{\text{loss}}\left(v|F\right) = \kappa_\nu v_0^{\nu}\left(t\right). \quad (42)$$

Comparison of (38) and (42) shows that both frequencies are related as,

$$\zeta_\nu(t) = 2\gamma_0 \frac{\kappa_{\nu+2}}{\kappa_\nu} \omega_\nu(t) \equiv 2\gamma_\nu \omega_\nu(t), \tag{43}$$

where $\gamma_\nu \kappa_\nu = \gamma_0 \kappa_{\nu+2}$ are time independent constants. Using (42) and (43) the equation for the r.m.s. velocity becomes, $\dot{v}_0 = -\gamma_\nu \kappa_\nu v_0^{\nu+1}$, yielding a solution, identical to (3) with γ replaced by γ_ν. Consequently the granular temperature at large times decays as $T = v_0^2 \sim t^{-2/\nu}$. For $\nu = 1$ one recovers the well known law of Haff [9], describing the long time decay of the granular temperature in the homogeneous cooling state of inelastic hard spheres. A homogeneous cooling law similar to (3) with $\nu = 6/5$ has been derived in [55], not on the basis of the Boltzmann equation (24) with an energy dependent collision rate $\propto g^{5/6}$, but by using the Boltzmann equation for inelastic hard spheres ($\nu = 1$) with an energy dependent coefficient of restitution $\alpha(g) = 1 - Ag^{1/5}$. For Maxwell models ($\nu \to 0$) the r.m.s. velocity in (3) reduces to $v_0(t) = v_0(0) \exp[-\gamma_0 t]$, in agreement with (41).

The homogeneous cooling law for the general class of ISS-models discussed here, can be cast into a universal form by changing to a new time variable, the *collision counter* or *internal* time τ of a particle, which represents the total number of collisions that a particle has suffered in the (external) time t. It is defined through the mean collision frequency,

$$d\tau = \omega_\nu(t)\, dt, \tag{44}$$

where $\omega_\nu(t) \sim v_0^\nu(t)$. After inserting (3) in this equation the differential equation can be solved to yield,

$$\nu\gamma_\nu\tau = \ln\left[1 + \nu\gamma_\nu\omega_\nu(0)\, t\right], \tag{45}$$

valid for all ISS-models. Combination of (45) and (3) shows that the r.m.s. velocity follows the *universal* homogeneous cooling law,

$$v_0(\tau) = v_0(0) \exp\left[-\gamma_\nu \tau\right], \tag{46}$$

where γ_ν is according to (43) and (39) proportional to the inelasticity γ_0, and depends on ν through the collision integrals κ_ν. We further observe that the relations (45) and (46) also holds for the inelastic BGK-model in Sect. 2 with γ_ν replaced by $\gamma = \frac{1}{2}\left(1 - \alpha^2\right)$ in (6).

So far we have been dealing with freely cooling systems. Next we address the *balance equation* for the granular temperature in driven cases, where the external input of energy counterbalances the collisional cooling, and may lead to a NESS. We proceed in the same manner as for the free case, and apply $\left(\int d\vec{v} v^2\right)$ to the Boltzmann equation in (30) with the result,

$$\partial_t \langle v^2 \rangle = -\zeta_\nu(t) \langle v^2 \rangle + 2 \langle \vec{v} \cdot \vec{a} \rangle + 2dD, \tag{47}$$

where the first term, $\int d\vec{v}v^2 I(v|F) = -\zeta_\nu(t)\langle v^2\rangle$, is obtained from (37), (38) and (25). The next two terms are obtained from the driving term in (30), i.e. $\int d\vec{v}v^2 \mathcal{F}F(v)$, by performing partial integrations.

The most common ways of driving dissipative fluids [22, 26, 29, 31] is by Gaussian white noise (WN) ($\vec{a}=0; D\neq 0$), or by a Gaussian thermostat (GT) ($\vec{a}=\gamma\vec{v}; D=0$), yielding for the balance equations,

$$\frac{dv_0^2}{dt}(t) = \begin{cases} (2\gamma - \zeta_\nu(t))\, v_0^2(t) & \text{(GT)} \\ 4D - \zeta_\nu(t)\, v_0^2(t) & \text{(WN)}. \end{cases} \tag{48}$$

Here the collisional loss, $-\zeta_\nu v_0^2$, is counterbalanced by the heat, $2\gamma v_0^2$, generated by the negative friction of the Gaussian thermostat, or by the heat, $4D$, generated by randomly kicking the particles.

As $t\to\infty$ the granular temperature will reach a NESS with $T(\infty) = v_0^2(\infty)$, where the r.m.s. velocity follows from (48) using (38),

$$v_0(\infty) = \begin{cases} (\gamma/\gamma_0\kappa_{\nu+2})^{1/\nu} & \text{(GT)} \\ (2D/\gamma_0\kappa_{\nu+2})^{1/(\nu+2)} & \text{(WN)}. \end{cases} \tag{49}$$

With reference to the discussion in Sect. 2.3 we note that the GT-fixed point solution $v_0(\infty)$ is *attracting* for $\nu>0$, leading to a stable NESS, and *unstable* for $\nu<0$ with $v_0(t)$ vanishing at $t\to\infty$ if $v_0(0)<v_0(\infty)$, and diverging if $v_0(0)>v_0(\infty)$. The case $\nu=0$ is *marginally stable.* For the WN-fixed point similar observations apply with the stability threshold $\nu=0$ replaced by $\nu=-2$.

4.2 Scaling and Non-equilibrium Steady States

Free Cases:

To investigate the existence of scaling solutions of the Boltzmann equation for freely cooling inelastic systems, we substitute the scaling ansatz (26) into (24), to obtain an integral equation for the scaling form $f(c)$. With the help of (44), (46) and (28) the left hand side of (24) becomes,

$$\begin{aligned} \text{l.h.s.} &= -\dot{v}_0 v_0^{-d-1}\left(\frac{d\tau}{dt}\right)\left\{df(c) + c\frac{d}{dc}f(c)\right\} \\ &= \gamma_\nu\omega_\nu(\tau)\, v_0^{-d}(\tau)\, \nabla_{\vec{c}}\cdot\vec{c}f(c) = \gamma_0\kappa_{\nu+2}\nabla_{\vec{c}}\cdot\vec{c}f(c)\; v_0^{\nu-d}(\tau). \end{aligned} \tag{50}$$

In the last equality we have used the relation $\gamma_\nu\kappa_\nu = \gamma_0\kappa_{\nu+2}$, implied by (43). The resulting integral equation for $f(c)$ becomes,

$$I(c|f) = \gamma_0\kappa_{\nu+2}\nabla_{\vec{c}}\cdot\vec{c}f = \frac{1}{d}\varpi_2\nabla_{\vec{c}}\cdot\vec{c}f. \tag{51}$$

With the help of (37) and (38) the second moment of the collision term, ϖ_2, can be expressed as,

$$\varpi_2 \equiv -\int d\vec{c} c^2 I(c|f) = d\gamma_0 \kappa_{\nu+2}. \tag{52}$$

Driven Cases:
We consider equation (30) for the NESS distribution in its rescaled form (31) with $v_0 = v_0(\infty)$, driven by a Gaussian thermostat (GT), $\{\vec{a} = \gamma\vec{v}; D = 0\}$ or by white noise (WN), $\{\vec{a} = 0; D \neq 0\}$. The scaling equations for $f(c)$ take the form,

$$I(c|f) = \begin{cases} \dfrac{\gamma}{v_0^\nu} \nabla_{\vec{c}} \cdot \vec{c} f = \dfrac{1}{d} \varpi_2 \nabla_{\vec{c}} \cdot \vec{c} f & \text{(GT)} \\ -\dfrac{D}{v_0^{\nu+2}} \nabla_{\vec{c}}^2 f = -\dfrac{1}{2d} \varpi_2 \nabla_{\vec{c}}^2 f & \text{(WN)}. \end{cases} \tag{53}$$

The first equality in (GT) and (WN) suggests that $f(c)$ depends explicitly on γ or D. This is however not the case. The stationarity relation (49) combined with (52) shows in fact that the following expressions,

$$\varpi_2 = d\gamma v_0^{-\nu} = 2dD v_0^{-(\nu+2)} = d\gamma_0 \kappa_{\nu+2}, \tag{54}$$

are *independent* of γ or D. So, we have used (49) to eliminate γ and D, and to put it in the universal form, containing ϖ_2.

4.3 Comments

Equivalence Free Cooling – Gaussian Thermostat:
Comparison of integral equation (51) for free cooling and (53) for the Gaussian thermostat shows that both equations are identical, as first observed by Montanero and Santos [26]. This implies that the scaling form $f(c)$ in free cooling ($\mathcal{F} = 0$) is identical to the NESS distribution $f(c)$ of the same system, driven by a Gaussian thermostat, provided both forms are rescaled to the same constant width $\int d\vec{c} c^2 f(c) = d/2$. This also implies that the scaling form for the free case can be measured by performing Monte Carlo simulations in a steady state [25, 26]. The same idea of systematically rescaling the velocities has also been used in molecular dynamics simulations of a freely cooling system of N inelastic hard spheres [58].

Universality:
The scaling equations (53) for $f(c)$ in systems driven by a Gaussian thermostat and by white noise, have a universal form, because ϖ_2 is independent of the friction constant γ, the noise strength D and the width $v_0(\infty)$, which may depend on the initial distribution. The scaling equations reduce for $\nu = 1$ to those for hard spheres [22], and for $\nu = 0$ to those for Maxwell models [40].

Perturbative Approach:
The moment of the collision term $\varpi_2 = d\gamma_0\kappa_{\nu+2}$ in the scaling equations (53), and the cooling rate $\zeta_\nu = \gamma_0\kappa_{\nu+2}v_0^\nu$ in (48) contain for all models the quantity κ_ν, as given by (38). It depends on the unknown function $f(c)$, except for $\nu = 0$ where $\kappa_2 = 1$ and $\varpi_2 = d\gamma_0 = \frac{1}{4}\left(1-\alpha^2\right)$. For the case of inelastic hard spheres, a perturbative method has been developed in [11, 22] for small inelasticities to solve both integral equations in (53) by expanding $f(c)$ in a series of Sonine polynomials $S_p\left(c^2\right)$, i.e.

$$f(c) = \phi(c)\{1 + \sum_{p=2}^{\infty} a_p S_p\left(c^2\right)\}, \tag{55}$$

where $\phi(c) = \pi^{-d/2}\exp\left(-c^2\right)$ is the Maxwellian. Here a_2 is essentially the fourth cumulant which has been calculated explicitly in [22]. For dimensions $d > 1$ it turns out to be proportional to the inelasticity $\left(1-\alpha^2\right)$. However, the one-dimensional case is exceptional because a_2 approaches a finite value for $\alpha \to 1$ [23]. The same method has been successfully applied by Cercignani et al. [31] to an inelastic Maxwell model, and by several authors [25–27, 55] to inelastic hard spheres and related problems. The method can be applied to the inelastic ISS-models as well.

The method focuses on the lower moments of $f(c)$, and the polynomial approximation, cut off after $S_2\left(c^2\right)$, gives a fair representation of $f(c)$ for velocities in the thermal range, $c \leq 2$. The expansion (55) can then be used to calculate κ_ν in (52) in the form $\kappa_\nu = \kappa_\nu^0 + \left(1-\alpha^2\right)\kappa_\nu^1 + \cdots$. For more details we refer to [22]. As an illustration we calculate the lowest approximation κ_ν^0 to κ_ν in (38) using $f(c) \simeq \phi(c)$. The result is,

$$\begin{aligned}
\kappa_\nu^0 &= \int_{\vec{n}}\int d\vec{c}d\vec{c}_1 \left|g_\parallel\right|^\nu \phi(c)\,\phi(c_1) \\
&= \frac{1}{\sqrt{2\pi}}\int_{-\infty}^{\infty} dx\,|x|^\nu e^{-x^2/2} = \frac{2^{\nu/2}}{\sqrt{\pi}}\Gamma\left(\frac{\nu+1}{2}\right).
\end{aligned} \tag{56}$$

The lowest approximation to (54) is then $\varpi_2^0 = d\gamma_0\kappa_\nu^0$. The damping rate γ_ν in (43) of the r.m.s. velocity becomes then $\gamma_\nu = (\nu+1)\gamma_0$, in agreement with the results for $\nu = 1$ in [22].

4.4 High Energy Tails

The polynomial expansion (55) describes $f(c)$ only in the thermal range, but contains no meaningful information about velocities c in the asymptotic range $(c \gg 1)$. However the high energy tails in the ISS-models with $\nu > 0$ can be determined by a procedure similar to the one used successfully for inelastic hard spheres systems [22], as well as in inelastic Maxwell models driven by white noise [40]. To do so we make the ansatz of stretched Gaussian behavior for the high energy tail, i.e. $f(c) \simeq B \exp\left[-\beta c^b\right]$ with $0 < b < 2$ and $\beta > 0$, and determine b and β by inserting this ansatz in the scaling equation (53) and requiring self-consistency. The border line case $b \to 2$ corresponds to Gaussian tails, and $b \to 0$ suggests power law tails with negative exponents.

An estimate of the rescaled collision term $I(c|f)$ in (24) is made in [22, 40]. This suggests that the loss term is asymptotically dominant over the gain term as long as the exponent b in $\exp\left[-\beta c^b\right]$ is restricted to $b > 0$. This estimate also applies to the ISS-models as long as the exponent $\nu > 0$ and the inelasticity γ_0 is non-vanishing. So the gain term is neglected. Moreover the loss term can be simplified in the asymptotic velocity range. Its dominant contribution comes from collisions with particles having velocities c_1 that are typically in the thermal range ($c_1 = \mathcal{O}(1)$). Consequently the collision rate $|(\vec{c}-\vec{c}_1)\cdot\vec{n}|^\nu$ for asymptotic dynamics may be replaced by $c^\nu \left|\hat{\vec{c}}\cdot\vec{n}\right|^\nu$, and the total collision term simplifies to,

$$I(c|f) \sim I_{\text{loss}}(c|f) \sim -c^\nu \beta_\nu f(c), \tag{57}$$

with β_ν given by (40). After these preparations we insert (57) and the stretched Gaussian form, $f(c) \sim \exp\left[-\beta c^b\right]$, into the Boltzmann equation (53), and follow the procedure sketched above. This gives the following universal results (see item 2 below (54)) for the asymptotic high energy tail in d–dimensional inelastic ν–models, $f(c) \sim \exp\left[-\beta c^b\right]$ with

$$\begin{array}{lll} b = 0 & \text{inconsistent} & (\text{GT};\ \nu = 0) \\ b = \nu & \beta = d\beta_\nu/\nu\varpi_2 & (\text{GT};\ \nu > 0) \\ b = \frac{1}{2}(\nu+2) & \beta = \frac{2}{\nu+2}\sqrt{\frac{2d\beta_\nu}{\varpi_2}} & (\text{WN};\ \nu > 0\ ), \end{array} \tag{58}$$

where the scaling functions and high energy tails of GT-driven and freely cooling systems are equivalent.

We conclude this subsection with some comments.

Driven Systems:

For $0 < b < 2$ both GT- and WN-driving lead to consistent asymptotic solutions of the scaling equations for $\nu > 0$ with overpopulated high energy tails of stretched Gaussian type. We also note, the larger the interactions (larger ν-values) at large impact energies, the smaller the overpopulation of the high energy tails. For $\nu = 2$, corresponding to the inelastic version of the Very Hard Particle model (see end of Sect. 3.2) we obtain Gaussian behavior ($b = 2$) for both types of thermostats, and there are no longer over-populated tails. In the case of stretched Gaussian tails *all* moments $\int d\vec{c}c^n f(c) < \infty$. This would not be the case for power law tails.

In the case of white noise driving, the above results with $b = 1 + \frac{1}{2}\nu$, include inelastic hard spheres ($\nu = 1$), as well as Maxwell models ($\nu = 0$), and the results coincide with the detailed predictions for $\nu = 1$ [22] and $\nu = 0$ [40]. For GT-driven or freely cooling ISS-models with $\nu > 0$ the tail distributions have a stretching exponent $b = \nu$. The property $\beta b \to 1/\gamma_0$ as

$\nu \downarrow 0$ would suggest power law behavior at the stability threshold, $\nu = 0$ (Maxwell models), with an exponent $a = 1/\gamma_0$ (see however next item). The result (58) also includes the exponential decay, $f(c) \sim \exp[-\beta c]$ for inelastic hard spheres, found in [22].

Stability Thresholds:
In the previous comments we have speculated on power law tails, in agreement with the exact solution [35] for the one-dimensional freely cooling Maxwell models with $f(c) \sim c^{-4}$ for $c \gg 1$. As will be shown in Sect. 5, the scaling form for d-dimensional Maxwell models in free cooling does indeed have an algebraic tail $f(c) \sim 1/c^{d+a}$, *but* with an exponent $a(\alpha) \neq 1/\gamma_0$. However, neither the existence of algebraic tails, nor the possible value of the power law exponents can be obtained legitimately from the results (58). The reason is that the result is not consistent with the *a priori* assumption $I_{\text{gain}} \ll I_{\text{loss}}$ at $c \gg 1$. In fact, gain and loss terms are of the same order of magnitude for $\nu = 0$.

Maxwell Models as Approximations to Hard Spheres:
As discussed in item 2 of Sect. 3.5, inelastic Maxwell models have also been introduced as a sensibly looking mathematical simplification of the Boltzmann equation for inelastic hard spheres. The results of the analysis in the previous section show that the effect of this simplification on the shape of the tail may be very drastic. In the free cooling case the hard sphere tail is exponential, $f(c) \sim \exp[-\beta c]$, whereas in the Maxwell model it is a power law tail, $f(c) \sim 1/c^{d+a}$. A smaller difference exists in the WN-driven case, where the hard sphere tail is $f(c) \sim \exp\left[-\beta c^{3/2}\right]$, and the Maxwell tail is $f(c) \sim \exp[-\beta c]$.

MC-Simulations at High and Low Inelasticity:
The results (58) predict more than the stretching exponent b. In fact the coefficient β for $\nu > 0$ in $f(c) \sim \exp\left[-\beta c^b\right]$ is given in terms of ϖ_2 in (54). For these $\nu-$values we have described a perturbative calculation for ϖ_2, which converges rapidly for $\alpha \to 1$, but gives poor results for $\alpha < 0.6$ [22, 26]. Nevertheless, it is possible to test the predictions of (58) for the GT- and the WN-thermostats, including the coefficient β for *all* values of the restitution coefficient α. This can be done by measuring the mean square velocity $v_0^2(\infty)$ in MC simulations on driven systems in a NESS for a given γ or D. From these data ϖ_2 can be calculated using (54).

5 Inelastic Maxwell Models

5.1 Fourier Transform Method

As discussed in Sect. 4.3, the behavior of the high energy tails of the scaling form in the ISS-models is only controlled by the loss term in the Boltzmann collision term, and generates the stretched Gaussian tails. However, in the borderline case ($\nu = 0$) the tail behavior is determined by an interplay between gain and loss term, which leads to algebraic tails. This makes the analysis more complicated. In this article we only discuss the borderline case in *free cooling*, formed by the Maxwell models, and we refer to [34] for a more comprehensive discussion.

In this section we want to demonstrate that the Boltzmann equation (24) for inelastic Maxwell models ($\nu = 0$),

$$\partial_t F(v,t) = -F(v,t) + \frac{1}{\alpha} \int_{\vec{n}} \int d\vec{w} F(v^{**},t) F(w^{**},t), \tag{59}$$

has a scaling solution with a power law tail. To do so we first consider the Fourier transform of the distribution function, $\Phi(k,t) = \left\langle \exp\left[-i\vec{k}\cdot\vec{v}\right]\right\rangle$, which is the characteristic function or generating function of the velocity moments. Because $F(v,t)$ is isotropic, $\Phi(k,t)$ is isotropic as well.

As an auxiliary step we first Fourier transform the gain term in (59), i.e.

$$\begin{aligned}&\int d\vec{v} \exp\left[-i\vec{k}\cdot\vec{v}\right] I_{\text{gain}}(v|F) = \\ &\int_{\vec{n}} \int d\vec{v} d\vec{w} \exp\left[-i\vec{k}\cdot\vec{v}^*\right] F(v,t) F(w,t) = \int_{\vec{n}} \Phi(k\eta_+,t)\Phi(k\eta_-,t). \end{aligned}\tag{60}$$

The transformation needed to obtain the first equality is the same as in (37). Then we use (14) to write the exponent as $\vec{k}\cdot\vec{v}_1^* = \vec{k}_-\cdot\vec{v}_1 + \vec{k}_+\cdot\vec{v}_2$, where

$$\begin{aligned}\vec{k}_+ &= p\vec{k}\cdot\vec{n}\vec{n} & \left|\vec{k}_+\right| &= kp\left|\left(\hat{\vec{k}}\cdot\vec{n}\right)\right| = k\eta_+(\vec{n}) \\ \vec{k}_- &= \vec{k} - \vec{k}_+ & \left|\vec{k}_-\right| &= k\sqrt{1 - z\left(\hat{\vec{k}}\cdot\vec{n}\right)^2} = k\eta_-(\vec{n}),\end{aligned}\tag{61}$$

with $p = 1 - q = \frac{1}{2}(1+\alpha)$ and $z = 1 - q^2$. The Fourier transform of (59) then becomes,

$$\partial_t \Phi(k,t) = -\Phi(k,t) + \int_{\vec{n}} \Phi(k\eta_+(\vec{n}),t)\Phi(k\eta_-(\vec{n}),t), \tag{62}$$

where $\Phi(0,t) = 1$ because of (25). In one-dimension $\eta_+(\vec{n}) = p$ and $\eta_-(\vec{n}) = q$, and $\int_{\vec{n}}$ can be replaced by unity. Moreover this equation simplifies to [29],

$$\partial_t \Phi(k,t) = \Phi(pk,t)\Phi(qk,t) - \Phi(k,t). \tag{63}$$

Because $F(v,t)$ is isotropic, only the even moments are non-vanishing, and the moment expansion of the characteristic function takes the form,

$$\Phi(k,t) = {\sum_n}' \frac{(-ik)^n}{n!} \left\langle \left(\hat{\vec{k}}\cdot\vec{v}\right)^n \right\rangle = {\sum_n}' (-ik)^n m_n(t), \tag{64}$$

where the prime indicates that $n = even$, and the moment $m_n(t)$ is defined as,

$$m_n(t) = \beta_n \langle v^n \rangle / n!, \tag{65}$$

where $\beta_n = \int_{\vec{n}} \left(\hat{\vec{k}} \cdot \hat{\vec{v}}\right)^n$ is given in (40). Moreover, the normalizations (25) give $m_0(t) = 1$ and $m_2(t) = \frac{1}{2}\beta_2 \langle v^2 \rangle = \frac{1}{4} v_0^2$.

Maxwell models have the unusual property that the system of moment equations for $m_n(t)$ is closed, and can be solved *sequentially*. The reason is that the rate $\dot{m}_n$ is only a function of lower moments $m_s(t)$ with $s \leq n$.

The moment equations are readily obtained by inserting the expansion (64) in the Fourier transformed Boltzmann equation (62), and equating the coefficients of equal powers of k. The result is,

$$\dot{m}_n + \lambda_n m_n = \sum_{l=2}^{n-2} h(l, n-l)\ m_l\, m_{n-l} \quad (n > 2), \tag{66}$$

where all labels $\{n, l, s\}$ take *even* values only. Following [38] or Appendix A of [41] the functions can be calculated for real positive values of ℓ and s with the result,

$$\begin{aligned} h(l, s) &= \int_{\vec{n}} \eta_+^l(\vec{n})\, \eta_-^s(\vec{n}) = p^l \beta_l\ {}_2F_1\left(-\frac{s}{2}, \frac{l+1}{2}; \frac{l+d}{2} \mid z\right) \\ \lambda_s &= 1 - h(s,0) - h(0,s) = \int_{\vec{n}} \left[1 - \eta_+^s(\vec{n}) - \eta_-^s(\vec{n})\right] \\ &= 1 - p^s \beta_s - {}_2F_1\left(-\frac{s}{2}, \frac{1}{2}; \frac{d}{2} \mid 1 - q^2\right). \end{aligned} \tag{67}$$

Here ${}_2F_1(\alpha, \beta; \gamma \mid z)$ is a hypergeometric function, and $\lambda_2 = 2pq/d = 2\gamma_0$ is easily obtained from the above results.

Next we consider the Fourier transform of the scaling relation (26), yielding $\Phi(k,t) = \phi(kv_0(t))$ with $v_0(t) = v_0(0)\exp(-\gamma_0 t)$ according to (41). Inserting Φ in (62) gives the integral equation for the scaling form $\phi(k)$, which reads

$$-\gamma_0 k \frac{d}{dk}\phi(k) + \phi(k) = \int_{\vec{n}} \phi(k\eta_+)\, \phi(k\eta_-). \tag{68}$$

For $d = 1$ it reduces to,

$$-pqk\frac{d}{dk}\phi(k) + \phi(k) = \phi(pk)\, \phi(qk). \tag{69}$$

The scaling form $\phi(k)$ is the generating function for the moments of $f(c)$, i.e.

$$\begin{aligned} \phi(k) &= \sum_n{}' \frac{(-ik)^n}{n!} \beta_n \langle c^n \rangle \equiv \sum_n{}' (-ik)^n \mu_n \\ &\simeq 1 - \tfrac{1}{4}k^2 + k^4 \mu_4 - k^6 \mu_6 + \cdots, \end{aligned} \tag{70}$$

where n =even, $\mu_0 = 1$ and $\mu_2 = \frac{1}{2}\beta_2 \langle c^2 \rangle = 1/4$ on account of the normalizations (27) and $\beta_2 = 1/d$ (see (40)). By inserting (70) into (68) one obtains the recursion relation,

$$\mu_n = \frac{1}{\lambda_n - \frac{1}{2}n\lambda_2} \sum_{l=2}^{n-2} h(l, n-l)\, \mu_l \mu_{n-l} \qquad (n > 2) \tag{71}$$

with initialization $\mu_2 = 1/4$ and all labels $n, l = even$. How these moments behave as a function of α has been calculated in [41] by numerically solving the recursion relation.

5.2 Small-k Singularity of the Characteristic Function

The asymptotic analysis in Sect. 4.4 of the high energy tail $f(c) \sim \exp\left[-\beta c^b\right]$ for Maxwell models ($\nu = 0$) yields $b = 0$, which suggests a power law tail $f(c) \sim 1/c^{d+a}$ as the leading large$-c$ behavior in $d-$dimensional systems. If this is indeed the case, then the moments μ_n of the scaling form $f(c)$ are convergent if $n < a$ and are divergent if $n > a$. As we are interested in physical solutions which can be *normalized*, and have a *finite energy*, a possible value of the power law exponent must obey $a > 2$.

The characteristic function is in fact a very suitable tool for investigating this problem. Suppose the moment μ_n with $n > a$ diverges, then the n-th derivative of the corresponding generating function also diverges at $k = 0$, i.e. $\phi(k)$ has a singularity at $k = 0$. Then a simple rescaling argument of the inverse Fourier transform shows that $\phi(k)$ has a dominant small-k singularity of the form $\phi(k) \sim |k|^a$, where $a \neq even$. On the other hand, whenever the exponent b in $f(c) \sim \exp\left[-\beta c^b\right]$ is positive, – as is the case for the inelastic Maxwell model *driven by white noise*, where $b = 1$ and $\beta = \sqrt{2/pq}$ – , then all moments are finite, and the characteristic function $\phi(k)$ is regular at the origin, i.e. can be expanded in powers of k^2.

We first illustrate our analysis for the one-dimensional case. As the requirement of finite total energy imposes the lower bound $a > 2$ on the exponent, we make the ansatz, consistent with (70) that the dominant small-k singularity has the form,

$$\phi(k) = 1 - \tfrac{1}{4}k^2 + A\,|k|^a\,, \tag{72}$$

insert this in (69), and equate the coefficients of equal powers of k. This yields,

$$\tfrac{1}{2}a\lambda_2 = \lambda_a \quad \text{or} \quad apq = 1 - p^a - q^a\,. \tag{73}$$

The equation has two roots, $a = 2, 3$, of which $a = 3$ is the one larger than 2. Here A is left undetermined. Consequently the one-dimensional scaling solution has a power law tail, $f(c) \sim 1/c^4$, in agreement with the exact solution in [35].

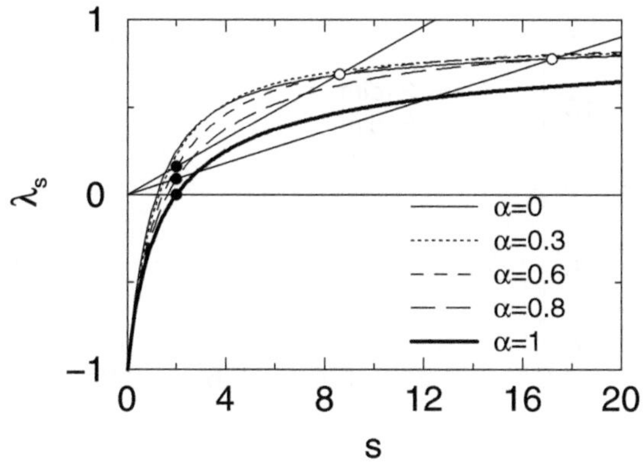

Fig. 2. Graphical solution of (74) for different values of the parameter α. The eigenvalue λ_s is a concave function of s, plotted for different values of the restitution coefficient α for the 2-D inelastic Maxwell model. The line $y = s\gamma_0$ is plotted for $\alpha = 0.6, 0.8$ and $\alpha = 1$(top to bottom). The intersections with λ_s determine the points s_0 (filled circles) and s_1 (open circles). Here $s_1 = a$ determines the exponent of the power law tail. For the elastic case ($\alpha = 1, \gamma_0 = 0$, energy conservation) there is only one intersection point

For general dimension we proceed in the same way as in the one-dimensional case, insert the ansatz (72) into (68), and equate the coefficients of equal powers of k. This yields for the coefficient of k^2 the identity $2\gamma_0 = \lambda_2$, and for the coefficient of k^a the transcendental equation,

$$\tfrac{1}{2}a\lambda_2 = \lambda_a = \textstyle\int_{\vec{n}} \left[1 - \eta_+^a - \eta_-^a\right] . \tag{74}$$

The equation above obviously has the solution $a = 2$. We are however interested in the solution with $a > 2$. In the elastic limit ($\alpha \to 1$) the solution is simple. There $\gamma_0 \to 0$ and a diverges. The contributions of $\eta_\pm^a$ on the right hand side vanish because $\eta_\pm < 1$, and the result is,

$$a \simeq \frac{1}{\gamma_0} = \frac{4d}{1-\alpha^2} . \tag{75}$$

For general values of α one can conveniently use an integral representation of ${}_2F_1$ to evaluate λ_a and solve the transcendental equation (74) numerically. We illustrate the solution method of (74) with the graphical construction in Fig. 2, where we look for intersections of the line $y = \frac{1}{2}\lambda_2 = \gamma_0 s$ with the curve $y = \lambda_s$ for different values of α.

The relevant properties of λ_s are: (i) $\lim_{s\to 0} \lambda_s = -1$; (ii) λ_s is a concave function, monotonically increasing with s, and *(iii)* all eigenvalues for positive *integers* n are positive (see Fig. 2). As can be seen from the graphical construction, the transcendental equation (74) has two solutions, the trivial one ($s_0 = 2$) and the solution $s_1 = a$ with $a > 2$. The numerical solutions

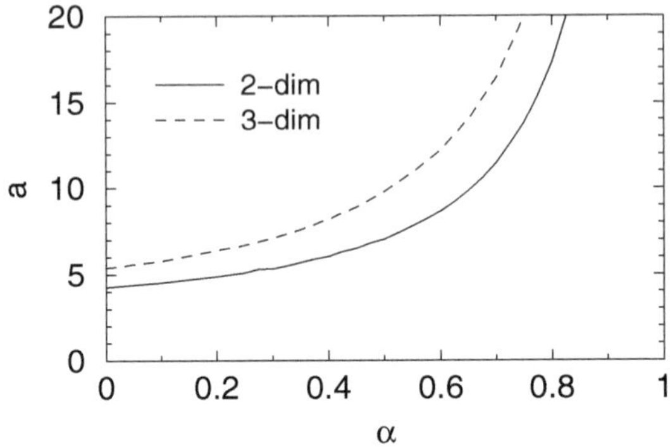

Fig. 3. Exponent $a(\alpha)$, which is the non-trivial root of (74), shown as a function of the coefficient of restitution α, and denoted in the plot by α. It determines the high energy tail $1/c^{a+d}$ of the scaling solution $f(c)$ of the 2-D and 3-D Maxwell models

for $d = 2, 3$ are shown in Fig. 3 as a function of α, and the α-dependence of the root $a(\alpha)$ can be understood from the graphical construction. In the elastic limit as $\alpha \uparrow 1$ the eigenvalue $\lambda_2(\alpha) \to 0$ because of energy conservation. In that limit the transcendental equation (74) no longer has a solution with $a > 2$, and $a(\alpha) \to \infty$ according to (75), as it should be. This is consistent with a Maxwellian tail distribution in the elastic case. Krapivsky and Ben-Naim have in fact solved the transcendental equation asymptotically for large d, which gives qualitatively the same results as shown in Fig. 3 for two and three dimensions.

These results establish the existence of scaling solutions $f(c) \sim 1/c^{d+a}$ with algebraic tails, where the exponent a is the solution of the transcendental equation (74) with $a > 2$. The exponent $a(\alpha)$ behaves as a function of α qualitatively the same as in the simple inelastic BGK model of Sect. 2.

Using a somewhat different analysis Krapivsky and Ben-Naim [37] independently obtained the same results for the algebraic tails in freely cooling Maxwell models.

5.3 Beyond Asymptotic Analysis

Free Cooling:
As mentioned already in Sect. 3.3, Baldassarri et al. [35] have obtained an exact solution $f(c)$ of the scaling equation for the one-dimensional Maxwell model in free cooling, and they demonstrated the importance of this scaling solution by means of Monte Carlo simulations. In doing so they found that $F(v,t)$ for different classes of initial distributions can be collapsed for *long* times t on this exact scaling solution $f(c)$, when $v_0(t)\,F(v,t)$ is plotted versus $c = v/v_0(t)$.

We briefly illustrate here how this solution is obtained from the one-dimensional scaling equation (69). One first verifies that the following function,

$$\phi(k) = (1 + \vartheta|k|)\exp(-\vartheta|k|), \tag{76}$$

with arbitrary positive ϑ is a solution, that can be Fourier inverted. It gives the scaling form,

$$f(c) = \frac{2}{\pi\vartheta}\frac{1}{\left(1 + c^2/\vartheta^2\right)^2}. \tag{77}$$

To determine the scaling form that satisfies the normalizations (27), we expand (76) in powers of $|k|$ to obtain,

$$\phi(k) = 1 - \tfrac{1}{2}\vartheta^2 k^2 + \tfrac{1}{3}\vartheta^3|k|^3 + \cdots. \tag{78}$$

Comparison of this result with (70) shows that $\vartheta = 1/\sqrt{2}$. Moreover, it confirms the ansatz (72) used to find solutions with small-k singularities. Comparison also shows the value of the coefficient $A = 1/\left[6\sqrt{2}\right]$ in (72), and the high energy tail is $f(c) \sim \mathcal{A}/c$ with $\mathcal{A} = 1/\left[\pi\sqrt{2}\right]$. The coefficients A and $\mathcal{A}$ can not be determined within the asymptotic method.

White Noise Driving:
For the one-dimensional Maxwell model driven by white noise the steady state solution has also been found exactly with the help of the Fourier transform method [29]. The characteristic function satisfies in that case a nonlinear finite difference equation, which can be solved by iteration. However, the analytic structure is rather complex, and makes it difficult to extract analytic information from that solution.

The observation that $f(c) \sim \exp[-\beta|c|]$ has an exponential high energy tail, was probably first made in numerical work of van der Hart and Nienhuis [43]. A more detailed analytic prediction about the asymptotic tail was given in [40], where it was shown that $\beta = \sqrt{8/(1-\alpha^2)}$ for all Maxwell models, independent of the dimensionality ($d = 1, 2, \cdots$), at least with the normalization, $\langle c^2 \rangle = \frac{1}{2}d$, used in this article. This result follows directly from (58) and the value $\varpi_2 = d\gamma_0 = \frac{1}{4}\left(1-\alpha^2\right)$, given in item 3 below (54). Furthermore, additional numerical and analytical work was also published by Marconi and Puglisi [59], and by Antal et al. [44]. Only recently more detailed analytic results have been extracted from the rather complex structure of the exact solution [45, 53].

6 Conclusions and Perspectives

We have studied asymptotic properties of scaling or similarity solutions, $F(v,t) = (v_0(t))^{-d} f(v/v_0(t))$, of the nonlinear Boltzmann equation in spatially homogeneous systems composed of particles with *inelastic* interactions for large times and large velocities. The large $t-$ and $v-$scales are relevant because on such scales the universal features of the solutions survive, while details of the initial distributions, of interaction strength, and degree of inelasticity, are mostly lost. The behavior of these scaling states, which describe nonequilibrium steady states (NESS), is less universal than the state of thermal equilibrium, because the form of the NESS distribution $f(c)$ depends on the way of driving the dissipative systems. Scaling solutions are very well suited to expose the universal features of the velocity distribution functions, because the velocities, $c = v/v_0(t)$, are measured in units of the r.m.s. velocity or instantaneous width $v_0(t)$ of the distribution.

The real importance of the scaling solutions is that the actual solutions $F(v,t)$ for large classes of initial distributions $F(v,0)$ (essentially initial data without over-populated tails -see [41]) rapidly approach these scaling solutions in the sense that after a short transient time the data $v_0^d(t)\,F(v,t)$ can be collapsed on a single scaling form $f(c)$, as first observed by Baldassarri et al. in MC simulations of the nonlinear Boltzmann equation for a one-dimensional inelastic Maxwell model. This rapid approach to a universal scaling function applies both to systems driven by a Gaussian or by a white noise thermostat, as well as for freely cooling systems, which show scaling behavior, identical to systems driven by a Gaussian thermostat.

Originally this scenario had the status of a conjecture for systems of inelastic hard spheres ($\nu = 1$) and inelastic Maxwell models ($\nu = 0$), as formulated in [41] where also some analytical evidence for the approach to a scaling form has been presented. For Maxwell models the conjecture has been rigorously proven in the mean time by Bobylev et al. [42]. The analysis in Sect. 4.1 also suggests what the basic criterion is for the approach of ISS-models to the scaling form. If the energy balance equation (48) has a stable/attractive fixed point solution, $v_0(\infty)$, then the distribution function $F(v,t)$ approaches – in the sense detailed in the paragraphs above – a stable NESS described by the scaling solution with a high energy tail that is generically of stretched Gaussian type.

The analysis in Sect. 4.1 of the energy balance equation also suggests that cases of marginal stability – which are the exceptional border line cases – are candidates for power law tails, where the freely evolving inelastic Maxwell model is a well known example. At what $\nu-$value an ISS-model is *marginal* depends on the type of energy input or thermostat. A further case of power law tails is the ISS-model with $\nu = -2$, which is the marginal case for white noise forcing. This, and the relation between marginal stability and power law tails, as well as additional cases of power law tails are further explored and verified by MC simulations in [34]. The general conclusion is here that among ISS-models stretched Gaussian tails represent the generic form of high energy tails, and that power law tails are exceptions.

In this article we have focused on the properties of the scaling solution, and in particular on its high energy tail. Here we have introduced the Boltzmann equation for new classes of inelastic interactions, named Inelastic Soft Spheres or ISS-models, corresponding to pseudo-repulsive power law potentials, with collision probabilities proportional to $\text{A} \sim |\vec{g} \cdot \vec{n}|^{\nu}$, covering hard scatterers like inelastic hard spheres ($\nu = 1$) and soft scatterers like pseudo-Maxwell molecules ($\nu = 0$). The energy loss in an inelastic interaction is proportional to the inelasticity, $\gamma_0 \sim (1 - \alpha^2)$, where α is the coefficient of restitution. We have studied two typical cases: a freely evolving system in homogeneous cooling state, without energy supply, and systems with energy supply, driven by Gaussian thermostats with negative friction or driven by Gaussian white noise.

The homogeneous cooling laws in these systems are described by the granular temperature $T \propto v_0^2$, with a long time decay as $t^{-2/\nu}$ or $\exp[-2\gamma_\nu \tau]$, where t is the external (laboratory) time, and τ is the internal time or collision counter, and γ_ν is proportional to the inelasticity γ_0.

An interesting feature of all ISS-models is that their scaling forms generically have a stretched Gaussian high energy tail, $f(c) \sim \exp[-\beta c^b]$. In *freely cooling* ISS-models with $\nu > 0$ the stretching exponent $b = \nu$ satisfies $0 < b = \nu \leq 2$. Among the ISS-models at the stability threshold only the freely cooling Maxwell model has been analyzed in the present paper, and it does yield power law tails, $f(c) \sim 1/c^{d+a}$, and the exponent a has been calculated from a transcendental equation. In the ISS-models with $\nu > 0$, driven by white noise, there exist again stretched Gaussian tails with $b = \frac{1}{2}(\nu + 2)$. The general conclusion is here: the larger the $\nu-$values, i.e. the larger the scattering cross-sections or collision frequencies at high energy, the closer the stretching exponent is to $b = 2$, i.e. the closer the tails are to Gaussian tails. All known results for higher dimensional inelastic hard spheres ($\nu = 1$) and inelastic Maxwell models ($\nu = 0$) are recovered in the present analysis.

In ISS-models with positive ν the tails are always stretched Gaussians, and they are determined only by the loss term in the Boltzmann equation. In the freely cooling Maxwell model, and more generally at a stability threshold – which are the cases leading to power law tails [34] – the loss and gain term in the nonlinear Boltzmann equation are of comparable size, and partially balancing each other.

We have also analyzed in Sect. 2 an inelastic BGK-model, introduced by Brey et al [49], and generalized it to an energy dependent collision frequency, and we added energy source terms to this kinetic equation as well. In all cases the scaling solution can be calculated exactly. The freely cooling model shows an algebraic tail, $f(c) \sim 1/c^{d+a}$ with an exponent $a \sim 1/\gamma \sim 2/(1 - \alpha^2)$. For white noise driving one finds asymptotically $f(c) \sim \exp[-\beta c]$ with $\beta \sim (1 - \alpha^2)^{-1/2}$. In this BGK-model the existence of algebraic tails, as well as the value of the tail exponents are independent of the energy dependence of the collision frequency in this model.

It is interesting to observe that the overpopulated high energy tails in the inelastic BGK-models, both for free cooling as well as for white noise driving, are essentially the same as for the more complicated Maxwell models. However, the present extension to an energy-dependent collision frequency does not capture the generic features of the less-singular stretched Gaussian tails for the ISS-models with $\nu > 0$, as predicted by the nonlinear Boltzmann equation for these models.

Acknowledgements

The authors want to thank J.W. Dufty, A. Santos, E. Trizac, A. Barrat and K. Shundyak for helpful discussions and correspondence. This work is supported by DGES (Spain), Grant No BFM-2001-0291.

References

1. H. Waldmann, in: *Handbuch der Physik*, Vol 12, S. Flügge (Ed.) (Springer-Verlag, 1958)
2. S. Chapman and T.G. Cowling, *Mathematical Theory of Non-Uniform Gases*, (Cambridge Univ. Press, 1970)
3. P. Résibois and M. de Leener, *Classical Kinetic Theory of Fluids* (Wiley, New York, 1977)
4. C. Cercignani, *The Boltzmann Equation and its Applications* (Springer Verlag, New York, 1988)
5. G.E. Uhlenbeck and G.W. Ford, *Lectures in Statistical Mechanics* (Americam Math. Society, Providence, Rhode Island, 1963)
6. C.S. Campbell, Annu. Rev. Fluid Mech. **22**, 57 (1990)
7. J.T. Jenkins and M.W. Richman, J. Fluid Mech. **192**, 313 (1988)
8. J.T. Jenkins and S.B. Savage, J. Fluid Mech. **130**, 187 (1985)
9. P.K. Haff, J. Fluid. Mech. **134**, 40 (1983)
10. N. Sela and I. Goldhirsch, Phys. Fluids **7**, 507 (1995)
11. A. Goldshtein and M. Shapiro, J. Fluid Mech. **282**, 75 (1995)
12. I. Goldhirsch and G. Zanetti, Phys. Rev. Lett. **70**, 1619 (1993) I. Goldhirsch, M.L. Tan and G. Zanetti, J. Scientific Computing **8**, 1 (1993).
13. S.R. Nagel, Rev. Mod. Phys. **64**, 321 (1992)
14. S. Luding, R. Cafiero and H.J. Herrmann, this volume, page 293
15. A. Baldassarri, U. Marini Bettolo Marconi, and A. Puglisi, this volume, page 95
16. J.J. Brey, J.W. Dufty and M.J. Ruiz-Montero, this volume, page 227
17. I. Goldhirsch and T. P. C. van Noije, Phys. Rev. E **61**, 3241 (2000)
18. M.O. Cáceres, Phys. Rev. E **67**, 016102 (2003)
19. D.R.M. Williams and F.C. MacKintosh, Phys. Rev. E **54**, R9 (1996)
20. J.J. Brey, J.W. Dufty and A. Santos, J. Stat. Phys. **87**, 1051 (1997)
21. S.E. Esipov and T. Pöschel, J. Stat. Phys. **86**, 1385 (1997)
22. T.P.C. van Noije and M.H. Ernst, Granular Matter **1**, 57 (1998), and cond-mat/9803042
23. A. Barrat, T. Biben, Z. Rácz, E. Trizac, and F. van Wijland, J. Phys. A: Math. Gen. **35**, 463 (2002) and cond-mat/0110345
24. Th. Biben, Ph. A. Martin and J. Piasecki, Physica A **310**, 308 (2002)

25. J.J. Brey, M.J. Ruiz-Montero and D. Cubero, Phys. Rev. E **54** 3664 (1996). J.J. Brey, D. Cubero and M.J. Ruiz-Montero, Phys. Rev. E **59**, 1256 (1999)
26. J.M. Montanero and A. Santos, Granular Matter **2**, 53 (2000) and cond-mat/0002323
27. M. Huthmann, J.A.G. Orza and R. Brito, Granular Matter **2**, 189 (2000), and cond-mat/0004079
28. I. Pagonabarraga, E. Trizac, T.P.C. van Noije and M.H. Ernst, Phys. Rev. E **65**, 011303 (2002) and cond-mat/0107570
29. E. Ben-Naim and P. Krapivsky, Phys. Rev. E **61**, R5 (2000)
30. A.V. Bobylev, J.A. Carrillo, and I.M. Gamba, J. Stat. Phys. **98**, 743 (2000); see also erratum: J. Stat. Phys. **103**, 1137 (2001)
31. J.A. Carrillo, C. Cercignani, and I.M. Gamba, Phys. Rev. E **62**, 7700 (2000); C. Cercignani, J. Stat. Phys. **102**, 1407 (2001)
32. G.P. Collins, Sci. Am. **284**(1), 17 (2001)
33. F. Rouyer and N. Menon, Phys. Rev. Lett. **85**, 3676 (2000); D.L. Blair and A. Kudrolli, Phys. Rev. E **64**, 050301(R) (2001); C. Bizon, M.D. Shattuck, J.B. Swift, and H.L. Swinney, Phys. Rev. E **6**, 4340 (1999); J.S. Olafsen and J.S. Urbach, Phys. Rev. Lett. **81**, 4369 (1998); Phys. Rev. E **60**, R2468 (1999); W. Losert, D.G.W. Cooper, J. Delour, A. Kudrolli and J.P. Gollub, Chaos **9**, 682 (1999); I.S. Aranson and J.S. Olafsen, Phys. Rev. E **66**, 061302 (2002)
34. M.H. Ernst, E. Trizac and A. Barrat, to appear
35. A. Baldassarri, U. Marini Bettolo Marconi, and A. Puglisi, Europhys. Lett. **58**, 14 (2002), and this volume, page 95
36. A. Baldassarri, U. Marini Bettolo Marconi, and A. Puglisi, Phys. Rev. E **65**, 051301 (2002)
37. P.L. Krapivsky and E. Ben-Naim, J. Phys. A: Math Gen. **35**, L147 (2002), and this volume, page 65
38. E. Ben-Naim and P. Krapivsky, Phys. Rev. E **66**, 11309 (2002)
39. M.H. Ernst and R. Brito, cond-mat/0111093, 6 Nov. 2001, and Europhys. Lett. **58**, 182 (2002)
40. M.H. Ernst and R. Brito, Phys. Rev. E **65**, 040301(R) (2002)
41. M.H. Ernst and R. Brito, J. Stat. Phys. **109**, 407 (2002)
42. A.V. Bobylev and C. Cercignani, J. Stat. Phys. **106**, 547 (2002); A.V. Bobylev, C. Cercignani and G. Toscani, J. Stat. Phys. **111**, 403 (2003)
43. B. Nienhuis and R. van der Hart, unpublished; R. van der Hart, *Road Models and Granular Gases*, Master's thesis, Universiteit van Amsterdam, May 24, 2002
44. T. Antal, M. Droz and A. Lipowski, Phys. Rev. E **66**, 062301(2002)
45. A. Santos and M.H. Ernst, cond-mat/0302285
46. A.V. Bobylev, Sov. Phys. Dokl. **20**, 820 (1975); A.V. Bobylev, Sov. Phys. Dokl. **21**, 632 (1976)
47. T. Krook and T.T. Wu, Phys. Rev. Lett. **36**, 1107 (1976)
48. P.L. Bhatnagar, E.P. Gross and M. Krook, Phys. Rev. A **94**, 511 (1954)
49. J.J. Brey, F. Moreno, J.W. Dufty , Phys. Rev. A **54**, 445 (1996)
50. S. Ulam, Adv. Apll. Math. **1**, 7 (1980)
51. M.H. Ernst, Phys. Rep. **78**, 1 (1981)
52. T.P.C. van Noije and M.H. Ernst, in: *Granular Gases*, T. Pöschel and S. Luding (Eds), Lecture Notes in Physics (Springer-Verlag, Berlin, 2001), p. 1–30. M.H. Ernst, in: *Dynamics: models and kinetic methods for non-equilibrium many body*

systems, J. Karkheck, (Ed.), Nato Science Series E-Vol. 371, (Kluwer Academic Publs, Dordrecht, 1998) p. 239–266
53. E. Ben-Naim and P.L. Krapivski, this volume, page 65
54. J.R. Dorfman, *An Introduction to Chaos in Nonequilibrium Statistical Mechanics*, Cambridge Lecture Notes in Physics, Vol. 14, section 13.4 (Cambridge University Press, 1999)
55. N.V. Brilliantov and T. Pöschel, in: *Granular Gases*, T. Pöschel and S. Luding (Eds), Lecture Notes in Physics (Springer-Verlag, Berlin, 2001), p. 100–124; R. Ramírez, T. Pöschel, N.V. Brilliantov and T. Schwager, Phys. Rev. E **60**, 4465 (1999)
56. S. Chen, Y. Deng, X. Nie and Y. Tu, Phys. Lett. A **269**, 218 (2000)
57. H. Hwang and K. Hutter, Continuum Mech. and Thermodynamics **7**, 357 (1995)
58. R. Soto, M. Mareschal and M. Malek Mansour, Phys. Rev. E **62**, 3836 (2000)
59. U.M. Bettolo Marconi, and A. Puglisi, Phys. Rev. E 66, 011301 (2002)

The Homogeneous Cooling State Revisited

Isaac Goldhirsch[1], S. Henri Noskowicz[1], and Oded Bar-Lev[2]

[1] Department of Fluid Mechanics and Heat Transfer, Faculty of Engineering, Tel-Aviv University, Ramat-Aviv, Tel-Aviv 69978, Israel

[2] School of Mathematics, Faculty of Exact Sciences, Tel-Aviv University, Ramat-Aviv, Tel-Aviv 69978, Israel

Abstract. The distribution function for the homogeneous cooling state (HCS) of a monodisperse collection of inelastically colliding spheres has more structure than hitherto thought. Define the following notation: e is the (fixed) coefficient of normal restitution, $\epsilon \equiv 1 - e^2$ the 'degree of inelasticity', T the granular temperature, and u the speed variable. There is an initial-state-dependent speed, u^*, such that for $u^* < u$ the HCS distribution function retains memory of the initial distribution. For $u < u^*$ the initial distribution converges to a universal scaling function. In the latter range of speeds, there is a near-Maxwellian subrange for $u < \mathcal{O}\left(\sqrt{T}/\sqrt{\epsilon}\right)$, a 'transition' subrange for $\mathcal{O}\left(\sqrt{T}/\sqrt{\epsilon}\right) < u < \mathcal{O}\left(\sqrt{T}/\epsilon\right)$, and a near-exponential subrange for $u > \mathcal{O}\left(\sqrt{T}/\epsilon\right)$. These speed ranges are more distinct for near-elastic collisions, but still observable for any value of $0 \le e < 1$. These results follow from numerical studies, heuristic considerations and each of three levels of analytic reduction of the Boltzmann equation: (i). an exact reduction to an equation in the speeds; (ii). an asymptotic reduction to a relatively simple one dimensional integrodifferential equation, and (iii). an asymptotic reduction of the latter to a simple transcendental equation; the latter leads to an approximate analytic solution for the distribution function of the HCS. The distribution functions obtained from the reduced descriptions are practically indistinguishable from the corresponding numerical solutions of the full Boltzmann equation. Possible implications of the results beyond the realm of the HCS are briefly mentioned.

1 Introduction

Granular gases lack a steady unforced homogeneous (of uniform macroscopic density) and isotropic (the directions of the particle velocities are isotropically distributed) state, i.e., an equilibrium-like state, since the kinetic energy of a free granular gas decreases with time as a result of the inelasticity of the grain collisions. A free, homogeneous and isotropic granular gas, whose energy does decay (hence its state is not stationary) is conceptually possible. This state, which is probably the 'simplest' state of a granular gas one can conceive, is known as the homogeneous cooling state (HCS). The Boltzmann equation corresponding to a granular gas indeed possesses an HCS solution. However, this solution is known to be unstable to clustering [1] and collapse [2], hence it is only a theoretical construct. Nevertheless, it turns out that the HCS is a useful concept in many respects, part of which are mentioned below.

First, a direct study of the HCS enables one to learn some properties of the inelastic Boltzmann equation. The HCS serves (more accurately, the local HCS does) as a zeroth order in some perturbation theories (based on the Chapman-Enskog expansion) that result in constitutive relations for granular gases [3]. Furthermore, the HCS has been employed in derivations of Green-Kubo relations for granular gases [4]. In addition, the study of the HCS reveals the breaking of equipartition [5, 6] in granular gases, in its simplest manifestation.

In spite of its basic significance in the field of granular gases, it seems that the HCS has not been thoroughly studied. Haff [7] was probably the first to show that the energy (or granular temperature) of an HCS of smooth inelastic spheres (with a fixed coefficient of normal restitution) decays as $\frac{1}{t^2}$, where t denotes time (corrections to Haff's law were observed numerically in ([1]-iv,v,vi) and theorized upon in ([8]-i,iii) and ([1]-vi,v,vi); as we only consider the HCS, we shall not dwell on this problem below). The velocity distribution of the HCS is not Maxwellian. Its departure from a Maxwellian distribution (in the bulk, i.e., not the high speed tail) has been studied perturbatively [8], and its tail has been shown [9] to decay exponentially (in the speed). It has also been discovered that the HCS distribution function can be appropriately scaled so that it is described (asymptotically in time) by a time-independent scaling function [8]. This property has been exploited in the above mentioned derivations of constitutive relations and Green-Kubo relations. All of the above results pertain to collections of smooth spheres. The HCS for e.g., frictional spheres has not been studied in great detail: mean field approaches [6] have revealed that it is (expectedly) more complex than the corresponding HCS for smooth spheres (see also ([8]-i) for an early account of the breakdown of equipartition in these systems).

Clearly the HCS deserves careful and detailed study. Due to the above mentioned instabilities, it is impractical to study it by molecular dynamics methods except in very small and nearly elastic systems (although Monte-Carlo methods, such as the DSMC allow for such a study, e.g. ([1]-v), they solve the underlying Boltzmann equation and do not provide a more basic validation as MD would have). The goal of the work presented below is to provide a detailed analytic description of the HCS for smooth spheres in three dimensions. Among other results, it is shown that the pertinent Boltzmann equation can be reduced to a one-dimensional integrodifferential equation in the speed, which quantitatively captures essentially all properties of the HCS.

Prior to embarking on the main goal of this article, it is perhaps important to stress that the coefficient of normal restitution for real grains is truly velocity dependent and that this property has important consequences, such as the elimination of the collapse phenomenon [2] and the destabilization of clusters [10]. This and other facts pertaining to realistic grains do not diminish the importance of the HCS described here as the simplest (and perhaps of academic nature) case of a granular gas.

2 Setting Up the Problem

2.1 The Collision Model

The model considered in the present work is that of a monodisperse collection of spheres of mass $m = 1$ and radius a. The system is assumed to be of infinite extent. The spheres interact by (binary) instantaneous collisions characterized by a fixed coefficient of normal restitution, denoted by e, with $0 \leq e \leq 1$. The basic collision process between two spheres, whose precollisional ('incoming') velocities are $\vec{v}_{\text{in},1}$ and $\vec{v}_{\text{in},2}$, and whose corresponding postcollisional ('outgoing') velocities are $\vec{v}_{\text{out},1}$ and $\vec{v}_{\text{out},2}$, respectively, is given by: $\vec{v}_{\text{out},21} \cdot \hat{\vec{n}} = -e\, \vec{v}_{\text{in},21} \cdot \hat{\vec{n}}$, where $\vec{v}_{\text{out},21} \equiv \vec{v}_{\text{out},2} - \vec{v}_{\text{out},1}$, $\vec{v}_{\text{in},21} \equiv \vec{v}_{\text{in},2} - \vec{v}_{\text{in},1}$, and $\hat{\vec{n}}$ is the unit vector pointing from the center of sphere "1" to the center of sphere "2". Combining the collision rule with momentum conservation one obtains:

$$\vec{v}_{\text{out},i} = \vec{v}_{\text{in},i} + \frac{1+e}{2} \left(\vec{v}_{\text{in},ji} \cdot \hat{\vec{n}} \right) \hat{\vec{n}} \,, \tag{1}$$

and

$$\vec{v}_{\text{in},i} = \vec{v}_{\text{out},i} + \frac{1+e}{2e} \left(\vec{v}_{\text{out},ji} \cdot \hat{\vec{n}} \right) \hat{\vec{n}} \,, \tag{2}$$

where $i, j \in \{1, 2\}$, and $\vec{v}_{\text{in},12} \cdot \hat{\vec{n}} \geq 0$ must hold.

2.2 The Boltzmann Equation

Let $f(\vec{r}, \vec{v}, t)$ denote the single particle (velocity) distribution function at point $\vec{r}$ at time t, where $\vec{v}$ denotes the (particle) velocity. By definition, $\int f(\vec{r}, \vec{v}, t) d\vec{v} = n(\vec{r}, t)$, where $n(\vec{r}, t)$ is the (particle) number density at point $\vec{r}$ at time t. For low densities the distribution function, f, for a granular gas, is believed to satisfy the inelastic Boltzmann equation [8, 11]:

$$\frac{\partial f(\vec{r}, \vec{v}, t)}{\partial t} + \vec{v} \cdot \nabla f(\vec{r}, \vec{v}, t) + \vec{F} \cdot \frac{\partial f(\vec{r}, \vec{v}, t)}{\partial \vec{v}} = \left(\frac{\partial f(\vec{r}, \vec{v}, t)}{\partial t} \right)_c \,, \tag{3}$$

where $\vec{F}$ is an external (velocity independent) force and $(\partial f / \partial t)_c$ represents the effect of the collisions. It is convenient [12] to separate the right hand side of (3) into a 'gain term', $(\partial f(\vec{r}, \vec{v}, t)/\partial t)_g \geq 0$, which represents all collisions that increase the number density of particles having velocity $\vec{v}$, and a 'loss term', $\left(\frac{\partial f(\vec{r}, \vec{v}, t)}{\partial t} \right)_\ell \geq 0$, which represents all collisions that decrease this number density, such that $\left(\frac{\partial f(\vec{r}, \vec{v}, t)}{\partial t} \right)_c = \left(\frac{\partial f(\vec{r}, \vec{v}, t)}{\partial t} \right)_g - \left(\frac{\partial f(\vec{r}, \vec{v}, t)}{\partial t} \right)_\ell$. A standard procedure [6, 11, 12] yields:

$$\left(\frac{\partial f(\vec{r}, \vec{v_1}, t)}{\partial t} \right)_\ell = \sigma_T f(\vec{r}, \vec{v}_1, t) \int |\vec{v}_{12}| \; f(\vec{r}, \vec{v}_2, t) \; d\vec{v}_2, \tag{4}$$

where $\sigma_T = 4\pi a^2$ is the total collision cross section. Notice that the form of the loss term is not affected by the inelasticity of collisions. The gain term is given by:

$$\left(\frac{\partial f\left(\vec{r},\vec{v_1},t\right)}{\partial t}\right)_g = \frac{1}{e^2}\int |\vec{v}_{12}|\ f(\vec{r},\vec{v}_1',t)f(\vec{r},\vec{v}_2',t)\ d\vec{v}_2\ b\ db\ d\phi\ , \tag{5}$$

where b is the (outgoing) impact parameter ($0 \le b \le 2a$), ϕ the (outgoing) azimuthal angle of the collision and $\{\vec{v}_1', \vec{v}_2'\}$ are the precollisional ('incoming') velocities corresponding to the postcollisional ('outgoing') velocities $\{\vec{v}_1, \vec{v}_2\}$, respectively. Notice that the dependence of (5) on the coefficient of restitution, e, is both explicit, through the factor $1/e^2$, and implicit, through the dependence of $\vec{v}_1'$ and $\vec{v}_2'$ on $\vec{v}_1$ and $\vec{v}_2$.

2.3 The Boltzmann Equation for the Homogeneous and Isotropic Case

In the homogeneous and isotropic case, the distribution function, f, at time t, depends only on the speed, v (and t), and the number density is a constant. Below, the external force field, $\vec{F}$, is assumed to vanish. In this case the Boltzmann equation can be reduced to an equation involving speeds alone, as shown below.

Consider first the loss term. Let μ be the cosine of the angle between $\vec{v}_2$ and $\vec{v}_1$ in (4). Using $v_{21}^2 = v_1^2 + v_2^2 - 2\mu v_1 v_2$, and integrating over μ and the corresponding azimuthal angle (the latter yielding a factor of 2π), one obtains from (4):

$$\left(\frac{\partial f(v_1,t)}{\partial t}\right)_\ell = \frac{2\pi\sigma_T}{3v_1} f(v_1,t)\int_0^\infty \left[(v_1+v_2)^3 - |v_1-v_2|^3\right] v_2 f(v_2,t)dv_2 \tag{6}$$

Further straightforward algebra yields from (6):

$$\left(\frac{\partial f(u,t)}{\partial t}\right)_\ell = 4\pi\sigma_T f(u)\left[\int_0^\infty \left(uv^2 + \frac{v^4}{3u}\right) f(v)dv \right.$$
$$\left. + \int_u^\infty \left(v^3 + \frac{1}{3}u^2 v - uv^2 - \frac{v^4}{3u}\right) f(v)dv\right]\ , \tag{7}$$

The angular integrations for the $(\partial f/\partial t)_g$ term are rather tedious. A detailed derivation will be presented elsewhere. The result is that the gain term can be expressed as a sum of nine expressions, $\{G_i(u); 1 \le i \le 9\}$ (presented in Appendix A-1): $(\partial f(u,t)/\partial t)_g = \sum_{i=1}^9 G_i(u,t)$.

3 Heuristic Analysis

In this section, a heuristic study of the HCS distribution function is presented and validated by numerical solutions of the pertinent Boltzmann equation.

It turns out that the values of the speed can be divided into a range in which memory of the the initial condition is retained, and one in which it is not. The latter range can be described by a scaling function and it consists of a near-Maxwellian, an intermediate and an exponential subrange.

3.1 Large Speeds

Assuming that for asymptotically large values of u, the distribution $f(u)$, decays faster than algebraically in u (a rather weak assumption, corroborated by the results below), it follows from (7) that the second integral on the r.h.s. of (7) is negligible with respect to the first integral for sufficiently large speeds. Hence, asymptotically:

$$\left(\frac{\partial f(u,t)}{\partial t}\right)_\ell \approx n\sigma_T f(u,t)\left[u+\frac{\langle v^2;t\rangle}{3u}\right], \tag{8}$$

where $\langle v^k;t\rangle$ denotes the value of the k-th moment of the velocity (at time t): $\langle v^k;t\rangle = \frac{4\pi}{n}\int_0^\infty {v'}^{k+2} f(v',t)\,dv'$. The asymptotic properties of the gain term are considered next.

Notice that both for a granular and elastic gas, 'high speed' particles are rare and they collide chiefly with particles of typical speed. When the faster particle emerging from such a collision is slower than the faster 'parent' particle, one can say that the collision contributes to a speed flux from high to low speeds, i.e., a 'downward' flux. Had there been no opposite flux in the elastic case, a state of equilibrium could not have existed. Therefore there must be (at least in the elastic case) collisions which contribute to an upward flux in the speed space to compensate for the downward flux. In contrast, in the inelastic case, the distribution function eventually shrinks to a delta function in the speed. Therefore, the downward flux must dominate over the upward flux. These observations are put on a more quantitative basis, by considering the kinematics of collisions, in the next paragraph.

Consider a collision $(\vec{v}_1', \vec{v}_2') \to (\vec{v}_1, \vec{v}_2)$ of a 'fast' particle '1' with a typical particle '2': $v_1' \gg v_2' = \mathcal{O}(\sqrt{\langle v^2\rangle})$. This collision can contribute to the upward flux, i.e., 'produce' a particle whose speed is larger than v_1', if either $v_1 > v_1'$ (henceforth, Case I) or $v_2 > v_1'$ (henceforth, Case II). Using the kinematic equations, (1,2), as well as $\vec{v}_{12}' \cdot \hat{\vec{n}} > 0$, this is equivalent to

Case I: $$\vec{v}_2' \cdot \hat{\vec{n}} < \vec{v}_1' \cdot \hat{\vec{n}} < -\frac{1+e}{3-e}\left(\vec{v}_2' \cdot \hat{\vec{n}}\right)$$

Case II: $$\left(\vec{v}_1' \cdot \hat{\vec{n}} - \vec{v}_2' \cdot \hat{\vec{n}}\right)\left(\vec{v}_1' \cdot \hat{\vec{n}} + \frac{3-e}{1+e}\vec{v}_2' \cdot \hat{\vec{n}}\right) > \left(\frac{2}{1+e}\right)^2 \left(v_1'^2 - v_2'^2\right)$$

Since, by assumption, $\left|\vec{v}_2' \cdot \hat{\vec{n}}\right| \le v_2' = \mathcal{O}\left(\sqrt{\langle v^2\rangle}\right)$, inequality I can be satisfied only if $\left|\vec{v}_1' \cdot \hat{\vec{n}}\right| = \mathcal{O}\left(\sqrt{\langle v^2\rangle}\right)$, hence $\left|\vec{v}_{12}' \cdot \hat{\vec{n}}\right| = \mathcal{O}\left(\sqrt{\langle v^2\rangle}\right)$. This is possible only for (the relatively rare) near grazing collisions (impact parameter close to $2a$). This result holds equally for both the elastic and inelastic cases. Consider next case II. When $e = 1$, the corresponding inequality is satisfied by $v_{1t}'^2 < v_{2t}'^2$, where the subscript t denotes the tangential component of a vector: $v_t \equiv \sqrt{v^2 - (\vec{v} \cdot \hat{\vec{n}})^2}$. The latter inequality can be rewritten as $v_{12t} \times (v_{1t} + v_{2t}) < 0$; this inequality is clearly satisfied by a finite fraction of the collisions. On the other hand, when $e < 1$, one obtains from condition **II**:

$$\frac{(3+e)(1-e)}{4}\left(\vec{v}_1' \cdot \hat{\vec{n}}\right)^2 - \frac{(1-e^2)}{2}\left(\vec{v}_2' \cdot \hat{\vec{n}}\right)\left(\vec{v}_1' \cdot \hat{\vec{n}}\right) + \frac{(3-e)(1+e)}{4}\left(\vec{v}_2' \cdot \hat{\vec{n}}\right)^2 < v_2'^2 - \left(\vec{v}_{1t}'\right)^2 \quad (9)$$

A straightforward analysis of the inequality, (9), reveals that it can be satisfied only when v_1' is smaller than $v_{1,\max}' = \sqrt{\frac{3-e}{2(1-e)}}v_2' = \sqrt{\frac{(3-e)(1+e)}{2(1-e^2)}}v_2'$ and that this maximal value is attained when $v_{1t}' = 0$ and $\vec{v}_2' \cdot \hat{\vec{n}} = \sqrt{\frac{1-e^2}{2(3-e)(1+e)}}v_2'$. It follows that in the inelastic case there is virtually no upward speed flux for $v_1' > v_{1,\max}$, i.e., when $(1-e^2)\, v_1'^2 > \mathcal{O}(\langle v^2 \rangle)$. As mentioned, the downward flux is hardly affected by inelasticity. Therefore, the gain term must be subdominant to the loss term at values of the speed which exceed the order of $v_{1\max}'$. Hence, there must be a value of the speed, $u^* \geq \mathcal{O}\left(\sqrt{\frac{\langle v^2 \rangle}{1-e^2}}\right)$, such that for $u > u^*$ the gain term is subdominant to the loss term. In this range of speeds the Boltzmann equation reduces to: $\frac{\partial f(u,t)}{\partial t} \sim -n\sigma_T u\, f(u,t)$ (upon keeping the first term in the r.h.s. of (8)), or $f(u,t) \sim f(u,0)e^{-n\sigma_T u t}$. Therefore, for $u > u^*$ the HCS distributions 'remembers' the initial condition. Notice that as the initial distribution, $f(u,0)$, is u dependent, the (large u) part of the distribution, which retains memory of the initial condition, is not necessarily exponential (although, as $t \to \infty$, the exponential dominates).

3.2 The Speed Ranges and Universality

As is well known [13, 14], the distribution function for an unforced system of elastically colliding spheres (or any other elastic system) converges very rapidly (in a time corresponding to a few collisions per particle) from practically any initial distribution to a near-Maxwellian distribution (except in the high speed tail [13], where convergence is much slower). The corresponding convergence rate can be estimated, for example, from the value of the leading nonzero eigenvalue of the linearized Boltzmann operator, which is $\mathcal{O}(1)$ in units of the inverse mean free time. Next, consider a quasielastic system. If the degree of inelasticity, $\epsilon \equiv 1 - e^2$, is sufficiently small, the outcome of a small number of collisions per particle is very close to the outcome of a small number of elastic collisions. Therefore a near elastic system is expected to

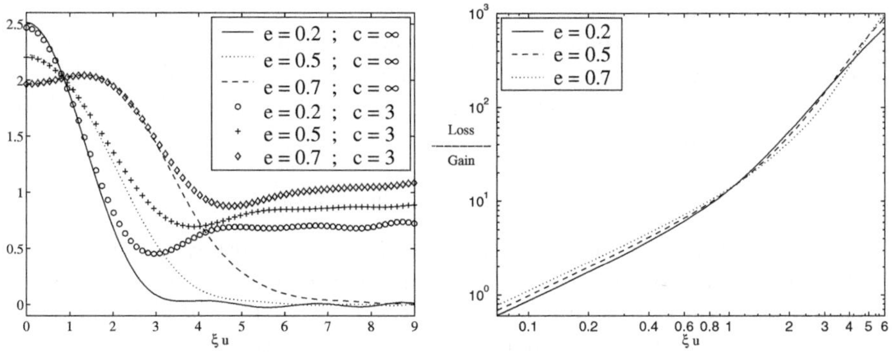

Fig. 1. Left: Development of the distribution function as a function of the number of accumulated collisions per particle, c. Shown are the negatives of the second derivative of the logarithm of the rescaled distributions versus the rescaled speed ξu for three values of the coefficient of restitution, e. Notice the crossover from the Maxwellian value, 2, to the value 0, corresponding to the exponential decay. Also notice the fast convergence of the bulk versus the slow convergence of the tail. The initial distribution is Maxwellian. **Right**: The ratio of the loss term to the gain term extracted from a converged numerical solution of the Boltzmann equation for three values of e, versus the rescaled speed. For $u\xi < 0.1$ the ratio is smaller than unity, indicating a dominance of the gain, when $0.1 < u\xi < 1/\sqrt{1-e^2}$ the ratio is proportional to $u\xi$ (the loss dominates the gain for $u\xi > 1$), and for $u\xi > 1/\sqrt{\epsilon}$ the ratio is proportional to $(\xi u)^{\gamma(\epsilon)}$, with $\gamma \approx 3$, in conformity with the theoretical calculation (Appendix C-3)

develop a near Maxwellian distribution for speeds of the order of $\sqrt{\langle v^2 \rangle}$ (or smaller) with a granular temperature (i.e., $\langle v^2 \rangle$) slowly decaying with time. As the energy loss in a single collision is proportional to $\epsilon(\vec{v}_{12} \cdot \hat{\vec{n}})^2$, one expects the near-Maxwellian distribution to extend to speeds of the order of $1/\sqrt{\epsilon}$. This implies that the statistical weight of the non-Maxwellian (tail) part of the distribution function is small. Similarly, one expects the memory of the details of the initial distribution to fade following a few collisions per particle, even when ϵ is not small, although it is not a-priori clear that in this case the bulk of the distribution function becomes near-Maxwellian. As a matter of fact we have checked (by solving the pertinent Boltzmann equation numerically) that this is the case down to quite low values of e (e.g., for $e = 0.2$). The development of the distribution function in time is demonstrated in Fig. 1. As the form of the distribution function emerging after a few collisions per particle is independent of the form of the initial distribution (except in the tail), it does not change in time any more, i.e., it becomes *universal* (but still dependent on the value of e). The universal distribution, being normalized, is characterized by e and its (shrinking) width alone. Define the inverse width, ξ, as follows: $\xi^2(t) \equiv \frac{3}{2\langle v^2;t \rangle}$. In the universal range, $d\xi/dt$ has dimensions of inverse length; the only available parameter having this dimension is $n\sigma_T$

(the inverse mean free path). It follows that $d\xi/dt = Cn\sigma_T$, where C is a constant. Mean field calculations (see remark in Appendix B) suggest that $C \propto \epsilon$, in conformity with systematic studies (cf. e.g., Sec. (4)). This is easy to understand since ξ does not change in time when $\epsilon = 0$. Defining λ by $C \equiv \epsilon\lambda\pi^{-\frac{1}{2}}$, one obtains:

$$\xi(t) = \epsilon\lambda n\sigma_T\pi^{-\frac{1}{2}}(t - t_0) + \xi(t_0) \; , \tag{10}$$

where t_0 is a time beyond which the distribution is largely (except at the far tail) universal. With this definition, one can express the bulk of the distribution function as a scaling function: $f(u,t) = n\xi^3(t)\tilde{f}(u\xi(t))$, where the prefactor is determined by dimensional considerations, and where $\tilde{f}$ is universal (its dependence on e has been suppressed).

Let u be a speed in the range corresponding to the universal (and scaling) form of f, at time t. Clearly, after a sufficient time elapses, u will be in the tail of the distribution, since the entire distribution shrinks. Let $t(u)$ be a time at which u is sufficiently larger than $\langle v^2; t(u)\rangle$, so that its subsequent dynamics is dominated by the loss term. Following the result presented at the end of Sec. (3.1): $f(u,t) = \exp\left[-n\sigma_T u(t - t(u))\right] f(u, t(u))$ for $t > t(u)$, hence, using (10): $f(u,t) = \exp\left[-\sqrt{\pi}u\frac{\xi(t)-\xi(t(u))}{\epsilon\lambda}\right] f(u,t(u))$. Substituting now the scaling form of f, multiplying both sides of the equation by u^3 and rearranging terms, one obtains:

$$u^3\xi^3(t)\tilde{f}(u\xi(t))\exp\left[\frac{\sqrt{\pi}u\xi(t)}{\epsilon\lambda}\right] = \exp\left[\frac{u\xi(t(u))\sqrt{\pi}}{\epsilon\lambda}\right] u^3\xi^3(t(u))\tilde{f}(u\xi(t(u))) \; . \tag{11}$$

Equation (11) implies that there must be a constant, A, such that $u\xi(t(u)) = A$. Clearly, $A \gg 1$ since u is in the tail of the distribution at $t = t(u)$. It follows that, for $t > t(u)$ the distribution, $f(u)$, is given by:

$$f(u,t) = n\left(A^3\tilde{f}(A)\exp\left[\frac{A\sqrt{\pi}}{\lambda\epsilon}\right]\right)\frac{1}{u^3}\exp\left[-\frac{\xi(t)u\sqrt{\pi}}{\lambda\epsilon}\right] \; . \tag{12}$$

Figure 1 depicts the convergence of a Maxwellian initial condition to a universal distribution for several values of e (a similar result is obtained for non-Maxwellian initial conditions). The crossover between the Maxwellian-like range and the exponential range is visible. As the values of ϵ used here are $\mathcal{O}(1)$ (for technical reasons) one cannot read the scaling of the crossover 'points', in terms of ϵ, off Fig. 1.

To reiterate, the exponential tail is, strictly speaking, an intermediate tail (developing out of the scaling part of the distribution at an earlier time), since the true tail retains the memory of the initial condition.

Assume for the moment that the near-Maxwellian range of speeds crosses over smoothly to the exponential range. Equating $\exp\left(-\xi^2u^2\right)$ with the exponential form, one obtains that the crossover must occur at $u\xi \propto 1/\epsilon$.

However, as explained above (see also Sec. (4)), the Maxwellian range is limited to $u\xi < 1/\sqrt{\epsilon}$. Therefore there must be a crossover region, 'starting' at $u\xi \propto 1/\sqrt{\epsilon}$, and ending where the exponential decay 'starts'. The crude matching of the Maxwellian and exponential ranges may be taken to suggest that the exponentially decaying range starts at $u\xi \propto 1/\epsilon$, a conclusion corroborated below.

4 The Near-Maxwellian Range of Speeds

This section is devoted to a study of the near-Maxwellian range of speeds. To this end it is convenient to perform a perturbative expansion of the pertinent Boltzmann equation in powers of the degree of inelasticity, $\epsilon = 1 - e^2$. Only the universal distribution is sought.

Denote the (universal) scaled distribution function by $\tilde{f}(\tilde{u})$, where $\tilde{u} \equiv \xi(t)u$, $f(u,t) = n\xi^3(t)\tilde{f}(\tilde{u})$, and $\xi(t)$ is given by (10). Notice that the parameter λ, cf. (11), is left to be determined by the perturbation theory. With this scaling, the Maxwellian solution transforms to $\tilde{f}_M \equiv \pi^{-3/2}\exp\left(-\tilde{u}^2\right)$. Upon scaling the impact parameter, b, by its maximal value, $2a$ ($\tilde{b} \equiv \frac{b}{2a}$), the (rescaled) Boltzmann equation reads:

$$3\tilde{f}(\tilde{v}_1) + \tilde{v}_1\frac{d\tilde{f}(\tilde{v}_1)}{d\tilde{v}_1} = \frac{1}{\epsilon\lambda\sqrt{\pi}}\int d\tilde{\vec{v}}_2\tilde{b}\,d\tilde{b}\,d\phi\left|\tilde{\vec{v}}_{12}\right|\left(\frac{1}{e^2}\tilde{f}(\tilde{v}_1')\tilde{f}(\tilde{v}_2') - \tilde{f}(\tilde{v}_1)\tilde{f}(\tilde{v}_2)\right)$$
$$\equiv \frac{1}{\epsilon\lambda\sqrt{\pi}}\,\tilde{B}(\tilde{f},\tilde{f};e)\ , \quad (13)$$

thereby defining the (bilinear) operator $\tilde{B}$. Define the function ϕ by $\tilde{f}(\tilde{v}) = \tilde{f}_M(\tilde{v})(1+\phi(\tilde{v}))$. The correction ϕ is calculated perturbatively, in powers of ϵ; it can be shown that this is the only choice of a small parameter which is a function of ϵ, up to a multiplicative constant, that yields a consistent perturbation theory. Notice that when $\epsilon = 0$, $\phi = 0$. The zeroth order (Maxwellian) distribution is conveniently chosen to possess the correct kinetic energy (average of $\tilde{u}^2$). Therefore ϕ must satisfy: $\int_0^\infty v^4\exp\left(-v^2\right)\phi(v)dv = 0$ (so that it does not affect the average kinetic energy), as well as $\int_0^\infty v^2\exp\left(-v^2\right)\phi(v)dv = 0$ (so as to preserve the normalization of $\tilde{f}$). Define now the following expansions: $\phi(u,e) = \sum_{k=1}^\infty \epsilon^k\phi_k(u)$, and $\lambda(e) = \sum_{k=0}^\infty \epsilon^k\lambda_k$. It is also convenient to define the expansion of the nonlinear (Boltzmann) operator $\tilde{B}$ in powers of ϵ: $\tilde{B}(f,g,e) = \sum_{k=0}^\infty \epsilon^k B_k(f,g)$. The last definition does not include an expansion of f or g in powers of ϵ; the only sources of ϵ dependence in the expansion of $\tilde{B}$ are the explicit $1/e^2$ term in the definition of this operator and the ϵ dependence of $\{\tilde{\vec{v}}_1', \tilde{\vec{v}}_2'\}$ on $\{\tilde{\vec{v}}_1, \tilde{\vec{v}}_2\}$. Therefore the full expansion of $\tilde{B}$ in powers of ϵ, when $\tilde{f}_M(1+\sum_k \epsilon^k\phi_k)$ is substituted for $\tilde{f}$, requires a further expansion. This, slightly cumbersome definition, is calculationally convenient, as seen below. The tilde superscripts are removed below for sake of notational simplicity. The zeroth order in ϵ, now including the expansion of f in powers of ϵ, of the Boltzmann equation, reads:

$$\frac{1}{\lambda_0\sqrt{\pi}}\left[B_0(f_M\phi_1, f_M) + B_0(f_M, f_M\phi_1)\right] = 3f_M + u\frac{df_M}{du} - \frac{1}{\lambda_0\sqrt{\pi}}B_1(f_M, f_M). \tag{14}$$

It is convenient to multiply both sides of the above equation by $u\lambda_0/4f_M$, thus obtaining:

$$\begin{aligned}\widehat{L} * \phi_1(u) \equiv \frac{u}{4f_M\sqrt{\pi}}&\left[B_0(f_M\phi_1, f_M) + B_0(f_M, f_M\phi_1)\right] \\ &= \frac{u\lambda_0}{4f_M}\left(3f_M + u\frac{df_M}{du} - \frac{1}{\lambda_0\sqrt{\pi}}B_1(f_M, f_M)\right).\end{aligned} \tag{15}$$

The left hand side of (15) defines the action of the linearized Boltzmann operator, $\widehat{L}$, and the right hand side is a source term, s_1, whose functional dependence on u is easy to calculate, cf. (55), as f_M is given and B_1 is easily obtained from $B(f, f; e)$. Only the prefactor, λ_0, is not a-priori known (see below). The integrations defining the operator $\widehat{L}$ can be straightforwardly carried out, yielding:

$$\begin{aligned}\widehat{L} * \phi = \sqrt{\pi}\int_0^u dv v\phi(v)\,\mathrm{erf}(v) + \sqrt{\pi}e^{u^2}\mathrm{erf}(u)\int_u^\infty dv v\phi(v)e^{-v^2} \\ -\left(\frac{u}{4}e^{-u^2} + \frac{\sqrt{\pi}}{4}\left(\frac{1}{2} + u^2\right)\mathrm{erf}(u)\right)\phi(u) - \int_0^u dv\left(v^2u^2 + \frac{v^4}{3}\right)e^{-v^2}\phi(v) \\ -\int_u^\infty dv\left(uv^3 + \frac{vu^3}{3}\right)e^{-v^2}\phi(v)\,.\end{aligned} \tag{16}$$

Thus, the correction ϕ_1 satisfies: $\widehat{L} * \phi_1(u) = s_1(u)$. It is easy to show that the n-th order in perturbation theory is of the form: $\widehat{L} * \phi_n = s_n$.

The operator $\widehat{L}$ is self-adjoint in a Hilbert space in which the scalar product of two real functions, $a(v)$ and $b(v)$, is: $(a, b) \equiv \int_0^\infty v f_M(v)a(v)b(v)dv$. Also, $\widehat{L} * v^k = 0$ for $k = 0, 2$ [14]: the functions 1 and v^2 are the only normalizable zero eigenfunctions of $\widehat{L}$; they clearly correspond to the conservation of particle number and energy (in the elastic limit). Hence:

$$\left(v^k, \widehat{L} * \phi_n\right) = \left(v^k, s_n\right) = \int_0^\infty dv v^{k+1} f_M(v)s_n(v) = 0 \qquad k = 0, 2. \tag{17}$$

These are the two solubility conditions at order n. In addition, the two normalization conditions that ϕ_n must satisfy, read:

$$\left(v^k, \phi_n\right) = \int_0^\infty dv v^{k+1} f_M(v)\phi_n(v) = 0 \qquad k = 1, 3. \tag{18}$$

The general solution of $\widehat{L} * \phi_n = s_n$ is $\phi_n(u) = a_n + b_n u^2 + \Phi_n(u)$, where, $a_n + b_n u^2$ is the solution of the homogeneous equation, $\widehat{L} * \phi_n = 0$. Without loss of generality one may impose $\Phi_n(0) = \Phi_n''(0) = 0$ on the inhomogeneous solution.

All in all ϕ_n must satisfy two conditions, (18), and these determine the parameters a_n and b_n. The source term, s_n, which depends on one free parameter, λ_{n-1} (e.g., s_1 depends on λ_0), must also satisfy two conditions, (17). This does not lead to an overdetermination of λ_{n-1}, because, as shown in Appendix C-2, the condition corresponding to $k = 0$ in (17) is identically satisfied by s_n. In the corresponding elastic case the condition $k = 2$ is also identically satisfied by the source term, in conformity with the fact that there is no additional free parameter in this case.
The condition $k = 2$, applied to s_n, reads:

$$\left(v^2, s_1\right) = \pi^{-3/2} \int_0^\infty dv v^3 e^{-v^2} \left[\frac{\lambda_0}{4}\left(3v - 2v^3\right) - I_2(v)\right] = 0, \tag{19}$$

where I_2 is given by (54). This integral can be carried out analytically yielding: $\lambda_0 = \frac{\sqrt{2}}{3}$, in conformity with the mean field result, (50). The correction ϕ_1 can be straightforwardly found from the equation $\widehat{L} * \phi_1 = s_1$ in terms of a polynomial expansion: $\phi_1(u) = a_1 + b_1 u^2 + A_4 u^4 + A_6 u^6 + ...$, where $a_1 = -0.320$, $b_1 = 0.574$, $A_4 = -0.246$, $A_6 = 6.6075 \times 10^{-2}$, $A_8 = -1.337 \times 10^{-2}$ and $A_{10} = 1.001 \times 10^{-3}$. One can check that this perturbative solution agrees well with the numerical solution of the Boltzmann equation up to $u = \mathcal{O}\left(1/\sqrt{\epsilon}\right)$ and slightly beyond (for values of ϵ which are $\mathcal{O}(1)$). Notice the rather rapid decrease of these coefficients. A straightforward asymptotic analysis of the equation $\widehat{L} * \phi_1 = s_1$ (using (16) for $u \gg 1$), reveals that: $\phi_1(u) \sim -u^2 \ln(u) - 2\sqrt{2}u/\sqrt{\pi} + 3.9538u^2$, i.e., the 'asymptotic' solution of the Boltzmann equation for large u is, to linear order in ϵ: $\pi^{-3/2} \exp\left(-u^2\right)\left(1 - \epsilon u^2 \ln\, u\right)$; this limit makes sense only for $\epsilon \ll 1$ and $u < \mathcal{O}\left(1/\sqrt{\epsilon}\right)$. This, and the above results imply that the perturbation expansion is valid for $u < 1/\sqrt{\epsilon}$, which is also the range of the near-Maxwellian distribution. The reduced equation presented in the next section shares this property.

5 Reduction of the Boltzmann Equation for the HCS

The goal of the present section is to present a derivation of a reduced equation for the HCS, on the basis of the Boltzmann equation. The reduction is achieved by an asymptotic analysis of the Boltzmann equation at high speeds, $u \gg 1$ (in the rescaled variables). The error allowed in the non-linear Boltzmann operator is algebraically subdominant to the loss term at large speeds. It turns out that the error of the resulting equation at low speeds is formally $\mathcal{O}(\epsilon)$. The solutions of the resulting equation, much like those of other equations derived by asymptotic methods, correctly reproduce the HCS distribution function, far beyond the nominal limits of its derivation. As a matter of fact, the solutions of the reduced equation are practically indistinguishable from those of the full Boltzmann equation for all speeds, even at values of the coefficient of restitution which are as low as 0.4. A formal perturbative analysis of the reduced equation in powers of ϵ yields a leading order correction which is precisely that obtained in the previous section for $1 \gg u \gg 1/\sqrt{\epsilon}$. Possible explanations for these features are proposed. Needless to say, the three universal speed ranges are fully reproduced by the reduced equation (analytically and numerically). This section ends with a second reduction of the Boltzmann equation, which is of the form of a transcendental equation. The latter yields solutions which are in excellent agreement with the numerical solutions (and, of course, with the first reduced equation). Thus this section provides a nearly analytic and uniform solution for the HCS. Finally, we wish to stress that the methods presented below are non-perturbative.

5.1 Derivation of the Reduced Equation

The goal of the present subsection is to derive a simplification of the Boltzmann equation, which is formally valid at large values of the speed.

As in the above, it is convenient to employ a non-dimensional representation with $\widetilde{u} = \xi u$, $f(u,t) = n\xi^3 \widetilde{f}(\widetilde{u})$ and all lengths rescaled by the particle radius. The integrals, $\{G_i\}$, are rescaled as follows: $\widetilde{G}_i \equiv G_i/\left(4\pi\sigma_T n^2 \xi^2\right)$. It follows that the gain term is given by:

$$\left(\frac{\partial f}{\partial t}\right)_g = \sum_{i=1}^{9} G_i = 4\pi\sigma_T n^2 \xi^2 \cdot \sum_{i=1}^{9} \widetilde{G}_i,$$

and the loss term is:

$$\left(\frac{\partial f}{\partial t}\right)_l = 4\pi\sigma_T n^2 \xi^2 \cdot \widetilde{f}(u) \left[\int_0^{\widetilde{u}} \left(\widetilde{u}\widetilde{v}^2 + \frac{\widetilde{v}^4}{3\widetilde{u}}\right) \widetilde{f}(\widetilde{v})\, d\widetilde{v} \right.$$
$$\left. + \int_{\widetilde{u}}^{\infty} \left(\widetilde{v}^3 + \frac{\widetilde{u}^2\widetilde{v}}{3}\right) \widetilde{f}(\widetilde{v})\, d\widetilde{v}\right] \equiv 4\pi\sigma_T n^2 \xi^2 \widetilde{\left(\frac{\partial f}{\partial t}\right)_l} .$$

so that

$$n\dot{\xi}\xi^2 \left(3\widetilde{f} + \widetilde{u}\frac{d\widetilde{f}}{d\widetilde{u}}\right) = 4\pi\sigma_T n^2 \xi^2 \cdot \left[\sum_{i=1}^{9} \widetilde{G}_i - \widetilde{\left(\frac{\partial f}{\partial t}\right)_l}\right].$$

Recalling that $\dot{\xi} = \epsilon\lambda n \sigma_T \pi^{-1/2}$ (following (19): $\lambda = \sqrt{2}/3 + \mathcal{O}(\epsilon)$), one obtains

$$\epsilon\frac{\lambda\pi^{-\frac{3}{2}}}{4}\left[3\widetilde{f} + \widetilde{u}\frac{d\widetilde{f}}{d\widetilde{u}}\right] = \left[\sum_{i=1}^{9} \widetilde{G}_i - \widetilde{\left(\frac{\partial f}{\partial t}\right)_l}\right], \qquad (20)$$

In the sequel all entities are assumed dimensionless and the tilde signs are dropped for notational simplicity.

As shown in Appendix A-2, only G_2 is non-negligible with respect to the loss term in the large-u limit. It is convenient to define (the dimensionless) G_2^* as follows:

$$G_2^*(\epsilon,u) \equiv \frac{8}{(1+e)^{\frac{3}{2}}(3-e)u}\int_0^u dv_1 v_1 f(v_1)$$
$$\times \int_{\sqrt{(u^2-v_1^2)(3-e)/(1+e)}}^{\infty} dv_2 \sqrt{(3-e)(v_1^2-u^2)+(1+e)v_2^2}\; v_2\; f(v_2), \quad (21)$$

where the dependence of G_2^* on ϵ has been spelled out. The difference between G_2^* and G_2 is in the ranges of integration. Denote this difference by: $G_2 - G_2^* = -\delta_1 G_2 - \delta_2 G_2$, where:

$$\delta_1 G_2(\epsilon,u) \equiv \frac{8}{(1+e)^{\frac{3}{2}}(3-e)u}\int_0^{u(1-e)/2} dv_1 v_1 f(v_1)$$
$$\times \int_{\sqrt{(u^2-v_1^2)(3-e)/(1+e)}}^{[2u-(1-e)v_1]/(1+e)} dv_2 \sqrt{(3-e)(v_1^2-u^2)+(1+e)v_2^2}\; v_2\; f(v_2), \quad (22)$$

and

$$\delta_2 G_2(\epsilon,u) \equiv \frac{8}{(1+e)^{\frac{3}{2}}(3-e)u}\int_0^u dv_1 v_1 f(v_1)$$
$$\times \int_{[2u-(1-e)v_1]/(1+e)}^{\infty} dv_2 \sqrt{(3-e)(v_1^2-u^2)+(1+e)v_2^2}\; v_2\; f(v_2). \quad (23)$$

It is shown in Appendix A-2-2 that δG_1 and δG_2 are algebraically subdominant to the loss term for $u \gg 1$. It follows that the gain term can be substituted for by G_2^*, the error being subdominant to the loss term. Consider the latter next. The large-u form of the loss term is: $uf(u)\int_0^\infty dv\, v^2 f(v) = \frac{1}{4\pi}uf(u)$, where the normalization of the dimensionless form of the distribution function has been used. Therefore, for large u, the Boltzmann equation for the HCS reduces to:

$$\frac{\epsilon\lambda}{4\pi^{\frac{3}{2}}}\left[3f + u\frac{df}{du}\right] = G_2^* - \frac{1}{4\pi}uf(u), \quad (24)$$

Interestingly, (24) becomes an identity for $\epsilon = 0$ and $f(u) = \pi^{-3/2}\exp\left(-u^2\right)$, as can be checked by direct substitution. Thus, (24) can be regarded to be a uniform approximation of the Boltzmann equation, whose error is algebraically small (with respect to the loss term) for $u \gg 1$, and $\mathcal{O}(\epsilon)$ for $u \le \mathcal{O}(1)$. Since the bulk of the distribution is near-Maxwellian for practically all physical values of e, one expects (24) to comprise a rather faithful approximation of the pertinent Boltzmann equation. It is further reduced below.

Define the following variable transformations for (21), $x \equiv v_1^2$, $y \equiv v_2^2$ and $z \equiv u^2$. Define a shift: $t \equiv y - \frac{3-e}{1+e}(z-x)$ and let $\mu \equiv \frac{3-e}{1+e}$. Also define: $\widehat{h}(v^2) \equiv f(v)$. With these transformations, one obtains from (21):

$$G_2^*(\epsilon, u) = \frac{2}{(3-e)(1+e)\sqrt{z}} \int_0^z \widehat{h}(z-x)dx \int_0^\infty \widehat{h}(t+\mu x)\sqrt{t}dt \tag{25}$$

Next, (25) is further simplified. Let $\widehat{h}(z) \equiv e^{-\psi(z)}$. We wish to show that $\widehat{h}(t+\mu x)$ in (25) can be replaced by $\exp[-\psi(\mu x) - t\psi'(\mu x)]$, where ψ' denotes the derivative of ψ with respect to its argument. To this end note that for $z = \mathcal{O}(1)$ and $t = \mathcal{O}(1)$, $\widehat{h}(t+\mu x)$ is near-exponential in its argument (i.e., near-Maxwellian in the original variables) and the above expansion is justified. The contribution of $t > \mathcal{O}(1)$ is at least exponentially small with respect to that of $t = \mathcal{O}(1)$ and will therefore be neglected. Consider next the case when $z > \mathcal{O}(1)$. Upon splitting the integration over x, from 0 to $\Lambda = O(1)$ and from Λ to z one obtains two contributions. In the first range one is justified in expanding ψ, and the contribution of the second range is at least exponentially smaller than that of the first range, for every value of t. Using this expansion in (25), one obtains:

$$G_2^*(\epsilon, u) \simeq \frac{\sqrt{\pi}}{(3-e)(1+e)\sqrt{z}} \int_0^z \widehat{h}(z-x)\widehat{h}(\mu x) \frac{1}{(\psi'(\mu x))^{\frac{3}{2}}} dx \tag{26}$$

Further straightforward considerations (taking into account that the distribution crosses over from a near-Maxwellian to an exponential for $u > \mathcal{O}(\frac{1}{\epsilon})$ and that $\widehat{h}(\mu z) < \widehat{h}(z)$, since $\mu > 1$), reveal (in agreement with numerical tests) that the integration over x in (26) is dominated by the 'small' x range, where the distribution is near-Maxwellian, hence $\widehat{h}$ is near exponential and $\psi'(\mu x) = 1$ in this range. One can therefore replace ψ' in (26) by unity, obtaining:

$$G_2^*(\epsilon, u) \simeq \frac{\sqrt{\pi}}{(3-e)(1+e)\sqrt{z}} \int_0^z \widehat{h}(z-x)\widehat{h}(\mu x) dx \tag{27}$$

Next, define: $\widehat{h}(z) \equiv \frac{(3-e)(1+e)}{4}\pi^{(-3/2)}h(z)$. Substituting this definition into (27) and using (24), one obtains:

$$\epsilon\frac{2\lambda}{\sqrt{\pi}}z^{\frac{3}{2}}h'(z) + \left(z + \frac{3\epsilon\lambda}{\sqrt{\pi}}\sqrt{z}\right)h(z) = \int_0^z h(z-x)h(\mu x)dx \tag{28}$$

As a final step, we neglect $\epsilon\sqrt{z}$ with respect to z since this contribution is only important when $z < \epsilon^2$, i.e., for values of z at which the asymptotic evaluation

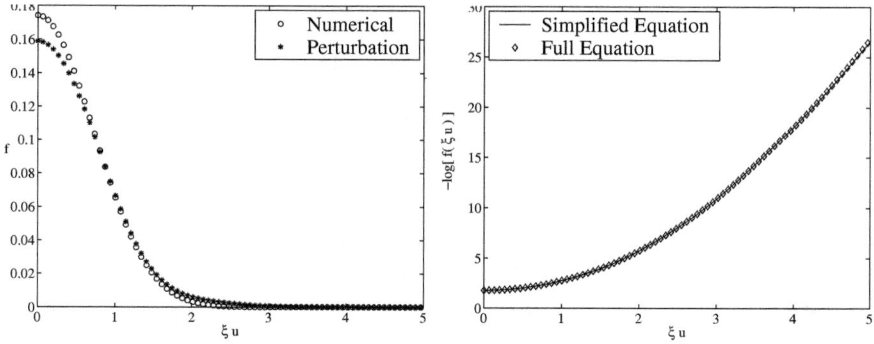

Fig. 2. Left: A comparison of the first order perturbation theory result for the HCS versus the numerical solution of the Boltzmann equation for $e = 0.8$. The (rescaled) distributions are shown versus the rescaled speed, ξu. **Right**: A comparison of the numerical solution of the Boltzmann equation for $e = 0.8$ with the numerical solution of the reduced equation (29). Shown are the logarithms of the distribution functions. Notice the close correspondence, even for small speeds and not-so-small ϵ

is not valid. Also the equation is valid only to within $\mathcal{O}(\epsilon)$. Therefore the final form of the reduced equation is:

$$\epsilon \frac{2\lambda}{\sqrt{\pi}} z^{\frac{3}{2}} h'(z) + zh(z) = \int_0^z h(z-x)\, h(\mu x)\, dx \tag{29}$$

One can easily check that the reduced equation admits a Maxwellian solution (i.e., $h(z) = e^{-z}$) when $\epsilon = 0$, and that the leading term (in $z = u^2$) of its perturbative expansion to first power in ϵ reads: $h(z) = e^{-z}(1 - \frac{1}{2}\epsilon z \ln z)$, in agreement with the large speed limit ($u \gg 1$ and $u\sqrt{\epsilon} < 1$, for $\epsilon \ll 1$) of the perturbative solution of the full equation. Thus, although the error of the equation is $\mathcal{O}(\epsilon)$ at low speeds, it is correct to order ϵu^2. As the numerical results (and some further analysis not presented here) show, cf. Fig. 2, right, the equation faithfully reproduces the distribution function for the HCS.

5.2 Second Reduction of the Boltzmann Equation

Although (29) represents an enormous simplification of the full Boltzmann equation it still requires a certain numerical effort to solve it. A further reduction is therefore useful.

Let $\phi(z) \equiv -\ln h$. With this definition, (29) transforms to:

$$z - \epsilon \frac{2\lambda}{\sqrt{\pi}} z^{\frac{3}{2}} \phi'(z) = \int_0^z e^{\phi(z) - \phi(\mu x) - \phi(z-x)} dx \tag{30}$$

The arguments presented in the previous subsection, as well as numerical studies (cf. Fig. 3), show that the integral in (30), like the integral in (26), is

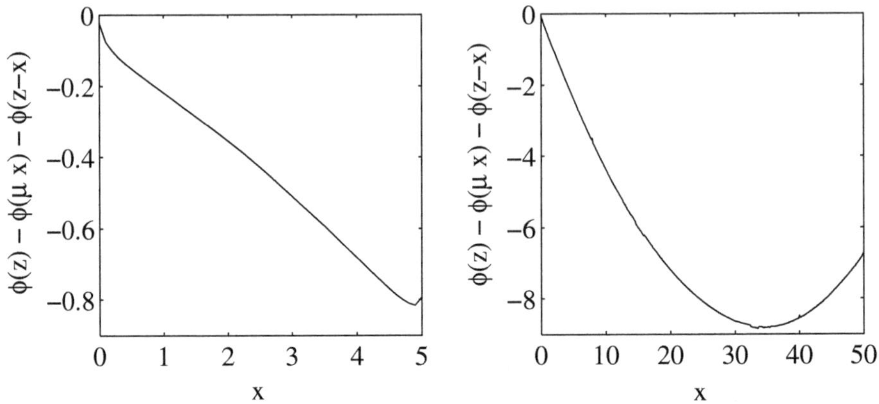

Fig. 3. Plot of the argument of the exponent in (30) versus x. The value of ϵ is 0.1. The left plot corresponds to $z = 5$ and the right plot corresponds to $z = 50$. These results are extracted from a numerical solution of the reduced equation, (29)

dominated by the small values of x when $z > \mathcal{O}(1/\epsilon)$, hence once can affect the following (approximate) replacement for $z > \mathcal{O}(1/\epsilon)$:

$$\phi(z) - \phi(z - x) - \phi(\mu x) \approx \phi'(z)x - \phi(\mu x) \approx \phi'(z)x - \phi(0) - \mu x, \tag{31}$$

where the second approximate equality follows from the fact that the distribution is near Maxwellian for small values of x (i.e., $\phi(z) \approx \phi(0) + z$ for small z). Consider now the range $z < \mathcal{O}(1/\epsilon)$. In this range the (approximate) linearity of ϕ in its argument, renders (31) valid as well. Finally, it is straightforward to deduce from (29) that $h(0)$ is close to unity (to within $O(\epsilon)$, hence one can substitute $\phi(0) = 0$. With these approximations, the integral on the right hand side of (30) (i.e., the integral of an exponential) can be immediately performed and (30) reduces to:

$$z - \epsilon \frac{2\lambda}{\sqrt{\pi}} z^{\frac{3}{2}} \phi'(z) = \frac{1 - e^{-q(z)}}{q(z)} z, \tag{32}$$

where $q(z) \equiv z\,(\mu - \phi'(z))$. Substituting $\phi'(z) = \mu - q(z)/z$, which follows from the definition of q, into (32), one obtains the following transcendental equation for $q(z)$:

$$\frac{2\epsilon\lambda}{\sqrt{\pi}} \left(\mu\sqrt{z} - \frac{q(z)}{\sqrt{z}} \right) - 1 + \frac{1 - e^{-q(z)}}{q(z)} = 0 \tag{33}$$

Once $q(z)$ is known, the value of $\phi(z)$ can be obtained by direct integration, using the definition of $q(z)$. A straightforward analysis of (33) shows that for $z < \mathcal{O}(1/\epsilon)$, $q(z)$ is approximately given by:

$$q(z) \approx \frac{\frac{4\epsilon\lambda\mu}{\sqrt{\pi}} z}{\sqrt{z} + \frac{4\epsilon\lambda}{\sqrt{\pi}}} \tag{34}$$

It follows that in this range $\phi(z) = \mu z + \mathcal{O}(\sqrt{z})$, i.e. the distribution is near-Maxwellian. When $z > \mathcal{O}(1/\epsilon^2)$, the solution of (33) is given by: $q(z) \approx \mu z - \sqrt{\pi z}/(2\epsilon\lambda)$, hence $\phi(z) \approx \sqrt{\pi z}/(\epsilon\lambda)$, which corresponds (as expected) to an exponential distribution. A full numerical evaluation of $q(z)$ from (33) and the calculation of $\phi(z)$ on the basis of this result yield a distribution function (i.e., $\exp[-\phi(z)]$), which is essentially indistinguishable from the numerical solution of the Boltzmann equation. Therefore there is no point in presenting a graph showing this correspondence. It is somewhat surprising that the second reduction still yields such good agreement with the numerical solution.

6 Concluding Remarks

A study of the properties of the HCS has been presented. Among the surprising features is the fact that the reduced descriptions so faithfully agree with the numerical solution (even in terms of perturbative corrections), in spite of the fact that the asymptotic analyses leading to the reductions are based on the large speed properties of the Boltzmann equation. It is possible that this good correspondence is due to the fact that the bulk of the distribution function is not significantly removed from a Maxwellian (except at values of e which are near zero) and that the asymptotics sets in (as numerical results show) at values of u which are basically of the thermal speed (for e not close to 1); when e is close to unity the approximate equation is a faithful approximation of the Boltzmann equation anyway. Another interesting feature is the nontrivial structure of the transition region, between the near-Maxwellian and the exponential ranges. This region is prominent for near-elastic values of the coefficient of restitution but it does exist (for $\mathcal{O}(1/\sqrt{\epsilon}) < \tilde{u} < \mathcal{O}(1/\epsilon)$ at every value of e. It is perhaps important to mention that the dominance of the loss over the gain term in the Boltzmann equation is 'just' algebraic (proportional to u^3 for the tail of the distribution function and linear in u in the transition region, cf. Appendix C-3 and Fig. 1). We believe that our detailed studies of the properties of the non-linear Boltzmann operator for the HCS should be useful to other, more complex granular flows, both in terms of results (e.g. concerning the linearized Boltzmann operator) and techniques. It should be noted that granular hydrodynamics depends on low order moments of the distribution function, and, as is well known, the latter are provided, to a good degree of approximation, by perturbation theory (except for coefficients of restitution which are smaller than about 0.7). Therefore the above study is not essential for the derivation of hydrodynamic descriptions (this does not necessarily imply that a thorough study of the distribution function pertaining to an important system, such as the HCS, is not warranted; distribution functions are actually measured experimentally these days and their non-Gaussian nature is of great interest). It may however be the starting point for the study of very inelastic systems or systems with large gradients, such as encountered in shocks, where the Chapman-Enskog theory is no longer valid. Whether this is indeed the case remains to be seen.

Acknowledgment

This work has been partially supported by the the US - Israel Binational Science Foundation (BSF), the Israel Science Foundation (ISF) and INTAS (the EU).

Appendix A: The Gain Term for the HCS

A-1: Reduction of the Gain Term

Since the sought distribution function is isotropic, one has:

$$\left(\frac{\partial f(u,t)}{\partial t}\right)_g = \frac{1}{4\pi}\int d\Omega_{\vec{u}} \left(\frac{\partial f(\vec{u},t)}{\partial t}\right)_g \tag{35}$$

Using the identity $\frac{1}{4\pi}\int d\Omega_{\vec{u}}\ \delta(\vec{v}_1-\vec{u}) = \frac{1}{2\pi u}\ \delta(v_1^2-u^2)$, one obtains from (35):

$$\int d\vec{v}_1 \int \frac{d\Omega_{\vec{u}}}{4\pi}\ \delta(\vec{v}_1-\vec{u})\left(\frac{\partial f(v_1,t)}{\partial t}\right)_g = \frac{1}{2\pi u}\int d\vec{v}_1\ \delta(v_1^2-u^2)\left(\frac{\partial f(v_1,t)}{\partial t}\right)_g . \tag{36}$$

Substituting this result into (5) and transforming the integration to incoming velocities, one obtains:

$$\left(\frac{\partial f(\vec{u},t)}{\partial t}\right)_g = \frac{1}{2\pi u}\int v'_{12} f(v'_1) f(v'_2)\delta(v_1^2-u^2) d\vec{v}'_1 d\vec{v}'_2 b' db' d\phi' , \tag{37}$$

where v_1 is the outgoing velocity of particle '1' and it represents a function of the incoming velocities, $\vec{v}'_1$ and $\vec{v}'_2$. The impact parameter, b', and the azimuthal angle, ϕ', relate to the incoming (precollisional) variables. In (37) the only ϕ' and b' dependent part of the integrand is the δ-function. The integration over the angles in the above expression is tedious. The result is that $\left(\frac{\partial f(\vec{u},t)}{\partial t}\right)_g = \sum_{i=1}^{9} G_i(u)$, where:

$$G_1 = \frac{2(3-e)}{1+e}\int_u^\infty dv_1 \int_{(2u+(1-e)v_1)/(1+e)}^\infty H(v_1,v_2) dv_2 \tag{38}$$

$$G_2 = \frac{2}{\sqrt{1+e}u}\int_{u(1-e)/2}^{u} dv_1 \int_{\sqrt{(u^2-v_1^2)(3-e)/(1+e)}}^{(2u-(1-e)v_1)/(1+e)} \sqrt{w^2-(3-e)u^2}\ \times \times H(v_1,v_2) dv_2 \tag{39}$$

$$G_3 = \frac{2(3-e)}{u(1+e)} \int_0^u dv_1 \int_{(2u+(1-e)v_1)/(1+e)}^{\infty} v_1 H(v_1, v_2) dv_2 \quad (40)$$

$$G_4 = \frac{2}{u} \int_u^{2u/(1-e)} dv_1 \int_0^{(2u-(1-e)v_1)/(1+e)} v_2 H(v_1, v_2) dv_2 \quad (41)$$

$$G_5 = \frac{1}{u\sqrt{1+e}} \int_0^{2u/(1-e)} dv_1 \int_{(2u-(1-e)v_1)/(1+e)}^{(2u+(1-e)v_1)/(1+e)} \Big[\sqrt{1+e}\, v_2 + \sqrt{w^2-(3-e)u^2}\Big] H(v_1, v_2) dv_2 \quad (42)$$

$$G_6 = \frac{1}{u\sqrt{1+e}} \int_{2u/(1-e)}^{\infty} dv_1 \int_{((1-e)v_1-2u)/(1+e)}^{((1-e)v_1+2u)(1+e)} \Big[\sqrt{1+e}\, v_2 + \sqrt{w^2-(3-e)u^2}\Big] H(v_1, v_2) dv_2 \quad (43)$$

$$G_7 = \frac{3-e}{u(1+e)} \int_u^{2u/(1-e)} dv_1 \int_{(2u-(1-e)v_1)/(1+e)}^{(2u+(1-e)v_1)/(1+e)} (u-v_1) H(v_1, v_2) dv_2 \quad (44)$$

$$G_8 = \frac{3-e}{u(1+e)} \int_{2u/(1-e)}^{\infty} dv_1 \int_{((1-e)v_1-2u)/(1+e)}^{((1-e)v_1+2u)/(1+e)} (u-v_1) H(v_1, v_2) dv_2 \quad (45)$$

$$G_9 = \frac{3-e}{u(1+e)} \int_0^u dv_1 \int_{(2u-(1-e)v_1)/(1+e)}^{(2u+(1-e)v_1)/(1+e)} (v_1-u) H(v_1, v_2) dv_2 \quad (46)$$

where $w^2 \equiv (3-e)v_1^2 + (1+e)v_2^2$ and $H(v_1, v_2) \equiv \frac{16\pi\sigma_T}{(3-e)(1+e)} v_1 v_2 f(v_1) f(v_2)$.

A-2: The Gain Term for Large Speeds

Below we present an analysis of the gain term for values of the speed that are far larger than the typical (thermal) speed. In this limit it turns out that $\{G_i\ ; i \neq 2\}$ are algebraically small with respect to the loss term. More specifically: $[G_i(u)] / [n\sigma_T u f(u)] < C/u$, for $i \neq 2$ and $u \gg 1$, irrespective of the value of e. Here C is an $\mathcal{O}(1)$ constant. Numerical results actually show that the ratio of the gain to the loss term for large u is smaller than the above bound; however this bound suffices for the purposes of our analysis.

A-2-1: Analysis of the $\{G_i\ ; i \neq 2\}$

It is convenient to study the dimensionless $\tilde{G}_i$ and compare to the dimensionless loss term. For notational simplicity we shall omit below the tilde signs and assume that all velocities are scaled by ξ, G_i are scaled by $4\pi\sigma_T n^2\xi^2$, and that the same scaling is applied to the loss term (see Sect. 5.1). In addition, $f(u)$ denotes below the non-dimensionalized universal distribution as a function of the non-dimensionalized speed.

The bounds below are based on two assumed properties. The first is that asymptotically any small power of u times $f(u)$ can be bounded by a constant. This will be referred to as 'property 1'. The second property is developed next. Define $I_n \equiv \int_u^\infty v^n f(v)dv$. Let $\psi(u) \equiv -\ln f(u)$. Since $f(u)$ is a decreasing function of u, $\psi'(u)$ is positive (the prime superscript denotes a derivative with respect to u). For small u: $\psi'(u) \propto u$ and $\psi''(u) \propto 1$, whereas for large u: $\psi'(u) \propto 1$ and $\psi''(u) \to 0$. We therefore assume that ψ' is bounded from below for $u > u_c > 0$, where we conveniently choose u_c to be larger than unity so that $1/\psi'$ is bounded by an $\mathcal{O}(1)$ value, denoted by C_0. From the properties of ψ'' we conjecture that $\psi'' > 0$ for all u. Using straightforward integration by parts, applied to the definition of I_n, one obtains:

$$I_n = \frac{u^n}{\psi'(u)} f(u) + \int_u^\infty \frac{n v^{n-1}}{\psi'(v)} f(v)dv - \int_u^\infty v^n \frac{\psi''(v)}{\psi'^2(v)} f(v)dv.$$

It follows that $I_n \leq C_0 u^n f(u) + C_0 n I_{n-1}$. Consequently, for small values of n and $u > u_c$ there is an $\mathcal{O}(1)$ constant, C, such that: $I_n < C u^n f(u)$. We refer to the latter inequality as 'property 2'. Below we use the symbol C to denote *all* $\mathcal{O}(1)$ bounding prefactors, irrespective of their values.

Consider first G_1. It is easy to check that $G_1 < C I_1^2$, hence using property 2: $G_1 < C u^2 f^2(u)$. For $u \gg 1$ it follows from property '1' that $G_1 < C f(u)$, hence it is subdominant to the loss term (the latter being $C u f(u)$ for large u).

Consider next G_3. Using its definition and the definition of I_n one obtains: $G_3 < \frac{C}{u} \int_0^u dv_1 v_1^2 f(v_1) I_1 \left([2u + (1-e)v_1]/(1+e) \right)$. Since I_1 is a decreasing function of its argument, I_1 in the latter integral can be bounded by $I_1(u)$. Next, since the integral $\int_0^u v^n f(v)dv$ is bounded by a constant for any positive value of n, it follows that: $G_3 < \frac{C}{u} I_1(u)$. Now, using property 2, it follows that: $G_3 < C f(u)$. A similar analysis yields $G_4 < C f(u)$, hence G_3 and G_4 are subdominant to the loss term for $u \gg 1$.

It is convenient to analyze G_5 and G_9 together. Define G_5^a as the part of G_5 in which the integration over v_1 ranges from 0 to u and G_5^b as the part of G_5 in which the integration over v_1 ranges from u to $2u/(1-e)$. Clearly: $G_5 = G_5^a + G_5^b$. Consider next the sum $G_5^a + G_9$. It follows from their definitions that:

$$G_5^a + G_9 = \frac{C}{u} \int_0^u dv_1 v_1 f(v_1) \int_{[2u-(1-e)v_1]/(1+e)}^{[2u+(1-e)v_1]/(1+e)} dv_2 v_2 f(v_2) \times \left[\frac{3-e}{1+e} (v_1 - u) + v_2 + \sqrt{\frac{w^2}{1+e} - \frac{3-e}{1+e} u^2} \right].$$

The term in the square brackets increases with v_2, hence it can be bounded from above by substituting the maximal value of v_2 in the range of integration.

Since $f(v_2)$ is a decreasing function of v_2 one can bound it by substituting the minimal value of v_2 in the range of integration for its argument. In addition, one can substitute for the explicit factor v_2, its maximal value in the range of integration. It follows that:

$$G_5^a + G_9 \le (1-e)\frac{C}{u}\int_0^u dv_1 v_1^3 f(v_1)\left[2u + (1-e)\,v_1\right] f\left(\frac{2u-(1-e)\,v_1}{1+e}\right)$$

By replacing v_1 in the argument of $f\left(\left[2u-(1-e)v_1\right]/(1+e)\right)$ by u one further increases the result, hence $G_5^a + G_9 \le C(1-e)f(u)$ and this sum is again subdominant to the loss term.

Next consider G_5^b. One can subdivide the range of integration over v_1 to $[u, \Lambda u]$, where Λ is a constant satisfying : $1 < \Lambda < 2/(1-e)$, and $[\Lambda u, 2u/(1-e)]$. The contribution of the second range is proportional (using property '2') to $(1-e)u^3 f\left(\frac{2-(1-e)\Lambda}{1+e}u\right) f(\Lambda u)$; using property '1' one concludes that this sum is asymptotically smaller than and clearly subdominant to $(1-e)Cf(\Lambda u)$, hence it is smaller than $(1-e)Cf(u)$. In the first range on can now use property '2', obtaining by straightforward analysis that $G_5^b \le (1-e)\Lambda u^3 f\left(\frac{2-(1-e)\Lambda}{1+e}u\right) f(u)$. Next, using property '1', this contribution is bounded by $C(1-e)f(u)$. Thus, for $u \gg 1$ it follows $G_5 + G_9 < C(1-e)f(u)$. An analysis similar to that for G_5^b yields: $G_7 \le (1-e)Cf(u)$. Since in the expressions for G_6 and G_8, the lower bound for v_1 is $2u/(1-e)$, the corresponding integrals are clearly subdominant to the loss term. A straightforward analysis shows that they too can be bounded from above by $(1-e)Cf(u)$. Note that we have not absorbed the constant $(1-e)$ in C for $\{G_i\,; 5 \le i \le 9\}$, as we wish to emphasize that these contributions to the gain term vanish in the elastic limit.

A-2-2: Large Speed Bounds on $\delta_1 G_2(\epsilon, u)$ and $\delta_2 G_2(\epsilon, u)$

The goal here is to prove that for $u \gg 1$, $\delta_1 G_2(\epsilon, u)$ and $\delta_2 G_2(\epsilon, u)$, cf. (22,23), are algebraically (in u) subdominant to the loss term (which is $\propto u\ f(u)$) for large u. As before, C represents a generic prefactor denoting $\mathcal{O}(1)$ numbers. Below, all functions and speeds are dimensionless.

Consider first $\delta_1 G_2$, (22). The lowest value of v_1 in the range of integration is zero and the lowest value of v_2 is $\frac{3-e}{2}u = \left(1+\frac{1-e}{2}\right)u$. Therefore, replacing $f(v_1)f(v_2)$ in the integrand by $f(0)f\left((1+\frac{1-e}{2})u\right)$ increases the value of the integral. With this substitution, the integral over v_1 and v_2 can be straightforwardly performed, yielding the following upper bound: $\delta_1 G_2 < \frac{C}{u}\left(((1-e)u)^5\right) f\left((1+\frac{1-e}{2})u\right)$. When $(1-e)u$ is of order unity or less, the bound is clearly subdominant to the loss term. When $(1-e)u$ is larger than order unity, the value of $f\left((1+\frac{1-e}{2})u\right)$ is (at least) exponentially subdominant to $f(u)$ and again the bound on $\delta_1 G_2$ is subdominant to the loss term.

Consider next $\delta_2 G_2$, cf. (23). Using $(1+e)/(3-e) \leq 1$ and replacing the v_1-dependent lower limit of the integration over v_2 by its smallest value, u, one obtains the following bound:

$$\delta_2 G_2 \leq \frac{C}{u} \int_0^u dv_1 \; v_1 \; f(v_1) \int_u^\infty dv_2 \sqrt{v_1^2 + v_2^2 - u^2} \; v_2 \; f(v_2). \qquad (47)$$

Let $\eta(u)$ be defined by $f(\eta(u)) = 1/u^2$. Clearly, $\eta(u)$ is proportional to a power of $\log(u)$ for large u. Divide now the range of integration over v_1 in (47) into a 'range A', in which $0 < v_1 < \eta(u)$, and let 'range B' be defined by: $\eta(u) < v_1 < u$. Denote the corresponding values of the integrals by $\delta_2 G_2^A$ and $\delta_2 G_2^B$, respectively. Now, in range B, replace $f(v_1)$ by its largest value (recall that f is a decreasing function of its argument), i.e. by $f(\eta(u)) = 1/u^2$, and replace the argument in the square root by $2v_2^2$, thereby increasing it. Next, perform the integration over v_1. One obtains: $\delta_2 G_2^B \leq \frac{C}{u} f(\eta(u)) \int_u^\infty dv_2 v_2^2 f(v_2)$. Using property 2 one obtains: $\delta_2 G_2^B \leq \frac{C}{u} f(\eta(u)) u^2 f(u)$, and using the definition of $f(\eta(u))$: $\delta_2 G_2^B \leq \frac{C}{u} f(u)$, hence this contribution is subdominant to the loss term. Next, consider $\delta_2 G_2^A$. In this range it is convenient to majorize $f(v_1)$ by $f(0)$ and perform the integration over v_1. One obtains:

$$\delta_2 G_2^A \leq \frac{C}{u} \int_u^\infty dv_2 \; v_2 \; f(v_2) \left[\left(\eta^2(u) + v_2^2 - u^2\right)^{\frac{3}{2}} - \left(v_2^2 - u^2\right)^{\frac{3}{2}} \right]. \qquad (48)$$

Next divide the range of integration in (48) into $u < v_2 < u_1$ (range A1) and $u_1 < v_2 < \infty$ (range A2), where u_1 satisfies $u_1^2 - u^2 = \alpha \eta^2(u)$, with $\alpha = 10$, for example. In the range A1 the argument in the square root in (48) is majorized by an $\mathcal{O}(1)$ number times $\eta^2(u)$. It follows that $\delta_2 G_2^{A1} \leq \frac{C}{u} \eta^3(u) \int_u^{u_1} dv_2 \; v_2 f(v_2)$. Using property 2 one now obtains: $\delta_2 G_2^{A1} \leq C\eta^3(u) f(u)$ and this contribution is also subdominant to the loss term. Finally, consider $\delta_2 G_2^{A2}$. In the range of v_2 corresponding to this contribution, $\eta^2(u) \ll v_2^2 - u^2$, hence one can bound this contribution by:

$$\delta_2 G_2^{A2} \leq \frac{C\eta^2(u)}{u} \int_{u_1}^\infty dv_2 \; v_2 \; f(v_2)(v_2^2 - u^2)^{\frac{1}{2}}. \qquad (49)$$

The latter integral can be shown to be subdominant to $u^{3/2} f(u)$ (provided $f(v_2)$ decays at least exponentially at large v_2), hence $\delta_2 G_2^{A2} \leq C\sqrt{u} f(u)$, and this contributions is also algebraically subdominant to the loss term. Thus, $\delta_1 G_2 + \delta_2 G_2$ is subdominant to the loss term.

Appendix B: Remarks on the Mean Field Approach

There are several possible approaches to developing a mean field theory for the HCS problem. A phenomenological way is presented in [7], see also [15]. Here we simply assume (as explained in Sec. (3-2)) that the distribution function remains Maxwellian. In this case, a substitution of the Maxwellian distribution in the r.h.s of the Boltzmann equation, followed by a multiplication of the result by v_1^2 and integration over v_1, yields:

$$\dot{\xi} = \epsilon \frac{\sqrt{2}}{3} n \sigma_T \pi^{-\frac{1}{2}}, \tag{50}$$

i.e., $\lambda = \sqrt{2}/3$ in the mean field approximation.

Appendix C: Perturbation Theory

C-1: Calculation of $1/\left(\epsilon\lambda\pi^{1/2}\right)\tilde{B}(\tilde{f}_M, \tilde{f}_M; e)(\tilde{v}_1)$

The following relation results from the kinematics of the collisions, (1) and (2), and the definition of $\tilde{b}$ (the outgoing rescaled impact parameter): $\tilde{v}_1'^2 + \tilde{v}_2'^2 - \tilde{v}_1^2 - \tilde{v}_2^2 = (\tilde{v}_{21})^2(1-\tilde{b}^2)\left(1-e^2\right)/\left(2e^2\right)$. Substituting this result in (13) and performing the integrations involved in the calculation of the right hand side of (13), one obtains:

$$\frac{1}{\epsilon\lambda\pi^{\frac{1}{2}}}\tilde{B}(\tilde{f}_M, \tilde{f}_M; e)(\tilde{v}_1) = \frac{4\tilde{f}_M(\tilde{v}_1)}{\tilde{v}_1\epsilon^2\lambda} I(\tilde{v}_1), \tag{51}$$

where the function I is defined as:

$$\begin{aligned} I(u,\epsilon) \equiv \int_0^\infty dv v \exp\left(-v^2\right) \Bigg[& (u+v) - |u-v| - \frac{\epsilon}{6}(u+v)^3 - |u-v|^3 \\ & - \frac{1}{2}\int_{(u-v)^2}^{(u+v)^2} \frac{dy}{\sqrt{y}} \exp\left(-\frac{\epsilon}{2(1-\epsilon)} y\right)\Bigg], \end{aligned} \tag{52}$$

Equation (52) can be further simplified to:

$$\begin{aligned} I(u) = & \int_0^u e^{-t^2} dt \left(1 - \frac{\epsilon u^2}{2} - \frac{\epsilon}{4}\right) - \frac{\epsilon u}{4} e^{-u^2} \\ & - \frac{\exp\left(-u^2 \frac{\epsilon K}{1+\epsilon K}\right)}{1+\epsilon K} \int_0^u \exp\left(-\frac{t^2}{1+\epsilon K}\right) dt, \end{aligned} \tag{53}$$

where $K \equiv (2(1-\epsilon))^{-1}$. For $u < \mathcal{O}(1/\sqrt{\epsilon})$, one is justified in expanding $I(u)$ in powers of ϵ: $I(u) = I_0(u) + \epsilon I_1(u) + \epsilon^2 I_2(u) + \mathcal{O}(\epsilon^3)$. One obtains, to second order in ϵ: $I_0 = I_1 = 0$ and

$$I_2(u) = \frac{\sqrt{\pi}}{2}\left(\frac{5}{32} + \frac{u^2}{8} - \frac{u^4}{8}\right)\mathrm{erf}(u) + \left(\frac{3u}{32} - \frac{u^3}{16}\right)e^{-u^2}. \tag{54}$$

Since $\tilde{B}(f_M, f_M; 1) = 0$, it follows that:

$$\lim_{\epsilon\to 0} \frac{1}{\epsilon\lambda\pi^{1/2}}\tilde{B}(\tilde{f}_M, \tilde{f}_M, e) = \frac{1}{\lambda_0\pi^{\frac{1}{2}}}\tilde{B}_1(\tilde{f}_M, \tilde{f}_M),$$

hence $1/\left(\lambda_0\pi^{1/2}\right)\tilde{B}_1(\tilde{f}_M,\tilde{f}_M) = I_2(u)$. It follows that the source term, s_1, i.e., the right hand side of (15), is given by:

$$s_1 = \frac{\lambda_0}{4}(3u - 2u^3) - I_2(u), \tag{55}$$

with $I_2(u)$ given by (54). Notice that, unlike in this Appendix, the tildes denoting dimensionless variables and functions are omitted in (15).

C-2: Proof of the First Solubility Condition

Recall that the general form of the n-th order in perturbation theory is $\widehat{L} * \phi_n = s_n$, where s_n is a 'source' term, which depends on previous orders and on λ_{n-1}. Following (17), s_n is required to satisfy both $(1, s_n) = 0$ and $(u^2, s_n) = 0$, where we recall that the scalar product, (f,g), is defined by $(f,g) \equiv \int_0^\infty v f_M(v) f(v) g(v) dv$, with $f_M(v) = \pi^{-3/2}\exp\left(-v^2\right)$. It is shown below that the condition $(1, s_n) = 0$ is identically satisfied to all orders in perturbation theory (in powers of ϵ). All entities in this Appendix are assumed to be dimensionless (and the tilde signs are omitted).

It follows from (13) that: $B(f,f;e) = \epsilon\lambda\sqrt{\pi}\left(3f + u\frac{df}{du}\right)$, where u is the speed variable. Formally expanding $B(f,f;e)$ in powers of ϵ (as explained in Sec. (4), i.e., the function f is not expanded), one obtains:

$$B_0(f,f) = \epsilon\lambda\sqrt{\pi}\left(3f + u\frac{df}{du}\right) - \sum_{n=1}^{\infty}\epsilon^n B_n(f,f). \tag{56}$$

Let $f = f_M(1+\phi)$ be the sought solution of the Boltzmann equation. Clearly, since B (cf. (13)) is a bilinear operator, one has:

$$\begin{aligned} B(f,f;e) = {} & B(f_M, f_M; e) + B(f_M\phi, f_M; e) + B(f_M, f_M\phi; e) \\ & + B(f_M\phi, f_M\phi; e). \end{aligned} \tag{57}$$

It therefore follows from (56) that:

$$\begin{aligned} B_0(f_M\phi, f_M) + B_0(f_M, f_M\phi) = {} & -B_0(f_M, f_M) - B_0(f_M\phi, f_M\phi) \\ & + \epsilon\lambda\sqrt{\pi}\left(3f + u\frac{df}{du}\right) - \sum_{n=1}^{\infty}\epsilon^n B_n(f,f). \end{aligned} \tag{58}$$

The left hand side of (58), multiplied by $u/\left(4f_M\sqrt{\pi}\right)$ is the definition of $\widehat{L} * \phi$, cf. (15). Therefore:

$$\begin{aligned} \widehat{L} * \phi = \frac{u}{4f_M\sqrt{\pi}}\Bigg(& -B_0(f_M, f_M) - B_0(f_M\phi, f_M\phi) + \epsilon\lambda\sqrt{\pi}\left(3f + u\frac{df}{du}\right) \\ & - \sum_{n=1}^{\infty}\epsilon^n B_n(f,f)\Bigg). \end{aligned} \tag{59}$$

Equation (59) can clearly be written as $\widehat{L} * \phi = s$, where $s = \sum_{n=1}^{\infty} \epsilon^n s_n$. It therefore suffices to show that $(1, s) = 0$. Noting that $B_0(f_M, f_M) = 0$, and using the definition of the scalar product in Sec. (4), one obtains:

$$(1,s) = \int_0^\infty du \frac{u^2}{4\sqrt{\pi}} \Bigg(- B_0(f_M\phi, f_M\phi) + \epsilon\lambda\sqrt{\pi}\left(3f + u\frac{df}{du}\right) - \sum_{n=1}^{\infty} \epsilon^n B_n(f,f)\Bigg). \quad (60)$$

First, notice that $\int_0^\infty du \left(3u^2 f(u) + u^3 \frac{df(u)}{du}\right) = 0$, as a simple integration by parts reveals. Next, since $u^2 du = 1/(4\pi) d\vec{u}$ (all functions depend on the speed, u, here) and since $\int d\vec{u} B_0(f,f) = 0$ for any function, f, it also follows that $\int du u^2 B_0(f_M\phi, f_M\phi) = 0$. Finally, notice that $\int d\vec{u} B(g, g; e) = 0$ for any function g. Expanding the last relation in powers of ϵ one obtains $\int d\vec{u} B_n(g,g) = 0$. It thus follows that $\int du u^2 B_n(f,f) = 0$. This completes the proof.

C-3: The Asymptotic Ratio of the Gain to the Loss Term

Consider G^* as given by (25). Since, as explained in the main text, this integral is dominated by the small x domain, one can replace $\widehat{h}(t + \mu x)$ by a Maxwellian transformed to the new variables, $\widehat{h}(t + \mu x) \approx \pi^{-3/2} e^{-(t+\mu x)}$. Next, replace $\widehat{h}(z - x)$ in this integral by its asymptotic form (an exponential distribution in u): $\widehat{h}(z - x) \approx C z^{-\eta} \exp\left(-\alpha\sqrt{z - x}/\epsilon\right)$, where η, α and C are $\mathcal{O}(1)$ constants (they are actually known, e.g., $\eta = 3$, but this is not important for the present derivation). With this substitution, a straightforward asymptotic analysis shows that for $z > \mathcal{O}\left(1/\epsilon^2\right)$ (i.e. for $u > \mathcal{O}\left(1/\epsilon\right)$), $G_2^*(u) \propto u^{-2} f(u)$, hence the ratio of the gain to the loss term is proportional to u^{-3}, in conformity with the numerical findings. When $1 \ll u \ll 1/\epsilon$, the gain term can be shown to be $\propto Cf$, and its ratio to the loss term is proportional to $1/u$, again in conformity with the numerical results.

References

1. (i). I. Goldhirsch, Clustering Instability in Granular Gases, Proc. DOE/NSF Meeting: Flow of Particulates and Fluids, 211–235, Worcester, MA (WPI: 1991). (ii). M.A. Hopkins and M.Y. Louge, Inelastic Microstructure in Rapid Granular Flows of Smooth Disks, Phys. Fluids A **3** (1), 47–57 (1991). (iii). I. Goldhirsch and G. Zanetti, Clustering Instability in Dissipative Gases, Phys. Rev. Lett. **70**, 1619–1622 (1993). (iv). I. Goldhirsch, M.L. Tan and G. Zanetti, A Molecular Dynamical Study of Granular Fluids I: The Unforced Granular Gas in Two Dimensions, J. Sci. Comp. **8** (1), 1–40 (1993). (v). J.J. Brey, M.J. Ruiz-Montero and D. Cubero, Homogeneous Cooling State of a Low Density Granular Flow, Phys. Rev. E **54** (4), 3664–3671 (1996). (vi). R. Brito and M. H. Ernst, Extension of Haff's law in granular flows, Europhys. Lett. **43** (5),

497–502 (1998). (vii). J.J. Brey, M. J. Ruiz-Monterro and D. Cubero, Origin of Density Clustering in Freely Evolving Granular Gas, Phys. Rev. E **60** (3), 3150–3157 (1999). (viii). S. Luding and H. J. Herrmann, Cluster-Growth in Cooling Granular Media. Chaos **9** (3), 673–681 (1999)

2. (i). S. McNamara and W.R. Young, Inelastic Collapse and Clumping in a One-Dimensional Granular Medium, Phys. Fluids A **4**, 496–504 (1992). (ii). S. McNamara and W.R. Young, Kinetics of a One-Dimensional Granular Medium in the Quasielastic Limit, Phys. Fluids A **5** (1), 34–35 (1993)
3. J.J. Brey, J.W. Dufty, C.S. Kim and A. Santos, Hydrodynamics for Granular Flow at Low Density, Phys. Rev. E **58**, 4638–4653 (1998)
4. (i). I. Goldhirsch and T.P.C. van Noije, Green-Kubo Relations for Granular Fluids, Phys. Rev. E **61** (3), 3241–3244 (2000). (ii). J.W. Dufty and J.J. Brey, Green-Kubo Expressions for a Granular Gas, J. Stat. Phys. **109** (3–4), 433–448 (2002)
5. (i). T.A. Knight and L.V. Woodcock, Test of the Equipartition Principle for Granular Spheres in a Saw-Tooth Shaker, J. Phys. A. **29** (15), 4365–4386 (1998). (ii). S. Luding, On the Relevance of "Molecular Chaos" for Granular Flows, ZAMM **80**, S9–S12 (2000)
6. (i). M. Huthmann and A. Zippelius, Dynamics of Inelastically Colliding Rough Spheres: Relaxation of Translational and Rotational Energy, Phys. Rev. E **56** (6), R6275–R6278 (1997). (ii). S. McNamara and S. Luding, Energy Nonequipartition in Systems of Inelastic Rough Spheres, Phys. Rev. E **58** (2), 2247–2250 (1998). (iii). S. Luding, M. Huthman, S. McNamara and A. Zippelius, Homogeneous Cooling of Rough, Dissipative Particles: Theory and Simulations. Phys. Rev. E **58** (3), 3416–3425 (1998). (iv). R. Cafiero and S. Luding, Mean Field Theory for a Driven Granular Gas of Frictional Particles, Physica A **280** (1–2), 142–147 (2000)
7. P. K. Haff, Grain Flow as a Fluid Mechanical Phenomenon, J. Fluid Mech. **134**, 401–948 (1983)
8. (i). A. Goldshtein and M. Shapiro, Mechanics of Collisional Motion of Granular Materials, Part I: General Hydrodynamic Equations, J. F. M. **282**, 75–114 (1995). (ii). J.J. Brey, J.W. Dufty and A. Santos, Dissipative Dynamics for Hard Spheres, J. Stat. Phys. **87** (5–6), 1051–1066 (1997). (iii). T.P.C. van Noije and M.H. Ernst, Velocity Distribution in Homogeneous Granular Fluids: the Free and the Heated Case, Granular Matter **1**, 57–64 (1998)
9. (i). S.E. Esipov and T. Pöschel, The Granular Phase Diagram, J. Stat. Phys. **86**, 1385–1395 (1997). (ii). J.J. Brey, D. Cubero and M.J. Ruiz-Montero, High Energy Tail in the Velocity Distribution of a Granular Gas, Phys. Rev. E **59**, 1256–1258 (1999)
10. (i). N.V. Brilliantov and T. Pöschel, Hydrodynamics of Granular Gases of Viscoelastic Particles, Phil. Trans. R. Soc. Lond. A **360**, 415–429 (2002) and refs therein. (ii). T. Pöschel, N.V. Brilliantov and T. Schwager, Long-time Behavior of Granular Gases with Impact-Velocity Dependent Coefficients of Restitution, Physica **A**, in press (2003), cond-mat/0212200, and refs therein
11. N. Sela and I. Goldhirsch, Hydrodynamic Equations for Rapid Flows of Smooth Inelastic Spheres, to Burnett Order, J. Fluid Mech., **361**, 41–74 (1998), and refs. therein
12. cf. e.g., K. Huang, Statistical Mechanics, Wiley, N.Y. (1963)

13. (i). I. Kuščer and M.M.R. Williams, Relaxation Constants of a Uniform Hard-Sphere Gas, Phys. Fluids. **10**, 1922–1927, (1967). (ii). F. Schürrer and G. Kügerl, The Relaxation of Single Hard Sphere Gases, Phys. Fluids. A **2**, 609–618 (1990)
14. C. Cercignani, Theory and Application of the Boltzmann Equation, Scottish Acad. Press, Edinburgh and London (1975)
15. I. Goldhirsch, Rapid Granular Flows, Annual Reviews of Fluid Mechanics, **35**, 267–293 (2003)

The Inelastic Maxwell Model

Eli Ben-Naim[1] and Paul L. Krapivsky[2]

[1] Theoretical Division and Center for Nonlinear Studies, Los Alamos National Laboratory, Los Alamos, NM 87545
[2] Center for Polymer Studies and Department of Physics, Boston University, Boston, MA 02215

Abstract. Dynamics of inelastic gases are studied within the framework of random collision processes. The corresponding Boltzmann equation with uniform collision rates is solved analytically for gases, impurities, and mixtures. Generally, the energy dissipation leads to a significant departure from the elastic case. Specifically, the velocity distributions have overpopulated high energy tails and different velocity components are correlated. In the freely cooling case, the velocity distribution develops an algebraic high-energy tail, with an exponent that depends sensitively on the dimension and the degree of dissipation. Moments of the velocity distribution exhibit multiscaling asymptotic behavior, and the autocorrelation function decays algebraically with time. In the forced case, the steady state velocity distribution decays exponentially at large velocities. An impurity immersed in a uniform inelastic gas may or may not mimic the behavior of the background, and the departure from the background behavior is characterized by a series of phase transitions.

1 Introduction

Granular gases consist of weakly deformable macroscopic particles that interact via contact interactions and dissipate energy during collisions [1]. In typical experimental situations the system is constantly supplied with energy to counter the loss occurring during inelastic collisions [2]. The most ubiquitous features of granular gases are their tendency to form clusters and their anomalous velocity statistics [3–6]. In turn, these characteristics lead to many interesting collective phenomena including density inhomogeneities [7], shocks [8, 9], size segregation [10], pattern formation [11], and phase transitions [12], to name a few.

Inelastic hard spheres provide a useful theoretical, computational, and experimental framework for studying granular gases [13–16]. Inelastic gases pose new theoretical challenges as their dynamics are governed by dissipative collisions and strong velocity correlations. Fundamental equilibrium characteristics such as ergodicity and equipartition of energy are typically violated by this nonequilibrium gas system. As a result, inelastic gases exhibit finite time singularities [17–19], chaotic behavior [20], breakdown of molecular chaos [21–25], and anomalous velocity statistics [26–30].

The significance of the energy dissipation is nicely demonstrated by considering a mean-field version of the hard sphere system, namely a random collision process. In this process, randomly chosen pairs of particles undergo inelastic collisions with a random impact direction. This process, often called the Maxwell model, is described by a Boltzmann equation with a velocity independent collision rate. As this model is random in all respects, it is analytically tractable. In fact, in classical kinetic theory of gases, the Maxwell model precedes the Boltzmann equation [31]. Historically, it played an important role in the development of kinetic theory [32–36], and it remains the subject of current research [37–39]. Recently, it has been realized that the Maxwell model is analytically tractable even for inelastic collisions [40–57]. In this review, we detail dynamics of uniform inelastic gases, isolated impurities in uniform gases, and binary mixtures.

First, we consider the one-dimensional gas where explicit analytic solutions are possible (Sec. 2). In the freely cooling case, the velocity distribution approaches a universal (dissipation independent) scaling form with an algebraic high-energy tail. Furthermore, the moments of the velocity distribution exhibit a multiscaling asymptotic behavior, and the velocity autocorrelation function decays algebraically with a non-universal exponent. In the uniformly heated case, the system approaches a steady state, and the velocity distribution has an exponential high-energy tail.

Many of these features extend to higher dimensions (Sec. 3). In the freely cooling case, however, the scaling function underlying the velocity distribution is no longer universal — its shape and its extremal behavior depend on the spatial dimension and the degree of dissipation. We obtain explicit expressions for low order moments and show the development of correlations between different velocity components. Such correlations are more pronounced in the freely cooling case, compared with the forced case.

When an impurity is immersed in a homogeneous cooling fluid, various scenarios are possible (Sec. 4). If the impurity mass is lighter than a critical mass, its velocity statistics are completely governed by the background. As the impurity mass increases, impurity to fluid moment ratios of sufficiently high order diverge asymptotically. A series of critical masses govern these phase transitions. Ultimately, the temperature ratio may diverge, and the impurity becomes so energetic that it effectively decouples from the fluid. In this case, its velocity distribution is distinct from the fluid. Although single-time velocity statistics of the impurity may be governed by the fluid, two-time statistics such as the autocorrelation function are always different.

For mixtures (Sec. 5), all components have the same asymptotic temperature decay, and a high-energy tail as in the uniform gas case. This behavior is shown in detail for one-dimensional freely cooling binary mixtures.

We finally describe a lattice generalization of the Maxwell model where particles are placed on a lattice and only nearest neighbors interact via inelastic collisions (Sec. 6). We show the development of spatial velocity correlations with a diffusively growing correlation length. These correlations significantly reduce the temperature cooling rate.

2 Uniform Gases: One Dimension

2.1 The Freely Cooling Case

Consider an infinite system of identical particles that interact via instantaneous binary collisions. When a pair of particles with velocities (u_1, u_2) collide, their post collision velocities (v_1, v_2) are given by the collision rule

$$v_{1,2} = u_{1,2} \mp (1-p)g, \tag{1}$$

with $g = u_1 - u_2$ the relative velocity, and p the dissipation parameter. The total momentum is always conserved, $u_1 + u_2 = v_1 + v_2$. Since the system is invariant under the Galilean transformation $v \to v - v_0$, we set the average velocity to zero, without loss of generality. In each collision, the relative velocity between the colliding particles is reduced by the restitution coefficient $r = 1-2p$. In an inelastic collision, there is an energy loss, $\Delta E = -p(1-p)g^2$ (particle mass is set to unity); energy loss is maximal for completely inelastic collisions ($p = 1/2$), and it vanishes for elastic collisions ($p = 0$).

In a random collision process, the collision rate is independent of the particles' velocities. Let $P(v,t)$ be the normalized velocity distribution at time t. It evolves according to the Boltzmann equation

$$\begin{aligned}\frac{\partial P(v,t)}{\partial t} = K \int du_1\, P(u_1,t) \int du_2\, P(u_2,t) \\ \times \left\{\delta\left[v - u_1 + (1-p)g\right] - \delta(v-u_2)\right\}.\end{aligned} \tag{2}$$

This Boltzmann equation, with a velocity independent collision kernel, is termed the Maxwell model in kinetic theory [35]. The overall collision rate $K = \sqrt{T}$, with the granular temperature $T = \int dv\, v^2 P(v,t)$, is chosen to represent the typical particle velocity. The quadratic integrand in (2) reflects the random and binary nature of the collision process, while the gain and loss terms reflect the collision rules (1). One can verify that the total number of particles, $\int dv P(v,t)$, and the total momentum, $\int dv\, vP(v,t)$, are conserved; hereinafter we set $\int dv P(v,t) = 1$ and $\int dv\, vP(v,t) = 0$.

For inelastic hard spheres, where the collision rate equals the relative velocity, the equation governing the temperature is part of an infinite hierarchy of equations. In contrast, in the Maxwell model, the temperature obeys a closed equation

$$\frac{dT}{dt} = -\lambda T^{3/2} \qquad \text{with} \qquad \lambda = 2p(1-p). \tag{3}$$

The temperature decays algebraically with time according to Haff's law [58]

$$T(t) = \frac{T_0}{(1+t/t_0)^2}, \tag{4}$$

with the time scale $t_0^{-1} = p(1-p)\sqrt{T_0}$. Eventually, all particles acquire the average velocity: $P(v,\infty) = \delta(v)$. Our goal is to characterize how the velocity distribution approaches this terminal state asymptotically. Since the overall collision rate is uniform, it is useful to characterize time by the collision counter $\tau = \int_0^t dt' K(t')$, equal to the average number of collisions experienced by a particle,

$$\tau = \frac{2}{\lambda}\ln(1 + t/t_0). \tag{5}$$

In terms of the number of collisions, $T(\tau) = T_0\, e^{-\lambda\tau}$.

The convolution structure of the Boltzmann equation suggests to apply the Fourier transform $F(k,\tau) = \int dv\, e^{ikv} P(v,\tau)$. This quantity evolves according to [40]

$$\frac{\partial}{\partial\tau} F(k,\tau) + F(k,\tau) = F(k - pk,\tau)\, F(pk,\tau). \tag{6}$$

This closed equation is both nonlinear and nonlocal, yet it is analytically tractable. Assuming that the velocity distribution approaches its final state in a self-similar fashion, we seek a scaling solution

$$P(v,t) = T^{-1/2}\mathcal{P}(w) \qquad \text{with} \qquad w = v\,T^{-1/2}. \tag{7}$$

The scaling form corresponding to the Fourier transform is $F(k,\tau) = f(z)$ with the variable $z = |k|\,T^{1/2}$. Substituting the scaling ansatz into (6) and using $dT/d\tau = -\lambda T$, the Fourier scaling function satisfies

$$-p(1-p)zf'(z) + f(z) = f(z - pz)f(pz). \tag{8}$$

This equation is supplemented by the small-z behavior $f(z) \cong 1 - \frac{1}{2}z^2$, which is dictated by the small wave number behavior, $F(k) \cong 1 - \frac{1}{2}k^2 T$. Subject to these conditions, the (unique) solution is [43]

$$f(z) = (1 + z)\,e^{-z}. \tag{9}$$

The scaled velocity distribution is obtained by performing an inverse Fourier transform

$$\mathcal{P}(w) = \frac{2}{\pi}\frac{1}{(1 + w^2)^2}. \tag{10}$$

Remarkably, the scaled velocity distribution is independent of the dissipation parameter p. Another important feature is the algebraic decay: $\mathcal{P}(w) \sim w^{-4}$ for $w \gg 1$. This behavior should be compared with the exponential high energy tails obtained for the traditional Boltzmann equation [26, 28]. The enhancement in the likelihood of finding energetic particles is due to the effective reduction in their collision rate.

Typically, in kinetic theory of molecular gases, the velocity distributions have sharp tails such that all moments of the distribution are finite[1]. Sonine expansions, where Maxwellian distributions are systematically modified by polynomials of increasing orders, are widely used to analyze the Boltzmann equation. The Maxwell model can be conveniently utilized to examine the applicability of this approach to situations where the velocity distributions have overpopulated high energy tails.

The scaling function does not characterize all features of the asymptotic time dependent behavior. Higher than third moments of the scaling function diverge, and since moments must be finite at all times, a direct calculation is necessary. Moments of the velocity distribution, $M_n(t) = \int dv\, v^n P(v,t)$, obey a closed set of equations

$$\frac{d}{d\tau} M_n + a_n M_n = \sum_{m=2}^{n-2} \binom{n}{m} p^m (1-p)^{n-m} M_m M_{n-m}, \tag{11}$$

with $a_n(p) = 1 - p^n - (1-p)^n$. These equations are solved recursively using $M_0 = 1$ and $M_1 = 0$. Assuming that all moments are initially finite we find that to leading order, the moments asymptotically decay as

$$M_n \sim e^{-a_n \tau} \sim t^{-2a_n/a_2}\,. \tag{12}$$

Indeed, the sum term in (11) is asymptotically negligible because the coefficients satisfy the inequality $a_n < a_m + a_{n-m}$ for all $1 < m < n-1$ [40]. Asymptotically, $M_n \sim e^{-a_n\tau}$ and $M_2 \sim e^{-a_2\tau}$, so $M_n \sim M_2^{\alpha_n}$ with $\alpha_n = a_n/a_2$. In this so-called multiscaling asymptotic behavior, the temperature does not characterize higher order moments (for ordinary scaling behavior $\alpha_n = n/2$). The indices α_n increase monotonically with p, so the stronger the dissipation, the more pronounced the multiscaling asymptotic behavior.

The scaling behavior characterizes only velocities of the scale of the typical velocity. For sufficiently large velocities, far outside the scaling region, the gain term in the Boltzmann equation is negligible, and $\left(\frac{\partial}{\partial\tau} + 1\right) P(v,\tau) = 0$. Consequently, a generic exponential decay, $P(v,\tau) \sim P_0(v)\exp(-\tau)$ characterizes such large velocities. Similarly, sufficiently large moments decay according to $M_n \sim e^{-\tau} \sim t^{-1/[p(1-p)]}$ for $n \to \infty$.

In several studies, inelasticity is treated as a small perturbation [29, 30, 59]. To first order in p, (6) reads $\left(\frac{\partial}{\partial\tau} + pk\frac{\partial}{\partial k}\right) F(k,\tau) = 0$. The solution to this equation, $F(k,\tau) = F_0(ke^{-p\tau})$, remembers the initial conditions forever, in contradiction with the exact asymptotic behavior (10). This example raises questions concerning the validity of perturbation analysis in the vicinity of $p = 0$. The $p \to 0$ limit is singular and indeed, in (8) the small parameter

[1] Algebraic high-energy tails may also characterize *nonequilibrium* states of elastic gases; for uniform shear flows of two-dimensional Maxwell molecules, this has been recently shown by Acedo, Santos, and Bobylev [38].

p multiplies the highest derivative. The singular nature of the quasi-elastic limit has profound general consequences.

The Fourier transform can be expressed as a series expansion. The linear term can be eliminated from (6) by making the transformation $F \to e^{\tau} F$ and $\tau \to 1 - e^{-\tau}$. Then a formal Taylor expansion solution can be found [42]:

$$F(k,\tau) = e^{-\tau} \sum_{n=0}^{\infty} \frac{(1-e^{-\tau})^n}{n!} F_n(k). \tag{13}$$

The expansion functions $F_n(k)$ are obtained recursively: $F_0(k) \equiv F(k, \tau = 0)$, and generally $F_{n+1}(k) = \sum_{m=0}^{n} \binom{n}{m} F_m(k - pk) F_{n-m}(pk)$. For instance, when the initial distribution is Maxwellian, the expansion functions, and hence the velocity distribution itself consist of sums of Maxwellians. Similarly, starting from a stretched exponential, all expansion functions consist of sums of stretched exponentials. In general, the expansion functions are products of F_0's with stretched arguments. This implies that starting from a compact initial distribution $P(v, 0)$, the velocity distribution $P(v, t)$ develops a set of singularities. For instance, a distribution with support in $[-v_0, v_0]$ becomes non-analytic at an infinite set of points $v_{l,m} = \pm p^l (1 - p)^m v_0$.

Thus far, we characterized velocity statistics at a specific point in time. The autocorrelation function

$$A\left(t', t\right) = \overline{v\left(t'\right)\, v(t)} \tag{14}$$

with the overline denoting an average over all particles quantifies (two-point) temporal correlations in the velocity of a tagged particle. We have for $\tau > \tau'$

$$\begin{aligned} \frac{d}{d\tau} A\left(\tau', \tau\right) &= \frac{d}{d\tau} \overline{v(\tau') v(\tau)} = \overline{v\left(\tau'\right)\, \left[dv(\tau)/d\tau\right]} \\ &= -(1-p)\, \overline{v\left(\tau'\right)\, \left[v(\tau) - u(\tau)\right]} = -(1-p)\, A\left(\tau', \tau\right). \end{aligned} \tag{15}$$

This derivation reflects averaging over all possible collisions between the tagged particle of velocity v and another particle of particle of velocity u, with the collision rule, $v \to v - (1-p)(v-u)$. Therefore, the autocorrelation, $A(\tau', \tau) = A(\tau', \tau') \exp\left[-(1-p)(\tau - \tau')\right]$, decays exponentially with the collision number, for $\tau > \tau'$ [45]. In terms of the original time variable:

$$A\left(t', t\right) = A_0 \left(1 + t'/t_0\right)^{1/p-2} \left(1 + t/t_0\right)^{-1/p}, \tag{16}$$

with $A_0 = T_0$. In particular, $A(t) \equiv A(0,t) \propto t^{-1/p}$, so memory of the initial conditions decays algebraically with time. The autocorrelation function decays faster than the temperature, $A(t) \leq T(t)$ (the two functions coincide in the completely inelastic case, $p = 1/2$). Generally, the autocorrelation is a function of the waiting time t' and the observation time t, and not simply of their difference, $t - t'$. This history dependence ("aging") merely reflects the fact that the collision rate keeps changing with time.

The spread in the position of a tagged particle, $\Delta^2(t) \equiv \langle |x(t) - x(0)|^2 \rangle$, is obtained from the autocorrelation function using $\Delta^2 = 2\int_0^t dt' \int_0^{t'} dt'' A(t'', t')$. Substituting the autocorrelation function (16), we find that asymptotically, the spread grows logarithmically with time

$$\Delta \propto \sqrt{\tau} \propto \sqrt{\ln t}. \tag{17}$$

This behavior reflects the t^{-1} decay of the overall velocity scale and is consistent with inelastic hard spheres results [26, 60–62].

2.2 The Forced Case

In experimental situations, granular ensembles are constantly supplied with energy, typically through the boundaries, to counter the dissipation occurring during collisions[2]. Theoretically, it is convenient to consider uniformly heated systems, as discussed by Williams and MacKintosh [14] in the case of one-dimensional inelastic hard rods and then extended to higher dimensions (see [28–30] and references therein). We therefore study a system where in addition to changes due to collisions, velocities also change due to an external forcing: $\frac{dv_j}{dt}|_{\text{heat}} = \xi_j$. We assume standard uncorrelated white noise: $\langle \xi_j \rangle = 0$ and $\langle \xi_i(t) \xi_j(t') \rangle = 2\overline{D}\delta_{ij}\delta(t - t')$.

Such white noise forcing amounts to diffusion with a 'diffusion' coefficient $\overline{D}$ in velocity space. Therefore, the Boltzmann equation (2) is augmented by a diffusion term

$$\left(\frac{\partial}{\partial t} - \overline{D}\frac{\partial^2}{\partial v^2}\right) P(v,t) = K \int du_1\, P(u_1,t) \int du_2\, P(u_2,t) \times \{\delta[v - u_1 + (1-p)g] - \delta(v - u_2)\} \tag{18}$$

The temperature changes according to $dT/dt + \lambda T^{3/2} = 2\overline{D}$, so the steady state temperature is $T_\infty = (2\overline{D}/\lambda)^{2/3}$. The relaxation toward the steady state is exponential, $|T - T_\infty| \sim \exp(-\text{const.} \times t)$.

At the steady state, the Fourier transform $F_\infty(k) \equiv F(k, \infty)$ satisfies

$$(1 + Dk^2)F_\infty(k) = F_\infty(k - pk)\, F_\infty(pk). \tag{19}$$

with $D = \overline{D}/\sqrt{T_\infty}$. Conservation of the total number of particles and the total momentum impose $F_\infty(0) = 1$ and $F'_\infty(0) = 0$, respectively. The solution is found recursively to give the following infinite product [40]

$$F_\infty(k) = \prod_{l=0}^{\infty} \prod_{m=0}^{l} \left[1 + p^{2m}(1-p)^{2(l-m)} Dk^2\right]^{-\binom{l}{m}}. \tag{20}$$

[2] When the mean-free path is comparable with the system size, the boundary effectively plays the role of a thermal heat bath.

Thus in one dimension the Fourier transform is determined analytically in the steady state. To extract the high-energy tail from (20) we note that $F_\infty(k)$ has an infinite series of poles located at $\pm i\left[p^{2m}(1-p)^{2(l-m)}\,D\right]^{-1/2}$. The simple poles at $\pm i/\sqrt{D}$ closest to the origin imply an exponential decay of the velocity distribution [49, 53],

$$P_\infty(v) \simeq \frac{A(p)}{v_*}\, e^{-|v|/v_*}, \qquad \text{with} \qquad v_* = \sqrt{D}, \tag{21}$$

when $|v| \to \infty$. A (lengthy) re-summation similar to that used in [40] yields the residue to this pole, and in turn, the prefactor

$$\begin{aligned} A(p) &= \frac{1}{2}\exp\left(\sum_{n=1}^{\infty}\frac{1}{n}\frac{1-a_{2n}(p)}{a_{2n}(p)}\right), \\ &= \frac{1}{2}\exp\left(\sum_{n=1}^{\infty}\frac{1}{n}\frac{p^{2n}+(1-p)^{2n}}{1-p^{2n}-(1-p)^{2n}}\right). \end{aligned} \tag{22}$$

In the interesting quasi-elastic limit ($p \to 0$) we expand the denominator. Keeping only the dominant terms simplifies the sum to $\sum_{n=1}^{\infty}\frac{1}{2pn^2} = \frac{\pi^2}{12p}$ and to leading order, the prefactor is $A(p) \propto \exp[\pi^2/(12p)]$.

A detailed analysis of the next poles shows that for sufficiently small velocities, $v \ll v_c$, the velocity distribution is essentially Maxwellian, $\exp(-v^2)$ [57]. For convenience, the velocities are normalized here by the rms velocity. Matching this with the exponential tail $\exp(-|v|/\sqrt{p})$ yields the crossover velocity $v_c \sim p^{-1/2}$ (this estimate holds up to a logarithmic correction [57]). Thus, although in general the asymptotic decay of the velocity distribution is exponential, in the quasi-elastic limit, the velocity distribution is Maxwellian over a growing velocity range. A similar interplay between a generic $\exp(-|v|^{3/2})$ tail and a $\exp(-|v|^3)$ decay in the quasi-elastic limit is found for inelastic hards spheres as well [30].

The leading high-energy behavior can also be derived by using a useful heuristic argument [26, 28, 49]. For sufficiently large velocities, the gain term in the collision integral in (18) is negligible. The resulting equation for the steady state distribution

$$D\,\frac{d^2}{dv^2}P_\infty(v) = -P_\infty(v) \tag{23}$$

yields the exponential high-energy tail (21). This argument applies to arbitrary collision rates. For example, if $K \propto v^\delta$, the right-hand side in (23) becomes $-v^\delta P_\infty$ implying that $P_\infty(v) \propto \exp(-|v|^\gamma)$ with $\gamma = 1+\delta/2$. For hard spheres ($\delta = 1$) one finds $\gamma = 3/2$ [28], and curiously, the Gaussian tail arises only for the so-called very hard spheres ($\delta = 2$) [35].

Finally, we notice that steady state properties in the heated case are intimately related with the relaxation properties in the cooling case. This can be seen via the cumulant expansion

$$F_\infty(k) = \exp\left[\sum_{n=1}^{\infty} \psi_n (-Dk^2)^n\right] . \tag{24}$$

Replacing the term $(1 + Dk^2)$ with $\exp[-\sum_{n=1} n^{-1}(-Dk^2)^n]$, and substituting the cumulant expansion into (19) yields $\psi_n = [na_{2n}(p)]^{-1}$. The cumulants κ_n are defined via $\ln F_\infty(k) = \sum_{n=1} \kappa_n (ik)^n/n!$. Therefore, the steady-state cumulants are directly related to the relaxation coefficients (12), $\kappa_{2n} = \frac{(2n)!}{n} D^n / a_{2n}$ (the odd cumulants vanish).

3 Uniform Gases: Arbitrary Dimension

3.1 The Freely Cooling Case

In general dimension, the colliding particles exchange momentum only along the impact direction. The post-collision velocities $\vec{v}_{1,2}$ are given by a linear combination of the pre-collision velocities $\vec{u}_{1,2}$,

$$\vec{v}_{1,2} = \vec{u}_{1,2} \mp (1-p)\,(\vec{g}\cdot\vec{n})\,\vec{n}. \tag{25}$$

Here $\vec{g} = \vec{u}_1 - \vec{u}_2$ is the relative velocity and $\vec{n}$ the unit vector connecting the particles' centers. The normal component of the relative velocity is reduced by the restitution coefficient $r = 1 - 2p$, and the energy dissipation is given by $\Delta E = -p(1-p)\,(\vec{g}\cdot\vec{n})^2$.

In random collision processes, both the impact direction and the identity of the colliding particles are chosen randomly. In such a process, the Boltzmann equation

$$\begin{aligned}\frac{\partial P(\vec{v},t)}{\partial t} &= K \int d\vec{n} \int d\vec{u}_1\, P(\vec{u}_1,t) \int d\vec{u}_2\, P(\vec{u}_2,t) \\ &\quad \times \left\{ \delta\left[\vec{v} - \vec{u}_1 + (1-p)\,(\vec{g}\cdot\vec{n})\,\vec{n}\right] - \delta\left(\vec{v} - \vec{u}_1\right) \right\}\end{aligned} \tag{26}$$

exactly describes the evolution of the velocity distribution function $P(\vec{v},t)$. The overall collision rate is chosen to represent the typical relative velocity, $K = \sqrt{T}$, with the granular temperature now being the average velocity fluctuation per degree of freedom, $T = \frac{1}{d}\int d\vec{v}\, v^2 P(\vec{v},t)$ with $v \equiv |\vec{v}|$. The evolution equation involves integration over all impact directions, and this angular integration should be normalized, $\int d\vec{n} = 1$. We tacitly ignored the restriction $\vec{g}\cdot\vec{n} > 0$ on the angular integration range in (26) since the integrand obeys the reflection symmetry $\vec{n} \to -\vec{n}$.

Several temporal characteristics such as the temperature and the autocorrelation behave as in the one-dimensional case. For example, the temperature satisfies (3) with prefactor $\lambda = 2p(1-p)/d$ reduced[3] by a factor d. The temperature therefore decays according to Haff's law (4), with the time scale $t_0 = d/\left[p(1-p)\sqrt{T_0}\right]$ set by the initial temperature. Similarly, the decay rate of the autocorrelation function is merely reduced by the spatial dimension, $\frac{d}{d\tau}A(\tau',\tau) = -\frac{1-p}{d}A(\tau',\tau)$. Consequently, the nonuniversal decay (16) and the logarithmic spread (17) hold in general.

The Fourier transform, $F(\vec{k},t) = \int d\vec{v}\, e^{i\vec{k}\cdot\vec{v}}\, P(\vec{v},t)$, satisfies

$$\frac{\partial}{\partial\tau} F(\vec{k},\tau) + F(\vec{k},\tau) = \int d\vec{n}\, F(\vec{k}-\vec{p},\tau)\; F(\vec{p},\tau) \tag{27}$$

with $\vec{p} = (1-p)(\vec{k}\cdot\vec{n})\,\vec{n}$ reflecting the momentum transfer occurring during collisions. This equation was obtained by multiplying (26) by $e^{i\vec{k}\cdot\vec{v}}$ and integrating over the velocities. The power of the Fourier transform is even more remarkable in higher dimensions[4] as it reduces the $(3d-1)$–fold integral in (26) to the $(d-1)$–fold integral in (27).

Hereinafter, we consider only isotropic velocity distributions. The Fourier transform depends only on $k \equiv |\vec{k}|$, so we write $F(\vec{k},\tau) = F(y,\tau)$ with $y = k^2$. To perform the angular integration, we employ spherical coordinates with the polar axis parallel to $\vec{k}$, so that $\hat{\vec{k}}\cdot\vec{n} = \cos\theta$. The θ-dependent factor of the angular integration measure $d\vec{n}$ is proportional to $(\sin\theta)^{d-2}d\theta$. Denoting angular integration with brackets, $\langle f\rangle \equiv \int d\vec{n} f$, and using $\mu = \cos^2\theta$ gives

$$\langle f\rangle = \int_0^1 d\mu\, \frac{\mu^{-\frac{1}{2}}(1-\mu)^{\frac{d-3}{2}}}{B\left(\frac{1}{2},\frac{d-1}{2}\right)}\, f(\mu), \tag{28}$$

where $B(a,b)$ is the beta function. This integration is properly normalized, $\langle 1\rangle = 1$. The governing equation (27) for the Fourier transform can now be rewritten in the compact from

$$\frac{\partial}{\partial\tau} F(y,\tau) + F(y,\tau) = \langle\, F(\xi y,\tau)\, F(\eta y,\tau)\,\rangle\,, \tag{29}$$

with the shorthand notations $\xi = 1-(1-p^2)\mu$ and $\eta = (1-p)^2\mu$. Unlike the one-dimensional case, explicit solutions of this nonlinear and nonlocal rate equation are cumbersome and practically useless. Nevertheless, most of the physically relevant features of the velocity distributions including the large velocity statistics and the time dependent behavior of the moments can be obtained analytically.

[3] The reduction in rate by a factor d is intuitive because of the d independent directions only the impact direction is relevant in collisions. Mathematically, the prefactor $\lambda = 2p(1-p)\int d\vec{n}\, n_1^2$ is computed by using the identity $n_1^2+\ldots+n_d^2 = 1$ that yields $\int d\vec{n}\, n_1^2 = 1/d$.

[4] For elastic Maxwell molecules, the Fourier transform was first used in unpublished theses by Krupp [34], and then rediscovered and successfully utilized by Bobylev (see [35, 36] for a review).

We seek a scaling solution: $P(\vec{v},t) = T^{-d/2}\mathcal{P}(w)$ with $w = vT^{-1/2}$, or equivalently $F(y,\tau) = \Phi(x)$ with $x = yT$. The scaling function $\Phi(x)$ satisfies

$$-\lambda x \Phi'(x) + \Phi(x) = \langle \Phi(\xi x)\, \Phi(\eta x)\rangle \tag{30}$$

and the boundary condition $\Phi(x) = 1 - \frac{1}{2}x + \cdots$ as $x \to 0$. In the elastic case, the velocity distribution is purely Maxwellian, $\Phi(x) = e^{-x/2}$. Indeed, $\lambda = 0$ and $\xi + \eta = 1$ in this case. A stochastic process of elastic collisions effectively randomizes the velocities and leads to a thermal distribution [31]. In practice, this collision algorithm is used in Molecular dynamics simulations to thermalize velocities.

From the one-dimensional case, we anticipate that the velocity distribution has an algebraic large velocity tail. Generally, the large velocity behavior can be determined from the small wave number behavior of the Fourier transform. For example, the small-x expansion of the one-dimensional solution (10) contains both regular and singular terms: $\Phi(x) = 1 - \frac{1}{2}x + \frac{1}{3}x^{3/2} + \cdots$, and the dominant singular $x^{3/2}$ term reflects the w^{-4} tail of $\mathcal{P}(w)$. In general, if $\mathcal{P}(w)$ has an algebraic tail,

$$\mathcal{P}(w) \sim w^{-\sigma} \qquad \text{as} \quad w \to \infty, \tag{31}$$

then $\Phi(x) \propto \int_0^\infty dw\, w^{d-1}\mathcal{P}(w)\, e^{iw\sqrt{x}}$ contains, apart from regular terms, the following dominant singular term: $\Phi_{\text{sing}}(x) \sim x^{(\sigma-d)/2}$ as $x \to 0$. The exponent σ can be now obtained by inserting $\Phi(x) = \Phi_{\text{reg}}(x) + \Phi_{\text{sing}}(x)$ into (30) and balancing the dominant singular terms. We find that σ is a root of the integral equation[5]

$$1 - \lambda\frac{\sigma - d}{2} = \left\langle \xi^{(\sigma-d)/2} + \eta^{(\sigma-d)/2} \right\rangle. \tag{32}$$

This equation can be recast as the eigenvalue problem $\lambda_\nu = \nu\lambda_1$ using $\lambda_\nu = \langle 1 - \xi^\nu - \eta^\nu\rangle$ and $\nu = (\sigma - d)/2$. It can also be expressed using the hypergeometric function ${}_2F_1(a,b;c;z)$ [63] and the Euler's gamma function:

$$1 - p(1-p)\frac{\sigma - d}{d} = {}_2F_1\left[\frac{d-\sigma}{2}, \frac{1}{2}; \frac{d}{2}; 1 - p^2\right] + (1-p)^{\sigma - d}\,\frac{\Gamma\left(\frac{\sigma-d+1}{2}\right)\Gamma\left(\frac{d}{2}\right)}{\Gamma\left(\frac{\sigma}{2}\right)\Gamma\left(\frac{1}{2}\right)}.$$

Clearly, the exponent $\sigma \equiv \sigma(d,p)$ depends in a nontrivial fashion on the spatial dimension d as well as the dissipation parameter p.

There are two limiting cases where the velocity distribution approaches a Maxwellian, thereby implying a divergence of the exponent σ. In the quasi-elastic limit ($p \to 0$) we have $\sigma \to d/p$. Even a minute degree of dissipation strongly changes the character of the system, and hence, energy dissipation is a singular perturbation [24, 29]. As the dimension increases, the relative

[5] This result was derived independently in [44, 47].

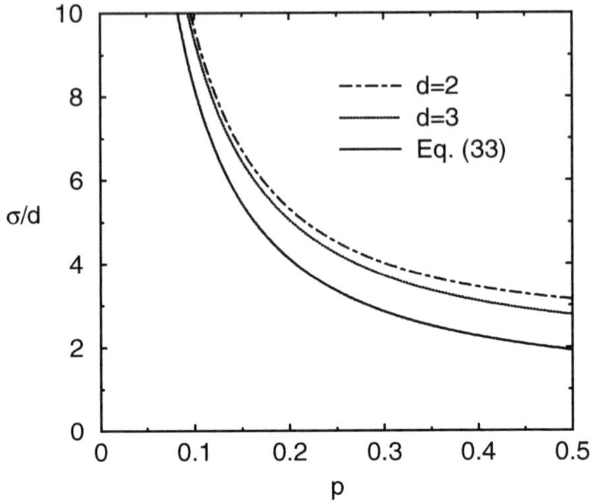

Fig. 1. The exponent σ versus the dissipation parameter p for $d = 2$, and 3. For comparison with the leading large-dimension behavior (33), σ is rescaled by d.

weight of the impact direction diminishes, and so does the role of dissipation. For large dimensions, the integral $\langle \eta^{(\sigma-d)/2} \rangle$ in (32) vanishes exponentially and asymptotic analysis of the remaining integral shows that the exponent grows linearly with the dimension:

$$\sigma \to d\,\frac{1+\frac{3}{2}p-p^3-p^{1/2}\left(1+\frac{5}{4}p\right)^{1/2}}{p(1-p^2)} \tag{33}$$

when $d \to \infty$ [44]. Equation (33) provides a decent approximation even at moderate dimensions (Fig. 1). Overall, the exponent $\sigma(d,p)$ increases monotonically with increasing d, and additionally, it increases monotonically with decreasing p. Both features are intuitive as they mirror the monotonic dependence of the energy dissipation rate $\lambda = 2p(1-p)/d$ on d and p. Remarkably, the exponent is very large. The completely inelastic case provides a lower bound for the exponent, $\sigma(d,p) \geq \sigma(d,1/2)$ with $\sigma(d,1/2) = 6.28753, 8.32937$, for $d = 2$, 3, respectively. For typical granular particles, $\sigma(d=3,p=0.1) \cong 30$, and such algebraic decays are impossible to measure in practice.

The algebraic tail of the velocity distribution implies that moments of the scaling function $\Phi(x)$ with sufficiently large indices diverge. In the scaling regime, moments of the velocity distribution can be calculated by expanding the Fourier transform in powers of x,

$$\Phi(x) \cong \sum_{n=0}^{[\nu]} \phi_n(-x)^n. \tag{34}$$

The order of terms in this expansion must be smaller than the order of the singular term, x^ν [47]. The coefficients ϕ_n, needed for calculating transport coefficients [54], yield the leading asymptotic behavior of the velocity moments, $M_k(t) = \int d\vec{v}\, v^k P(\vec{v}, t)$, via the relation $(2n)!\, T^n \phi_n \simeq \langle \mu^n \rangle\, M_{2n}$. Inserting the moment expansion into (30) yields a closed hierarchy of equations

$$(\lambda_n - n\lambda_1)\phi_n = \sum_{m=1}^{n-1} \lambda_{m,n-m} \phi_m \phi_{n-m}, \tag{35}$$

with $\lambda_n = \langle 1 - \xi^n - \eta^n \rangle$ and $\lambda_{m,l} = \langle \xi^m \eta^l \rangle$. Calculation of these coefficients requires the following integrals [45]

$$\langle \mu^n \rangle = \frac{\Gamma\left(\frac{d}{2}\right) \Gamma\left(n + \frac{1}{2}\right)}{\Gamma\left(\frac{1}{2}\right) \Gamma\left(n + \frac{d}{2}\right)} = \frac{1}{d} \frac{3}{d+2} \cdots \frac{2(n-1)+1}{2(n-1)+d}.$$

Of course, $\phi_0 = 1$ and $\phi_1 = 1/2$; further coefficients can be determined recursively from (35), e.g.,

$$\phi_2 = \frac{1}{8} \frac{1 - 3\,\frac{1-p^2}{d+2}}{1 - 3\,\frac{1+p^2}{d+2}}. \tag{36}$$

This coefficient is finite only when $\lambda_2 > 2\lambda_1$ or $d < d_2 = 1 + 3p^2$. Generally ϕ_n is finite only when the left hand side of (35) is positive, $\lambda_n > n\lambda_1$, or equivalently, when the dimension is sufficiently small, $d < d_n$. The crossover dimensions d_n are determined from $\lambda_n = n\lambda_1$.

For $d > d_n$, moments with index smaller than $2n$ are characterized by the temperature, while higher moments exhibit multiscaling asymptotic behavior. The time evolution of the moments can be studied using the expansion

$$F(y, \tau) = \sum_{n=0}^{\infty} f_n(\tau)\, (-y)^n. \tag{37}$$

The actual moments are related to the f_n coefficients via $(2n)! f_n = \langle \mu^n \rangle M_{2n}$. Substituting the expansion (37) into (29) yields the evolution equations

$$\frac{d}{d\tau} f_n + \lambda_n f_n = \sum_{m=1}^{n-1} \lambda_{m,n-m} f_m f_{n-m}. \tag{38}$$

We have $f_1 \propto e^{-\lambda_1 \tau}$, this is just the Haff's law. From $df_2/d\tau + \lambda_2 f_2 = \lambda_{1,1} f_1^2$ we see that f_2 is a linear combination of two exponentials, $e^{-\lambda_2 \tau}$ and $e^{-2\lambda_1 \tau}$. The two decay coefficients are equal $\lambda_2 = 2\lambda_1$ at the crossover dimension $d = d_2$. As expected, when $d < d_2$ the fourth moment is dictated by the second moment, $f_2 \propto f_1^2$. Otherwise, $f_2 \propto \exp(-\lambda_2 \tau)$. In general,

$$M_{2n} \propto \begin{cases} \exp(-n\lambda_1 \tau) & d < d_n; \\ \exp(-\lambda_n \tau) & d > d_n. \end{cases} \tag{39}$$

Fixing the dimension and the dissipation parameter, moments of sufficiently high order exhibit multiscaling asymptotic behavior. In practice, the exponent σ is large, and multiscaling occurs only for very high order moments.

3.2 The Forced Case

We consider white noise forcing as in the one-dimensional case. The steady state Fourier transform, $F_\infty(k) \equiv F_\infty(k^2)$, satisfies

$$(1 + Dk^2)F_\infty(k^2) = \langle F_\infty(\xi k^2)\, F_\infty(\eta k^2)\rangle . \tag{40}$$

This equation is solved recursively by employing the cumulant expansion (24). Writing $1 + Dk^2 = \exp\left[-\sum_{n\geq 1}(-Dk^2)^n/n\right]$, we recast (40) into

$$1 = \left\langle \exp\left[-\sum_{n=1}^{\infty}\left(\widetilde{\psi}_n - n^{-1}\right)\left(-Dk^2\right)^n\right]\right\rangle , \tag{41}$$

with the auxiliary variables $\widetilde{\psi}_n = \psi_n(1 - \xi^n - \eta^n)$. The coefficients ψ_n are obtained by evaluating recursively the angular integrals of the auxiliary variables, $\left\langle \widetilde{\psi}_n \right\rangle$, and then using the identities $\psi_n = \left\langle \widetilde{\psi}_n \right\rangle / \lambda_n$. The few first coefficients can be determined explicitly, e.g., $\psi_1 = 1/\lambda$ and

$$\psi_2 = \frac{3\,d^2}{4(d+2)(1-p^2) - 12(1-p)^2(1+p^2)} . \tag{42}$$

The nonvanishing second cumulant shows the steady state distribution is not Maxwellian. Moreover, the poles of the Fourier transform located at $k = \pm i/\sqrt{D}$ indicate that the large velocity tail of the distribution is exponential as in the one-dimensional case (21). Exponential decay is also suggested by the heuristic argument detailed above.

3.3 Velocity Correlations

Maxwellian velocity distributions were originally obtained for random elastic collision processes (see Ref. [32], p. 36). Maxwell's seminal derivation involves two basic assumptions: (1) The velocity distribution is isotropic, and (2) Correlations between the velocity components are absent. In inelastic gases, the velocity distributions are non-Maxwellian – therefore, there must be correlations between the velocity components[6]. The quantity

$$U = \frac{\langle v_x^2 v_y^2\rangle - \langle v_x^2\rangle \langle v_y^2\rangle}{\langle v_x^2\rangle \langle v_y^2\rangle} \tag{43}$$

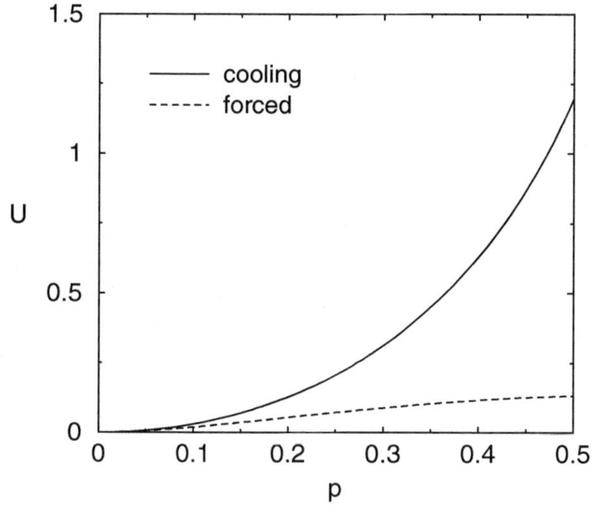

Fig. 2. The correlation measure U for spatial dimension $d = 3$. Shown are the freely cooling case and the forced case.

provides a natural correlation measure. A non-vanishing U indicates that velocity correlations do exist, and the larger U the larger the correlation.

To compute U, we apply the identities:

$$\langle v_x^2 \rangle = \frac{\partial^2}{\partial k_x^2} F \bigg|_{\vec{k}=\vec{0}} \quad \text{and} \quad \langle v_x^2 v_y^2 \rangle = \frac{\partial^2}{\partial k_x^2} \frac{\partial^2}{\partial k_y^2} F \bigg|_{\vec{k}=\vec{0}} . \tag{44}$$

The correlation measure is directly related to the fourth cumulant of the isotropic velocity distribution. In the freely cooling case, $U = 8\phi_2 - 1$; in the forced case, $U = \psi_2/\psi_1^2$. Substituting the corresponding coefficients yields

$$U_{\text{cooling}} = \frac{6p^2}{d - (1 + 3p^2)}, \tag{45a}$$

$$U_{\text{forced}} = \frac{6p^2(1-p)}{(d+2)(1+p) - 3(1-p)(1+p^2)} . \tag{45b}$$

Generally, correlations increase monotonically with p. Correlations are much weaker in the presence of an energy source because the nature of the driving is random (Fig. 2). The perfectly inelastic case provides an upper bound. For example, when $d = 3$ and $p = 1/2$ we have $U_{\text{cooling}} = 6/5$ and $U_{\text{forced}} = 2/15$. Correlations decay as $U \propto p^2$ and $U \propto d^{-1}$ in the respective limiting cases of $p \to 0$ and $d \to \infty$, where Maxwellian distributions are recovered.

[6] Inelastic collisions discriminate the impact direction, thereby generating correlations among the velocity components.

4 Impurities

The dynamics of an impurity immersed in a uniform granular fluid often defies intuition. The impurity does not generally behave as a tracer particle, and instead it may move either faster or slower than the fluid. Despite extensive studies, many theoretical and experimental questions regarding the dynamics of impurities remain open [64–72].

Impurities represent the simplest form of polydispersity, an important characteristic of granular media [73, 74]. Theoretically, the impurity problem is a natural first step in the study of mixtures as it involves fewer dissipation parameters. The fluid background is not affected by the presence of the impurity, and the impurity can be seen as "enslaved" to the fluid background.

We study dynamics of a single impurity particle in a uniform background of identical inelastic particles. We set the fluid particle mass to unity and the impurity mass to m. Collisions between two fluid particles are characterized by the dissipation parameter p as in (25), while collisions between the impurity and any fluid particle are characterized by the dissipation parameter q. When an impurity particle of velocity $\vec{u}_1$ collides with a fluid particle of velocity $\vec{u}_2$ its post-collision velocity $\vec{v}_1$ is given by

$$\vec{v}_1 = \vec{u}_1 - (1-q)\,(\vec{g}\cdot\vec{n})\,\vec{n}. \tag{46}$$

Two restitution coefficients, $r_p \equiv r = 1-2p$ and $r_q = m-(m+1)q$, characterize fluid-fluid and impurity-fluid collisions, respectively. The restitution coefficients obey $0 \leq r \leq 1$, and the dissipation parameters accordingly satisfy $0 \leq p \leq 1/2$ and $\frac{m-1}{m+1} \leq q \leq \frac{m}{m+1}$. The energy dissipated in an impurity-fluid collision is $\Delta E = -m(1-q)[2-(m+1)(1-q)]\,(\vec{g}\cdot\vec{n})^2$.

Let $Q\,(\vec{v},t)$ be the normalized velocity distribution of the impurity. In a random collision process, the impurity velocity distribution evolves according to the *linear* Lorentz-Boltzmann equation

$$\begin{aligned}\frac{\partial Q\,(\vec{v},t)}{\partial t} = K_q \int d\vec{n} \int d\vec{u}_1\, Q\,(\vec{u}_1,t) \int d\vec{u}_2\, P\,(\vec{u}_2,t) \\ \times\ \{\delta\left[\vec{v}-\vec{u}_1+(1-q)(\vec{g}\cdot\vec{n})\vec{n}\right] - \delta\,(\vec{v}-\vec{u}_1)\}\,,\end{aligned} \tag{47}$$

while the fluid distribution obeys (26). Two rates, $K_p \equiv K$ and K_q, characterize fluid-fluid and fluid-impurity collisions, respectively. These rates can be eliminated from the respective equations (26) and (47) by introducing the collision counters, $\tau_p \equiv \tau = \int_0^t dt' K_p\,(t')$, and $\tau_q = \int_0^t dt' K_q\,(t')$. The Lorentz-Boltzmann equation is further simplified by using the Fourier transform, $G(\vec{k},t) = \int d\vec{v}\, e^{i\vec{k}\cdot\vec{v}}\, Q(\vec{v},t)$

$$\frac{\partial}{\partial \tau_q}\, G(\vec{k}) + G(\vec{k}) = \int d\vec{n}\, G(\vec{k}-\vec{q})\, F\,(\vec{q})\,, \tag{48}$$

with $\vec{q} = (1-q)\,(\vec{k}\cdot\vec{n})\,\vec{n}$. This equation supplements the fluid equation (27).

We consider two versions of the Maxwell model: An idealized case with equal collision rates, $K_p = K_q$ (model A); and a more physical case with collision rates proportional to appropriate average relative velocities (model B). In the forced case, one can obtain the impurity velocity distribution using the cumulant expansion, and generally, velocity statistics of the impurity are similar to the background. Below, we discuss the more interesting freely cooling case.

4.1 Model A

When the two collision rates are equal, all pairs of particles are equally likely to collide with each other. For simplicity we set the overall rate to unity, $K_p = K_q = 1$; then both collision counters equal time, $\tau_p = \tau_q = t$.

In the freely cooling case, the fluid temperature decays exponentially, $T(t) = T_0 \exp[-2p(1-p)t/d]$. The impurity temperature, $\Theta(t)$, defined by $\Theta(t) = \frac{1}{d}\int d\vec{v}\, v^2 Q(\vec{v},t)$, is coupled to the fluid temperature via a linear rate equation

$$\frac{d}{dt}\Theta = -\frac{1-q^2}{d}\Theta + \frac{(1-q)^2}{d}T. \tag{49}$$

The solution to this equation is a linear combination of two exponentials

$$\Theta(t) = (\Theta_0 - cT_0)\, e^{-(1-q^2)t/d} + cT_0\, e^{-2p(1-p)t/d}. \tag{50}$$

The constant $c = (1-q)^2/[1-q^2-2p(1-p)]$ is simply the ratio between the impurity temperature and the fluid temperature in the long time limit. This generalizes the elastic fluid result $c = (1-q)/(1+q)$ [75, 76]. Generally, the impurity and the fluid have different energies, and this lack of equipartition is typical to granular particles [6, 77]

There are two different regimes of behavior. When $1-q^2 > 2p(1-p)$, the impurity temperature is proportional to the fluid temperature asymptotically, $\frac{\Theta(t)}{T(t)} \to c$ as $t \to \infty$. In the complementary region $2p(1-p) > 1-q^2$ the ratio of the fluid temperature to the impurity temperature vanishes. Since the system is governed by three parameters (m, r_p, r_q), it is convenient to consider the restitution coefficients as fixed and to vary the impurity mass. From the definition of the restitution coefficients the borderline case, $q^2 = 1-2p(1-p)$, defines a critical impurity mass

$$m_* = \frac{r_q + \sqrt{(1+r_p^2)/2}}{1-\sqrt{(1+r_p^2)/2}}. \tag{51}$$

This critical mass is always larger than unity. Asymptotically, the impurity to fluid temperature ratio exhibits two different behaviors

$$\frac{\Theta(t)}{T(t)} \to \begin{cases} c & m < m_*, \\ \infty & m \geq m_*. \end{cases} \tag{52}$$

We term these two regimes, the light impurity phase and the heavy impurity phase, respectively. In the light impurity phase, the initial impurity temperature becomes irrelevant asymptotically, and the impurity is governed by the fluid background. In the heavy impurity phase, the impurity is infinitely more energetic compared with the fluid and practically, it sees a static fluid. A qualitatively similar phase transition is observed for impurities in inelastic hard-spheres [71, 72]. Interestingly, the dependence on the dimension is secondary as both the critical mass, m_*, and the temperature ratio c are independent of d. Below we study velocity statistics including the velocity distribution and its moments primarily in one dimension, where explicit solutions are possible.

The Light Impurity Phase. In this phase, $m < m_*$, the impurity is governed by the fluid background. We therefore seek scaling solutions of the same form as the fluid's: $Q(v,t) = T^{-1/2}\mathcal{Q}\left(vT^{-1/2}\right)$ and $G(k,t) = g\left(|k|T^{1/2}\right)$. From the Lorentz-Boltzmann equation (48), the latter scaling function satisfies the *linear* equation

$$-p(1-p)zg'(z) + g(z) = g(qz)\left[1+(1-q)z\right]e^{-(1-q)z}. \tag{53}$$

Since the fluid scaling function is a combination of $z^n e^{-z}$ with $n=0$ and $n=1$, we try the series ansatz

$$g(z) = \sum_{n=0}^{\infty} g_n z^n e^{-z} \tag{54}$$

for the impurity. The first few coefficients, $g_0 = g_1 = 1$ and $g_2 = (1-c)/2$, follow from the small-z behavior $g(z) \cong 1 - \frac{1}{2}cz^2$. The rest are obtained recursively

$$g_n = \frac{q^{n-1}(1-q) - p(1-p)}{1 - q^n - np(1-p)}\, g_{n-1}. \tag{55}$$

The Fourier transform can be inverted to obtain the impurity velocity distribution function explicitly. The inverse Fourier transform of $e^{-\kappa z}$ is $\frac{1}{\pi}\frac{\kappa}{\kappa^2+w^2}$; the inverse transforms of $z^n e^{-z}$ can be obtained using successive differentiation with respect to κ. Therefore, the solution is a series of powers of Lorentzians

$$\mathcal{Q}(w) = \frac{2}{\pi}\sum_{n=2}^{\infty} Q_n \frac{1}{(1+w^2)^n} \tag{56}$$

The coefficients Q_n are linear combinations of the coefficients g_k's with $k \le n+1$, e.g., $Q_2 = 1 - 3g_2 + 3g_3$ and $Q_3 = 4g_2 - 24g_3 + 60g_4$. There are special values of q for which the infinite sum terminates at a finite order. Of course, when $q = p$, the impurity is identical to the fluid and $Q_n = 0$ for all $n > 2$. When $q^2(1-q) = p(1-p)$, one has $Q_n = 0$ for all $n > 3$, so the velocity distribution includes only the first two terms. Regardless of q, the first squared Lorentzian term dominates the tail of the velocity distribution $\mathcal{Q}(w) \sim \mathcal{P}(w) \sim w^{-4}$, as $w \to \infty$. One can show that this behavior extends to higher dimensions, $\mathcal{Q}(w) \sim \mathcal{P}(w) \sim w^{-\sigma}$. Therefore, the impurity has the same algebraic extremal velocity statistics as the fluid.

While the scaling functions underlying the impurity and the fluid are similar, more subtle features may differ. Moments of the impurity velocity distribution, $L_n(t) = \int dv\, v^n\, Q(v,t)$, obey the recursive equations

$$\frac{d}{dt}L_n + b_n L_n = \sum_{m=2}^{n-2} \binom{n}{m} q^m (1-q)^{n-m} L_m M_{n-m}, \tag{57}$$

with $b_n(q) = 1 - q^n$. Asymptotically, the fluid moments decay exponentially according to $M_n(t) \propto e^{-a_n(p)\,t}$. Using this asymptotic behavior, we analyze the (even) impurity moments. The second moment, i.e. the impurity temperature, was shown to behave similar to the fluid temperature when $a_2(p) < b_2(q)$. The fourth moment behaves similarly to the fourth moment of the fluid, $L_4 \propto M_4$, when $a_4(p) < b_4(q)$. However, in the complementary case, the fourth moment behaves differently, $L_n(t) \propto e^{-b_4(q)t}$. In general, when $a_n(p) < b_n(q)$, the nth impurity moment is proportional to the nth fluid moment, $L_n \propto M_n$. Otherwise, when $a_n(p) > b_n(q)$, the nth impurity moment is no longer governed by the fluid and

$$L_n(t) \propto e^{-(1-q^n)t}. \tag{58}$$

This series of transition affecting moments of decreasing order occurs at increasing impurity masses,

$$m_1 > m_2 > \cdots > m_\infty. \tag{59}$$

When $m \geq m_n$, the ratio M_{2k}/L_{2k} diverges asymptotically for all $k \geq n$. The transition masses

$$m_n = \frac{r_q + \frac{1}{2}\left[(1-r_p)^{2n} + (1+r_p)^{2n}\right]^{\frac{1}{2n}}}{1 - \frac{1}{2}\left[(1-r_p)^{2n} + (1+r_p)^{2n}\right]^{\frac{1}{2n}}} \tag{60}$$

are found from $q^{2n} = p^{2n} + (1-p)^{2n}$ and the definitions of the restitution coefficients. All of the transition masses are larger than unity, so the impurity must be heavier than the fluid for any transition to occur. The largest transition mass is $m_1 \equiv m_*$, and the smallest transition mass is $m_\infty = \lim_{n\to\infty} m_n = (1 + r_p + 2r_q)/(1 - r_p)$. Impurities lighter than the latter mass, $m < m_\infty$, mimic the fluid completely.

The Heavy Impurity Phase. In this phase, $m > m_*$, the velocities of the fluid particles are asymptotically negligible compared with the velocity of the impurity. Fluid particles become stationary as viewed by the impurity, and effectively, $\vec{u}_2 \equiv 0$ in the collision rule (46):

$$\vec{v} = \vec{u} - (1-q)\,(\vec{u}\cdot\vec{n})\,\vec{n}. \tag{61}$$

Mathematically, this process is reminiscent of a Lorentz gas [78]. Physically, the two processes are different. In the granular impurity system, a heavy particle scatters of a static background of lighter particles, while in the Lorentz gas the scatterers are infinitely massive.

We first consider the one-dimensional case. Setting $u_2 \equiv 0$ in the Lorentz-Boltzmann equation (47), integration over the fluid velocity u_2 is trivial, $\int du_2\, P(u_2,t) = 1$, and integration over the impurity velocity u_1 gives

$$\frac{\partial}{\partial t}Q(v,t) + Q(v,t) = \frac{1}{q}Q\left(\frac{v}{q}, t\right). \tag{62}$$

This equation can be solved directly by considering the stochastic process the impurity particle experiences. In a sequence of collisions, the impurity velocity changes according to $v_0 \to qv_0 \to q^2v_0 \to \cdots$ with v_0 the initial velocity. After n collisions the impurity velocity decreases exponentially, $v_n = q^n v_0$. Furthermore, the collision process is random, and therefore, the probability that the impurity undergoes exactly n collisions up to time t is Poissonian $t^n e^{-t}/n!$. Thus, the velocity distribution function reads

$$Q(v,t) = e^{-t}\sum_{n=0}^{\infty}\frac{t^n}{n!}\frac{1}{q^n}Q_0\left(\frac{v}{q^n}\right), \tag{63}$$

where $Q_0(v)$ is the initial velocity distribution of the impurity. The impurity velocity distribution function is a time-dependent combination of "replicas" of the initial velocity distribution. Since the arguments are stretched, compact velocity distributions display an infinite set of singularities, a generic feature of the Maxwell model.

In contrast to the velocity distribution, the moments $L_n(t)$ exhibit a much simpler behavior. Indeed, from (62) one finds that every moment is coupled only to itself, $\frac{d}{dt}L_n = -(1-q^n)L_n$. Solving this equation we recover (58); in the heavy impurity phase, however, it holds for all n. Therefore the moments exhibit multiscaling asymptotic behavior. The decay coefficients, characterizing the n-th moment, depend on n in a nonlinear fashion.

4.2 Model B

For hard sphere particles, the collision rates are proportional to the relative velocity. The overall collision rates in the Maxwell model represent the average relative velocity, and a natural choice is $\sqrt{\left\langle(\vec{v}_1-\vec{v}_2)^2\right\rangle} \propto \sqrt{(T_1+T_2)/2}$.

Therefore, the rates $K_p = \sqrt{T}$ and $K_q = \sqrt{(T+\Theta)/2}$, should be used in the Boltzmann equation (26) and the Lorentz-Boltzmann equation (47), respectively. This modification suppresses the heavy impurity phase, although the secondary transitions corresponding to higher order moments remain.

The fluid and the impurity temperatures obey

$$\frac{d}{dt}T = -\sqrt{T}\left[\frac{2p(1-p)}{d}T\right],$$
$$\frac{d}{dt}\Theta = \sqrt{\frac{T+\Theta}{2}}\left[-\frac{1-q^2}{d}\Theta + \frac{(1-q)^2}{d}T\right].$$

Consequently, the temperature ratio, $S = \Theta/T$, evolves according to

$$\frac{1}{\sqrt{T}}\frac{d}{dt}S = \sqrt{\frac{1+S}{2}}\left[-\frac{1-q^2}{d}S + \frac{(1-q)^2}{d}\right] + \frac{2p(1-p)}{d}S. \tag{64}$$

The loss term, which grows as $S^{3/2}$, eventually overtakes the gain term that grows only linearly with S. Therefore, the asymptotic temperature ratio remains finite, $S \to c$, where c is the root of the cubic equation

$$\sqrt{\frac{1+c}{2}}\left(c - \frac{1-q}{1+q}\right) = \frac{2p(1-p)}{1-q^2}c. \tag{65}$$

Consequently, there is only one phase, the light impurity phase. Intuitively, as the impurity becomes more energetic, it collides more often with fluid particles. This mechanism limits the growth rate of the temperature ratio. As in model A, energy equipartition does not generally occur, and the behavior is largely independent of the spatial dimension.

Qualitatively, results obtained for the light impurity phase in model A extend to model B. The impurity velocity distribution follows a scaling asymptotic behavior. The only difference is that the collision terms are proportional to $\beta^{-1} = \sqrt{(1+c)/2}$, and (53) generalizes as follows

$$-\beta p(1-p)zg'(z) + g(z) = (1+\alpha z)\,e^{-\alpha z}\,g(qz). \tag{66}$$

Seeking a series solution of the form (54) leads to the following recursion relations for the coefficients

$$g_n = \frac{(1-q)q^{n-1} - \beta p(1-p)}{1-q^n - n\beta p(1-p)}\,g_{n-1}. \tag{67}$$

Again, the velocity distribution is a combination of powers of Lorentzians as in (56). Moreover, the coefficients Q_n are linear combinations of the coefficients g_n's as in model A. Most importantly, the large-velocity tail is generic $\mathcal{Q}(w) \sim w^{-4}$.

The fluid moments evolve according to (11) with $\tau \equiv \tau_p$, while the impurity moments evolve according to

$$\frac{d}{d\tau_q} L_n + b_n L_n = \sum_{m=2}^{n-2} \binom{n}{m} q^m (1-q)^{n-m} M_m L_{m-j}, \tag{68}$$

Note that asymptotically $\tau_q \to \tau_p/\beta$, so the fluid moments decay according to $M_n \propto e^{-\beta a_n(p)\tau_q}$. Hence, when $\beta a_n(p) < b_n(q)$, the impurity moments are enslaved to the fluid moments, i.e., $L_n \propto M_n$ asymptotically. Otherwise, sufficiently large impurity moments behave differently than the fluid moments, viz. $M_n \propto e^{-b_n \tau_q}$. Although the primary transition affecting the second moment does *not* occur ($m_1 \equiv m_* = \infty$), secondary transitions affecting larger moments do occur at a series of masses, as in (59). The transition masses m_n are found by solving $\beta a_n(p) = b_n(q)$ simultaneously with (65). For example, for completely inelastic collisions ($r_p = r_q = 0$) one finds $m_2 = 1.65$. These transitions imply that some velocity statistics of the impurity, specifically large moments, are no longer governed by the fluid.

4.3 Velocity Autocorrelations

The impurity autocorrelation function satisfies $dA/d\tau_q = -(1-q)A$. Therefore, its decay is similar to (15)

$$A\left(\tau_q', \tau_q\right) \sim A_0 \exp\left[-(1-q)\left(\tau_q - \tau_q'\right)\right] \tag{69}$$

with q replacing p and τ_q replacing $\tau \equiv \tau_p$. For model A where the collision counters equal time, the decay remains exponential. However, the impurity autocorrelation decays with a different rate than the fluid. For model B, one can show that the algebraic decay (16) holds asymptotically with p replaced by q. We conclude that while one-point velocity statistics of the impurity are governed by the fluid, two-time statistics are different.

5 Mixtures

Granular media typically consists of mixtures of granular particles of several types. In contrast with the impurity problem, the different components of a mixture are coupled to each other. Moreover, there are numerous parameters, making mixtures much harder to treat analytically [51]. Remarkably, the impurity solution can be generalized to arbitrary mixtures[7]. We detail the freely cooling case in one dimension.

Consider a binary mixture where particles of type 1 have mass m_1 and concentration c_1, and similarly for particles of type 2. We set $c_1 + c_2 = 1$. Also, unit collision rates are considered for simplicity. Collisions between a

[7] Mixtures were treated analytically in the elastic Maxwell model [79, 80]

particle of type $i = 1,2$ and a particle of type $j = 1,2$ are characterized by the dissipation parameter $p_{ij} = (m_i - m_j r_{ij})/(m_i + m_j)$, with r_{ij} the restitution coefficient. We denote the normalized velocity distribution of component i by $P_i(v,t)$ and its Fourier transform by $F_i(k,t)$. The governing equations now couple the two distributions

$$\frac{\partial}{\partial t} F_i(k) + F_i(k) = \sum_{j=1}^{2} c_j F_j\left(k - p_{ij}k\right) F_i\left(p_{ij}k\right). \tag{70}$$

Let $T_i = \int dv\, v^2\, P_i(v,t)$ be the temperature of the ith component. Writing $F_i(k,t) \cong 1 - \frac{1}{2}k^2 T_i$, the temperatures evolve according

$$\frac{d}{dt}\begin{pmatrix} T_1 \\ T_2 \end{pmatrix} = -\begin{pmatrix} \lambda_{11} & \lambda_{12} \\ \lambda_{21} & \lambda_{22} \end{pmatrix}\begin{pmatrix} T_1 \\ T_2 \end{pmatrix}. \tag{71}$$

The diagonal matrix elements are $\lambda_{ii} = 2c_i p_{ii}(1 - p_{ii}) + c_j(1 - p_{ij}^2)$ with $j \neq i$, and the off-diagonal matrix elements are $\lambda_{ij} = -c_j(1-p_{ij})^2$. Therefore, the temperatures are sums of two exponential terms, $\exp(-\lambda_\pm t)$. The decay coefficients are the two (positive) eigenvalues

$$\lambda_\pm = \frac{1}{2}\left[\lambda_{11} + \lambda_{22} \pm \sqrt{(\lambda_{11} - \lambda_{22})^2 + 4\lambda_{12}\lambda_{21}}\,\right]. \tag{72}$$

Asymptotically, the term with the smaller decay rate $\lambda \equiv \lambda_-$ dominates

$$T_i(t) \simeq C_i e^{-\lambda t}, \qquad \text{with} \qquad C_i = \frac{(\lambda_+ - \lambda_{ii})T_i(0) - \lambda_{ij}T_j(0)}{\lambda_+ - \lambda_-}. \tag{73}$$

The ratio between the two temperatures approaches $C_2/C_1 = (\lambda - \lambda_{11})/\lambda_{12}$, and the two components have different temperatures. Furthermore, as long as the components are coupled, both temperatures are finite. For example, a vanishing C_2 implies that one of the concentrations vanishes as $\lambda = \lambda_{1,1}$.

As in the cases of homogeneous gases and impurities, we seek a scaling solution of the form $P_i(v,t) = e^{\lambda t/2}\,\mathcal{P}_i(w)$ with $w = v\, e^{\lambda t/2}$, or equivalently, $F_i(k,t) = f_i(z)$ with $z = |k|\, e^{-\lambda t/2}$. The scaling functions $f_i(z)$ are coupled via the non-local differential equations

$$-\frac{1}{2}\lambda z f_i(z) + f_i(z) = \sum_{j=1}^{2} c_j f_j(z - p_{ij}z)\, f_i(p_{ij}z). \tag{74}$$

Substituting the series solution (54) , $f_i(z) = \sum_{n=0}^{\infty} A_{i,n} z^n e^{-z}$, yields recursion relations for the coefficients $A_{i,n}$

$$\left(1 - \frac{n\lambda}{2}\right) A_{i,n} + \frac{\lambda}{2} A_{i,n-1} = \sum_{j=1}^{2}\sum_{m=0}^{n} c_j A_{j,m} A_{i,n-m} (1 - p_{ij})^m p_{ij}^{n-m}. \tag{75}$$

The small z behavior $f_i(z) = 1 - \frac{1}{2}C_i z^2$ implies the first three coefficients $A_{i,0} = A_{i,1} = 1$, and $A_{i,2} = (1-C_i)/2$. For $n \geq 3$, the coefficients $(A_{1,n}, A_{2,n})$ are solved recursively in pairs. Each such pair satisfies two inhomogeneous linear equations. Therefore, as in the impurity case, the velocity distribution is an infinite series of powers of Lorentzians (56). Although the two velocity distributions are different, they have the same extremal behavior, $\mathcal{P}_i(w) \sim w^{-4}$. It is straightforward to generalize the above to mixtures with arbitrary number of components, and to incorporate different collision rates, in particular, $K_{ij} = \sqrt{(T_i + T_j)/2}$.

6 Lattice Gases

Inelastic collisions generate spatial correlations and consequently, inelastic gases exhibit spatial structures such as shocks. Thus far, we considered only mean-field collision processes where there is no underlying spatial structure. Random collision processes can be naturally generalized by placing particles on lattice sites and allowing only nearest neighbors to collide.

Consider a one-dimensional lattice where each site is occupied by a single particle. Let v_j be the velocity of the particle at site j. The velocity of such a particle changes according to (1) due to interactions with either of its two neighbors. Time is conveniently characterized by the collision counter τ. In an infinitesimal time interval $d\tau$, the velocity of a particle changes as follows

$$v_j(\tau + d\tau) = \begin{cases} v_j(\tau) & \text{prob. } 1 - 2d\tau; \\ v_j(\tau) - (1-p)\left[v_j(\tau) - v_{j-1}(\tau)\right] & \text{prob. } d\tau; \\ v_j(\tau) - (1-p)\left[v_j(\tau) - v_{j+1}(\tau)\right] & \text{prob. } d\tau. \end{cases} \tag{76}$$

This process is stochastic and we are interested in averages over all possible realizations of the process, denoted by an overline. We consider random initial conditions where the average velocity vanishes and no correlations are present: $\overline{v_j(0)} = 0$ and $\overline{v_i(0)v_j(0)} = T_0\delta_{i,j}$.

Spatial velocity correlations satisfy closed equations as in the Ising-Glauber spin model [81]. For example, from the dynamical rules (76), the average velocity $V_j(\tau) = \overline{v_j(\tau)}$, obeys a discrete diffusion equation

$$\frac{dV_j}{d\tau} = (1-p)(V_{j-1} - 2V_j + V_{j+1}). \tag{77}$$

From the initial conditions, $V_j(0) = 0$, we obtain $V_j(t) = 0$. Consider the spatial correlation function $\overline{v_i v_j}$. The initial state is translationally invariant, so this property persists. The correlation functions $R_n = \overline{v_j v_{j+n}}$ satisfy

$$\begin{aligned}\frac{dR_n}{d\tau} &= 2(1-p)(R_{n-1} - 2R_n + R_{n+1}), \qquad n \geq 2;\\ \frac{dR_1}{d\tau} &= 2(1-p)\left[pR_0 - (1+p)R_1 + R_2\right];\\ \frac{dR_0}{d\tau} &= 4p(1-p)\left[R_1 - R_0\right].\end{aligned} \tag{78}$$

The initial conditions are $R_n(0) = T_0\delta_{n,0}$. Since we are interested in the asymptotic behavior, we employ the continuum approximation. The correlation function satisfies the diffusion equation $\partial R/\partial\tau = 2(1-p)\partial^2 R/\partial^2 n$, so the solution is the Gaussian

$$R_n(\tau) \simeq \frac{T_0}{\sqrt{8(1-p)\pi\tau}} \exp\left[-\frac{n^2}{8(1-p)\tau}\right]. \tag{79}$$

The temperature, $T \equiv R_0$ decays as $T(\tau) \simeq T_0\left[8(1-p)\pi\tau\right]^{-1/2}$ in the long time limit[8].

Although no correlations were present in the initial conditions, spatial correlations develop at later times. The corresponding correlation length ξ grows diffusively with the collision counter, $\xi \sim \tau^{1/2}$. The system consists of a network of domains of typical size ξ. Inside a domain, velocities are strongly correlated, and momentum conservation yields a relation between the average velocity and the domain size, $v_* \sim \xi^{-1/2} \sim \tau^{-1/4}$. This scale is consistent with the temperature behavior above, $T \sim v_*^2$. In arbitrary dimension, this scaling argument yields $T \sim \tau^{-d/2}$ [22, 43]. At least for scalar velocities, the correlation functions obey closed equations in arbitrary dimension [82], and this behavior can be also obtained analytically.

The actual time dependence is determined from the collision rate. We consider two choices. In model A, the rate is proportional to the typical velocity $K^2 = T$. In the physically more realistic model B, the rate is proportional to the average relative velocity $K^2 = \overline{(v_{j+1} - v_j)^2} = 2(R_0 - R_1) \propto -dT/d\tau$. Hence $K \sim \tau^{-\alpha}$ with $\alpha = d/4$ (model A) and $\alpha = (d+2)/4$ (model B). The time $t = \int_0^\tau d\tau'/K(\tau')$ grows as $t \sim \tau^{1+\alpha}$, and therefore,

$$T \sim t^{-\gamma} \qquad \text{with} \qquad \gamma = \begin{cases} \frac{2d}{d+4} & \text{model A;} \\ \frac{2d}{d+6} & \text{model B.} \end{cases} \tag{80}$$

In either case, Haff's cooling law is recovered only in the infinite dimension limit, $d \to \infty$. Otherwise, the appearance of spatial correlations slows down the temperature decay. As neighboring particles become correlated their relative velocity and consequently collision rate is reduced. Qualitatively similar behavior was shown for freely cooling inelastic gases, namely, breakdown of the mean-field cooling law due to the formation of strong spatial correlations among particles velocities [24, 25].

[8] An exact solution of the discrete equations (78) is possible. It yields an identical asymptotic expression for the temperature.

7 Conclusions

We presented analytic results for random inelastic collision processes. In general, when the collision rate is uniform, the convolution structure of the collision integrals translates to products in Fourier space. While the governing equations are both nonlinear and nonlocal, they are closed and amenable to analytical treatment. In the freely cooling case, a small wave number analysis of the Fourier transform displays both regular and singular terms. The regular terms yield the low order moments, while the leading singular term gives the high-energy tail. In the forced case, the Fourier transform is an analytic function and complex residue analysis yields the high-energy tail.

In the freely cooling case, the velocity distribution approaches a scaling form and displays an algebraic large velocity tail. In one dimension, the scaling function is a universal; otherwise, it depends on the degree of inelasticity. The exponent governing the high-energy tail is typically very large, and may be difficult to measure in practice. We have also shown that sufficiently large moments exhibit multiscaling behavior, and hence, are not characterized by the temperature. The autocorrelation function decays algebraically, with an exponent that depends on the dissipation parameters. The spread of a tagged particle exhibits a universal $\sqrt{\ln t}$ growth.

We demonstrated that an impurity immersed in a uniform fluid may or may not mimic the background. Fixing the dissipation parameters, for sufficiently low impurity mass, the impurity's velocity distribution, velocity moments and extremal velocity statistics are all governed by the fluid. However, there is a series of phase transitions occurring at a series of increasing masses, where impurity moments of decreasing orders decouple from the fluid. These transitions indicate that sufficiently heavy impurities are very energetic and effectively, they experience a static fluid background.

For binary mixtures, we examined only primary velocity statistics and showed that qualitatively, all components have the same temperature decay and extremal velocity statistics. It remains to be seen whether the different components may exhibit different asymptotic behaviors of more subtle velocity statistics such as the high order moments. The impurity case demonstrates how in certain limiting cases, mixtures may display anomalous behavior.

In the forced case, injection of energy counters the energy dissipation and the system relaxes toward a steady state. We considered white noise forcing, and showed that the steady state distribution has an exponential high-energy tail. Steady state characteristics in the forced case are directly related to time dependent relaxation characteristics in the freely cooling case.

Results obtained in the framework of the inelastic Maxwell model are exact for random collision processes where spatial structure is absent, and particles collide irrespective of their relative velocities. Such results are an uncontrolled approximation of the inelastic hard sphere problem. Nevertheless, as a conceptual tool, the inelastic Maxwell model is powerful. For example, it demonstrates the development of correlations among the velocity components, as well as spatial correlations. It also raises doubts concerning the suitability of several widely used techniques such as perturbation expansions in the quasi-elastic limit, and expansions in terms of Sonine polynomials.

We demonstrated that a lattice gas generalization of the Maxwell model remains tractable when the kinematic constraint (particles should collide only if they are moving toward each other) is ignored. Taking this constraint into account leads to shock-like structures in one dimension and to vortices in two dimensions [43]. Lattice models can therefore be used to study development of spatial correlations and spatial structures in inelastic gases.

We presented in detail the most basic realization of the Maxwell model. Several other generalizations are feasible. Experiments can now measure the distributions of impact angles and of the effective restitution coefficients. Such phenomenological information can be incorporated into the rate equations. The integration measure can be redefined to give different weights to different angles. Moreover, a distribution of restitution coefficients can be introduced by integrating the collision integrals with respect to the dissipation parameters. For example, random collision processes were successfully used to model the role of the boundary in driven gases [83].

We restricted our attention to the case where all moments of the initial velocity distribution are finite. However, there are other infinite energy solutions of the Boltzmann equation. For example, any Lorentzian, $F(k) = \exp(-Ck)$, is a steady state solution of (6). A remaining challenge is to classify the evolution of an arbitrary velocity distribution. Also, it will be interesting to characterize the full spectrum of extremal behaviors, characterizing velocities far larger than the typical velocity.

Finally, to model transport and other hydrodynamics problems one has to incorporate the spatial dependence explicitly. In the framework of the Maxwell model the simplest such problem – the shear flow – has been recently investigated by Cercignani [84]. For inelastic hard spheres, a number of unidirectional hydrodynamic flows were analyzed by Goldshtein and coworkers [85–87]. It would be interesting to investigate these problems in the framework of the Maxwell model.

Acknowledgments

We have benefited from discussions and correspondence with A. Baldassarri and M.H. Ernst. We also thank N. Brilliantov, A. Goldshtein, A. Puglisi, S. Redner, and H.A. Rose for useful discussions. This research was supported by DOE (W-7405-ENG-36) and NSF (DMR9978902).

References

1. T. Pöschel and S. Luding (editors), *Granular Gases* (Springer, Berlin, 2000)
2. H.M. Jaeger, S.R. Nagel, and R.P. Behringer, Rev. Mod. Phys. **68**, 1259 (1996)
3. W. Losert, D.G.W. Cooper, J. Delour, A. Kudrolli, and J.P. Gollub, Chaos **9**, 682 (1999)
4. J.S. Olafsen and J.S. Urbach, Phys. Rev. E **60**, R2468 (1999)
5. F. Rouyer and N. Menon, Phys. Rev. Lett. **85**, 3676 (2000)
6. R.D. Wildman and D.J. Parker, Phys. Rev. Lett. **88**, 064301 (2002)
7. A. Kudrolli, M. Wolpert and J.P. Gollub, Phys. Rev. Lett. **78**, 1383 (1997)
8. A. Samadani, L. Mahadevan, and A. Kudrolli, J. Fluid Mech. **452**, 293 (2002)
9. E. Rericha, C. Bizon, M.D. Shattuck, and H.L. Swinney, Phys. Rev. Lett. **88**, 014302 (2002)
10. K.M. Hill and J. Kakalios, Phys. Rev. E R3610 (1996)
11. F. Mello, P.B. Umbanhowar, and H.L. Swinney, Phys. Rev. Lett. **75**, 3838 (1995)
12. J.S. Olafsen and J.S. Urbach, Phys. Rev. Lett. **81**, 4369 (1998)
13. H.J. Herrmann, in: *Disorder and Granular Media*, edited by D. Bideau and A. Hansen (Elsevier Science Publishers, 1993)
14. D.R.M. Williams and F.C. MacKintosh, Phys. Rev. E **54**, R9 (1996)
15. E.L. Grossman, T. Zhou, and E. Ben-Naim, Phys. Rev. E **55**, 4200 (1997)
16. C. Bizon, M.D. Shattuck, J.B. Swift, W.D. McCormick, and H. L. Swinney, Phys. Rev. Lett. **80**, 57 (1998)
17. B. Bernu and R. Mazighi, J. Phys. A **23**, 5745 (1990)
18. S. McNamara and W.R. Young, Phys. Fluids A **4**, 496 (1992); Phys. Rev. E **50**, R28 (1994)
19. T. Zhou and L.P. Kadanoff, Phys. Rev. E **54**, 623 (1996)
20. Y. Du, H. Li, and L.P. Kadanoff, Phys. Rev. Lett. **74**, 1268 (1995)
21. I. Goldhirsch and G. Zanetti, Phys. Rev. Lett. **70**, 1619 (1993)
22. R. Brito and M.H. Ernst, Europhys. Lett. **43**, 497 (1998)
23. M.R. Swift, M. Boamfă, S.J. Cornell, and A. Maritan, Phys. Rev. Lett. **80**, 4410 (1998).
24. E. Ben-Naim, S.Y. Chen, G.D. Doolen, and S. Redner, Phys. Rev. Lett. **83**, 4069 (1999)
25. X. Nie, E. Ben-Naim, and S.Y. Chen, Phys. Rev. Lett. **89**, 204301 (2002)
26. S.E. Esipov and T. Pöschel, J. Stat. Phys. **86**, 1385 (1997)
27. P. Deltour and A. Barrat, J. Phys. I France **7**, 137 (1997)
28. T.P.C. van Noije and M.H. Ernst, Granular Matter **1**, 57 (1998)
29. A. Puglisi, V. Loreto, U.M.B. Marconi, and A. Vulpiani, Phys. Rev. E **59**, 5582 (1999)
30. A. Barrat, T. Biben, Z. Rácz, E. Trizac, and F. van Wijland, J. Phys. A **35**, 463 (2002)
31. J.C. Maxwell, Phil. Trans. R. Soc. **157**, 49 (1867)
32. P. Résibois and M. de Leener, *Classical Kinetic Theory of Fluids* (John Wiley, New York, 1977)
33. C. Truesdell and R.G. Muncaster, *Fundamentals of Maxwell's Kinetic Theory of a Simple Monoatomic Gas* (Academic Press, New York, 1980)
34. R.S. Krupp, *A nonequilibrium solution of the Fourier transformed Boltzmann equation*, M.S. Thesis, MIT (1967); *Investigation of solutions to the Fourier transformed Boltzmann equation*, Ph.D. Thesis, MIT (1970)

35. M.H. Ernst, Phys. Reports **78**, 1 (1981)
36. A. V. Bobylev, Sov. Sci. Rev. C. Math. Phys. **7**, 111 (1988)
37. D. Baganoff, Phys. Fluids **14**, 3403 (2002)
38. L. Acedo, A. Santos, and A.V. Bobylev, J. Stat. Phys. **109**, 1027 (2002)
39. H.-D. Kim and H. Hayakawa, *cond-mat*/0202003 and *cond-mat*/0202036
40. E. Ben-Naim and P.L. Krapivsky, Phys. Rev. E **61**, R5 (2000)
41. J.A. Carrillo, C. Cercignani, and I.M. Gamba, Phys. Rev. E **62**, 7700 (2000)
42. A.V. Bobylev, J.A. Carrillo, and I.M. Gamba, J. Stat. Phys. **98**, 743 (2000)
43. A. Baldassarri, U.M.B. Marconi, and A. Puglisi, Europhys. Lett. **58**, 14 (2002)
44. P.L. Krapivsky and E. Ben-Naim, J. Phys. A **35**, L147 (2002)
45. E. Ben-Naim and P.L. Krapivsky, Phys. Rev. E **66**, 011309 (2002)
46. E. Ben-Naim and P.L. Krapivsky, Eur. Phys. Jour. E **8**, 507 (2002)
47. M.H. Ernst and R. Brito, Europhys. Lett. **58**, 182 (2002)
48. M.H. Ernst and R. Brito, J. Stat. Phys. **109**, 407 (2002)
49. M.H. Ernst and R. Brito, Phys. Rev. E **65**, 040301 (2002)
50. A.V. Bobylev and C. Cercignani, J. Stat. Phys. **106**, 547 (2002)
51. U.M.B. Marconi and A. Puglisi, Phys. Rev. E **65**, 051305 (2002)
52. U.M.B. Marconi and A. Puglisi, Phys. Rev. E **66**, 011301 (2002)
53. T. Antal, M. Droz, and A. Lipowski, Phys. Rev. E **66**, 062301 (2002)
54. A. Santos, Physica A **321**, 442 (2003)
55. H. Hayakawa, *cond-mat*/0209630
56. V. Garzo, *cond-mat*/0211373
57. A. Santos and M.H. Ernst, *cond-mat*/0302285
58. P.K. Haff, J. Fluid Mech. **134**, 401 (1983)
59. E.L. Grossman and B. Rooman, Phys. Fluids **8**, 3218 (1996)
60. N.V. Brilliantov and T. Pöschel, Phys. Rev. E **61** 1716 (2000)
61. H.-D. Kim and H. Hayakawa, J. Phys. Soc. Japan **70**, 1954 (2001)
62. J. Lutsko, J.J. Brey, and J.W. Dufty, Phys. Rev. E **65**, 051304 (2002)
63. G.E. Andrews, R. Askey, and R. Roy, *Special Functions* (Cambridge University Press, New York, 1999)
64. A. Rosato, K.J. Strandburg, F. Prinz, and R.H. Swendsen, Phys. Rev. Lett. **58**, 1038 (1987)
65. R. Jullien, P. Meakin, and A. Pavlovitch, Phys. Rev. Lett. **69**, 640 (1992)
66. J.B. Knight, H.M. Jaeger, and S.R. Nagel, Phys. Rev. Lett. **70**, 3728 (1993)
67. J. Duran and R. Jullien, Phys. Rev. Lett. **80**, 3547 (1998)
68. D.C. Hong, P.V. Quinn, and S. Luding, Phys. Rev. Lett. **86**, 3423 (2001)
69. M.E. Mobius, B.E. Lauderdale, S.R. Nagel, and H.M. Jaeger, Nature **414**, 270 (2001)
70. J.W. Dufty, *cond-mat*/0108444
71. A. Santos and J.W. Dufty, Phys. Rev. Lett. **86**, 4823 (2001)
72. A. Santos and J.W. Dufty, Phys. Rev. E **64**, 051305 (2001)
73. J.C. Williams, Powder Technology **15**, 245 (1976)
74. C.S. Campbell, Annu. Rev. Fluid Mech. **22**, 57 (1990)
75. F.B. Pidduck, Proc. London Math. Soc. Ser. *2*, **15**, 89 (1915)
76. P.A. Martin and J. Piasecki, Europhys. Lett. **46**, 613 (1999)
77. K. Feitosa and N. Menon, Phys. Rev. Lett. **19**, 8301 (2002)
78. H.A. Lorentz, Arch. Néerl. **10**, 336 (1905); for a review, see E. H. Hauge, in: *Transport Phenomena*, eds. G. Kirczenow and J. Marro (Springer-Verlag, Berlin, 1974)

79. R. Dickman, J. Stat. Phys. **41**, 607 (1985)
80. C. Marin, V. Garzo, and A. Santos, Phys. Rev. E **51**, 3812 (1995)
81. R.J. Glauber, J. Math. Phys. **4**, 294 (1963)
82. E. Ben-Naim, L. Frachebourg, and P.L. Krapivsky, Phys. Rev. E **53**, 3078 (1996)
83. J.S. van Zon, F.C. MacKintosh, *cond-mat*/0205512
84. C. Cercignani, J. Stat. Phys. **102**, 407 (2001)
85. A. Goldshtein, M. Shapiro, and C. Gutfinger, J. Fluid Mech. **316**, 29 (1996)
86. V. Kamenetsky, A. Goldshtein, M. Shapiro, and D. Degani, Phys. Fluids **12**, 3036 (2000)
87. A. Goldshtein, V. Kamenetsky, A. Potapov, M. Shapiro, C. Campbell, and D. Degani, Granular Matter **4**, 115 (2002)

Velocity Fluctuations in Cooling Granular Gases

Andrea Baldassarri[1], Umberto Marini Bettolo Marconi[2], and Andrea Puglisi[3]

[1] INFM Udr Roma 1, University of Rome "La Sapienza", piazzale Aldo Moro 2, 00185, Roma, Italy
[2] Dipartimento di Matematica e Fisica and INFM Udr Camerino, University of Camerino, Via Madonna delle Carceri 62032 Camerino, Italy
[3] INFM Center for Statistical Mechanics and Complexity, University "La Sapienza", piazzale Aldo Moro 2, 00185 Roma, Italy

Abstract. We study the formation and the dynamics of correlations in the velocity field for 1D and 2D cooling granular gases with the assumption of negligible density fluctuations ("Homogeneous Velocity-correlated Cooling State", HVCS). It is shown that the predictions of mean field models fail when velocity fluctuations become important. The study of correlations is done by means of molecular dynamics and introducing an Inelastic Lattice Maxwell Models. This lattice model is able to reproduce all the properties of the Homogeneous Cooling State and several features of the HVCS. Moreover it allows very precise measurements of structure functions and other crucial statistical indicators. The study suggests that both the 1D and the 2D dynamics of the velocity field are compatible with a diffusive dynamics at large scale with a more complex behavior at small scale. In 2D the issue of scale separation, which is of interest in the context of kinetic theories, is addressed.

1 Introduction

The inelastic hard spheres model [1] without energy input, initially prepared in a homogeneous state, exhibits, after a rapid transient, a regime characterized by homogeneous density and a probability distribution of velocities that depends on time only through the total kinetic energy (global granular temperature $T_g(t)$), i.e. a scaling velocity distribution. This is the so called *Homogeneous Cooling State* (HCS) or Haff regime [2]. It has been shown by several authors [3–6] that this state is unstable with respect to shear and clustering instabilities: structures can form that seem to minimize dissipation, mainly in the form of velocity vortices and high density clusters. These instabilities grow on different space and time scales, so that one can investigate them separately. Several theories have been proposed to take into account the emergence of structures in granular gases. Some of these are more fundamental, because derive the correlation functions directly from the kinetic equations [7]; others, that assume the validity of the hydrodynamic description, deserve the name of mesoscopic theories [8]; others are phenomenological theories that suggest analogies with Burgers equation [9, 10] or spinodal decomposition models [11] or mode coupling theories [12]. Some of these theories can describe the behavior of the cooling granular gas far deeply into the correlated regime, giving predictions for the asymptotic decay of energy. In this paper, after a brief summary of the results of these theories, we show how the departure from the HCS can be well modeled by a class of models obtained placing on a lattice the original homogeneous Inelastic Maxwell Model. These models have the disadvantages of being conceived under the assumption of negligible density fluctuations and therefore can be useful only in the description of the first instability, i.e. the growth of velocity fluctuations. In Sect. 2 we review the HCS, its instabilities and the existing theories. In Sect. 3 we briefly discuss the Inelastic Maxwell Model [13–15], which is a starting point for the introduction of the Inelastic Lattice Maxwell Models (ILMM). In Sect. 4 and in Sect. 5 the analysis and results of the ILMM in one and two dimensions are reviewed [14, 16]. Finally in Sect. 6 conclusions are drawn.

2 Instabilities of the Homogeneous Cooling State

A cooling granular gas in d dimensions is defined as an ensemble of N grains, i.e. hard objects (rods if $d = 1$, disks if $d = 2$, spheres if $d = 3$), placed in a box of fixed volume with periodic boundary conditions. The grains evolve freely and interact with each other through instantaneous binary inelastic collisions. The rule that gives the velocities after the collision as functions of the velocities before the collision is the definition of the particular granular gas model. In this case we use the model with constant restitution coefficient without rotational degrees of freedom and set the mass of the grains $m = 1$. Other models have been considered in the literature [1]. The collision rule between a particle with velocity $\vec{v}$ and one with velocity $\vec{v}^*$ for this model is:

$$\vec{v}' = \vec{v} - \frac{1+r}{2}\left[(\vec{v}-\vec{v}^*)\cdot\hat{\vec{n}}\right]\hat{\vec{n}} \tag{1}$$

where the primed velocity is the post-collisional one, $\hat{\vec{n}} = (\vec{r}-\vec{r}^*)/|\vec{r}-\vec{r}^*|$ is the unit vector in the direction joining the centers of colliding particles, and r is the restitution coefficient and takes values between 0 and 1. When $r = 1$ the collision is elastic.

If granular gases are prepared in a homogeneous situation (uniform-random positions of grains, Gaussian or uniform-random initial velocity with no preferred direction), it happens that the imposed homogeneity is broken after a certain time, if the system or the inelasticity are large enough. The more accepted scenario is a two time homogeneity breaking: at a time t_s the velocity field becomes unstable to the formation of shear bands, then at a time $t_c > t_s$ the density field becomes unstable toward the formation of high density clusters.

2.1 The Homogeneous Cooling State

A granular gas evolving from a homogeneously random state loses memory of its initial condition after a time of the order of one collision per particle and rapidly enters the Homogeneous Cooling Regime. This regime is expected to be well described by the granular Boltzmann Equation (see for example [7]). This means that in this regime the total probability density function (p.d.f.) at collision is factorisable in one-particle p.d.f.'s: $p_N(\vec{r}_1...\vec{r}_N, \vec{v}_1...\vec{v}_N, t) = \prod_{i=1}^N p_1(\vec{r}_i, \vec{v}_i, t)$.

The kinetic definition of HCS is given by the homogeneity ansatz plus the scaling ansatz for the one particle distribution function:

$$p_1(\vec{r}, \vec{v}, t) = \frac{1}{V} P(\vec{v}, t) = \frac{1}{V v_0^d(t)} \tilde{P}(\vec{c}) \tag{2}$$

where $\vec{c} = \vec{v}/v_0(t)$ and $v_0(t)$ is the thermal velocity defined by $T(t) = v_0^2(t)/d$ with $T(t)$ the granular temperature; here we have assumed that $\int d\vec{v} \int d\vec{r} p_1 = 1$ and also that $\int d\vec{v} P = 1$. If the (2) is inserted in the Boltzmann Equation an equation for the temperature is obtained:

$$\frac{dT}{dt} = -2\omega\gamma T \tag{3}$$

where ω is the time dependent collision frequency, while γ is the time independent cooling rate. These two functions can be approximated, using the Maxwellian approximation $\tilde{P} \approx (\pi)^{-d/2} \exp\left(-dc^2/2\right)$, by ω_0 and γ_0:

$$\omega_0 \propto \sqrt{T} \tag{4a}$$

$$\gamma_0 = \frac{1 - r^2}{2d}. \tag{4b}$$

In this case the solution of the temperature equation (3) reads:

$$T(t) = \frac{T(0)}{(1 + \gamma_0 t/t_0)^2} = T(0) \exp(-2\gamma_0 \tau) \tag{5}$$

where $t_0 = 1/\omega(T(0))$ is the mean free time at the initial temperature $T(0)$ and

$$\tau = \frac{1}{\gamma_0} \ln(1 + \gamma_0 + t/t_0) \tag{6}$$

is the cumulated collision number obtained from the definition $d\tau = \omega(T(t))dt$. Equation (5) is known as Haff's law [2].

Corrections to the constants appearing in (5) stem from a more careful consideration of the HCS: when the volume fraction is non negligible the Enskog-Boltzmann equation should be employed instead of Boltzmann equation. This is identical to the Boltzmann equation but for a multiplicative constant in the collision integral that takes into account static density correlations due to the fact that the gas is not perfectly dilute. Very recently it has been shown [17] that there are also small velocity correlations which must be considered and that modify the molecular chaos hypothesis: with this correlations the law $T \sim t^{-2}$ is still valid but more precise corrections to the constant γ_0/t_0 appear.

The Haff law can be also derived in the framework of granular hydrodynamics [6, 18]: in the HCS hydrodynamics can be considered valid as a consequence of homogeneity of the density and velocity field and of the slow temperature decay. The evolution of the (space uniform) temperature field is exactly the same as in (5).

2.2 Instabilities of the Homogeneous Cooling State

Since the seminal work of Goldhirsch and Zanetti [3], is known that the HCS is unstable toward the formation of clusters and vortices. Kinetic arguments, linear stability analysis of hydrodynamic equations and numerical simulations show that large scale fluctuations of velocity and density can grow breaking the homogeneity of the system.

The linear stability analysis of hydrodynamics proposed by Ernst and coworkers [8, 19] gives, to our opinion, a good description of the mechanism of formation and growth of instabilities. The procedure is applied to the fields rescaled with respect to their values in the Homogeneous Cooling State. The stability analysis is performed in $\vec{k}$ space (the space of Fourier modes) and, when $d > 1$, the velocity field is decomposed in a parallel component $\tilde{\vec{u}}_{\parallel}(\vec{k}, t)$ and in $d-1$ orthogonal components $\tilde{\vec{u}}_{\perp}(\vec{k}, t)$ with respect to the vector $\vec{k}$. The most important result of this analysis is the existence of two threshold values $k_{\perp}^{*} \sim \sqrt{\gamma_0}$ and $k_{H}^{*} \sim \gamma_0$, indicating the presence of two linearly unstable modes, named the shear and heat instabilities respectively. The heat mode is a combination of density and temperature fluctuations and the corresponding instability is directly related to the so-called cluster instability. This analysis also shows that, at low inelasticities, the shear and heat instabilities are well separated.

It is important to note that if the linear total size L is small enough, the minimum wave number $k_{min} = 2\pi/L$ can be larger than both k_H^* and $k_{\perp}^*$, suppressing both the instabilities. More remarkably, one can tune L in order to make $k_H^* < k_{min} < k_{\perp}^*$, obtaining a system which displays shear instability but remains homogeneous in density.

When fluctuations grow, structures emerge. In molecular dynamics simulations these structure appear as shocks in $d = 1$, or vortices and clusters in $d = 2$. In $d > 2$, a detailed analysis of structure factors (from fluctuating hydrodynamics [8] or ring kinetic theory [7]) has established that all the main correlation lengths (which indicate the typical size of structures *in the rescaled fields*) should display a diffusive growth $L \sim \tau^{1/2}$.

After a brief transient the description suggested by linear analysis ceases to be meaningful. Molecular dynamics (MD), in the absence of experiments (the cooling granular gas being a reference model and not a real system), are the more reliable tools, but they are very cpu-time consuming: to obtain a decent statistics of the asymptotic dynamics of a granular gas, large enough to observe instabilities, MD simulations must be carried on for days. To our knowledge a definitive clear picture of the asymptotics of this model is still lacking (with month-lasting simulations Isobe [20] has obtained what he called the "final state", which resembles 2D turbulence, but this was done in a quasi-elastic limit).

On the other side there are effective models that can grasp the essence of the phenomena with particular assumptions. Here we rapidly review some of these, without the intention of being exhaustive:

- Brito and Ernst have shown by means of a mode-coupling [12] theory that the asymptotic energy decay should follow a diffusive form $\sim \tau^{-d/2}$.
- Wakou et al. [11] have demonstrated that the evolution of the flow field of a granular fluid, neglecting the convective term $\vec{u} \cdot \nabla \vec{u}$, can be cast in the form of a Time Dependent Ginzburg-Landau equation for a non-conserved order parameter. The energy functional has a *continuous* set of degenerate minima, having the shape of a Mexican hat. When the convective term is added, only subsets of admissible solutions are selected out of this infinite set of minima. In two dimensions only *two* distinct minima survive: therefore the $d = 2$ cooling granular gas, during the formation of instabilities, greatly resembles the spinodal decomposition for a non-conserved order parameter (Model A universality class, in the Hohenberg-Halperin classification [21]).
- Ben-Naim et al. [9] have studied the cooling granular gas in $d = 1$, discovering that, after the homogeneous (Haff) phase, it asymptotically becomes independent of the value of inelasticity ϵ and maps onto the sticky gas model. For such a sticky gas the asymptotic temperature decay is proportional to $t^{-2/3}$, when density clusters and velocity shocks form. The velocity field of the sticky gas [22] is described by the inviscid limit of the Burgers equation [23]. The relation to the Burgers equation is useful to establish an estimate of the tails of the asymptotic velocity distribution $P(v,t) \sim t^{1/3}\Phi(vt^{1/3})$ (independent of r), which reads $\Phi(z) \sim \exp(-|z|^3)$. The MD simulations have revealed only slight deviations from the Gaussian, and the authors have imputed the discrepancy from the expected behavior to the smallness of the constant in front of $|z|^3$. More recently it has been conjectured [10] that the Burgers equation describes the flow velocity $\vec{u}(\vec{r},t)$ of a cooling granular gas for arbitrary values of the inelasticity ϵ in generic d dimensions. This conjecture implies that the asymptotic behavior of the cooling inelastic gas is independent of ϵ (as it always falls in the universality class of the sticky gas) and that the upper critical dimension for the disappearance of the inelastic

collapse is $d_c = 4$. However MD simulations in $d > 4$ seem to show that the shear instability and the inelastic collapse do not disappear [24].

3 A Starting Point: The Homogeneous Inelastic Maxwell Model

To better illustrate the analysis of the Inelastic Lattice Maxwell Models (ILMM) we start with a brief description of the homogeneous Inelastic Maxwell Model (IMM). The definition of the model is given in terms of its Boltzmann Equation, which, for scalar velocities ($d = 1$), reads:

$$\frac{\partial P(v,\tau)}{\partial \tau} + P(v,\tau) = \frac{1}{1-\gamma^*} \int du P(u,\tau) P\left(\frac{v-\gamma^* u}{1-\gamma^*}, \tau\right) \tag{7}$$

with $\gamma^* = (1-r)/2$. The IMM can be introduced in different ways: (a) its Boltzmann equation can be regarded as the master equation of a simple stochastic process inspired to a model put forward by S.Ulam [25]: the process consists in the evolution of an ensemble of N velocities, at each time step (corresponding to $\Delta\tau = 2/N$, so that τ increases of 1 after averagely one collision per particle) two velocities are chosen at random with uniform distribution and are changed according to the inelastic collision rule; (b) the associated Boltzmann equation can be considered the inelastic generalization of the equation for elastic Maxwell Molecules [26, 27]; (c) its Boltzmann equation can be derived from the ordinary Boltzmann equation for inelastic hard spheres assuming that the term $|v - v'|$ in the collision integral can be approximated by $\sqrt{T}$, as suggested by Bobylev et al. [28].

The recent surge of interest for this model has been triggered by the discovery of an exact solution [13–15] for the $d = 1$ case (7) and the appearance of computable power law tails in the solutions of $d > 1$ cases [29, 30]. In a previous paper Ben-Naim and Krapivsky [31] showed that (7) could not have a scaling solution with finite moments: in fact, a scaling form

$$P(v,\tau) \to \frac{1}{v_0(\tau)} f\left(\frac{v}{v_0(\tau)}\right) \tag{8}$$

with $v_0^2(\tau) = \int v^2 P(v,\tau) dv = E(\tau) \sim \exp(-\tau a_2)$, imposes a temporal dependence of the higher moments (if finite):

$$\langle v^{2m} \rangle = v_0(\tau)^{2m} \mu_{2m} \tag{9a}$$

$$\text{where } \mu_{2m} = \int dy f(y) y^{2m} \tag{9b}$$

does not depend on time. But they found that

$$\lim_{\tau\to\infty} \frac{\langle v^{2m}(\tau)\rangle}{\left(\langle v^2(\tau)\rangle\right)^m} = \infty. \tag{10}$$

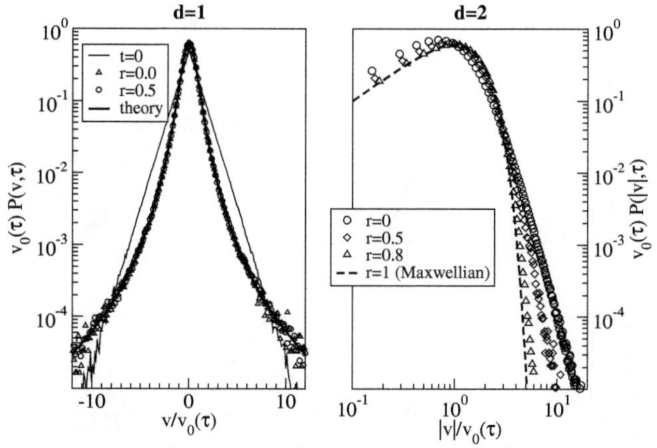

Fig. 1. Asymptotic velocity distributions $P(v,\tau)$ versus $v/v_0(\tau)$ for different values of r from the simulation of the Inelastic Maxwell Model in 1D (left) and 2D (right). In 1D the asymptotic distribution is independent of r and collapse to the (11). In 1D, the chosen initial distribution (exponential) is drawn (same result with uniform and Gaussian initial distribution). In 2D the distributions still present power-law tails, but the power depends upon r: for $r \to 1$ the pdf tends to a Maxwell distribution. Data refers to more than $N = 10^6$ particles

This result does not exclude the existence of a scaling solution (8) but requires a solution with diverging moments of order $2m \geq 4$, i.e. very large algebraic tails. In this case standard methods [27] based on Fourier transform fails due to the singularity in $k = 0$ corresponding to such algebraic tails.

Remarkably, a direct inspection shows that the following velocity distribution

$$P(v,\tau) = \frac{2}{\pi v_0(\tau)\left[1 + \left(\frac{v}{v_0(\tau)}\right)^2\right]^2} \tag{11}$$

is indeed a solution of the non-linear Boltzmann equation (7), for every value of r [13–15]. The solution (11) has the property that the moments of order $2m \geq 4$ diverge, and its Fourier transform has a singularity of the type $|k|k^2$ in $k = 0$.

We believe that (11) represents the asymptotic solution for a large class of initial distributions, for the following arguments:

- as shown by Ben Naim and Krapivsky [31], the dynamics of the moments for a generic starting distribution can be computed, giving the following limit $\lim_{t\to\infty}\langle v^{2m}(t)\rangle/\langle v^2(t)\rangle^m = \infty$ for $m > 1$;
- secondly we performed numerical simulations of the BK model, collecting evidence of the convergence to the solution (11) for several starting velocity distributions, namely uniform, exponential (see Fig. 1, left frame) or Gaussian.

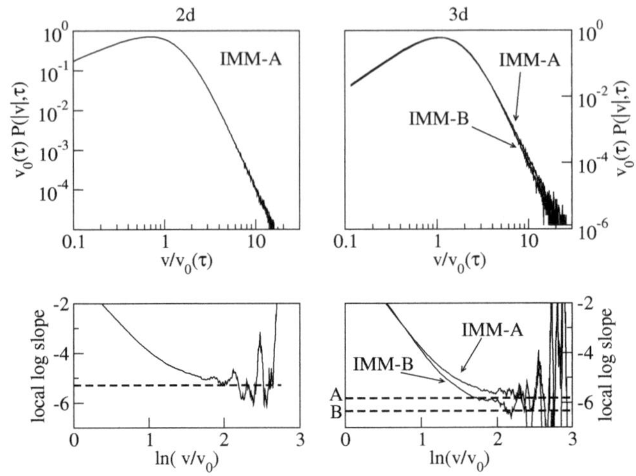

Fig. 2. Asymptotic velocity distributions $P(|v|, \tau)$ versus $|v|/v_0(\tau)$ for $r = 0$ from the simulation of the Inelastic Maxwell Model in 2D (left) and 3D (right). In 2D the asymptotic distribution tail converges to the exactly computed exponent (model IMM-A). In 3D the convergence is slower, but the difference of the predicted exponent for model IMM-A and IMM-B can be appreciated. In the bottom panels the convergences of the local logarithmic slope is put in evidence. (For the precise definition of the two models see [29].) Data refers to more than $N = 10^6$ particles

For higher dimensions, the problem of a scaling solution has been recently addressed in [10, 29]. Assuming a solution with large algebraic tails, it is possible to compute the asymptotic exponent as a solution of a transcendental equation. Such equation depends explicitly on the restitution coefficient, indicating that the tail exponent varies with r (and correctly diverges for $r \to 1$, corresponding to the recovering of the Maxwell distribution for the elastic case).

A numerical observation of such algebraic tails in 2 dimensions are shown in right frame of Fig. 1, where a dependence on r is put in evidence. A more precise comparison of the measured exponent and the theoretical predictions are shown in Fig. 2. In this figure we show the results of extensive numerical simulations for the completely inelastic case ($r = 0$) of several models, in two and three dimensions.

In fact, as noted in [29], the tail exponent is in general dependent on the details of the collisions, as for instance on the distribution of the impact parameter. Two different choices have been studied in 3 dimensions, corresponding to the models denoted IMM-A and IMM-B in reference [29]: in the first model grazing collisions are favored with respect to the second one. It is interesting to notice that the analytical and numerical study confirm the existence of a scaling solution with algebraic tails, irrespectively of the collision details, but that the precise value of the exponent is not universal.

The relevance of such a scaling solution for $d > 1$ has been addressed by Bobylev *et al.* [32], who have proved the existence and the uniqueness of such an asymptotic solution (giving details on the basin of attraction).

Interesting features emerge when the inelastic Maxwell model is extended to treat grains having different physical properties, such as unequal masses, different restitution coefficients, radii, etc. The study of a binary Maxwell mixture with scalar velocities showed few unexpected properties [33, 34]:

1. the kinetic temperatures of the two species are different, i.e. there is no energy equipartition, but the the ratio T_1/T_2 reaches asymptotically a constant value. Such a feature agrees with the results obtained in the framework of the Boltzmann-Enskog transport equation by Garzò and Dufty [35].
2. The velocity distributions in general differ in shape.
3. The velocity distributions have power law tails. These exponents can be calculated analytically, by a suitable generalization of a technique devised by Ernst and Brito [29]. The exponent of the power law tails in cooling binary Maxwell mixtures can attain values in between 2 and ∞, depending upon the parameters of the system.

4 The One-Dimensional Gas

The solution of the IMM given in (11) has nothing to do with the real dynamics of a cooling granular gas in the homogeneous regime, where no algebraic tails appear. Instead, a good agreement (in the HCS) is found between the solution of Boltzmann equation [36] and Molecular Dynamics [37]. However we will show in the following that the Inelastic Maxwell Model with the addition of a special topology (the 1d lattice) is able to reproduce the homogeneous regime and some essential features of the first instability (enhancement of velocity fluctuations).

4.1 Molecular Dynamics

An MD simulation of N hard rods on a ring of length L shows that, besides hydrodynamic instabilities discussed above, also a kinetic singularity appears that prevents long simulation runs, the so-called "inelastic collapse" [4, 38]. The inelastic collapse is the divergence of the collision rate of a small set of grains, analogous to the divergence of the collision rate of a ball left bouncing on the floor. This means that a group of particles will experience an infinite number of collisions in a finite time. To avoid such a problem, many authors [9, 39] have proposed the introduction of a velocity cut-off δ such that, whenever two particles collide with a relative velocity smaller, in absolute value, than δ, the collision is elastic. Taking $\delta^2 \ll T_{min}$, being T_{min} the minimum granular temperature that one expects to observe, one can be

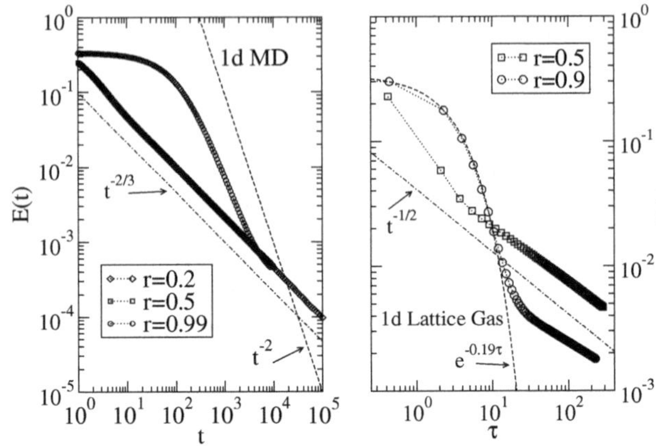

Fig. 3. Time behavior of kinetic energy for the Inelastic Hard Rods (left) and for the Inelastic Lattice Maxwell Model (right). The homogeneous Haff stage is evident only for quasi-elastic system, whereas the more inelastic ones enter almost immediately in the correlated regime. Note the different time units used (t is the physical time, defined only in MD simulations, while τ is the cumulated number of collisions per particle). The Haff law $E \sim t^{-2} \sim \exp(-\gamma_0 \tau)$ is verified for both systems. The correlated regime presents a behavior $t^{-2/3}$ independent of r for the Hard Rods, while appears diffusive (in collision units) $\tau^{-1/2}$ and r-dependent for the Lattice Model. Data refer to $N = 10^6$ particles (both models)

sure that the choice of δ will not influence the main results of the simulation. Performing an efficient Event Driven simulation scheme (based on a tree ordering of the event times) it is possible to study easily very large systems, with $N \sim 10^6$ for thousands of collisions per particle.

In the left panel of Fig. 3 the decay of kinetic energy $E(t) \equiv T(t) = \langle v^2 \rangle$ is shown. It displays, for different values of r, the same behavior with two regimes: in the HCS $E(t) \sim t^{-2}$ (Haff's law), then it decays differently, i.e. $E(t) \sim t^{-3/2}$. Noticeably all the curves seem to collapse in the second regime, i.e. the asymptotic regime is equivalent (not only in the exponent, but also in the coefficients) for every value of r.

The velocity pdf's are shown in Fig. 4: in the HCS it appears the peculiar "two-peaks" form, which is also predicted by the study of the Boltzmann equation in the quasi-elastic limit [36]. In the second regime (which cannot be described by the Boltzmann equation, because of the presence of correlations) the pdf becomes a Gaussian.

What can be said about correlations in the second regime? An inspection of the velocity profile v_i vs. x_i (see top panel of Fig. 5) suggests the presence of shocks, i.e. asymmetric "discontinuities" of the velocity profile in correspondence with peaks of density (clusters). A more quantitative information about the velocity field is given by the study of the structure function $S(k,t) = \hat{v}(k,t)\hat{v}(-k,t)$ where $\hat{v}(k,t)$ is the Fourier transform of the field

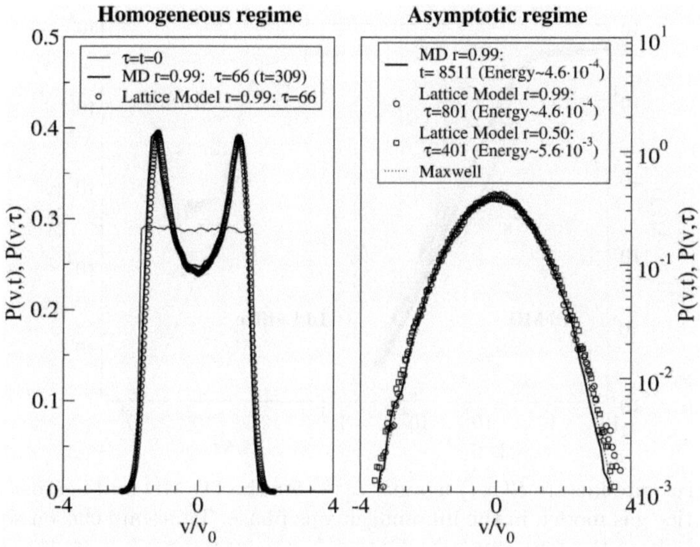

Fig. 4. Rescaled velocity distributions for the 1D MD and in the 1D ILMM, during the homogeneous (left) and the inhomogeneous phase (right). In the left frame is also shown the initial distribution (both models). The distributions refer to systems having the same energy. Data refer to $N = 10^6$ particles (both models) with $r = 0.99$ and $r = 0.5$ (for the lattice model in the inhomogeneous regime)

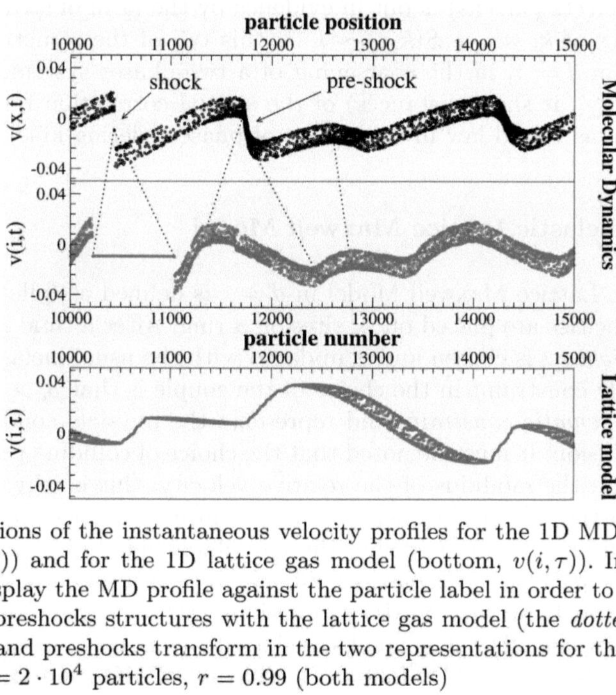

Fig. 5. Portions of the instantaneous velocity profiles for the 1D MD (top $v(x,t)$, middle $v(i,t)$) and for the 1D lattice gas model (bottom, $v(i,\tau)$). In the middle frame we display the MD profile against the particle label in order to compare the shocks and preshocks structures with the lattice gas model (the *dotted lines* show how shocks and preshocks transform in the two representations for the MD). Data refers to $N = 2 \cdot 10^4$ particles, $r = 0.99$ (both models)

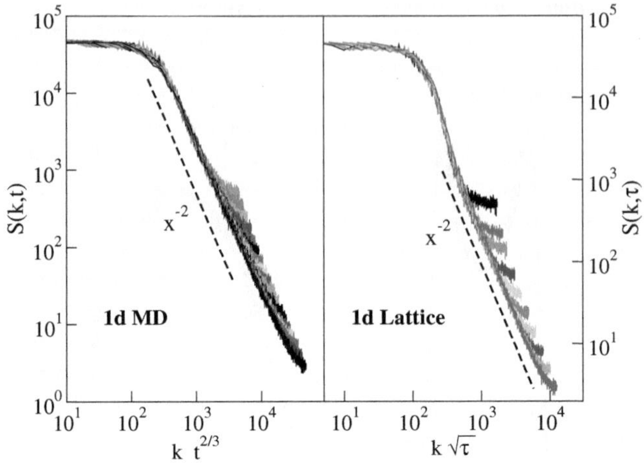

Fig. 6. Structure factors $S(k,t)$ against $kt^{2/3}$ for the 1D MD and against $k\tau^{1/2}$ for the 1D lattice gas model, in the inhomogeneous phase. Times are chosen so that the two systems have the same energies. Data refers to system with more than $N = 10^5$ particles, $r = 0.5$ (both models)

$v(i,t)$, presented in the left frame of Fig. 6. A good collapse is obtained if $S(k,t)$ is plotted at different times t versus $kt^{2/3} \sim k/E(t)$. The presence of short scale defects (shocks) is put in evidence by the form of structure factor at high values of k, where $S(k,t) \sim k^{-2}$: this tail of the structure function is expected, in $d = 1$, in the coarsening of a two-phases system, due to the non-analyticity (at short distances) of the spatial correlation function. This is also known as Porod law in the theory of phase ordering kinetics [40–42].

4.2 The Inelastic Lattice Maxwell Model

The Inelastic Lattice Maxwell Model in $d = 1$ is defined as follows. A set of N scalar velocities are placed on N sites on a ring. After a time $\Delta\tau = 2/N$ a couple of *neighbors* is chosen and is updated with the usual inelastic collision rule. The only constraint in the choice of the couple is that $v_i > v_{i+1}$: this is called the *kinematic constraint* and represents the physical condition necessary for a collision. It must be noted that the choice of colliding particles does not depend on the modulus of the relative velocity: this is why we consider this a "Inelastic Maxwell Model" embedded in a 1d lattice. In this model the only measure of time is τ, i.e. the cumulated number of collisions per particle. This makes many measures difficult to be compared with analogous MD measurements: in fact in MD simulations $\tau(t)$ depends upon the choice of δ (the elastic cut-off), while all observables depend on t independently of δ.

The analysis of the ILMM follows closely that of MD simulations and the results [13, 14] are presented, in parallel, in the same Figs. 3, 4, 5, 6.

The study of the decay of energy (Fig. 3) indicates that the model well reproduces the Haff regime ($E(\tau) \sim \exp(-\gamma_0 \tau)$) and displays a decay in the second regime of the form $\sim \tau^{-1/2}$.

The velocity pdf's reproduce exactly the ones measured in MD simulations (see Fig. 4), showing the "two-peaks" form in the Haff stage and the Gaussian in the late stage. It must be stressed again that these features are obtained simply applying to the Inelastic Maxwell Model (which displays tails $P(v) \sim v^{-4}$) a two-neighbors (instead of N-neighbors) topology.

The inspection of velocity profiles must be carefully carried: in this model the particles cannot move and therefore a comparison with MD should be made considering v_i vs. i instead of x_i. The good agreement between MD and ILMM can be appreciated in the two bottom panels of Fig. 5: with this choice of abscissa, the shocks observed in MD simulations appear reversed and smoothed, making the v profile resemble very similar to the one observed in the ILMM. Analogously the structure factor presents similar features with MD, mainly a k^{-2} behavior at high values of k (short scales) which is a signature of topological defects. In the lattice case the collapse can be obtained plotting $S(k,\tau)$ against $k\tau^{1/2}$, i.e. $\sim k/E(\tau)$ as in the MD case. We have also verified that, removing the kinematic constraint, this structure disappears and the model behaves as governed by a simple diffusion equation.

5 The Two-Dimensional Gas

The number of interesting and realistic results obtained with the one dimensional Inelastic Lattice Maxwell Model, in spite of its extreme simplicity, convinced us to extend our study to two dimensions.

However, a smaller number of results is known from Molecular Dynamics in $d = 2$, especially in the correlated regime, which requires very long simulations. Therefore the comparison between MD and the ILMM to the case $d = 2$ is more difficult. Only recently MD simulations start to give reliable measurements even for long times, and, where possible, the comparison with our ILMM is promising.

More speculatively, we tried a series of less standard measurements with two different aims: the first is to try to understand what kind of effective equation could describe the velocity field of the ILMM; the second is to discuss some basic hypothesis necessary to a hydrodynamic description of this granular fluid. After a brief review of the known results from MD, we present the results obtained with our 2D ILMM. First the same analysis of energy decay, velocity pdf and structure factors is carried on, as in the 1d case. Moreover, to better capture the nature of the topological defects generated by the dynamics, we consider the pdf of the velocity increments. Then we present an analysis of temporal correlations, mainly aimed to identify an effective equation for the velocity field. Finally we suggest a measurement of a hydrodynamic temperature and we sketch a discussion about the very presence of scale separation.

5.1 Known Results from Molecular Dynamics

To our knowledge few MD simulations have inspected the behavior of inelastic hard disks in the late stage, i.e. after the end of Haff regime and the growth of instabilities. The departure from the homogeneity has been studied by Ernst and co-workers [6–8] and their MD simulations have shown results compatible with their predicted $\tau^{1/2}$ scaling law for the linear size of structures (vortices and clusters). A figure in a work of Huthmann et al. [43] was the first suggestion that the global velocity pdf asymptotically returns to a Gaussian (while it has exponential tails in the HCS [44]) and recently different works on large MD simulations have confirmed this [10, 45]. Recently this "return-to-the-Gaussian" scenario was put in strict relation with the presence of a Burgers-like dynamics [10]. The debate on the asymptotic energy seems to be still open, even if there are strong indications of a universal (independent of r) decay of the form $E(t) \sim t^{-1}$ [10]. A recent work of Isobe [20] demonstrates that many features of the "last" stage resemble the dynamics of 2D turbulence.

5.2 The Inelastic Maxwell Lattice Model

The ILMM in $d = 2$ is identical to that in $d = 1$ except for the fact that the particles have a 2-components vector velocity $\vec{v}_i$ and that they are placed on a triangular $2D$ lattice, i.e. every particle has 6 neighbors. The dynamics is the same as in $d = 1$; the kinematic constraint here reads $(\vec{v}_i - \vec{v}_j) \cdot (\vec{r}_i - \vec{r}_j) < 0$.

The decay of energy $E(\tau)$ is shown in Fig. 7 for two different values of r. The Haff regime $E(\tau) \sim \exp(-\gamma_0 \tau)$ is well reproduced (when $r = 0.2$ it terminates too early to be appreciated). The asymptotic decay reads $E(\tau) \sim \tau^{-1}$. Both this measurement and the decay observed in the $d = 1$ case, are compatible with a universal law $E(\tau) \sim \tau^{-d/2}$ as for a purely diffusive field.

The rescaled velocity pdf's (see Fig. 8), which are initially Gaussian, become non-Gaussian in the Haff regime, with larger tails, but return Gaussian in the late stage. To the best of our knowledge this scenario is compatible with the MD simulations of Inelastic Hard Disks. The larger tails observed in the Haff regime can correspond to the exponential tails found in [44]. The analysis that follows shows that, in the late stage, the velocity field is far from being homogeneous: the global velocity pdf is strongly influenced by this homogeneity and its Gaussian form could be simply the signature of the presence of many independent domains (i.e. it is just the distribution of the average velocities of these domains).

In the Figs. 9 and 10 we show the presence of vortices in the velocity fields. The second figure represents the same system at a later time, demonstrating the coarsening of vortices (their number reduces and their size increases).

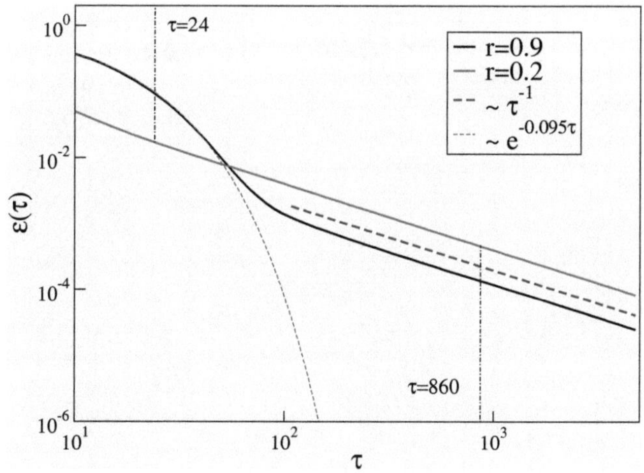

Fig. 7. Energy decay for $r = 0.9$ and $r = 0.2$ (1024^2 sites). The *bold dashed line* $\sim 1/\tau$ is a guide to the eye for the asymptotic energy decay, while the *light dashed line* is the exponential fit corresponding to the Haff law $\exp(-2\gamma_0\tau)$. The Haff regime is too short to be observed in the system with $r = 0.2$. The two indicated times $\tau = 24$ and $\tau = 860$ correspond to the plots of Fig. 15

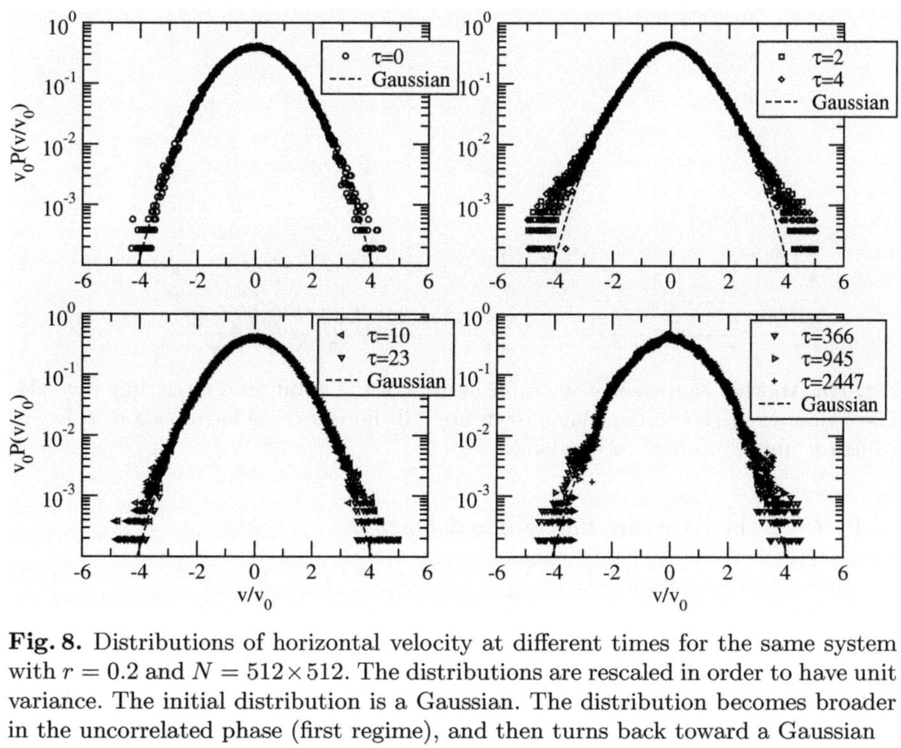

Fig. 8. Distributions of horizontal velocity at different times for the same system with $r = 0.2$ and $N = 512 \times 512$. The distributions are rescaled in order to have unit variance. The initial distribution is a Gaussian. The distribution becomes broader in the uncorrelated phase (first regime), and then turns back toward a Gaussian

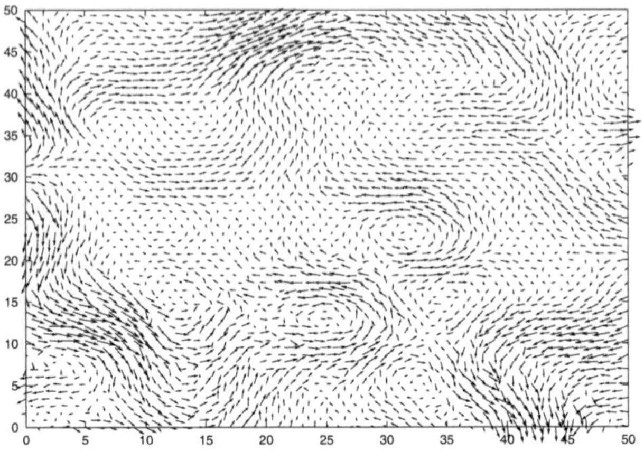

Fig. 9. A (zoomed) snapshot of the velocity field at time $\tau = 52$ for the Inelastic Lattice Gas, $d = 2$, with $r = 0.7$ and size $N = 512 \times 512$. The time has been chosen at the beginning of the correlated regime. It is evident the presence of vortices. All the velocities have been rescaled to arbitrary units, in order to be visible

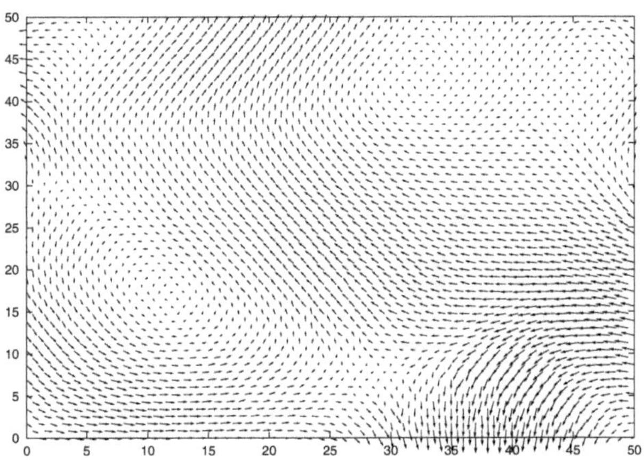

Fig. 10. Another snapshot of the same system of Fig. 9 but at a later time $\tau = 535$. The diameter of the vortices has grown up. All the velocities have been rescaled to arbitrary units, in order to be visible

In $d = 2$ the structure function is defined as

$$S^{t,l}(k,t) = \sum_{\hat{k}} \vec{v}^{t,l}\left(\vec{k},t\right) \vec{v}^{t,l}\left(-\vec{k},t\right) \tag{12}$$

where the superscripts t, l indicate the transverse and longitudinal components of the field with respect to the wave vector $\vec{k}$ and the sum $\sum_{\hat{k}}$ is over a circular shell of radius k. The analysis of the structure factor, shown in Fig. 11, is performed for two systems at different inelasticities. The common feature

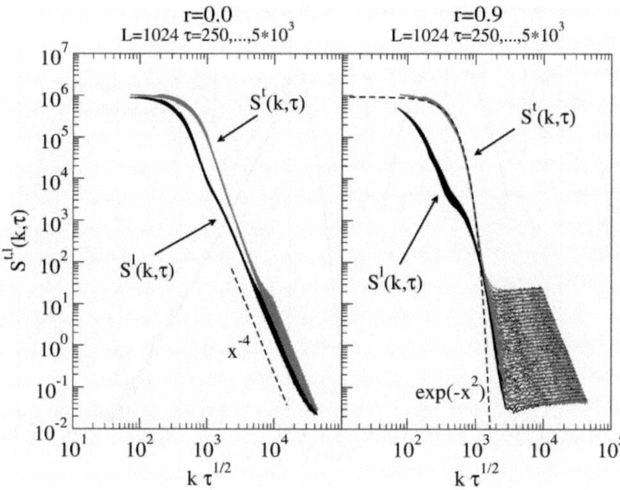

Fig. 11. Data collapse of the transverse (S^t) and longitudinal (S^l) structure functions for $r = 0.$ and $r = 0.9$ (system size 1024^2 sites, times ranging from $\tau = 500$ to $\tau = 10^4$). The wave-number k has been multiplied by $\sqrt{\tau}$. Notice the presence of the plateaux for the more elastic system. For comparison we have drawn the laws x^{-4} and $\exp(-x^2)$

is a good collapse of different time curves if plotted against $k\tau^{1/2}$ which is the signature of a domain growth (diameter of the vortices) $L(\tau) \sim \tau^{1/2}$. The elasticity however changes dramatically the other features: the more elastic system has a Gaussian structure factor at large scale (low values of k) and a plateau at small scales, i.e. quasi-elastic collisions randomize neighbour velocities while the effect of inelasticity induces a structure at large scale. The less elastic system does not present such a plateau, but a decay k^{-4} at small scales which is analogous to the Porod law [40, 41] expected from the presence of defects in a phase ordering process. The difference between transversal and longitudinal component (more appreciable in the quasi-elastic system) is coherent with the theoretical analysis discussed in the introduction, which predicts an earlier and stronger correlation for the transversal component.

In the $d = 2$ ILMM we have also observed shock-like phenomena. Even if in lattice models term "shock" is not appropriate, as real discontinuities cannot be observed. The inspection of the distribution of longitudinal and transversal velocity differences, defined as

$$\Delta_l\left(\vec{R},\vec{i}\right) = (\vec{v}_{i+R} - \vec{v}_i) \cdot \frac{\vec{R}}{R} \tag{13a}$$

$$\Delta_t\left(\vec{R},\vec{i}\right) = (\vec{v}_{i+R} - \vec{v}_i) \times \frac{\vec{R}}{R} \tag{13b}$$

is contained in Fig. 12. The main graph shows the case $R = 1$, i.e. the distributions of gradients. Both the longitudinal and transversal components

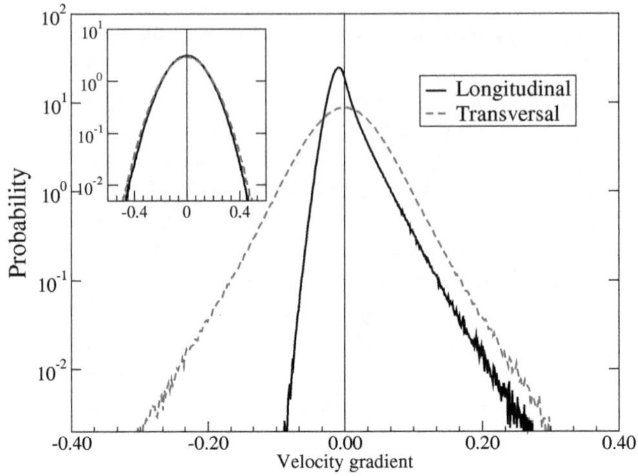

Fig. 12. Probability densities of the longitudinal and transverse velocity increments. The main figure shows the p.d.f. of the velocity differences for $R = 1$. The inset shows the Gaussian shape measured for $R = 40$ (larger than $L(t)$ for this simulation: $r = 0.2$, $t = 620$, system size 2048^2)

are distributed with non-Gaussian tails, but the longitudinal component presents also a strong asymmetry: this means that preferentially there are strong velocity differences in the direction of growing longitudinal velocity. This is compatible with the analysis of shocks in the bottom panels of Fig. 5 in the $d = 1$ case. When $R \gg 1$ the distribution of velocity differences is a Gaussian.

To better study the dynamics of the system we have calculated the two-times self-correlation of the velocity components

$$C(\tau_1, \tau_2) = \frac{\sum_i v_i(\tau_1) v_i(\tau_2)}{N} \tag{14}$$

During a short time transient, the self-correlation function of our model depends on $\tau_1 - \tau_2$, i.e. it is time translational invariant (*TTI*). Later, $C(\tau_1, \tau_2)$ reaches an "aging" regime and depends only on the ratio $x = \tau_1/\tau_2$. This *TTI* transient regime is similar to what occurs during the coarsening process of a quenched magnetic system: the self-correlation of the local magnetization $a(\tau_w, \tau_w + \tau)$ for $\tau \ll \tau_w$ shows a *TTI* decay toward a constant value $m_{eq}^2(T_{quench})$ that is the square of the equilibrium magnetization, indicating that the local magnetization is evolving in an ergodic-like fashion. Thereafter the self-correlation decays with the *aging* scaling law indicated above. Obviously, when $T_{quench} \to 0$, the *TTI* transient regime disappears. In our model the behavior of the self-correlation is even more subtle, as the cooling process imposes a (slowly) decreasing "equilibrium" temperature $T_{quench} \to 0$: this progressively erodes the *TTI* regime and better resembles a finite rate quench. The same dependence on the *TTI* manifests itself in the angular

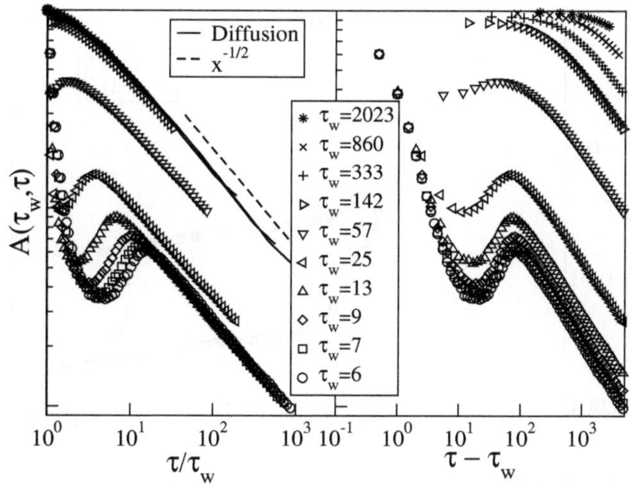

Fig. 13. Angular auto-correlation function $A(\tau, \tau_w)$ for different values of the waiting time τ_w and $r = 0.9$ (1024^2 sites). The graph on the left shows the convergence to the τ/τ_w diffusive scaling regime, for large τ_w. For small τ_w, a local minimum is visible (for such a quasi elastic dynamics). In the graph on the right the same data are plotted vs $\tau - \tau_w$: note that the small τ_w curves tend to collapse. For higher τ_w the position of the local minimum does not move sensibly, but its value grows and goes to 1 for large τ_w

auto-correlation, shown in Fig. 13:

$$A(\tau, \tau_w) = \frac{1}{N}\sum_i \cos(\theta_i(\tau_w + \tau) - \theta_i(\tau_w)). \tag{15}$$

The non-monotonic behavior of $A(t, t_w)$ suggests that the initial direction of the velocity induces a change in the velocities of the surrounding particles, which in turn generates, through a sequence of correlated collisions, a kind of retarded field oriented as the initial velocity. As t_w increases the maximum is less and less pronounced.

The behavior of the energy decay ($E(\tau) \sim \tau^{-d/2}$), the law of growth of domains ($L(\tau) \sim \tau^{1/2}$) and the "aging" form of the two-times self-correlations, as well as the Gaussian velocity pdf's and the Gaussian form at large scale in the structure factors, are strong evidence that the model behaves similarly to a system governed by the diffusion equation:

$$\frac{\partial \phi_i}{\partial \tau} = D \sum_j \frac{\partial_i^2 \phi}{\partial r_j^2} \tag{16}$$

where ϕ is a suitable field depending on the velocity field (it could be a scalar, for example the z-component of the vorticity field $\nabla \times \vec{v}$).

To perform a further check of this conjecture, we have measured the persistence exponent [46] of the model, which is very sensitive to the class of

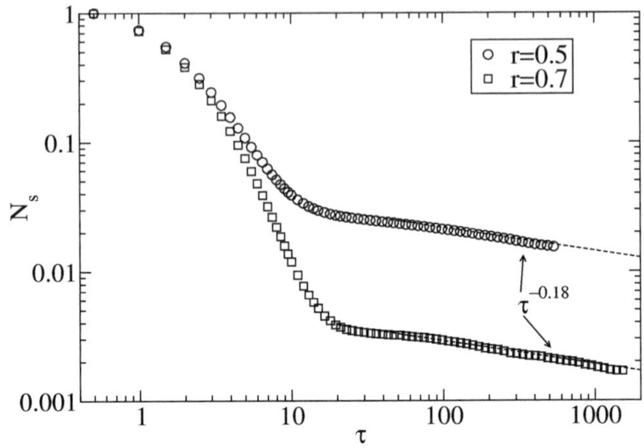

Fig. 14. Measure of the persistence of the Inelastic Lattice Gas in $d = 2$, for two different values of the restitution coefficient. The power-law regime corresponds to the correlated regime. The measured exponent $\theta = 0.18$ is, with this precision, equal to the persistence exponent numerically measured in the diffusive dynamics

universality. We have counted the number $N_s(\tau)$ of sites where the x velocity component never changed sign from the starting time of the dynamics up to time τ. In the correlated regime we observe $N_s(\tau) \sim \tau^{-\theta}$ with $\theta = 0.18$ which is equal, up to this precision, to the persistence exponent numerically measured for the diffusive dynamics.

In the spirit of verifying the possibility of a hydrodynamic (mesoscopic) description of this model, we have also analysed the average local granular temperature T_σ, defined as:

$$T_\sigma = \left\langle |\vec{v} - \langle \vec{v} \rangle_\sigma|^2 \right\rangle_\sigma \tag{17}$$

where $\langle ... \rangle_\sigma$ means an average on a region of linear size σ.

If we call $L(t)$ a characteristic correlation length of the system, since when $\sigma \gg L(t)$ the local average tends to the global (zero) momentum, then $\lim_{\sigma \to \infty} T_\sigma = E$. For $\sigma < L(t)$, instead, $T_\sigma < E$. The behavior of T_σ in the uncorrelated (Haff) regime and in the correlated (asymptotic) regime for two different values of r is presented in Fig. 15. For quasi elastic systems T_σ exhibits a plateau for $1 \ll \sigma \ll L(t)$ that identifies the strength of the internal noise (see also the plateau in the structure factor, Fig. 11) and indicates the mesoscopic scale necessary for a hydrodynamics description. The local temperature ceases to be well defined for smaller r, revealing an absence of scale separation between microscopic and macroscopic fluctuations in the strongly inelastic regime [47].

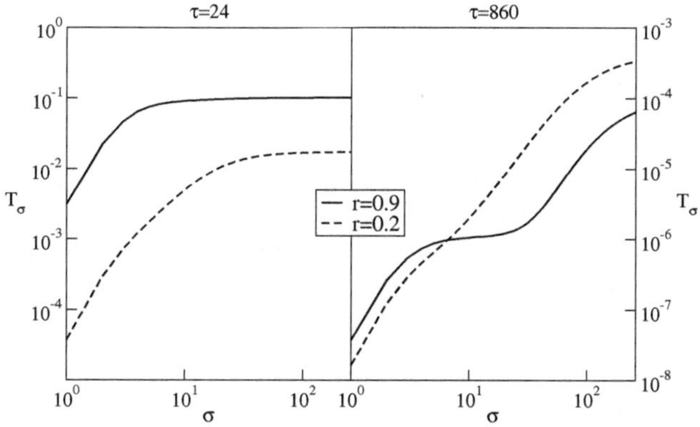

Fig. 15. The scale dependent temperature, T_σ, defined as function of the coarse graining size σ for $\tau = 24$ (incoherent regime) and $\tau = 860$ (correlated regime for both choices of r): the system is the same of Fig. 7. In the correlated regime the more elastic case presents a plateau at intermediate wavelengths, indicating a well defined *mesoscopic* temperature, and therefore a clear separation between the microscopic and the macroscopic scales

6 Conclusions

We have analysed essentially three different problems with a strong interconnection, in the framework of the physics of cooling granular gases. Our first step (IMM) was to demonstrate that a simplified 1D version of the Boltzmann equation for inelastic gases can be exactly solved by a $P(v)$ with power law tails. However, realistic granular systems never show such large tails: the key feature that must be added to this oversimplified model is a realistic topology. Therefore our second step (ILMM, $d = 1$) was to embed the IMM on a $1D$ lattice. The $P(v)$ changes dramatically and becomes identical to those measured in MD simulations, also in the velocity-correlated stage. Moreover, the ILMM allows for a study of spatial correlations, resulting in shock-like structures very similar to the one observed in MD and strong analogies with the MD structure functions. Our last step (ILMM, $d = 2$) has been the passage from $d = 1$ to $d = 2$, which mainly results in the appearance of vortices. Vortices are ubiquitous in granular gases experiments and simulations. The ILMM simply produces vortices as a result of the competition between the parallelizing effect of inelastic collisions and the constraint given by the conservation of total momentum. Moreover, a large set of evidences (energy decay, structure factors, Gaussian velocity pdf's, aging self-correlations, persistence exponent) indicate that the dynamical behavior of the system is compatible with a diffusion equation, even if short scale defects (appearing as shocks, internal noise and tails *à la* Porod in the structure functions) make this model richer.

References

1. T. Pöschel and S. Luding. *Granular Gases*. Springer Verlag, Berlin, 2000
2. P.K. Haff. Grain flow as a fluid-mechanical phenomenon. *J. Fluid Mech.*, 134:401, 1983
3. I. Goldhirsch and G. Zanetti. Clustering instability in dissipative gases. *Phys. Rev. Lett.*, 70:1619, 1993
4. S. McNamara and W.R. Young. Inelastic collapse and clumping in a one-dimensional granular medium. *Phys. Fluids A*, 4:496, 1992
5. P. Deltour and J.-L. Barrat. Quantitative study of a freely cooling granular medium. *J. Phys. I France*, 7:137, 1997
6. T.P.C. van Noije, M.H. Ernst, R. Brito, and J.A.G. Orza. Mesoscopic theory of granular fluids. *Phys. Rev. Lett.*, 79:411, 1997
7. T.P.C. van Noije and M.H. Ernst. Ring kinetic theory for an idealized granular gas. *Physica A*, 251:266, 1998
8. T.P.C. van Noije and M. H. Ernst. Cahn-Hilliard theory for unstable granular flows. *Phys. Rev. E*, 61:1765, 2000
9. E. Ben-Naim, S.Y. Chen, G.D. Doolent, and S. Redner. Shock-like dynamics of inelastic gases. *Phys. Rev. Lett.*, 83:4069, 1999
10. X. Nie, E. Ben-Naim, and S. Chen. Dynamics of freely cooling granular gases. *Phys. Rev. Lett.*, 89:204301, 2002
11. J. Wakou, R. Brito, and M.H. Ernst. Towards a Landau-Ginzburg-type Theory for Granular Fluids. *J. Stat. Phys.*, 3:107, 2002
12. R. Brito and M.H. Ernst. Extension of Haff's cooling law in granular flows. *Europhys. Lett.*, 43:497, 1998
13. A. Baldassarri, U. Marini Bettolo Marconi, and A. Puglisi. Influence of correlations on the velocity statistics of scalar granular gases. *Europhys. Lett.*, 58:14, 2002
14. A. Baldassarri, U. Marini Bettolo Marconi, and A. Puglisi. Models of freely evolving granular gases. *Advances Complex Systems*, 4:321, 2001
15. A. Baldassarri, U. Marini Bettolo Marconi, and A. Puglisi. Kinetics models of inelastic gases. *Math. Mod. Meth. Appl. S.*, 12:965, 2002
16. A. Baldassarri, U. Marini Bettolo Marconi, and A. Puglisi. Cooling of a lattice granular fluid as an ordering process. *Phys. Rev. E*, 65:051301, 2002
17. T. Pöschel, N.V. Brilliantov, and T. Schwager. Violation of molecular chaos in dissipative gases. *Int. J. Mod. Phys. C*, 13:1263, 2002
18. J.J. Brey, J.W. Dufty, C.S. Kim, and A. Santos. Hydrodynamics for granular flow at low density. *Phys. Rev. E*, 58:4638, 1998
19. J.A.G. Orza, R. Brito, T.P.C. van Noije, and M.H. Ernst. Patterns and long range correlations in idealized granular flows. *Int. J. of Mod. Phys. C*, 8:953, 1997
20. M. Isobe. Private communication
21. P.C. Hohenberg and B.I. Halperin. Theory of dynamic critical phenomena. *Rev.Mod.Phys*, 49:435, 1977
22. G.F. Carnevale, Y. Pomeau, and W.R. Young. Statistics of ballistic agglomeration. *Phys. Rev. Lett.*, 64:2913, 1990
23. S.F. Shandarin and Y.B. Zeldovich. The large-scale structure of the universe: Turbulence, intermittency, structures in a self-gravitating medium. *Rev. Mod. Phys.*, 61:185, 1989

24. A. Barrat E. Trizac. Free cooling and inelastic collapse of granular gases in high dimensions. *Eur. Phys. J. E*, 5:161, 2000
25. S. Ulam. On the operations of pair production, transmutations and generalized random walk. *Adv.Appl.Math.*, 1:7, 1980
26. J.C. Maxwell. On the dynamic theory of gases. *Phil. Trans. R. Soc.*, 157:49, 1867
27. M.H. Ernst. Nonlinear model-Boltzmann equations and exact solutions. *Phys. Rep.*, 78:1, 1981
28. A.V. Bobylev, J.A. Carrillo, and I.M. Gamba. On some properties of kinetic and hydrodynamic equations for inelastic interactions. *J. Stat. Phys.*, 98:743, 2000
29. M.H. Ernst. Scaling solutions of inelastic Boltzmann equations with over-populated high energy tails. *J. Stat. Phys.*, 109:407, 2002
30. E. Ben-Naim and P. Krapivsky. Scaling, multiscaling, and nontrivial exponents in inelastic collision processes. *Phys. Rev. E*, 66:011309, 2002
31. E. Ben-Naim and P.L. Krapivsky. Multiscaling in infinite dimensional collision processes. *Phys. Rev. E.*, 61:R5, 2000
32. A. Bobylev, C. Cercignani, and G. Toscani. Proof of an asymptotic property of self-similar solutions of the Boltzmann equation for granular materials. *J. Stat. Phys.*, 111:403, 2003
33. U. Marini Bettolo Marconi and A. Puglisi. Mean field model of free cooling inelastic mixtures. *Phys.Rev. E*, 65:051305, 2002
34. U. Marini Bettolo Marconi and A. Puglisi. Steady state properties of a mean field model of driven inelastic mixtures. *Phys. Rev. E*, 66:011301, 2002
35. Vicente Garzó and James Dufty. Homogeneous cooling state for a granular mixture. *Phys. Rev. E*, 60:5706, 1999
36. D. Benedetto, E. Caglioti, and M. Pulvirenti. Kinetic equations for granular media. *Math. Mod. and Num. An.*, 31:615, 1997
37. A. Barrat, T. Biben, Z. Racz, E. Trizac, and F. van Wijland. On the velocity distributions of the one-dimensional inelastic gas. *J. Phys. A*, 35:463, 2002
38. S. McNamara and W.R. Young. Inelastic collapse in two dimensions. *Phys. Rev. E*, 50:R28, 1994
39. S. McNamara S. Luding. How to handle the inelastic collapse of a dissipative hard-sphere gas with the tc model. *Granular Matter*, 1:113, 1998
40. G. Porod. X-ray small-angle scattering of close packed colloid systems. Part I. *Kolloid Z.*, 124:83, 1951
41. G. Porod. X-ray small-angle scattering of close packed colloid systems. Part II. *Kolloid Z.*, 125:51, 1952
42. A.J. Bray. Theory of phase ordering kinetics. *Adv. Phys.*, 357:43, 1994
43. M. Huthmann, J.A.G. Orza, and R. Brito. Dynamics of deviations from the gaussian state in a freely cooling homogeneous system of smooth inelastic particles. *Granular Matter*, 2:189, 2000
44. S.E. Esipov and T. Pöschel. The granular phase diagram. *J. Stat. Phys.*, 86:1385, 1997
45. H. Nakanishi. Velocity distribution of inelastic granular gas in homogeneous cooling state. Unpublished (cond-mat/0208477), 2002
46. A.J. Bray, B. Derrida, and C. Godrèche. Nontrivial algebraic decay in a soluble model of coarsening. *Europhys. Lett.*, 27:175, 1994
47. I. Goldhirsch. Scales and kinetics of granular flows. *Chaos*, 9:659, 1999

Self-Similar Asymptotics for the Boltzmann Equation with Inelastic Interactions

Alexander V. Bobylev[1] and Carlo Cercignani[2]

[1] Department of Mathematics, Karlstad University, Karlstad, Sweden
[2] Dipartimento di Matematica, Politecnico di Milano, Milano, Italy

Abstract. We consider a question related to the kinetic theory of granular materials. The model of hard spheres with inelastic collisions is replaced by a Maxwell model, characterized by a collision frequency independent of the relative speed of colliding particles. Our main result is that, in the space-homogeneous case, a self-similar asymptotics holds, as conjectured by Ernst-Brito. The proof holds for any initial distribution function with a finite moment of some order greater than two.

1 Introduction

In this paper we survey some questions related to the self-similar asymptotics in the theory of the Boltzmann equation for inelastic particles (IBE).

At the kinetic level, the usual description of a granular material is based on a model of hard spheres with inelastic collisions. For the sake of mathematical convenience this description is sometimes replaced by a Maxwell model, with a collision frequency independent of the relative speed of the colliding particles [1] (it appears that the one-dimensional version of the model was independently introduced in [2]).

There is an important difference between the elastic and inelastic Maxwell models. The elastic one is clearly related to the classical dynamics of gas molecules interacting with a force inversely proportional to the fifth power of the distance. In the inelastic case, we first consider a kinetic equation corresponding to the dynamics of macroscopic particles of finite size (say, inelastic hard spheres). Then the basic equation (in the low density limit) is the IBE for hard spheres. The inelastic Maxwell model does not have an underlying particle dynamics: it is just a mathematical model though with a clear probabilistic meaning (this makes it easy to perform numerical experiments). Thus one should be very careful when trying to draw any conclusion for hard sphere grains on the basis of results obtained from Maxwell model.

A mature technique (based on the use of Fourier transform) can be easily generalized to IBE to answer (this does not occur very often in kinetic theory) some relevant questions concerning the self-similar asymptotics in a quite rigorous form. Such asymptotics was recently proved by the authors for solutions of EBE with infinite energy in [3] [4]. Independently, it was conjectured recently by Ernst and Brito [5] [6] that a similar property holds also for solutions of IBE with finite energy, i. e. there is a self-similar solution providing the asymptotic behavior for more general solutions. The conjecture seems to be quite non-trivial since the corresponding self-similar solution would have a power-like tail. It was already known [1] that, for almost all values of the restitution coefficient, self-similar solutions with finite moments of all orders cannot be positive.

Ernst and Brito presented two main arguments to explain the reasons in favor of this conjecture:

(a) higher moments of “rescaled” solutions of IBE tend to infinity for long times;

(b) a whole family of self-similar solutions must be two-parametric (a fact well-known for EBE) having in the general case power-like tails with the “same” (see (a)) number of finite moments.

No method to “extract” this special self-similar solution satisfying their conjecture (which must be unique for a given class of initial conditions) from the whole two-parameter family was given by EB [5] [6]. In any case their conjecture seems to be correct; we have proved rigorously [7] that (a) such unique self-similar solution does exist and (b) the conjecture is true for any initial distribution function having a finite moment of some order greater than two (”something more” than finite energy) [7, 9].

There is a subtle point to be understood: the tails are “high-energy” in a special sense: the time-evolution cools a granular gas to a time dependent temperature $T = T(t)$. The conjecture to be proved says that, when $t \to \infty$ (and $T(t) \to 0$), the power-like behavior is exhibited at energies which are relatively large with respect to $T(t)$, but these energies may be very low with respect to $T(0)$. On the other hand, any moment of the solution remains bounded for all $t > 0$ provided it was finite at $t = 0$. Thus, the power-like behavior at intermediate energies $T(t) \leq E \leq T(0)$, $t \to \infty$, is not very important for the “truly” high energy tails.

Our main approach to Maxwell models is based on the Fourier transform. In other words, we deal directly not with the Inelastic Boltzmann Equation for the distribution function $f(\vec{v}, t)$, but with its characteristic function (Fourier transform in the velocity space $\mathbb{R}^n$) $\phi(\vec{k}, t) = (f, \exp(-i\vec{k} \cdot \vec{v}))$. Then it is well known [1] that $\phi(\vec{k}, t)$ satisfies the following equation:

$$\frac{\partial \phi}{\partial t} = \hat{Q}(\phi, \phi) = \int_{S^{n-1}} d\vec{n} B\left(\hat{\vec{k}} \cdot \vec{n}\right) \left[\phi\left(\vec{k}_+\right) \phi\left(\vec{k} - \vec{k}_+\right) - \phi(0)\phi\left(\vec{k}\right)\right], \quad (1)$$

where (below $0 < e < 1$ is the restitution coefficient)

$$\vec{k}_+ = z\left(\vec{k} \cdot \vec{n}\right) \vec{n}, \qquad z = (1 + e)/2, \qquad \hat{\vec{k}} = \vec{k} \Big/ \left|\vec{k}\right| . \quad (2)$$

The “cross-section” $B(\cos\theta)$ depends on the concrete model and dimensionality $n > 1$ (see [7] for details). All our results are valid for any non-negative $B(\dots)$ integrable over the sphere S^{n-1} (usually we assume that such integral is equal to one).

The main idea is to make use of relative simplicity of this equation compare to IBE in the velocity space. We begin, in Sect. 2, with a simpler isotropic case (this case is of particular importance since the asymptotic self-similar solution is isotropic). Then we generalize our results in Sect. 3-4.

2 Isotropic Equation and Preliminary Result

In the isotropic case the n-dimensional distribution function $f(|\vec{v}|, t)$ ($n = 2, \dots$) satisfies

$$\frac{\partial f}{\partial t} = Q(f, f), \quad f_{|t=0} = f_0(|\vec{v}|), \quad \vec{v} \in \mathbb{R}^n, t > 0, \tag{3}$$

where $Q(f, f)$ is defined in such a way that the characteristic function

$$\phi\left(\frac{|\vec{k}|^2}{2}, t\right) = \int_{\mathbb{R}^n} f(|\vec{v}|, t) \, e^{-i\vec{k}\cdot\vec{v}} d\vec{v}, \quad \vec{k} \in \mathbb{R}^n, \tag{4}$$

satisfies (1) in isotropic form

$$\frac{\partial \phi}{\partial t} = \int_0^1 dsG(s) \left\{\phi(z^2 sx)\phi[(1 - \beta s)x] - \phi(x)\right\}, \tag{5}$$

$$1/2 < z < 1, \quad \beta = z(2 - z), \quad x \geq 0,$$

where $G(s)$ is a non-negative function normalized to one on $[0, 1]$. Our first results on self-similar asymptotics for the Inelastic Boltzmann Equation are formulated below (see [7] for the proof).

Theorem 1. *(1) There exists a unique non-negative function $F(|\vec{v}|)$ satisfying the conditions*

$$\int_{\mathbb{R}^n} F(|\vec{v}|) \, d\vec{v} = \frac{1}{n} \int_{\mathbb{R}^n} |\vec{v}|^2 F(|\vec{v}|) d\vec{v} = 1. \tag{6}$$

such that

$$f_S(|\vec{v}|, t) = e^{n\mu t} F(|\vec{v}| e^{\mu t}), \quad \mu = z(1 - z) \int_0^1 dsG(s)s, \tag{7}$$

satisfies (3). The function $F(|\vec{v}|)$ is bounded and continuous on $\mathbb{R}^n$.

(2) There exists a non-empty class A of initial conditions $f_0(|\vec{v}|)$ satisfying conditions (6) such that the corresponding solutions $f(|\vec{v}|, t)$ have the following property

$$e^{-n\mu t} f(|\vec{v}| e^{-\mu t}, t) \to F(|\vec{v}|), \quad t \to \infty \tag{8}$$

where the convergence is understood in the sense of probability measures.

(3) If $f_0(|\vec{v}\,|) \geq 0$ satisfies (6) and

$$\int_{\mathbb{R}^n} f_0(|\vec{v}\,|)|\vec{v}\,|^{2+\epsilon} d\vec{v} \leq \infty \qquad \int_{\mathbb{R}^n} f_0(|\vec{v}\,|) e^{i\vec{k}\cdot\vec{v}} d\vec{v} \geq 0,$$

for some $\epsilon > 0$ and all $\vec{k} \in \mathbb{R}^n$, then $f_0 \in A$. In particular, the Maxwellian distribution

$$f_0(|\vec{v}\,|) = M(|\vec{v}\,|) = (2\pi)^{-n/2} e^{-|\vec{v}\,|^2/2} \tag{9}$$

belongs to the class A.

These results can be easily explained to a reader who is not interested in mathematical details and proofs. For simplicity, we shall do this for the case $n = 3$.

We consider the initial value problem (3), where $f_0\,(|\vec{v}\,|)$ is normalized as

$$(f_0, 1) = 1, \quad (f_0, |\vec{v}\,|^2) = 3. \tag{10}$$

It is well-known that

$$f\,(|\vec{v}\,|, t) \to \delta\,(|\vec{v}\,|)\,, \quad T(t) = \frac{1}{3}\left(f, |\vec{v}\,|^2\right) = e^{-2\mu t}, \ \mu = \frac{1}{8}\left(1 - e^2\right), \tag{11}$$

where $0 \leq e < 1$ is the restitution coefficient.

We have constructed (Theorem 1) the positive self-similar solution of (3)

$$f_S\,(|\vec{v}\,|, t) = e^{3\mu t} F\left(|\vec{v}\,| e^{\mu t}\right), \tag{12}$$

and proved that for any initial condition $f_0(\vec{v})$ satisfying (10) and two additional assumptions

$$\left(f_0, |\vec{v}\,|^{2+\epsilon}\right) < \infty, \quad \left(f_0, e^{-i\vec{k}\cdot\vec{v}}\right) \geq 0, \ \vec{k} \in \mathbb{R}^3, \tag{13}$$

the corresponding solution $f(|\vec{v}|, t)$ of the problem (3) has a self-similar asymptotics:

$$f\,(|\vec{v}\,|, t) \cong f_S(|\vec{v}\,|, t), \quad \text{if } t \to \infty, \, |\vec{v}\,| \to 0, \tag{14}$$

such that $|\vec{v}\,| e^{\mu t}$ remains finite. The exact meaning of this asymptotic equality is the following:

$$e^{-3\mu t} f\left(|\vec{v}\,| e^{-\mu t}, t\right) \to_{t \to \infty} F(|\vec{v}\,|), \tag{15}$$

where the convergence is understood in the sense of probability measures.

Thus, we have proved the Ernst-Brito conjecture for a wide class of initial data, including the Maxwellian

$$M(|\vec{v}\,|) = (2\pi)^{-3/2} \exp\left(-|\vec{v}|^2/2\right), \tag{16}$$

all satisfying the conditions (13). It can be also shown [7] that the function $F(|\vec{v}\,|)$ has a restricted number of moments not for all, but for almost all values of the restitution coefficient $0 \leq e < 1$; there exists a sequence $\{e_n\}$, $e_n \to 1$ such that the function $F(|\vec{v}\,|)$ has finite moments of all orders if $e = e_n$, $n = 1, 2, \ldots$ (this degeneracy is discussed in detail in Ref. [7]).

3 A Recent Extension: Preliminaries

It is clear, however, that the restrictions on initial conditions used are related to mathematical technicalities, not to physics. To this end the tools developed in [8] prove useful. In a recent paper written jointly with Toscani [9] we proved the conjecture for a very general class of initial conditions. The main result is the proof that the self-similar asymptotics (Ernst-Brito conjecture) holds for any initial distribution function having "something more" than finite energy (namely, a finite moment of some order greater than two). It seems plausible that this restriction, typical also for the classic (elastic) Boltzmann equation, cannot be weakened [8].

The idea of the proof is the same as in [7]. If the conjecture of the self-similar asymptotics is true for two solutions $\phi_{1,2}(\vec{k},t)$ of (1), satisfying

$$(f_0, 1) = 1, \quad (f_0, \vec{v}\,) = 0, \quad (f_0, |\vec{v}\,|^2) = d, \tag{17}$$

$$\phi(\vec{k},t) = 1 - \frac{|\vec{k}|^2}{2} e^{-\lambda t} - \frac{1}{2} q_{\alpha\beta}(t) \left(k_\alpha k_\beta - \frac{|\vec{k}|^2}{d} \delta_{\alpha\beta} \right) + \mathcal{O}\left(|\vec{k}|^{2(1+\delta)} \right), \tag{18}$$

$$\delta > 0, \qquad |\vec{k}| \to 0\,,$$

then

$$\lim_{t\to\infty} \left| \phi_1(\vec{k}e^{\mu t}, t) - \phi_2(\vec{k}e^{\mu t}, t) \right| = 0, \quad \vec{k} \in \mathbb{R}^d\,, \tag{19}$$

where $\mu = 2\lambda$ was defined in (7) in terms of $G(s)$ (its definition in terms of $B(\cos\theta)$ will be given below. Equation (18) means simply that we can separate terms of order $\mathcal{O}(|\vec{k}|^2)$ and evaluate them explicitly. In particular (see [9]),

$$q_{\alpha\beta}(t) = q_{\alpha\beta}(0)\, e^{-\lambda_1 t}\,, \qquad \lambda_1 > \lambda\,. \tag{20}$$

Our goal now is to prove the equality (19) which is equivalent to the self-similar asymptotics (note that the existence of the self-similar solution of (1)

$$\phi_S\left(\vec{k},t\right) = \psi\left(\vec{k}\,e^{-\mu t}\right), \qquad \psi = \left(F, e^{-i\vec{k}\cdot\vec{v}}\right), \tag{21}$$

was already established in Theorem 1). Therefore, we denote:

$$u = \phi_1 - \phi_2, \qquad \Phi = \frac{1}{2}(\phi_1 + \phi_2) \tag{22}$$

and consider the equation for $u(x,t)$. By the elementary identity

$$ab - cd = \frac{1}{2}[(a+c)(b-d) + (b+d)(a-c)] \tag{23}$$

and (see[7] for details)

$$\frac{\partial \phi}{\partial t} = \hat{Q}(\phi,\phi) = \int_{S^{n-1}} d\vec{n} B\left(\hat{\vec{k}} \cdot \vec{n}\right)\left[\phi\left(\vec{k}_+\right)\phi\left(\vec{k}-\vec{k}_+\right) - \phi(0)\phi\left(\vec{k}\right)\right], \tag{24}$$

(with the notation

$$\left\langle \Phi(\vec{k},\vec{n})\right\rangle = \int_{S^{n-1}} d\vec{n} B\left(\hat{\vec{k}} \cdot \vec{n}\right)\Phi\left(\vec{k},\vec{n}\right) \quad), \tag{25}$$

we obtain

$$\frac{\partial u}{\partial t} + u = \left\langle \Phi\left(\vec{k}_+\right)u\left(\vec{k}-\vec{k}_+\right) + \Phi\left(\vec{k}-\vec{k}_+\right)u(\vec{k}_+)\right\rangle, \tag{26}$$

$$u_{|t=0} = u_0\left(\vec{k}\right) = \phi_1\left(\vec{k},0\right) - \phi_2\left(\vec{k},0\right).$$

The obvious estimate

$$\Phi \le \frac{1}{2}\left(|\phi_1| + |\phi_2|\right) \le 1 \tag{27}$$

leads to the inequality

$$\left|\frac{\partial u}{\partial t} + u\right| \le \hat{L}^+|u|, \quad \hat{L}^+ = \hat{L} + \hat{I}, \quad u_{|t=0} = u_0 \tag{28}$$

where $\hat{I}$ is the identity, $\hat{L}$ is the linearized operator defined by

$$\hat{L}\phi\left(\vec{k},t\right) = \left\langle \phi\left[z\left(\vec{k}\cdot\vec{n}\right)\vec{n}\right] + \phi\left[\vec{k} - z\left(\vec{k}\cdot\vec{n}\right)\vec{n}\right] - \phi\left(\vec{k}\right)\right\rangle. \tag{29}$$

If $u = ye^{-t}$ then

$$\left|\frac{\partial y}{\partial t}\right| \le \hat{L}^+|y|, \quad y_{|t=0} = u_0\left(\vec{k}\right). \tag{30}$$

On the other hand for any complex function $y(\vec{k},t)$ ($\vec{k} \in \mathbb{R}^d$ is fixed)

$$|y|\left|\frac{\partial |y|}{\partial t}\right| = \frac{1}{2}\left|\frac{\partial}{\partial t}(yy^*)\right| = \frac{1}{2}\left|y\frac{\partial y^*}{\partial t} + y^*\frac{\partial y}{\partial t}\right| \le |y|\left|\frac{\partial y}{\partial t}\right|,$$

where y^* is the complex conjugate of y. Therefore (note that y and its time derivative are continuous functions of $\vec{k} \in \mathbb{R}^d$ and $t \in \mathbb{R}_+$):

$$\left|\frac{\partial |y|}{\partial t}\right| \le \left|\frac{\partial y}{\partial t}\right| \le \hat{L}^+|y|, \quad y_{|t=0} = u_0\left(\vec{k}\right). \tag{31}$$

The operator $\hat{L}^+$ is positive ($\hat{L}^+ y \geq 0$ for any $y \geq 0$). Hence

$$\left| y\left(\vec{k},t\right)\right| \leq e^{t\hat{L}^+}\left| u_0\left(\vec{k}\right)\right|, \tag{32}$$

or, equivalently,

$$\left| u\left(\vec{k},t\right)\right| \leq e^{t\hat{L}}\left| u_0\left(\vec{k}\right)\right|, \tag{33}$$

in the notation (29). This simple consideration already leads to the proof of (8) under the additional restriction

$$q_{\alpha\beta}(0) = \left(f_0, v_\alpha v_\beta - \frac{|\vec{v}\,|^2}{n}\delta_{\alpha\beta}\right) = 0. \tag{34}$$

In such a case

$$u_0\left(\vec{k}\right) = \mathcal{O}\left(|\vec{k}|^{2(1+\delta)}\right), \quad |\vec{k}| \to 0; \left| u_0\left(\vec{k}\right)\right| \leq \left|\phi_1\left(\vec{k},0\right)\right| + \left|\phi_2\left(\vec{k},0\right)\right| \leq 2. \tag{35}$$

Hence for any $0 < \epsilon < \delta$ there exists a constant A_ϵ such that

$$\left| u_0\left(\vec{k}\right)\right| \leq A_\epsilon |\vec{k}\,|^{2(1+\epsilon)}, \tag{36}$$

and therefore

$$\left(e^{t\hat{L}}|u_0|\right)\left(\vec{k}\right) |\leq A_\epsilon |\,\vec{k}|^{2(1+\epsilon)} e^{-\lambda(1+\epsilon)t}, \tag{37}$$

where

$$\lambda(p) = \int_{S^{n-1}} d\vec{n} B\left(\hat{\vec{k}}\cdot\vec{n}\right)\left\{1 - z^{2p}\left(\hat{\vec{k}}\cdot\vec{n}\right)^{2p} - \left[1 - \beta\left(\hat{\vec{k}}\cdot\vec{n}\right)^2\right]^p\right\}, \tag{38}$$

$$\hat{\vec{k}} \in S^{n-1},\ \beta = z(2-z),\ p \geq 1,$$

($\lambda(p)$ can be expressed through a 1-d integral, but we do not need it). Thus we obtain from the estimate (33)

$$\left| u\left(\vec{k},t\right)\right| \leq A_\epsilon |\vec{k}|^{2(1+\epsilon)} e^{-\lambda(1+\epsilon)t}, \tag{39}$$

for any $\vec{k} \in \mathbb{R}^n$, $t \in \mathbb{R}$ and $0 < \epsilon < \delta$ provided the condition (34) is fulfilled.

We note that

$$\mu = \frac{\lambda}{2} = \frac{1}{2}\lambda(1) \tag{40}$$

in the notation (38). Hence

$$\left| u\left(\vec{k}e^{\mu t},t\right)\right| \leq A_\epsilon |\vec{k}|^{2(1+\epsilon)} e^{-\Delta(\epsilon)t}, \quad \Delta(\epsilon) = \lambda(1+\epsilon) - (1+\epsilon)\lambda(1). \tag{41}$$

The inequality $\Delta(\epsilon) > 0$ was already proved (in a more general case) in [7] (Lemma 6.1). For the reader's convenience we also present a simplified proof here. It is sufficient to show that

$$\Delta'(0) = \lambda'(1) - \lambda(1) > 0. \tag{42}$$

A simple calculation leads to the equality

$$\Delta'(0) = \int_{S^{n-1}} d\vec{n} B\left(\hat{\vec{k}} \cdot \vec{n}\right) F\left[\left(\hat{\vec{k}} \cdot \vec{n}\right)^2, z\right], \tag{43}$$

where

$$F(x; z) = a(z^2 x) + a(1 - \beta x) - 1, \quad a(x) = x(1 - \log x), \quad 0 \le x \le 1, \frac{1}{2} \le z \le 1,$$

$$\beta = z(2 - z) = 1 - (1 - z)^2.$$

We consider $F(x; z)$ as a function of $x \in [0, 1]$ for a fixed $1/2 \le z \le 1$. It is easy to verify that

$$F(0; z) = 0, \quad F(1; z) = a(z^2) + a\left[(1 - z)^2\right] - 1;$$

$$F_x(x; z) \to_{x \to 0} \infty, \quad F_{xx}(x; z) < 0,$$

where the index x denotes a partial derivative with respect to x. Hence, $F(x; z)$ cannot have local minima in (0,1). Therefore $F(x; z) > 0$ for all $x \in (0, 1)$ provided $F(1; z) \ge 0$. Noting that

$$F(1; 1) = 0, \quad F_z(1; z) = 4[h(1 - z) - h(z)], \quad h(z) = z \log z,$$

we observe that $h'(z) > 0$ if $z > e^{-1}$. Hence $F_z(1; z) < 0$ for $z \in (1/2, 1]$ and $F_z(1; 1/2) = 0$. Therefore

$$F(1; z) > 0; \qquad \Rightarrow F(x; z) > 0, \quad \frac{1}{2} \le z < 1 \quad 0 < x \le 1.$$

Hence $\Delta' > 0$, (see (43)) provided $B(\cos\theta)$ is not concentrated at $\theta = \pi/2$. The latter case, however, is no interesting at all since it leads to the equality $\lambda = 2\mu = 0$ (see (40)). To exclude such case it is enough to assume $B(x) \in L^1_+([-1, 1])$.

Hence the inequality $\Delta(\epsilon) > 0$ holds in all cases when $\mu > 0$. Then we conclude, by the estimate(41) and the definition (40) that

$$\lim_{t \to \infty} \left| u\left(\vec{k} e^{\mu t}, t\right) \right| = 0, \quad \vec{k} \in \mathbb{R}^n, \tag{44}$$

where $u(\vec{k}, t)$ denotes the difference between any two solutions of the problem (24) with an initial distribution function satisfying the condition (34). Taking the self-similar solution $\psi(\vec{k} e^{\mu t})$ as one of the two we obtain the following result

Lemma 1. *Let $f(\vec{v},t)$ be any solution of the problem satisfying the conditions (17) and (34). Then its characteristic function $\phi(\vec{k},t) = (f, e^{-i\vec{k}\cdot\vec{v}})$ has the self-similar asymptotics*

$$\lim_{t\to\infty} \phi\left(\vec{k}e^{\mu t}, t\right) = \psi\left(\vec{k}\right) = \left(F, e^{-i\vec{k}\cdot\vec{v}}\right), \quad \vec{k} \in \mathbb{R}^n,$$

where $F(|\vec{v}|)$ is defined in Theorem 1.

In the next section we shall see that the extra condition (34) can be removed.

4 Complete Proof of the Conjecture

Let us consider the most general case. We return to the (exact!) (26) and represent $u(\vec{k},t)$ as a sum

$$u(\vec{k},t) = u_1(\vec{k},t) + w(\vec{k},t) \tag{45}$$

$$u_1 = \frac{1}{2} b_{\alpha\beta}(t) \left(k_\alpha k_\beta - \frac{|\vec{k}|^2}{d} \delta_{\alpha\beta} \right), \quad b_{\alpha\beta} = q^1_{\alpha\beta} - q^2_{\alpha\beta}, \quad w = \mathcal{O}\left(|\vec{k}|^{2(1+\delta)}\right).$$

We also note that, in the notation (28),

$$b_{\alpha\beta}(t) = b_{\alpha\beta}(0) e^{-\lambda_1 t}, \quad \frac{\partial u_1}{\partial t} + u_1 = \hat{L}^+ u_1, \tag{46}$$

Then we obtain from (26)

$$\frac{\partial w}{\partial t} + w = \left\langle \Phi\left(\vec{k}_+\right) w\left(\vec{k} - \vec{k}_+\right) + \Phi\left(\vec{k} - \vec{k}_+\right) w\left(\vec{k}_+\right) \right\rangle \tag{47}$$

$$\left\langle \left[1 - \Phi\left(\vec{k}_+\right)\right] u_1\left(\vec{k} - \vec{k}_+\right) + \left[1 - \Phi\left(\vec{k} - \vec{k}_+\right)\right] u_1\left(\vec{k}_+\right) \right\rangle, w_{|t=0} = w_0(\vec{k}).$$

In order to estimate the source term we note that

$$\left|\Phi\left(\vec{k}\right)\right| \le 1, \quad \left|1 - \Phi\left(\vec{k}\right)\right| \le 2; \quad 1 - \Phi\left(\vec{k}\right) = \mathcal{O}\left(|\vec{k}|^2\right), \quad \vec{k} \to 0.$$

Hence for any $0 < \epsilon < 2$ there exists a constant B_ϵ such that

$$\left|1 - \Phi\left(\vec{k}\right)\right| \le B_\epsilon |\vec{k}|^{2\epsilon}, \quad \vec{k} \in \mathbb{R}^d.$$

On the other hand,

$$\left|u_1\left(\vec{k},t\right)\right| \le \text{const.} |\vec{k}|^2 \exp\left(-\lambda_1 t\right), \quad \vec{k} \in \mathbb{R}^d.$$

Therefore

$$\left|\frac{\partial w}{\partial t} + w\right| \le \hat{L}^+ |w| + C_\epsilon |\vec{k}|^{2(1+\epsilon)} \exp\left(-\lambda_1 t\right).$$

The same arguments as in Sect. 4 (see derivation of (33) from (28)) lead to the estimate

$$\frac{\partial |w|}{\partial t} \leq \hat{L}|w| + C_\epsilon |\vec{k}|^{2(1+\epsilon)} \exp\left(-\lambda_1 t\right) .$$

Hence (note that $e^{t\hat{L}}$ is a positive operator) we obtain

$$\left|w\left(\vec{k},t\right)\right| \leq e^{t\hat{L}} \left|w_0\left(\vec{k}\right)\right| + C_\epsilon \int_0^t d\tau e^{-\lambda_1(t-\tau)} e^{\tau \hat{L}} |\vec{k}|^{2(1+\epsilon)} \qquad (48)$$

The first term was already estimated in (37) (we assume that the condition (36) holds for $w_0(\vec{k})$). The second term, which we denote by $I_\epsilon(\vec{k},t)$, reads

$$I_\epsilon\left(\vec{k},t\right) = C_\epsilon \int_0^t d\tau e^{-\lambda_1(t-\tau)} e^{-\lambda(1+\epsilon)\tau} |\vec{k}|^{2(1+\epsilon)}$$

$$= C_\epsilon \frac{|\vec{k}|^{2(1+\epsilon)}}{\lambda_1 - \lambda(1+\epsilon)} \left[e^{-\lambda(1+\epsilon)t} - e^{-\lambda_1 t}\right] .$$

We note that $\lambda(1+\epsilon) \to \lambda(1) = \lambda$, as $\epsilon \to 0$, whereas $\lambda_1 > \lambda$ does not depend on ϵ. Therefore we obtain (for sufficiently small $\epsilon > 0$) the same upper estimate

$$\left|I_\epsilon\left(\vec{k},t\right)\right| \leq D_\epsilon |\vec{k}|^{2(1+\epsilon)} e^{-\lambda(1+\epsilon)t},$$

as we got already in (37) for the first term in the right hand side of (48). Hence

$$\left|w\left(\vec{k},t\right)\right| \leq E_\epsilon |\vec{k}|^{2(1+\epsilon)} e^{-\lambda(1+\epsilon)t},$$

and (see (45), (46))

$$\left|u\left(\vec{k},t\right)\right| = \left|u_1\left(\vec{k},t\right) + w\left(\vec{k},t\right)\right| \leq E_\epsilon |\vec{k}|^{2(1+\epsilon)} e^{-\lambda(1+\epsilon)t} + C_\epsilon |\vec{k}|^2 e^{-\lambda_1 t}.$$

Therefore

$$\left|u(\vec{k}e^{\mu t},t)\right| \leq E_\epsilon |\vec{k}|^{2(1+\epsilon)} e^{-\Delta(\epsilon)t} + C_\epsilon |\vec{k}|^2 e^{-(\lambda_1-\lambda)t},$$

in the notation (41). The inequality $\Delta(\epsilon) > 0$ was proved in Sect. 4, and we already mentioned that $\lambda_1 > \lambda = 2\mu$. Thus the equality

$$\lim_{t\to\infty} \left|\phi_1\left(\vec{k}e^{\mu t},t\right) - \phi_2\left(\vec{k}e^{\mu t},t\right)\right| = 0, \quad \vec{k} \in \mathbb{R}^d \qquad (49)$$

is proved for any two solutions of the problem (24) without the assumption $q^1_{\alpha\beta} = q^2_{\alpha\beta} = 0$ used in Sect. 5. If we replace $\phi_2(\vec{k},t)$ by the self-similar solution $\psi(\vec{k}e^{-\mu t})$ and $\phi_1(\vec{k},t)$ by $\phi(\vec{k},t)$, then our result reads

$$\lim_{t\to\infty} \phi\left(\vec{k}e^{\mu t},t\right) = \psi\left(\vec{k}\right), \quad \vec{k} \in \mathbb{R}^d$$

where $\phi_0(\vec{k}) = \phi(\vec{k}, 0)$ satisfies the most general assumptions.

We already mentioned in Sect. 4 that the pointwise convergence of the characteristic functions implies the weak convergence of distribution functions. Therefore we can formulate the final result in terms of solutions of the kinetic equation:

Theorem 2. *Let $f(\vec{v}, t)$ be the solution of the problem, where*

$$(f_0, 1) = 1, \quad (f_0, \vec{v}) = 0, \quad (f_0, |\vec{v}|^2) = d, \left(f_0, |\vec{v}|^{2+\delta}\right) < \infty, \qquad \delta > 0.$$

Then

$$e^{-d\mu t} f\left(\vec{v} e^{-\mu t}, t\right) \to F\left(|\vec{v}|\right), \quad t \to \infty, \quad n \geq 2,$$

where the convergence is understood in the sense of probability measures. $F(|\vec{v}|)$ and μ are defined in Theorem 1.

Proof. It is already given above. We just note that the term "solution" is understood in the same sense as in Ref. [9].

Thus, the Ernst-Brito conjecture is proved for practically all initial conditions having interest in the applications.

Acknowledgments

The research described in the paper was supported by MURST of Italy and NFR of Sweden. We thank Thorsten Pöschel for discussion.

References

1. A.V. Bobylev, J.A. Carrillo, and I.A. Gamba. On some properties of kinetic and hydrodynamic equations for inelastic interaction. *J. Stat. Phys.*, 98: 743–773, 2000
2. E. Ben-Naim and P. Krapivsky. Multiscaling in inelastic collisions. *Phys. Rev. E*, 61: R5–R8, 2000
3. A.V. Bobylev and C. Cercignani. Exact eternal solutions of the Boltzmann equation. *J. Stat. Phys.* , 106: 1019–1038, 2002
4. A.V. Bobylev and C. Cercignani. Self-similar solutions of the Boltzmann equation and their applications. *J. Stat. Phys.*, 106: 1039–1071, 2002
5. M.H. Ernst and R. Brito. High energy tails for inelastic Maxwell models. *Europhys. Lett.*, 58: 182, 2002
6. M.H. Ernst and R. Brito. Scaling Solutions of Inelastic Boltzmann Equations with over-populated High Energy Tails. *J. Stat. Phys.*, 109: 407, 2002.
7. A.V. Bobylev and C. Cercignani. Self-similar asymptotics for the Boltzmann equation with inelastic and elastic interactions. *J. Stat. Phys.*, 110: 333–375, 2003
8. E. Gabetta, G. Toscani and B. Wennberg. Metrics for probability distributions and the trend to equilibrium for solutions of the Boltzmann equation. *J. Stat. Phys.*, 81: 901–934, 1995
9. A.V. Bobylev, C. Cercignani and G. Toscani. Proof of an asymptotic property of self-similar solutions of the Boltzmann for granular materials. *J. Stat. Phys.*, 111: 403–416, 2003

Kinetic Integrals in the Kinetic Theory of Dissipative Gases

Thorsten Pöschel and Nikolai V. Brilliantov

Institut für Biochemie, Charité, Humboldt-Universität zu Berlin,
Monbijoustraße 2, 10117 Berlin, Germany

Abstract. The kinetic theory of gases, including Granular Gases, is based on the Boltzmann equation. Many properties of the gas, from the characteristics of the velocity distribution function to the transport coefficients may be expressed in terms of functions of the collision integral which we call kinetic integrals. Although the evaluation of these functions is conceptually straightforward, technically it is frequently rather cumbersome. We report here a method for the analytical evaluation of kinetic integrals using computer algebra. We apply this method for the computation of some properties of Granular Gases, ranging from the moments of the velocity distribution function to the transport coefficients. For their technical complexity most of these quantities cannot be computed manually.

1 Introduction

The collision of elastic hard spheres of initial velocities $\vec{v}_1$ and $\vec{v}_2$ is described by the collision law

$$\vec{v}_1^{\,\prime} = \vec{v}_1 - \left(\vec{v}_{12} \cdot \vec{e}\,\right) \vec{e}\,, \qquad \vec{v}_2^{\,\prime} = \vec{v}_2 + \left(\vec{v}_{12} \cdot \vec{e}\,\right) \vec{e} \tag{1}$$

where $\vec{v}_{1/2}^{\,\prime}$ denote their velocities after the collision, $\vec{v}_{12} \equiv \vec{v}_1 - \vec{v}_2$ and $\vec{e} \equiv \vec{r}_{12}/r_{12}$ with $\vec{r}_{12} \equiv \vec{r}_1 - \vec{r}_2$ at the instant of the collision. For dissipatively colliding particles the coefficient of restitution $\varepsilon \in (0, 1)$ characterizes the loss of relative normal velocity of the particles; $\varepsilon = 1$ represents elastic collisions, $\varepsilon = 0$ stands for sticky collisions where the relative velocity after a collision drops to zero. Consequently, the collision rule for dissipatively colliding particles reads

$$\vec{v}_1^{\,\prime} = \vec{v}_1 - \frac{1+\varepsilon}{2} \left(\vec{v}_{12} \cdot \vec{e}\,\right) \vec{e}\,, \qquad \vec{v}_2^{\,\prime} = \vec{v}_2 + \frac{1+\varepsilon}{2} \left(\vec{v}_{12} \cdot \vec{e}\,\right) \vec{e}\,. \tag{2}$$

The kinetic theory uses the concepts of direct and inverse collisions: In a *direct collision*, as described above, the particle velocities change by $\{\vec{v}_1, \vec{v}_2\} \to \{\vec{v}_1^{\,\prime}, \vec{v}_2^{\,\prime}\}$. An *inverse collision* is a collision which results in the velocities $\vec{v}_{1/2}$, i.e. $\{\vec{v}_1^{\,\prime\prime}, \vec{v}_2^{\,\prime\prime}\} \to \{\vec{v}_1, \vec{v}_2\}$.

The starting point of the kinetic theory of gases of elastic particles [1, 2] as well as dissipatively colliding particles (Granular Gases) [3] is the Boltzmann equation. It describes the evolution of the distribution function $f\left(\vec{r}, \vec{v}, t\right)$ of the number of gas particles at time t in the infinitesimal phase-space volume $d\vec{r}$, $d\vec{v}$ located at $\vec{r}$, $\vec{v}$. For dissipative gases it reads (e.g. [3–5])

$$\begin{aligned}\left(\frac{\partial}{\partial t}+\vec{v}_1\cdot\vec{\nabla}\right)f\left(\vec{v}_1,t\right) &= \sigma^{d-1}\int d\vec{v}_2\int d\vec{e}\,\Theta\left(-\vec{v}_{12}\cdot\vec{e}\right)\left|\vec{v}_{12}\cdot\vec{e}\right| \times \\ &\quad\times\left[\chi f\left(\vec{v}_1^{\,\prime\prime},t\right)f\left(\vec{v}_2^{\,\prime\prime},t\right)-f\left(\vec{v}_1,t\right)f\left(\vec{v}_2,t\right)\right] \\ &\equiv I\left(f,f\right)\,,\end{aligned} \tag{3}$$

which defines the collision integral $I\left(f,f\right)$. Here σ is the particle diameter and d is the dimension of the system. The integration is performed over the velocity $\vec{v}_2$ of the particle 2 of a colliding pair $\{1-2\}$ and over the unit vector $\vec{e}$. $\Theta(x)$ denotes the Heaviside step-function. The Boltzmann equation describes the change of the population in the phase space due to the flux of particles against the density gradient (second term in the lhs of (3)) and due to collisions (rhs of (3)).

The factor χ in the gain term of the collision integral is specific for Granular Gases. It appears as a result of the transformation from the velocities $\vec{v}_1^{\,\prime\prime}, \vec{v}_2^{\,\prime\prime}$ of the inverse collision to the velocities $\vec{v}_1, \vec{v}_2$. Frequently it is assumed $\varepsilon=$ const. (we will call this *simplified collision model*, see Sect. 3), the factor χ is also a constant for this model, $\chi=1/\varepsilon^2$. The case of molecular gases is obtained in the elastic limit $\varepsilon=1$.

The collision integral has an important property: Given

$$\langle\psi\left(t\right)\rangle\equiv\int d\vec{v}_1\psi\left(\vec{v}\right)f\left(\vec{v},t\right) \tag{4}$$

is the average value of some function $\psi\left(\vec{v}\right)$ and $\Delta\psi\left(\vec{v}_i\right)\equiv\left[\psi\left(\vec{v}_i^{\,\prime}\right)-\psi\left(\vec{v}_i\right)\right]$ denotes the change of $\psi\left(\vec{v}_i\right)$ due to a direct collision. Then [1, 2]

$$\begin{aligned}\frac{d}{dt}\langle\psi\left(t\right)\rangle &= \int d\vec{v}_1\psi\left(\vec{v}_1\right)\frac{\partial}{\partial t}f\left(\vec{v}_1,t\right)=\int d\vec{v}_1\psi\left(\vec{v}_1\right)I\left(f,f\right) \\ &= \frac{\sigma^{d-1}}{2}\int d\vec{v}_1 d\vec{v}_2\int d\vec{e}\,\Theta\left(-\vec{v}_{12}\cdot\vec{e}\right)\left|\vec{v}_{12}\cdot\vec{e}\right|f\left(\vec{v}_1,t\right)f\left(\vec{v}_2,t\right) \\ &\quad\times\Delta\left[\psi\left(\vec{v}_1\right)+\psi\left(\vec{v}_2\right)\right]\,.\end{aligned} \tag{5}$$

Hence, the time derivative of any avarage quantity $\psi\left(\vec{v}\right)$ is expressed in terms of a function of the collision integral. The integration is performed over two velocities $\vec{v}_1$, $\vec{v}_2$ and over the unit vector $\vec{e}$. The integrand is the product of two distribution functions and of the expression $\Delta\left[\psi\left(\vec{v}_1\right)+\psi\left(\vec{v}_2\right)\right]$, which depends on the above three vectors since the after-collision velocities are given by (2).

We will show below that all important properties of dilute homogeneous and non-homogeneous Granular Gases may be formulated in terms of expressions of the collision integral of the same structure as in (5). We will call these functions *kinetic integrals.* The structures of the kinetic integrals do not depend on the collision model, which allows to elaborate a general approach for gases of elastic particles, for dissipative gases with a simplified collisional model $\varepsilon = \text{const.}$, as well as for gases of particles which collide via a realistic collision model as discussed in Sect. 3.

The aim of this paper is to describe a general approach for the analytical evaluation of kinetic integrals by means of symbolic algebra. To illustrate its application we compute various characteristics of homogeneous and inhomogeneous Granular Gases.

2 A Simple Example

We consider a homogeneous Granular Gas where the velocity distribution function is independent of $\vec{r}$. The granular temperature T is defined by the second moment of the velocity distribution function:

$$\frac{d}{2} nT(t) = \int d\vec{v} \frac{mv^2}{2} f(\vec{v}, t) \,, \tag{6}$$

which evolves due to [4–7]

$$\frac{dT}{dt} = -\zeta T \qquad \text{with} \qquad \zeta \equiv \frac{2}{d} n\sigma^{d-1} \sqrt{\frac{2T}{m}} \mu_2 \,, \tag{7}$$

where ζ is the cooling coefficient of the homogeneous gas, n is the number density, m is the particle mass and the coefficient μ_2 is the second moment of the dimensionless collision integral, with the general definition [4, 5, 7]

$$\mu_p = -\frac{1}{2} \int d\vec{c}_1 \int d\vec{c}_2 \int d\vec{e}\, \Theta\left(-\vec{c}_{12} \cdot \vec{e}\right) \left|\vec{c}_{12} \cdot \vec{e}\right| \tilde{f}\left(c_1\right) \tilde{f}\left(c_2\right) \Delta\left(c_1^p + c_2^p\right) . \tag{8}$$

Here $\tilde{f}(c)$ is the dimensionless distribution function

$$f(v, t) = \frac{n}{\vec{v}_T^d} \tilde{f}\left(\frac{v}{\vec{v}_T}\right) , \tag{9}$$

which depends on the reduced velocity $\vec{c} \equiv \vec{v}/\vec{v}_T$, with

$$\vec{v}_T(t) \equiv \sqrt{\frac{2T(t)}{m}} \tag{10}$$

being the thermal velocity of the gas.

For Granular Gases the velocity distribution function deviates from the Maxwell distribution. This deviation is characterized by the coefficients of the Sonine polynomial expansion:

$$\tilde{f}(\vec{c}) = \phi(c) \left[1 + \sum_{p=1}^{\infty} a_p S_p \left(c^2\right) \right] \tag{11}$$

where the leading term $\phi(c) \equiv \pi^{-d/2} \exp\left(-c^2\right)$ is the reduced Maxwell distribution, a_p are numerical coefficients and the Sonine polynomials read for general dimension d:

$$\begin{aligned}
S_0(x) &= 1 \\
S_1(x) &= -x + \frac{d}{2} \\
S_2(x) &= \frac{1}{2}x^2 - \frac{1}{2}(d+2)\,x + \frac{1}{8}(d+2)\,d \\
&\text{etc.}
\end{aligned} \tag{12}$$

For small dissipation ($\varepsilon \lesssim 1$) the deviations of the velocity distribution function from the Maxwell distribution are small. Therefore, we approximate

$$\tilde{f}(c) \simeq \phi(c) \left[1 + a_2 S_2\left(c^2\right)\right] , \tag{13}$$

where we take into account that $a_1 = 0$ due to the definition of temperature [4, 5, 7].

If we further neglect terms $\mathcal{O}\left(a_2^2\right)$, the second moment of the dimensionless collision integral reads

$$\begin{aligned}
\mu_2 = \int d\vec{c}_1 \int d\vec{c}_2 \int d\vec{e}\, \Theta\left(-\vec{c}_{12}\cdot\vec{e}\right) \left|\vec{c}_{12}\cdot\vec{e}\right| \phi(c_1)\, \phi(c_2) \times \\
\times -\frac{1}{2} \left\{1 + a_2 \left[S_2\left(c_1^2\right) + S_2\left(c_2^2\right)\right]\right\} \Delta\left(c_1^2 + c_2^2\right) .
\end{aligned} \tag{14}$$

This expression is of the form

$$\mu_2 = \int d\vec{c}_1 \int d\vec{c}_2 \int d\vec{e}\, \Theta\left(-\vec{c}_{12}\cdot\vec{e}\right) \left|\vec{c}_{12}\cdot\vec{e}\right| \phi(c_1)\, \phi(c_2)\, \mathrm{Expr}(\,\ldots\,) \tag{15}$$

with $\mathrm{Expr} = -\frac{1}{2}\left\{1 + a_2\left[S_2\left(c_1^2\right) + S_2\left(c_2^2\right)\right]\right\} \Delta\left(c_1^2 + c_2^2\right)$

We apply the following Maple program to obtain μ_2 and, consequently, the cooling rate of the Granular Gas due to (7):

```
> restart;
> libname:=libname,'/home/kies/KineticIntegral/':
> with(KineticIntegral);
        [DefDimension, DefJ, DefS, KIinit, getJexpr, unDefJ]
> KIinit();
> DefDimension(3);

Dimension = 3

Sonine polynomials defined
> DefJ();

Basic Integrals will be evaluated
> Expr:=Delta2*(-1/2*(1+a2*(S(2,c1p)+S(2,c2p)))):
> mu2:=getJexpr(0,Expr,0);

expressing 16 terms in form of Basic Integrals
```

$$\mu 2 := -\frac{1}{16}\sqrt{\pi}\sqrt{2}\,(\varepsilon - 1)\,(1+\varepsilon)\,(16 + 3\,a2)$$

This example is intended to demonstrate the application of symbolic algebra using the package `KineticIntegral.m` for the evaluation of kinetic integrals without explaining the details. The mathematical method as well as the according Maple package are subject of the next sections.

Keeping terms $\mathcal{O}\left(a_2^2\right)$ we obtain the result

$$\mu_2 = \sqrt{2\pi}\left(1-\varepsilon^2\right)\left(1+\frac{3}{32}a_2\right)^2 , \tag{16}$$

which yields with (7) the temperature evolution law (Haff's law)

$$T(t) = \frac{T_0}{\left(1+t/\tau_0\right)^2} \tag{17}$$

with the characteristic time τ_0

$$\tau_0^{-1} = \frac{\mu_2}{3\sqrt{8\pi}}\tau_c(0)^{-1} = \frac{1}{6}\left(1-\varepsilon^2\right)\left(1+\frac{3}{16}a_2+\frac{9}{1024}a_2^2\right)\tau_c(0)^{-1} , \tag{18}$$

where the mean collision time is given by

$$\tau_c(t)^{-1} \equiv 4\sqrt{\pi}\sigma^2 n\sqrt{\frac{T(t)}{m}} . \tag{19}$$

This example shows that with few lines only we can derive the temperature decay of a Granular Gas. Since the first part of the program is identical in all applications, only the main part is given below in the text. The program above then reads

```
DefDimension(3);
DefJ();
Expr:=Delta2*(-1/2**(1+a2*(S(2,c1p)+S(2,c2p)))):
mu2:=getJexpr(0,Expr,0);
```

3 Granular Gases of Viscoelastic Particles

It has been shown experimentally, e.g. [8], as well as theoretically [9] that the coefficient of restitution of realistic particles is not a material constant but it depends on the impact velocity. For the case of viscoelastic interaction the coefficient of restitution can be derived from the detailed analysis of the interaction of colliding spheres. The force between viscoelastic spheres of radii R_i and R_j which deform by $\xi \equiv |\vec{r}_i - \vec{r}_j| - R_i - R_j$ at the rate $\dot{\xi}$ is given as a superposition of Hertz's contact force [10] and a viscous force [11]:

$$F = \frac{2Y\sqrt{R^{\rm eff}}}{3(1-\nu^2)}\left(\xi^{3/2} + \frac{3}{2}A\sqrt{\xi}\dot{\xi}\right) \equiv \alpha\xi^{3/2} + \frac{3}{2}A\alpha\sqrt{\xi}\dot{\xi}\,, \tag{20}$$

where Y, ν are respectively the Young modulus and the Poisson ratio and $R^{\rm eff} \equiv R_iR_j/(R_i+R_j)$. The dissipative constant A is a function of the bulk viscosities of the particles material (see [11] for details). Integrating Newton's equation of motion for a particle collision the coefficient of restitution is obtained as a function of the normal component of the impact velocity $g \equiv |\vec{e}\cdot(\vec{v}_1 - \vec{v}_2)|$ [9, 12]

$$\varepsilon = 1 - C_1 A\alpha^{2/5} g^{1/5} + \frac{3}{5}C_1^2 A^2 \alpha^{4/5} g^{2/5} \mp \dots \tag{21}$$

The numerical coefficient C_1 reads

$$C_1 = \frac{\sqrt{\pi}}{2^{1/5}5^{2/5}}\,\frac{\Gamma(3/5)}{\Gamma(21/10)}\,. \tag{22}$$

Using the definitions of the thermal velocity, (10), and the reduced velocity $\vec{c} \equiv \vec{v}/\vec{v}_T$ the impact-velocity dependent coefficient of restitution, (21), may be written in terms of the reduced velocity:

$$\varepsilon = 1 - C_1\delta'(t)\,|\vec{c}_{12}\cdot\vec{e}|^{1/5} + \frac{3}{5}C_1^2\delta'^2(t)\,|\vec{c}_{12}\cdot\vec{e}|^{2/5}\,, \tag{23}$$

where

$$\delta'(t) \equiv A\alpha^{2/5}\left[2T(t)\right]^{1/10} \equiv \delta\left(\frac{2T(t)}{T_0}\right)^{1/10}\,, \tag{24}$$

with $\delta \equiv A\alpha^{2/5}T_0^{1/10}$ and with T_0 being the initial temperature. We will need this form below.

The derivation of the kinetic properties of Granular Gases of viscoelastic particles is significantly more complicated than for the case of the simplified coefficient of restitution (ε = const.), see e.g. [5, 13]. The main reason of these complications is the factor χ in (3) which is not a constant anymore but a complicated function of the impact velocity [5]. The elaboration of the kinetics of Granular Gases of viscoelastic particles becomes very technical and was the main motivation for the application of computational symbolic algebra as presented here.

In this paper we consider systems of three-dimensional spheres, i.e., for the case of viscoelastic spheres the particles interact by the force (20). Hence, for $d = 2$ we describe a gas of three-dimensional viscoelastic spheres whose motion is restricted to a plane. The coefficient of restitution for $d = 2$, i.e. for viscoelastic disks, has been derived recently [14, 15].

4 Evaluation of Kinetic Integrals

4.1 Definition of the Basic Integrals

The evaluation of kinetic integrals requires the computation of a sum of simpler integrals which we call *basic integrals.* These basic integrals may be computed analytically. Kinetic integrals are then decomposed into basic integrals. Although this decomposition is straightforward, the number of basic integrals may be as large as several thousands. Thus in practice it is not feasible to evaluate kinetic integrals manually. Let us consider a simple kinetic integral and show how it can be expressed in terms of basic integrals. In what follows we elaborate the computation of kinetic integrals for general dimension d. Below the application of the method will be illustrated only for the most important case $d = 3$. The corresponding results for $d = 2$ can be obtained by changing the command `DefDimension(3)` into `DefDimension(2)` in the beginning of the programs.

We assume that the dissipation is small enough to keep only first nonvanishing term in the Sonine polynomials expansion (11), i.e. we assume that the approximation

$$\tilde{f}(c) \simeq \phi(c)\left[1 + a_2 S_2\left(c^2\right)\right] . \tag{25}$$

is adequate. With this approximation the moments of the dimensionless collision integral μ_p reads

$$\mu_p = -\frac{1}{2}\int d\vec{c}_1 \int d\vec{c}_2 \int d\vec{e}\,\Theta\left(-\vec{c}_{12}\cdot\vec{e}\right)\left|\vec{c}_{12}\cdot\vec{e}\right|\phi\left(c_1\right)\phi\left(c_2\right)\times$$
$$\times\left\{1 + a_2\left[S_2\left(c_1^2\right) + S_2\left(c_2^2\right)\right] + a_2^2 S_2\left(c_1^2\right) S_2\left(c_2^2\right)\right\}\Delta\left(c_1^p + c_2^p\right) . \tag{26}$$

To evaluate the kinetic integral in (26) it is convenient to use the center of mass velocity $\vec{C}$ and the relative velocity $\vec{c}_{12}$:

$$\vec{c}_1 = \vec{C} + \frac{1}{2}\vec{c}_{12}\,, \qquad \vec{c}_2 = \vec{C} - \frac{1}{2}\vec{c}_{12}\,. \tag{27}$$

It is easy to check that the Jacobian of the transformation (27) equals unity. The product of the Maxwell distributions $\phi\left(\vec{c}_1\right)\phi\left(\vec{c}_2\right)$ transforms into a product of a Maxwell distributions for the center of mass (an effective particle of double mass) and of a Maxwell distribution for the relative motion (an effective particle of half mass)

$$\begin{aligned}\phi\left(\vec{c}_1\right)\phi\left(\vec{c}_2\right) \rightarrow & \frac{1}{(2\pi)^{d/2}}\exp\left(-\frac{1}{2}c_{12}^2\right)\left(\frac{2}{\pi}\right)^{d/2}\exp\left(-2C^2\right) \\ \equiv & \,\phi\left(\vec{c}_{12}\right)\phi\left(\vec{C}\right)\,.\end{aligned} \tag{28}$$

The terms in squared brackets in (26) hence transform into

$$\begin{aligned}&\left[S_2\left(c_1^2\right) + S_2\left(c_2^2\right)\right] = \\ &\qquad C^4 + \left(\vec{C}\cdot\vec{c}_{12}\right)^2 + \frac{1}{16}c_{12}^4 + \frac{1}{2}C^2c_{12}^2 - 5C^2 - \frac{5}{4}c_{12}^2 + \frac{15}{4}\,.\end{aligned} \tag{29}$$

To find the quantities $\Delta\left(c_1^p + c_2^p\right)$ $(p = 2, 4)$ we rewrite the collision law (2) in terms of the $\vec{C}$ and $\vec{c}_{12}$ too:

$$\begin{aligned}\vec{c}_1^{\,\prime} =& \vec{C} + \frac{1}{2}\vec{c}_{12} - \frac{1}{2}(1+\varepsilon)\left(\vec{c}_{12}\cdot\vec{e}\,\right)\vec{e} \\ \vec{c}_2^{\,\prime} =& \vec{C} - \frac{1}{2}\vec{c}_{12} + \frac{1}{2}(1+\varepsilon)\left(\vec{c}_{12}\cdot\vec{e}\,\right)\vec{e}\,.\end{aligned} \tag{30}$$

Then straightforward algebra yields

$$\Delta\left(c_1^2 + c_2^2\right) = -\frac{1}{2}\left(1-\varepsilon^2\right)\left(\vec{c}_{12}\cdot\vec{e}\,\right)^2 \tag{31}$$

and

$$\begin{aligned}\Delta\left(c_1^4 + c_2^4\right) =& 2(1+\varepsilon)^2\left(\vec{c}_{12}\cdot\vec{e}\,\right)^2\left(\vec{C}\cdot\vec{e}\,\right)^2 + \frac{1}{8}\left(1-\varepsilon^2\right)^2\left(\vec{c}_{12}\cdot\vec{e}\,\right)^4 \\ & - \frac{1}{4}\left(1-\varepsilon^2\right)\left(\vec{c}_{12}\cdot\vec{e}\,\right)^2\vec{c}_{12}^{\,2} - \left(1-\varepsilon^2\right)\vec{C}^2\left(\vec{c}_{12}\cdot\vec{e}\,\right)^2 \\ & - 4(1+\varepsilon)\left(\vec{C}\cdot\vec{c}_{12}\right)\left(\vec{C}\cdot\vec{e}\,\right)\left(\vec{c}_{12}\cdot\vec{e}\,\right)\,.\end{aligned} \tag{32}$$

Combining the terms in (29), (31), and (32) the integrands for μ_2 and μ_4 in (26) contain products of the Maxwell distributions for C and c_{12} and the factors C^k, c_{12}^l, $(\vec{c}\cdot\vec{c}_{12})^m$ with exponents k, l, m, and similar other factors. Hence we define the *basic integral* by

$$J_{k,l,m,n,p,\alpha} \equiv \int d\vec{c}_{12} \int d\vec{C} \int d\vec{e}\,\Theta\left(-\vec{c}_{12}\cdot\vec{e}\right)\left|\vec{c}_{12}\cdot\vec{e}\right|^{1+\alpha}$$

$$\times\ \phi\left(c_{12}\right)\phi\left(C\right)C^k c_{12}^l\left(\vec{C}\cdot\vec{c}_{12}\right)^m\left(\vec{C}\cdot\vec{e}\right)^n\left(\vec{c}_{12}\cdot\vec{e}\right)^p\ , \quad (33)$$

and express μ_p via these quantities. In particular μ_2 reads in linear approximation with respect to a_2:

$$\mu_2 = \frac{1}{4}\left(1-\varepsilon^2\right)\left[J_{0,0,0,0,2,0} + a_2\left(J_{4,0,0,0,2,0} + J_{0,0,2,0,2,0} + \frac{1}{16}J_{0,4,0,0,2,0}\right.\right.$$

$$\left.\left. + \frac{1}{2}J_{2,2,0,0,2,0} - 5J_{2,0,0,0,2,0} - \frac{5}{4}J_{0,2,0,0,2,0} + \frac{15}{4}J_{0,0,0,0,2,0}\right)\right]\ . \quad (34)$$

The corresponding expression for μ_4 contains already 35 basic integrals. Keeping the second order terms with respect to a_2 yields expressions which contain 60 basic integrals for μ_2 and 328 basic integrals for μ_4. To handle such expressions one should obviously use computer algebra.

4.2 Computation of the Basic Integrals

Solution of the Angular Integrals

The solution of the basic integral as defined by (33) is straightforward for $n = 0$, however, for $n = 1, 2$ it requires some tricks (see, e.g.,. [16]). The presented approach may be generalised to $n > 2$, here we need only $n = 0, 1, 2$.

In the d-dimensional space the Cartesian coordinates of a vector $\vec{r} = (x_1, \ldots, x_d)$ are related with the spherical coordinates, expressed by the absolute value of the vector $r \geq 0$ and $d-1$ angles θ_1, θ_2, $\ldots\theta_{d-1}$ by the transformation

$$\begin{aligned}
x_1 &= r\sin\theta_1\sin\theta_2\ldots\sin\theta_{d-1} \\
x_m &= r\cos\theta_{m-1}\prod_{k=m}^{d-1}\sin\theta_k\ , \qquad m = 2,3,\ldots d-1 \\
x_d &= r\cos\theta_{d-1}\ ,
\end{aligned} \quad (35)$$

where

$$\begin{aligned}
&0 \leq \theta_1 \leq 2\pi \\
&0 \leq \theta_m \leq \pi\ , \qquad m = 2,3,\ldots d-1\ .
\end{aligned} \quad (36)$$

Hence, the infinitesimal solid angle reads

$$d\vec{e} = \prod_{k=1}^{d-1}\sin^{k-1}\theta_k d\theta_k\ . \quad (37)$$

Correspondingly we obtain for the surface area of a d-dimensional unit sphere

$$\begin{aligned}\Omega_d &= \int d\vec{e} = \int_0^{2\pi} d\theta_1 \int_0^{\pi} \sin\theta_2 d\theta_2 \int_0^{\pi} \sin^2\theta_3 d\theta_3 \ldots \int_0^{\pi} \sin^{d-2}\theta_{d-1} d\theta_{d-1} \\ &= 2\pi \, \frac{\Gamma(1)\,\Gamma\left(\frac{1}{2}\right)}{\Gamma\left(\frac{3}{2}\right)} \, \frac{\Gamma\left(\frac{3}{2}\right)\Gamma\left(\frac{1}{2}\right)}{\Gamma(2)} \, \frac{\Gamma(2)\,\Gamma\left(\frac{1}{2}\right)}{\Gamma\left(\frac{5}{2}\right)} \cdots \frac{\Gamma\left(\frac{d-1}{2}\right)\Gamma\left(\frac{1}{2}\right)}{\Gamma\left(\frac{d}{2}\right)} \\ &= 2\pi \left[\Gamma\left(\frac{1}{2}\right)\right]^{d-2} \frac{1}{\Gamma\left(\frac{d}{2}\right)} = \frac{2\pi^{d/2}}{\Gamma\left(\frac{d}{2}\right)} ,\end{aligned} \tag{38}$$

where we take into account that $\Gamma(1) = 1$, $\Gamma(1/2) = \sqrt{\pi}$ and we use the trigonometric integrals

$$\int_0^{\pi/2} \sin^l\theta \cos^n\theta d\theta = \frac{1}{2} \frac{\Gamma\left(\frac{l+1}{2}\right)\Gamma\left(\frac{n+1}{2}\right)}{\Gamma\left(\frac{l+n}{2}+1\right)} . \tag{39}$$

and

$$\int_0^{\pi} \sin^l\theta d\theta = \frac{\Gamma\left(\frac{l+1}{2}\right)\Gamma\left(\frac{1}{2}\right)}{\Gamma\left(\frac{l}{2}+1\right)} . \tag{40}$$

Before calculating the Basic Integral we evaluate some more simple expressions. We will need the angular integral

$$\beta_m = \int d\vec{e}\,\Theta\left(\vec{e}\cdot\hat{\vec{g}}\right)\left(\vec{e}\cdot\hat{\vec{g}}\right)^m = (-1)^m \int d\vec{e}\,\Theta\left(-\vec{e}\cdot\hat{\vec{g}}\right)\left(\vec{e}\cdot\hat{\vec{g}}\right)^m , \tag{41}$$

where vectors with a hat denote unit vectors (e.g. $\hat{\vec{g}} \equiv \vec{g}/g$). In (41) we change the variables $\vec{e} \to -\vec{e}$ to get the expression for β_m which will be mainly used in the calculations. Let us choose the coordinate axis OX_d along $\vec{g}$, so that $\vec{e}\cdot\hat{\vec{g}} = \cos\theta_{d-1}$. Using (40) (39) the above properties of the Gamma function and (37) for the infinitesimal solid angle $d\vec{e}$ we obtain

$$\begin{aligned}\beta_m &= \int d\vec{e}\,\Theta\left(\vec{e}\cdot\hat{\vec{g}}\right)\left(\vec{e}\cdot\hat{\vec{g}}\right)^m \\ &= \int_0^{2\pi} d\theta_1 \int_0^{\pi} \sin\theta_2 d\theta_2 \int_0^{\pi} \sin^2\theta_3 d\theta_3 \ldots \int_0^{\pi/2} \sin^{d-2}\theta_{d-1} \cos^m\theta_{d-1} d\theta_{d-1} \\ &= 2\pi \, \frac{\Gamma(1)\,\Gamma\left(\frac{1}{2}\right)}{\Gamma\left(\frac{3}{2}\right)} \, \frac{\Gamma\left(\frac{3}{2}\right)\Gamma\left(\frac{1}{2}\right)}{\Gamma(2)} \cdots \frac{1}{2} \frac{\Gamma\left(\frac{d-1}{2}\right)\Gamma\left(\frac{m+1}{2}\right)}{\Gamma\left(\frac{d+m}{2}\right)} \\ &= \pi^{(d-1)/2} \frac{\Gamma\left(\frac{m+1}{2}\right)}{\Gamma\left(\frac{m+d}{2}\right)} ,\end{aligned} \tag{42}$$

where we take into account that the integration over θ_{d-1} is to be performed in the interval $[0, \pi/2]$ due to the factor $\Theta\left(\vec{e}\cdot\hat{\vec{g}}\right)$. Note that β_m does not depend on the vector $\hat{\vec{g}}$. Similarly we obtain

$$\int d\vec{e}\left(\vec{e}\cdot\hat{\vec{g}}\right)^m = \int d\vec{e}\left[\Theta\left(\vec{e}\cdot\hat{\vec{g}}\right)+\Theta\left(-\vec{e}\cdot\hat{\vec{g}}\right)\right]\left(\vec{e}\cdot\hat{\vec{g}}\right)^m = \left[1+(-1)^m\right]\beta_m$$
$$= \int d\hat{\vec{g}}\left(\hat{\vec{g}}\cdot\hat{\vec{C}}\right)^m , \tag{43}$$

where we take into account that $\vec{e}$ is the integration variable and that β_m is independent of $\hat{\vec{g}}$. This allows to change the notation for the integration variable and for $\hat{\vec{g}}$: $\vec{e}\to\hat{\vec{g}}$ and $\hat{\vec{g}}\to\hat{\vec{C}}$.

Reduction of the Basic Integrals

We also need the integrals

$$\int d\vec{C}C^r\phi(C) = \int_0^\infty c^{d-1+r}\left(\frac{2}{\pi}\right)^{d/2} e^{-2C^2}\,dC\int d\hat{\vec{C}}$$
$$= \left(\frac{2}{\pi}\right)^{d/2}\frac{1}{4}\frac{\sqrt{2}\Gamma\left(\frac{d+r}{2}\right)}{2^{\frac{d+r-1}{2}}}\Omega_d = 2^{-r/2}\frac{\Gamma\left(\frac{d+r}{2}\right)}{\Gamma\left(\frac{d}{2}\right)} \equiv \gamma_r \tag{44}$$

and

$$\int d\vec{g}g^r\phi(g)$$
$$= \int_0^\infty g^{d-1+r}\left(\frac{1}{2\pi}\right)^{d/2} e^{-g^2/2}\,dg\int d\hat{\vec{g}} = 2^{r/2}\frac{\Gamma\left(\frac{d+r}{2}\right)}{\Gamma\left(\frac{d}{2}\right)} = 2^r\gamma_r\,, \tag{45}$$

where again we have used $\int d\hat{\vec{g}} = \int d\hat{\vec{C}} = \int d\vec{e} = \Omega_d$ according to (38).

For $n=0$ the definition of the basic integral, (33), reads (with $\vec{g}\equiv\vec{c}_{12}$)

$$J_{k,l,m,0,p,\alpha} = \int d\vec{g}\int d\vec{C}\phi(g)\phi(C)C^k g^l(\vec{C}\cdot\vec{g})^m K_{\alpha,p}(g)\,, \tag{46}$$

where we introduce the integral

$$K_{\alpha,p}(g) \equiv \int d\vec{\mu}\,\left|\vec{e}\cdot\vec{g}\right|^\alpha\left(\vec{e}\cdot\vec{g}\right)^p\,, \tag{47}$$

with the short-hand notation $d\vec{\mu}\equiv d\vec{e}\,\Theta\left(-\vec{e}\cdot\vec{g}\right)\left|\vec{e}\cdot\vec{g}\right|$. Similarly, for $n=1$ one can write

$$J_{k,l,m,1,p,\alpha} = \int d\vec{g}\int d\vec{C}\phi(g)\phi(C)C^k g^l\left(\vec{C}\cdot\vec{g}\right)^m\left(\vec{C}\cdot\vec{I}_{\alpha,p}(g)\right)\,, \tag{48}$$

with the vectorial integral

$$\vec{I}_{\alpha,p}(g) \equiv \int d\vec{\mu}\, \vec{e}\, |\vec{e}\cdot\vec{g}\,|^{\alpha} \left(\vec{e}\cdot\vec{g}\,\right)^{p} . \tag{49}$$

Finally, for $n=2$:

$$J_{k,l,m,2,p,\alpha} = \int d\vec{g} \int d\vec{C} \phi(g)\phi(C) C^k g^l \left(\vec{C}\cdot\vec{g}\right)^m \vec{C}\cdot\hat{H}_{\alpha,p}(g)\cdot\vec{C} \tag{50}$$

where the dyad $\hat{H}_{\alpha,p}(g)$ is given by

$$\hat{H}_{\alpha,p}(g) \equiv \int d\vec{\mu}\, \vec{e}\,\vec{e}\, |\vec{e}\cdot\vec{g}\,|^{\alpha} \left(\vec{e}\cdot\vec{g}\,\right)^{p} \tag{51}$$

and where we take into account that according to the vector-dyadic product

$$\vec{C}\cdot\vec{e}\,\vec{e}\cdot\vec{C} = \left(\vec{C}\cdot\vec{e}\right)\left(\vec{e}\cdot\vec{C}\right) = \left(\vec{C}\cdot\vec{e}\right)^2 . \tag{52}$$

With the quantities introduced above we can express the basic integral step by step.

Solution of the Auxiliary Vectorial and Tensorial Integrals

The integral $K_{\alpha,\,p}(g)$ reads then

$$\begin{aligned} K_{\alpha,p}(g) &= \int d\vec{e}\,\Theta\left(-\vec{e}\cdot\vec{g}\,\right) |\vec{e}\cdot\vec{g}\,|^{\alpha+1} \left(\vec{e}\cdot\vec{g}\,\right)^{p} \\ &= (-1)^p g^{\,p+\alpha+1} \int d\vec{e}\,\Theta\left(\vec{e}\cdot\hat{\vec{g}}\right)\left(\vec{e}\cdot\hat{\vec{g}}\right)^{p+\alpha+1} \\ &= (-1)^p g^{\,p+\alpha+1} \beta_{p+\alpha+1} \end{aligned} \tag{53}$$

with the definition (41) of β_m. Consider now the vectorial integral $\vec{I}_{\alpha,p}(g) = \vec{g}G(g)$. The result of the integration over $\vec{e}$ must be a vector. Due to the symmetry of the problem this vector should be directed along the vector $\vec{g}$ since there is no other preferred direction. Thus we write

$$\vec{I}_{\alpha,p}(g) = \vec{g}\,G(g) . \tag{54}$$

The function $G(g)$ may be found by multiplying both sides of (49) by $\vec{g}$:

$$\vec{g}\cdot\vec{I}_{\alpha,p}(g) = g^2 G(g) = \int d\vec{e}\,\Theta\left(-\vec{e}\cdot\vec{g}\,\right) |\vec{e}\cdot\vec{g}\,|^{\alpha+1} \left(\vec{e}\cdot\vec{g}\,\right)^{p+1} = K_{\alpha,p+1}(g) . \tag{55}$$

Therefore, $G(g) = g^{-2}K_{\alpha,p+1}(g)$ and thus with (53) for $K_{\alpha,p}(g)$ we obtain

$$\vec{I}_{\alpha,p}(g) = (-1)^{p+1} \beta_{p+\alpha+2}\, g^{\,p+\alpha}\vec{g} . \tag{56}$$

The integrand in (51) is a dyad proportional to the dyad $\vec{e}\,\vec{e}$. Since integration over $\vec{e}$ does not change the symmetry, we expect that the components of the resultant dyad $\hat{H}_{\alpha,p}(g)$ are composed of the components of the vector $\vec{g}$, the only vectorial quantity which is contained. Since the result may also have a non-vectorial part, the following form of $\hat{H}_{\alpha,p}(g)$ is suggested:

$$\hat{H}(g) = A(g)\,\vec{g}\,\vec{g} + B(g)g^2\hat{U}\,, \tag{57}$$

where $\hat{U}$ is a unit dyad (i.e. the unit diagonal matrix) and the functions $A(g)$ and $B(g)$ are to be determined. Multiplying $\hat{H}$ scalarly from both sides by $\vec{g}$ we obtain the first equation for $A(g)$ and $B(g)$:

$$\begin{aligned}\vec{g}\cdot\hat{H}\cdot\vec{g} &= A(g)g^4 + B(g)g^4\\ &= \int d\vec{e}\,\Theta\left(-\vec{e}\cdot\vec{g}\,\right)\left|\vec{e}\cdot\vec{g}\,\right|^{\alpha+1}\left(\vec{e}\cdot\vec{g}\,\right)^{p+2} = K_{\alpha,p+2}(g)\,,\end{aligned} \tag{58}$$

where we use

$$\vec{g}\cdot\vec{g}\,\vec{g}\cdot\vec{g} = \left(g^2\right)^2 = g^4 \qquad \text{and} \qquad \vec{g}\cdot\hat{U}\cdot\vec{g} = \vec{g}\cdot\vec{g} = g^2\,. \tag{59}$$

Then taking the trace of $\hat{H}$ (i.e. computing the sum of diagonal elements) we arrive at the second equation:

$$\mathrm{Tr}\;\hat{H} = A(g)g^2 + d\,B(g)g^2 = \int d\vec{e}\,\Theta\left(-\vec{e}\cdot\vec{g}\,\right)\left|\vec{e}\cdot\vec{g}\,\right|^{\alpha+1}\left(\vec{e}\cdot\vec{g}\,\right)^{p} = K_{\alpha,p}(g)\,. \tag{60}$$

In (60) (with the summation convention) we take into account

$$\mathrm{Tr}\;\vec{g}\,\vec{g} = g_i g_i = g^2\,, \qquad \mathrm{Tr}\;\hat{U} = \delta_{ii} = d\,, \qquad \mathrm{Tr}\;\vec{e}\,\vec{e} = e_i e_i = e^2 = 1\,. \tag{61}$$

Solving the set of equations (58), (60) for the functions $A(g)$ and $B(g)$ and using (53) for $K_{\alpha,p}(g)$ we obtain

$$\hat{H} = \frac{(-1)^p\,g^{p+\alpha-1}}{(d-1)}\left[\left(d\beta_{p+\alpha+3} - \beta_{p+\alpha+1}\right)\vec{g}\,\vec{g} + \left(\beta_{p+\alpha+1} - \beta_{p+\alpha+3}\right)g^2\hat{U}\right]\,. \tag{62}$$

Solution for the Basic Integrals

With the results for the scalar integral $K_{\alpha,p}(g)$, for the vectorial integral $\vec{I}_{\alpha,p}(g)$ and for the tensorial integral $\hat{H}_{\alpha,p}(g)$ we can evaluate the basic integrals. For the first basic integral with $n=0$ we find with (53)

$$
\begin{aligned}
&J_{k,l,m,0,p,\alpha} \\
&= (-1)^p \beta_{p+\alpha+1} \int d\vec{g} \int d\vec{C} \phi(g)\phi(C) C^{k+m} g^{\,l+m+p+\alpha+1} \left(\hat{\vec{C}} \cdot \hat{\vec{g}}\right)^m \\
&= (-1)^p \beta_{p+\alpha+1} \int d\vec{C} C^{k+m} \phi(C) \int d\vec{g} g^{l+m+p+\alpha+1} \phi(g) \int d\hat{\vec{g}} \left(\hat{\vec{g}} \cdot \hat{\vec{C}}\right)^m \Omega_d^{-1} ,
\end{aligned}
\tag{63}
$$

where we write the vectors $\vec{g}$ and $\vec{C}$ in the form $\vec{g}=g\hat{\vec{g}}$ and $\vec{C}=C\hat{\vec{C}}$ and for the integration over $d\vec{g}$ use

$$
\begin{aligned}
&\int d\vec{g} F_1(g) F_2\left(\hat{\vec{g}}\right) = \int_0^\infty g^{d-1} F_1(g) dg \int d\hat{\vec{g}} F_2\left(\hat{\vec{g}}\right) \\
&= \int_0^\infty g^{d-1} F_1(g) dg \int d\hat{\vec{g}} \, \frac{\int d\hat{\vec{g}} F_2\left(\hat{\vec{g}}\right)}{\int d\hat{\vec{g}}} = \int d\vec{g} F_1(g) \int d\hat{\vec{g}} F_2\left(\hat{\vec{g}}\right) \Omega_d^{-1}
\end{aligned}
\tag{64}
$$

(recall that $\int d\hat{\vec{g}} = \Omega_d$). The integral over $d\hat{\vec{g}}$ in (63) is $\left[1+(-1)^m\right]\beta_m$, according to (43), while the integrals over $d\vec{C}$ and over $d\vec{g}$ are given by (44) and (45), respectively. As the result we obtain the basic integral for $n=0$:

$$
\begin{aligned}
&J_{k,l,m,0,p,\alpha} \\
&= (-1)^p \left[1+(-1)^m\right] 2^{\,l+m+p+\alpha+1} \Omega_d^{-1} \beta_{p+\alpha+1} \beta_m \gamma_{k+m} \gamma_{l+m+p+\alpha+1} \,.
\end{aligned}
\tag{65}
$$

The evaluation of the basic integral for $n=1$ and $n=2$ may be performed by substituting (56) and (62) correspondingly into (48) and (50). The integrals which are obtained by this substitute are precisely of the same structure as in (63). Hence, the calculations for $n=1$ and $n=2$ are analogous to the case $n=0$. The final result reads for $n=1$:

$$
\begin{aligned}
J_{k,l,m,1,p,\alpha} &= (-1)^{p+1} \left[1+(-1)^{m+1}\right] 2^{\,l+m+p+\alpha+1} \Omega_d^{-1} \\
&\qquad \times \beta_{p+\alpha+2} \beta_{m+1} \gamma_{k+m+1} \gamma_{l+m+p+\alpha+1}
\end{aligned}
\tag{66}
$$

and for $n=2$:

$$
\begin{aligned}
J_{k,l,m,2,p,\alpha} &= (-1)^p \left[1+(-1)^m\right] 2^{\,l+m+p+\alpha+1} \left[(d-1)\Omega_d\right]^{-1} \\
&\qquad \times \gamma_{k+m+2} \gamma_{l+m+p+\alpha+1} \left[\left(d\beta_{p+\alpha+3} - \beta_{p+\alpha+1}\right) \beta_{m+2}\right. \\
&\qquad\qquad \left. + \left(\beta_{p+\alpha+1} - \beta_{p+\alpha+3}\right) \beta_m\right] .
\end{aligned}
\tag{67}
$$

The constants β_m, γ_m and Ω_d which are contained in the basic integrals have been defined above.

We gave the solution of the basic integrals for $n \le 2$ with arbitrary other indexes. As illustrated below, these basic integrals are mainly used in the kinetic gas theory. The present approach, however, may be straightforwardly extended. For example, the integral for $n = 3$ may be written (similar to (51) in terms of the tensor $H_{\alpha,p}^{(i,j,k)}(g) = \int d\mu e_i e_j e_k \left|\vec{e}\cdot\vec{g}\right|^{\alpha} \left(\vec{e}\cdot\vec{g}\right)^p$. Then this tensor and finally the basic integral may be solved using the same symmetry-based approach.

5 Computational Formula Manipulation to Evaluate Kinetic Integrals

Kinetic integrals have the structure

$$\int d\vec{c}_1 \int d\vec{c}_2 \int d\vec{e}\, \Theta\left(-\vec{c}_{12}\cdot\vec{e}\right) \left|\vec{c}_{12}\cdot\vec{e}\right| \phi\left(c_1\right) \phi\left(c_2\right) \mathrm{Expr}\left(\dots\right) , \tag{68}$$

where Expr $(\dots)$ specifies a mathematical expression in terms of the pre- and aftercollision velocities, the Sonine polynomials and their coefficients and other variables (see below). The evaluation of such integrals by means of basic integrals $J_{k,l,m,n,p,\alpha}$ is straightforward, although for practical applications even simple looking expressions may result in sums of thousands of terms. Since the algebra follows an algorithm, we can use a formula manipulation program, such as Maple to evaluate kinetic integrals.

Before giving a systematic description we wish to describe the example from Sect. 2 in detail, where we computed the second moment of the collision integral μ_2 in linear approximation with respect to a_2.

The kernel of the kinetic integral (14) reads

$$\mathrm{Expr} = -\frac{1}{2}\left\{1 + a_2\left[S_2\left(c_1^2\right) + S_2\left(c_2^2\right)\right]\right\} \Delta\left(c_1^2 + c_2^2\right) \tag{69}$$

The computation in the listing in Sect. 2 is started with `restart` to clean the memory of the computer from earlier results, then with `libname...` we extend the library path of Maple and load the library (`with...`). The next line is an output produced by maple, indicating that the commands `DefDimension`, `DefJ`, `DefS`, `KIinit`, `getJexpr`, and `unDefJ` can be used now. Here starts the evaluation of our integral: First we initialize the computation by `KIinit()` which defines some internal variables. Then the dimension of the problem is specified. Here we wish to evaluate the Integral in three dimensions. With the line `DefJ()` we require that the Basic Integrals will be evaluated, i.e., in the final result we generate real numbers according to (65)-(67), e.g., $J_{0,0,2,0,2,0} = 6\sqrt{2\pi}$. The line `Expr:=...` defines the kernel of the integral due to (68). Finally `mu2:=getJexpr(0,Expr,0)` starts the computation. The arguments of `getJexpr()` as well as the syntax of `Expr` are described in detail below.

We explain the Maple package `KineticIntegral.m` in detail. To this end we comment the program line by line, bold numbers refer to line numbers in the listing. The package as well as the examples can be obtained at [17]. The program uses internal variables whose names follow a regular scheme: `c1p`, `c2p`, `c1a`, and `c2a` stand for the pre-collision and after-collision velocities of the particles. The unit vector `e`$\equiv E \equiv \vec{r}_{12}/r_{12}$ and the dimensionality `DIM` are further variables. Scalar products are named by `xDOTy`, where `x` and `y` are vectors. An example is `c1pDOTc2p`$\equiv \vec{c}_1 \cdot \vec{c}_2$. A full list of the internal variables is given in the table below. All these symbols can be used to specify the kernel `Expr` of the kinetic integral.

`c1p`	$\vec{c}_1$: pre-collision velocity of particle 1
`c2p`	$\vec{c}_2$: pre-collision velocity of particle 2
`c1a`	$\vec{c}_1'$: after-collision velocity of particle 1
`c2a`	$\vec{c}_2'$: after-collision velocity of particle 2
`c12`	$\vec{c}_{12}$: relative velocity
`C`	$\vec{C}$: centre of mass velocity
`c12DOTe=eDOTc12`	$\vec{c}_{12} \cdot \vec{e}$
`CDOTc12=c12DOTC`	$\vec{C} \cdot \vec{c}_{12}$
`c1pDOTc1p`	$(\vec{c}_1)^2 = \vec{C}^2 + \vec{c}_{12}^{\,2}/4 + \vec{C} \cdot \vec{c}_{12}$
`c2pDOTc2p`	$(\vec{c}_2)^2 = \vec{C}^2 + \vec{c}_{12}^{\,2}/4 - \vec{C} \cdot \vec{c}_{12}$
`c1pDOTc2p =c2pDOTc1p`	$\vec{c}_1 \cdot \vec{c}_2 = \vec{C}^2 - \vec{c}_{12}^{\,2}/4$
`c1aDOTc1a`	$(\vec{c}_1')^2 = \vec{C}^2 + \vec{c}_{12}^{\,2}/4 - \left(1-\varepsilon^2\right)/4\,(\vec{c}_{12}\cdot\vec{e})^2 + {}$ $+\vec{C}\cdot\vec{c}_{12} - (1+\varepsilon)\left(\vec{C}\cdot\vec{e}\right)(\vec{c}_{12}\cdot\vec{e})$
`c2aDOTc2a`	$(\vec{c}_2')^2 = \vec{C}^2 + \vec{c}_{12}^{\,2}/4 - \left(1-\varepsilon^2\right)/4\,(\vec{c}_{12}\cdot\vec{e})^2 - {}$ $-\vec{C}\cdot\vec{c}_{12} + (1+\varepsilon)\left(\vec{C}\cdot\vec{e}\right)(\vec{c}_{12}\cdot\vec{e})$
`c1aDOTc2a=c2aDOTc1a`	$\vec{c}_1' \cdot \vec{c}_2' = \vec{C}^2 - \vec{c}_{12}^{\,2}/4 + \left(1-\varepsilon^2\right)/4\,(\vec{c}_{12}\cdot\vec{e})^2$
`c1pDOTc1a=c1aDOTc1p`	$\vec{c}_1 \cdot \vec{c}_1' = \vec{C}^2 + \vec{c}_{12}^{\,2}/4 + \vec{C}\cdot\vec{c}_{12} -$ $-(1+\varepsilon)/2\left(\vec{C}\cdot\vec{e}\right)(\vec{c}_{12}\cdot\vec{e}) - (1+\varepsilon)/4\,(\vec{c}_{12}\cdot\vec{e})^2$
`c2pDOTc2a=c2aDOTc2p`	$\vec{c}_2 \cdot \vec{c}_2' = \vec{C}^2 + \vec{c}_{12}^{\,2}/4 - \vec{C}\cdot\vec{c}_{12} +$ $+(1+\varepsilon)/2\left(\vec{C}\cdot\vec{e}\right)(\vec{c}_{12}\cdot\vec{e}) - (1+\varepsilon)/4\,(\vec{c}_{12}\cdot\vec{e})^2$
`c1aDOTc2p=c2pDOTc1a`	$\vec{c}_1' \cdot \vec{c}_2 = \vec{C}^2 - \vec{c}_{12}^{\,2}/4 -$ $-(1+\varepsilon)/2\left(\vec{C}\cdot\vec{e}\right)(\vec{c}_{12}\cdot\vec{e}) + (1+\varepsilon)/4\,(\vec{c}_{12}\cdot\vec{e})^2$
`c1pDOTc2a=c2aDOTc1p`	$\vec{c}_1 \cdot \vec{c}_2' = \vec{C}^2 - \vec{c}_{12}^{\,2}/4 +$ $+(1+\varepsilon)/2\left(\vec{C}\cdot\vec{e}\right)(\vec{c}_{12}\cdot\vec{e}) + (1+\varepsilon)/4\,(\vec{c}_{12}\cdot\vec{e})^2$
`Delta2`, `Delta4`	see (31), (32)
`DIM`	dimension
`C1`, `dprime`	describes the coefficient of restitution of viscoelastic particles, see (23) !!!!
`absc12DOTe5`	$\left\lvert\vec{c}_{12}\cdot\vec{e}\right\rvert^{1/5}$

The variables `Delta2` and `Delta4` are defined in the initialization `KIinit()`:

```
1  KIinit:=proc()
2    global Delta2,Delta4:
3    unprotect('Delta2','Delta4'):
4    Delta2:=(1/2*epsilon^2-1/2)*c12DOTe^2:
5    Delta4:=(-4*epsilon-4)*CDOTe*c12DOTe*CDOTc12
6     +(1/8-1/4*epsilon^2+1/8*epsilon^4)*c12DOTe^4
7     +(-1/4+1/4*epsilon^2)*c12^2*c12DOTe^2
8     +(2*epsilon^2+2+4*epsilon)*c12DOTe^2*CDOTe^2
9     +(epsilon^2-1)*C^2*c12DOTe^2:
10   protect('Delta2','Delta4'):
11 end:
```

The kernel Expr may contain Sonine polynomials. They are defined by DefS() according to (12).

```
12 DefS:=proc()
13  global S:
14  if(assigned(DIM)) then
15   unprotect('S'):
16   S:=proc(a::integer,x::algebraic)
17    local r,xx:
18    xx:=subs(c1p=C+1/2*c12,x):
19    xx:=subs(c2p=C-1/2*c12,xx):
20    xx:=subs(c1a=C+1/2*c12-1/2*(1+epsilon)*c12DOTe*e,xx):
21    xx:=subs(c2a=C-1/2*c12+1/2*(1+epsilon)*c12DOTe*e,xx):
22    xx:=expand(xx^2):
23    xx:=algsubs(C*c12=CDOTc12,xx):
24    if(a=0)then
25     r:=1:
26    elif (a=1) then
27     r:=-xx+DIM/2:
28    elif (a=2) then
29     r:=xx^2/2 - (DIM+2)*xx/2 +(DIM+2)*DIM/8:
30    else
31     ERROR(`invalid argument`,a):
32    fi:
33    r:=factor(r):
34    end:
35   protect('S'):
36   printf("Sonine polynomials defined\n"):
37  else
38   ERROR(`Dimension undefined`):
39  fi:
40 end:
```

First at line **14** it is checked whether the dimensionality DIM is specified, otherwise DefS reports an error at line **38**. The syntax of the Sonine polynomials is S(a,expr), a specifies the order of the polynomial and expr is its algebraic argument which may contain the pre- and after-collision particle velocities c1p, c1a, c2p, and c2a. In lines **18-21** these velocities are translated into expressions of the center of mass velocity C, the relative velocity c12, the coefficient of restitution epsilon, the unit vector e and the scalar product c12DOTe, due to (30). There is an important difference in the notations of the Sonine coefficients due to (12) and in the Maple program:

$$\texttt{S(a,expr)} \equiv S_a\left(\text{expr}^2\right) . \tag{70}$$

This difference of the notation is taken into account by line **22** where the entire term Expr with all the described substitutions is squared. In the squared expressions there appears the term $\vec{C} \cdot \vec{c}_{12}$ which is again substituted by CDOTc12. The program evaluates Sonine polynomials of the order $a = \{0, 1, 2\}$ which is done in line **24-29**. Any incorrect input a is acknowledged by an error message, line **31**.

All of the defined internal variables, including the Sonine polynomials are accessible within the Maple program. In general, it is of not much value to list them since according to the substitution of variables even small terms become impressive expressions. For debugging purposes, however, this feature may be useful.

There are two different output modes: first we may be interested in the result of an integration in terms of basic integrals $J_{k,l,m,n,p,\alpha}$ and second (usually) we are interested in obtaining real numbers. The translation of the basic integrals into real numbers is given by (65)–(67). These equations are implemented in DefJ():

```
41 DefJ:=proc()
42  global J:
43  unprotect(J):
44  if(assigned(DIM)) then
45   J:=proc(k::integer,l::integer,m::integer,n::integer,
46    p::integer,a::integer)
47    local J0, J1, J2, beta, gamma, omega:
48    beta:=m->Pi^((DIM-1)/2)*GAMMA((m+1)/2)/GAMMA((m+DIM)/2):
49    gamma:=m->2^(-m/2)*GAMMA((m+DIM)/2)/GAMMA(DIM/2):
50    omega:=2*Pi^(DIM/2)/GAMMA(DIM/2):
51    J0:=(k,l,m,p,a)->(-1)^p*(1+(-1)^m)*2^(l+m+p+a+1)/omega*beta(p+a+1)
52     *beta(m)*gamma(k+m)*gamma(l+m+p+a+1):
53    J1:=(k,l,m,p,a)->(-1)^(p+1)*(1+(-1)^(m+1))*2^(l+m+p+a+1)/omega*beta(p+a+2)
54     *beta(m+1)*gamma(k+m+1)*gamma(l+m+p+a+1):
55    J2:=(k,l,m,p,a)->(-1)^p*(1+(-1)^m)*2^(l+m+p+a+1)/omega*gamma(k+m+2)
56     *gamma(l+m+p+a+1)*((DIM*beta(p+a+3)-beta(p+a+1))*beta(m+2)
57     +(beta(p+a+1)-beta(p+a+3))*beta(m))/(DIM-1):
58    simplify(1/2*J0(k,l,m,p,a/5)*(n-1)*(n-2)-J1(k,l,m,p,a/5)*n*(n-2)
59     +1/2*J2(k,l,m,p,a/5)*n*(n-1)):
60    end:
61    protect('J'):
62    printf("Basic Integrals will be evaluated\n"):
63  else
64   ERROR(`Dimension undefined`):
65  fi:
66 end:
```

After calling defJ all results are given as real numbers by evaluating the basic integrals, e.g.,

```
> DefJ();

Basic Integrals will be evaluated
> BIdemo:=J(0,0,1,1,1,1);
```

$$BIdemo := \frac{11}{100} \frac{2^{(3/5)}\, \pi^{(3/2)}}{\sin(\frac{1}{10}\,\pi)\, \Gamma(\frac{9}{10})}$$

```
> evalf(BIdemo);
```

$$2.811423645$$

The counterpart of `DefJ()` is `unDefJ()`. After calling `unDefJ` any result is given in terms of basic integrals.

```
67 unDefJ:=proc()
68   global J:
69   if (assigned(J)) then
70     unprotect(J):
71     unassign('J');
72   fi;
73   printf("Basic Integrals will not be evaluated\n"):
74 end:
```

In the described programs `DefS` and `DefJ` we have already used the variable `DIM` which contains the dimensionality of the Basic Integral. This variable is set by `DefDimension()`:

```
75 DefDimension:=proc(d::integer)
76   global DIM:
77   unprotect('DIM'):
78   DIM:=d:
79   printf("Dimension = %d\n",d):
80   protect('DIM'):
81   DefS():
82   if (assigned(J)) then
83     DefJ():
84   fi:
85 end:
```

After changing the variable `DIM` in line **78** the Sonine polynomials which are specific for any dimension have to be redefined (line **81**). For the case that `DefJ()` has been called before, the basic integrals have to be redefined as well, line **83**.

The heart of the Maple package is the program `getJexpr()`. This program has three arguments: `epstype` specifies the type of the particle interaction. `epstype`=0 stands for constant coefficient of restitution and `epstype`=1 stands for viscoelastic particles. The second argument is the kernel of the integral `expr` which is a function of the particle velocities before and after the collision, the dimension, the Sonine polynomials and their arguments, the Sonine coefficients, and scalar products of the vectorial values. In fact `expr` may contain all variables which are listed in the table on page 146. The third argument describes the output mode of the program. It will be described later. The text of the program `getJexpr()` is listed below:

```
86 getJexpr:=proc(epstype::integer,expr::algebraic,
87  ou::integer)
88  global di,ddi:
89  local Expr,termJ,it,term,ko,epow,aa,k,l,m,n,p,a2pow,
90  vareps,C1expl,C1omega0,C12omega1:
91  Expr:=subs(c2pDOTc1p=c1pDOTc2p,expr);
92  Expr:=subs(c2aDOTc1a=c1aDOTc2a,Expr);
93  Expr:=subs(c1pDOTc1a=c1aDOTc1p,Expr);
94  Expr:=subs(c2pDOTc2a=c2aDOTc2p,Expr);
95  Expr:=subs(c2pDOTc1a=c1aDOTc2p,Expr);
96  Expr:=subs(c2aDOTc1p=c1pDOTc2a,Expr);
97  Expr:=subs(c1pDOTc1p=(C*C+1/4*c12*c12+CDOTc12),Expr);
98  Expr:=subs(c2pDOTc2p=(C*C+1/4*c12*c12-CDOTc12),Expr);
```

```
 Expr:=subs(c1pDOTc2p=(C*C-1/4*c12*c12),Expr);
 Expr:=subs(c1aDOTc1a=(C*C+1/4*c12*c12-(1-epsilon*epsilon)/4*c12DOTe*c12DOTe+CDOTc12
  -(1+epsilon)*CDOTe*c12DOTe),Expr);
 Expr:=subs(c2aDOTc2a=(C*C+1/4*c12*c12-(1-epsilon*epsilon)/4*c12DOTe*c12DOTe-CDOTc12
  +(1+epsilon)*CDOTe*c12DOTe),Expr);
 Expr:=subs(c1aDOTc2a=(C*C-1/4*c12*c12+(1-epsilon*epsilon)/4*c12DOTe*c12DOTe),Expr);
 Expr:=subs(c1aDOTc1p=(C*C+1/4*c12*c12+CDOTc12-(1+epsilon)/2*CDOTe*c12DOTe
  -(1+epsilon)/4*c12DOTe*c12DOTe),Expr);
 Expr:=subs(c2aDOTc2p=(C*C+1/4*c12*c12-CDOTc12+(1+epsilon)/2*CDOTe*c12DOTe
  -(1+epsilon)/4*c12DOTe*c12DOTe),Expr);
 Expr:=subs(c1aDOTc2p=(C*C-1/4*c12*c12-(1+epsilon)/2*CDOTe*c12DOTe
  +(1+epsilon)/4*c12DOTe*c12DOTe),Expr);
 Expr:=subs(c1pDOTc2a=(C*C-1/4*c12*c12+(1+epsilon)/2*CDOTe*c12DOTe
  +(1+epsilon)/4*c12DOTe*c12DOTe),Expr);
 if(epstype=0) then
  Expr:=expand(Expr):
 elif(epstype=1) then
  vareps:=1-C1*dprime*absc12DOTe5+3/5*C1^2*dprime^2*absc12DOTe5^2:
  Expr:=expand(algsubs(epsilon=vareps,Expr)):
 else
  ERROR(`invalid argument`,epstype):
 fi:
 Expr:=expand(simplify(Expr)):
 termJ:=0:
 printf("expressing %d terms in form of Basic Integrals\n",
  nops(Expr)):
 for it from 1 to nops(Expr) do
  term:=op(it,Expr):
  ko:=coeffs(term,{epsilon,absc12DOTe5,C,c12,CDOTc12,CDOTe,c12DOTe,a2}):
  epow:=degree(term,epsilon):
  aa:=degree(term,absc12DOTe5):
  k:=degree(term,C):
  l:=degree(term,c12):
  m:=degree(term,CDOTc12):
  n:=degree(term,CDOTe):
  p:=degree(term,c12DOTe):
  a2pow:=degree(term,a2):
  termJ:=termJ+epsilon^epow*ko*a2^a2pow*J(k,l,m,n,p,aa):
  di[it]:=ko*epsilon^epow*c12DOTe^aa*C^k*c12^l*CDOTc12^m*CDOTe^n*c12DOTe^p*a2^a2pow:
  ddi[it]:=term:
  if(simplify(di[it]-ddi[it])!=0) then
   ERROR(`error evaluating term`,it,`. see di[], ddi[]`):
  fi:
 od:
 termJ:=simplify(termJ):
 if (ou=1) then
  C1expl:=GAMMA(3/5)*sqrt(Pi)/(2^(1/5)*5^(2/5)*GAMMA(21/10));
  termJ:=subs(C1=C1expl,termJ):
  expand(evalf(termJ)):
 elif (ou=0) then
  factor(simplify(termJ)):
 elif (ou=2) then
  C1omega0:=omega0/(2*sqrt(2*Pi)*2^(1/10)*GAMMA(21/10)):
  C12omega1:=omega1/(sqrt(2*Pi)*2^(1/5)*GAMMA(16/5)):
  termJ:=taylor(termJ,dprime=0,3):
  termJ:=normal(simplify(termJ,GAMMA,trig));
  termJ:=subs(C1^2=C12omega1,termJ);
  termJ:=subs(C1=C1omega0,termJ);
  termJ:=normal(simplify(termJ));
 else
  ERROR(`invalid argument`,ou):
 fi:
end:
```

First, in lines **91-112** the kernel of the integral `expr` is processed. All variables are expressed in terms of `epsilon`, `C`, `c12`, `CDOTc12`, `CDOTe`, `c12DOTe`, `a2`. If we deal with viscoelastic particles (`epstype`=1), the coefficient of restitution is a function of the relative velocity. It is expressed in our variables by (23), where the coefficients δ' describes the material properties of the viscoelastic particles and C_1 is a numerical constant, see (21). Therefore, for this case ε has to be replaced by an expression in our variables which is done in lines **116-117**. We introduce a new internal variable `absc12DOTe5`$\equiv \left|\vec{c}_{12}\cdot\vec{e}\right|^{1/5}$.

The entire kernel of the integral is then in line **121** represented as a sum of polynomials in the variables `C`, `c12`, `CDOTc12`, `CDOTe`, `c12DOTe`, `a2`, `epsilon` (for $\varepsilon = \text{const.}$), and `absc12DOTe5` (for viscoelastic particles). These are precisely the variables which define our basic integrals for which we know the solution (see (65)–(67)). For any of these terms we determine in the loop (lines **125-142**) the indices of the according basic integral. The terms are then successively picked out of the sum by line **126** and stored in `term`. The next line determines the coefficients of the variables in this term, they are stored temporarily in the list `coeffs`. In lines **128-134** the powers of the variables in the present term are determined, i.e., `epow` is the power of `epsilon` in `|term|`,`aa` is the power of `absc12DOTe5`, etc. These powers determine the indices of the basic integral. In this way all terms of the sum are transformed in line **136** into expressions of the basic integrals and the according prefactors. By processing the terms in the loop the variable `termJ` grows and finally it contains the full sum in terms of basic integrals.

The vector `di[]` in line **137** is for testing purposes. It reconstructs the present term by using the powers of the variables as computed above and its prefactors. If the reconstructed term differs from the original term (line **139**) an error has occurred. This may happen by an typing error for the kernel of the integral. The program then reports the error and the number of the concerned term.

In line **143** the integral is completely expressed in terms of basic integrals. There are different output modes, specified by the third argument `ou` of the program `getJexpr()`. If `ou=0` the output is given in symbolic form, using the Gamma-function and its properties. The result is analytically exact, however, the output may become very large. For `ou=1` the result is given by a floating point number. Before, the numerical constant C_1 is replaced in line **146** by its analytical expression (22). For the choice `ou=2` the result is given as a Taylor expansion using the numerical constants $\omega_{1/2}$, which make the expression more compact. These constants read:

$$\omega_0 \equiv 2\sqrt{2\pi}2^{1/10}\Gamma\left(\frac{21}{10}\right)C_1 \approx 6.485\,, \qquad \omega_1 \equiv \sqrt{2\pi}2^{1/5}\Gamma\left(\frac{16}{5}\right)C_1^2 \approx 9.285\,. \tag{71}$$

Finally the defined commands have to be bundled to a package and stored with the name `KineticIntegral.m`. This is done by the last lines of the program:

```
KineticIntegral:=table():
KineticIntegral[KIinit]:=eval(KIinit):
KineticIntegral[DefJ]:=eval(DefJ):
KineticIntegral[unDefJ]:=eval(unDefJ):
KineticIntegral[getJexpr]:=eval(getJexpr):
KineticIntegral[DefDimension]:=eval(DefDimension):
KineticIntegral[DefS]:=eval(DefS):
save KineticIntegral, `KineticIntegral.m`;
```

The presented program generates the maple library `KineticIntegral.m`. To evaluate integrals of the type (68) the library can be invoked and used as demonstrated in the example in Sect. 2.

6 Kinetic Integrals in the Kinetic Theory of Granular Gases

6.1 Homogeneous Cooling State

Temperature Decay and Velocity Distribution Function for the Simplified Collision Model (ε = const.)

In the first stage of the evolution of an initially homogeneous gas of dissipative particles the gas preserves its homogeneity, however, its temperature decays persistently. This state of the evolution is called the homogeneous cooling state. In this state the velocity distribution function does not depend on the space coordinate $\vec{r}$. Hence, the homogeneous cooling state is characterized by the gas temperature T as defined by (6).

In our introducing example in Sect. 2 we have derived the evolution of temperature of a Granular Gas. The deviation of the velocity distribution function from the Maxwell distribution is characterized by the Sonine coefficients a_i (see (11)). If the deviations from the Maxwell distributions are small only the first nonvanishing term in the expansion (11) contributes considerably to the distribution function, i.e. we use approximation (13). For the simplified collision model (ε = const.) the second Sonine coefficient a_2 is related with the moments of the collision integral by [4, 7]

$$5\mu_2 (1 + a_2) - \mu_4 = 0 \,. \tag{72}$$

This equation is solved by the program below:

```
DefDimension(3);
DefJ();
Expr:=Delta2*(-1/2*(1+a2*(S(2,c1p)+S(2,c2p)) + a2^2*S(2,c1p)*S(2,c2p))):
mu2:=getJexpr(0,Expr,0);
Expr:=Delta4*(-1/2*(1+a2*(S(2,c1p)+S(2,c2p)) + a2^2*S(2,c1p)*S(2,c2p))):
mu4:=getJexpr(0,Expr,0);
A2:=solve(5*mu2*(1+a2)-mu4=0,a2):
AA2:=unapply(Re(A2[2]),epsilon):
A2NE:=epsilon->(16*(1-epsilon)*(1-2*epsilon^2))
  /(81-17*epsilon+30*epsilon^2*(1-epsilon)):
plot([AA2,A2NE],0..1,linestyle=[1,4],
 legend=["nonlin.theory","lin.theory (v. Noije, Ernst)"],
 font=[COURIER,20],thickness=2,labels=[e,a2],color=black);
```

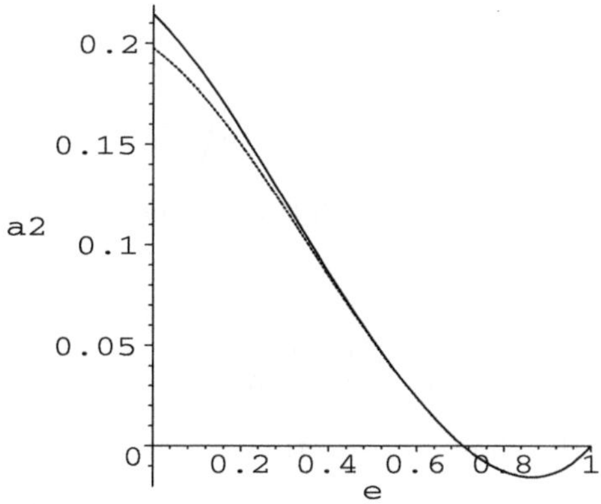

Fig. 1. Second Sonine coefficient a_2 over the coefficien of restitution ε. *Full line:* solution of (72) (nonlinear theory [4]), *dashed line:* linear solution of (72), i.e. only linear terms with respect to a_2 are kept in this equation [7]

As a result we obtain the second Sonine coefficient as a function of the coefficient of restitution as drawn in Fig. 1. The full analytic expression for $a_2(\varepsilon)$ can be found in [4].

Temperature Decay and Velocity Distribution Function for Gases of Viscoelastic Particles

For Granular Gases of viscoelastic particles under the adiabatic approximation [18] the cooling rate and the second Sonine coefficient are computed in the same way as for $\varepsilon = \text{const}$. The according program reads

```
1 DefDimension(3):
2 DefJ():
3 Expr:=Delta2*(-1/2*(1+a2*(S(2,c1p)+S(2,c2p)) + a2^2*S(2,c1p)*S(2,c2p))):
4 mu2:=getJexpr(1,Expr,2);
5 Expr:=Delta4*(-1/2*(1+a2*(S(2,c1p)+S(2,c2p)) + a2^2*S(2,c1p)*S(2,c2p))):
6 mu4:=getJexpr(1,Expr,2):
7 A2:=solve(5*mu2*(1+a2)-mu4=0,a2);
```

which yields μ_2 on line **5** in terms of a_2 and then a_2 on line **9**. Note that on line **9** we need to take a physical root $a_2 \propto \delta'$. The result for a_2 reads

$$a_2 = a_{21}\delta' + a_{22}{\delta'}^2 + \ldots \tag{73}$$

$$a_{21} = -\frac{3\omega_0}{20\sqrt{2\pi}} \approx -0.388\,, \qquad a_{22} = \frac{12063}{640000}\frac{\omega_0^2}{\pi} + \frac{27}{40}\frac{\omega_1}{\sqrt{2\pi}} \approx 2.752\,. \tag{74}$$

Correspondingly the cooling coefficient, expressed via μ_2 (see (7)) reads

$$\zeta = \frac{2}{3}\sqrt{\frac{2T}{m}}n\sigma^2\left(\omega_0\delta' - \omega_2{\delta'}^2 + \ldots\right)\,, \tag{75}$$

where

$$\omega_2 \equiv \omega_1 + \frac{9}{500}\omega_0^2\sqrt{\frac{2}{\pi}}, \tag{76}$$

which yields for the temperature evolution in linear approximation with respect to δ

$$\frac{T(t)}{T_0} = \left(1 + \frac{t}{\tau_0}\right)^{-5/3} \tag{77}$$

where we introduce the characteristic time

$$\tau_0^{-1} = \frac{16}{5} q_0 \delta \tau_c(0)^{-1} = \frac{48}{5} q_0 \delta\, 4\sigma^2 n \sqrt{\frac{\pi T_0}{m}}, \tag{78}$$

with the initial mean collision time $\tau_c(0)$ and the constant

$$q_0 = 2^{1/5} \Gamma\left(\frac{21}{10}\right) \frac{C_1}{8} \approx 0.173\,. \tag{79}$$

The evaluation of a_2 requires already the solution of 786 basic integrals. The above results for the temperature evolution and second Sonine coefficient have been reported in [4, 5].

Coefficient of Self-Diffusion for the Simplified Collision Model ($\varepsilon =$ const.)

If we assume Molecular Chaos in the homogeneous cooling state the velocity correlation function

$$\langle \vec{v}(t)\vec{v}(t')\rangle = \langle \vec{v}(t')^2\rangle \exp\left(-\frac{|t-t'|}{\tau_v}\right) \tag{80}$$

is related with the coefficient of self-diffusion

$$D(t) = \frac{T}{m}\tau_v(t) \tag{81}$$

via the velocity relaxation time τ_v. This quantity reads [19]

$$\begin{aligned} \tau_v^{-1}(t) &= \frac{1}{12\sqrt{2\pi}}\tau_c^{-1}(t) \\ &\times \int d\vec{c}_1 d\vec{c}_2 \int d\vec{e}\, \Theta\left(-\vec{c}_{12}\cdot\vec{e}\right)\left|\vec{c}_{12}\cdot\vec{e}\right| \tilde{f}\left(\vec{c}_1,t\right)\tilde{f}\left(\vec{c}_2,t\right)\left(1+\varepsilon\right)\left(\vec{c}_{12}\cdot\vec{e}\right)^2 . \end{aligned} \tag{82}$$

Again we recognize the kinetic integral in (82) which allows for the application of the Maple package. Using the program

```
DefDimension(3);
unDefJ();
Expr:=tau_c_inv/(12*sqrt(2*Pi))*(1+epsilon)*c12DOTe*c12DOTe*(1+a2*S(2,c1p))
      *(1+a2*+S(2,c2p))):
DefJ();
tau_v_inv:=getJexpr(0,Expr,0);
```

we obtain the velocity relaxation time

$$\tau_v^{-1}(t) = \frac{(1+\varepsilon)}{3}\left(1+\frac{3a_2}{32}\right)^2 \tau_c^{-1}(t) \tag{83}$$

which yields with (81) the self-diffusion coefficient

$$D(t) = \frac{2D_0(t)}{(1+\varepsilon)\left(1+\frac{3}{32}a_2\right)^2}, \tag{84}$$

where

$$D_0^{-1}(t) = \frac{8}{3}\sigma^2 n \sqrt{\frac{\pi m}{T(t)}}. \tag{85}$$

is the Enskog self-diffusion coefficient of the elastic gas.

The self-diffusion coefficient in the linear approximation with respect to a_2 has been computed for the simplified collision model in [20].

Coefficient of Self-Diffusion for Gases of Viscoelastic Particles

A similar analysis can be performed for gases of viscoelastic particles:

```
1  DefDimension(3):
2  DefJ():
3  Expr:=(1+epsilon)*c12DOTe*c12DOTe*(1+a2*(S(2,c1p)+S(2,c2p))+a2^2*S(2,c1p)*S(2,c2p)):
4  hattauV:= getJexpr(1,Expr,2):
5  tauVinv:=1/(2*DIM)*sqrt(2*T/m)*n*sigma^(DIM-1)*hattauV:
6  Expr:=Delta2*(-1/2*(1+a2*(S(2,c1p)+S(2,c2p))+a2^2*S(2,c1p)*S(2,c2p))):
7  mu2:=getJexpr(1,Expr,2):
8  Expr:=Delta4*(-1/2*(1+a2*(S(2,c1p)+S(2,c2p))+a2^2*S(2,c1p)*S(2,c2p))):
9  mu4:=getJexpr(1,Expr,2):
10 tmpA2:={solve(mu4=mu2/DIM*4*DIM/4*(DIM+2)*(1+a2),a2)}:
11 a2:=tmpA2[1]:
12 a2:=convert(a2,polynom):
13 tauVinv:=convert(taylor(tauVinv,dprime,3),polynom):
14 Dvis:=T/m/tauVinv:
15 DEnsk:=DIM*GAMMA(DIM/2)*sqrt(T/m) / (4*Pi^((DIM-1)/2)*n*sigma^(DIM-1)):
16 simplify(convert(taylor(Dvis/DEnsk,dprime,3),polynom));
```

At line **14** we need to take a root $a_2 \propto \delta'$. The coefficient of self-diffusion for gases of viscoelastic particles in linear approximation with respect to a_2 has been reportet in [19]. Here we give the complete solution with respect to a_2 which becomes feasible due to symbolic algebra:

$$D(t) = D_0(t)\left[1+\frac{89}{640}\sqrt{\frac{2}{\pi}}\omega_0\,\delta' + \left(\frac{625537}{20480000}\frac{\omega_0^2}{\pi} - \frac{1851}{14080}\sqrt{\frac{2}{\pi}}\omega_1\right)\delta'^2 + \cdots\right] \tag{86}$$

6.2 Inhomogeneous Granular Gases

Transport Coefficients

So far we have considered Granular Gases in the homogeneous cooling state. To describe the non-homogeneous gas a number of perturbative approaches has been developed. The most successive are the Chapman-Enskog approach [1, 2] and the Grad method [21]; here we consider the former one.

In the Chapman-Enskog approach it is assumed that (after a short relaxation time) the time and space dependence of the velocity distribution function $f(\vec{r}, \vec{v}, t)$ occurs only via density $n(\vec{r}, t)$, flow velocity $\vec{u}(\vec{r}, t)$ and temperature $T(\vec{r}, t)$. These quantities are, respectively, the zeroth, first and second order moments of the velocity distribution function; $n(\vec{r}, t)$ and $\vec{u}(\vec{r}, t)$ are expressed in terms of $f(\vec{r}, \vec{v}, t)$ in the same way is temperature in (6) [1, 2]. These three fields evolve according to the hydrodynamic equations (e.g. [13, 18]),

$$
\begin{aligned}
\frac{\partial n}{\partial t} + \vec{\nabla} \cdot (n\vec{u}) = 0 \\
\frac{\partial \vec{u}}{\partial t} + \vec{u} \cdot \vec{\nabla}\vec{u} + (nm)^{-1}\vec{\nabla} \cdot \hat{P} = 0 \\
\frac{\partial T}{\partial t} + \vec{u} \cdot \vec{\nabla}T + \frac{2}{3n}\left(P_{ij}\nabla_j u_i + \vec{\nabla} \cdot \vec{q}\right) + \zeta T = 0 \, ,
\end{aligned}
\tag{87}
$$

where $\hat{P}$ denotes the pressure tensor and $\vec{q}$ is the vector of the heat flux. In the Navier-Stokes hydrodynamics the pressure tensor and the heat flux depend linearly on the field gradients:

$$
\begin{aligned}
P_{ij} &= p\delta_{ij} - \eta\left(\nabla_i u_j + \nabla_j u_i - \frac{2}{3}\delta_{ij}\vec{\nabla} \cdot \vec{u}\right) \\
\vec{q} &= -\kappa\vec{\nabla}T - \mu\vec{\nabla}n \, ,
\end{aligned}
\tag{88}
$$

with the hydrostatic pressure $p = nT$ in the dilute limit. Here η is the viscosity coefficient, κ is the coefficient of thermal conductivity and coefficient μ characterises the heat flux due to the density gradients and does have an analogue for elastic gases.

The transport coefficients η, κ, μ may be calculated using the Chapman-Enskog approach adopted for the dissipative gases. The most complete analysis of this problem for the gas with a simplified collisional model with a constant ε have been performed by Brey and coauthors [22, 23]. They have expressed transport coefficients for the dissipative gases in terms of functions of the collision integral. The generalization of this approach for the dissipative gases with the velocity-dependent restitution coefficient has been done in [18], where it has been shown that the transport coefficient may be expressed in terms of the same functions of the collision integral as for the case of $\varepsilon = \text{const}$.

Transport Coefficients for the Simplified Collision Model

Using the notations of [18] the viscosity coefficient reads for the case of $\varepsilon =$ const. (see also [22]),

$$\eta = -\frac{1}{\sigma^2}\sqrt{\frac{mT}{2}}\frac{1}{\frac{1}{3}\mu_2 + \frac{2}{5}\Omega_\eta} \,. \tag{89}$$

where μ_2 has been found above (see (16)) and with the function of the collision integral Ω_η

$$\Omega_\eta \equiv \int d\vec{c}_1 \int d\vec{c}_2 \int d\vec{e}\, \Theta\left(-\vec{c}_{12}\cdot\vec{e}\right)\left|\vec{c}_{12}\cdot\vec{e}\right| \tilde{f}^{(0)}(c_1) \times \\ \times\, \phi(c_2) D_{ij}(\vec{c}_2)\Delta\left[D_{ij}(\vec{c}_1) + D_{ij}(\vec{c}_2)\right] \,, \tag{90}$$

where $D_{ij}\left(\vec{c}\right)$ is the traceless tensor,

$$D_{ij}\left(\vec{c}\right) \equiv c_i c_j - \frac{1}{3}\delta_{ij}c^2 \,. \tag{91}$$

In (90) we recognize the kinetic integral. For the simplified collision model ($\varepsilon =$ const.) it is solved by the program

```
DefDimension(3);
DefJ();
Expr :=(-1/DIM*c2pDOTc2p*Delta2+c1aDOTc2p*c1aDOTc2p+c2aDOTc2p*c2aDOTc2p
       -c2pDOTc2p*c2pDOTc2p-c1pDOTc2p*c1pDOTc2p)*(1+a2*S(2,c1p)):
OmegaETA:=getJexpr(0,Expr,0);
```

yielding

$$\Omega_\eta = -\sqrt{2\pi}\,(1+\varepsilon)(3-\varepsilon)\left(1 - \frac{a_2}{32}\right) \,, \tag{92}$$

Similarly the coefficient of thermal conductivity may be expressed as

$$\kappa = -\frac{15}{8\sigma^2 g_2(\sigma)}\sqrt{\frac{T}{2m}}\frac{1+2a_2}{\mu_2 + \frac{1}{5}\Omega_\kappa} \,, \tag{93}$$

with the function Ω_κ defined as

$$\Omega_\kappa \equiv \int d\vec{c}_1 \int d\vec{c}_2 \int d\vec{e}\, \Theta\left(-\vec{c}_{12}\cdot\vec{e}\right)\left|\vec{c}_{12}\cdot\vec{e}\right| \tilde{f}^{(0)}(c_1) \times \\ \times\, \phi(c_2)\vec{S}\left(\vec{c}_2\right)\cdot\Delta\left[\vec{S}\left(\vec{c}_1\right) + \vec{S}\left(\vec{c}_2\right)\right] \,, \tag{94}$$

and with the vectorial quantity,

$$\vec{S}(\vec{c}) \equiv \left(c^2 - \frac{5}{2}\right)\vec{c}. \tag{95}$$

The coefficient μ may be also written in terms of Ω_κ:

$$\mu = \frac{5}{2n\sigma^2}\sqrt{\frac{T}{2m}}\frac{a_2}{\mu_2 + \frac{4}{15}\Omega_\kappa} . \tag{96}$$

The function Ω_κ is computed by the program

```
DefDimension(3);
DefJ();
Expr:=(c2pDOTc2p-(DIM+2)/2)*(c1aDOTc2p*c1aDOTc1a+c2aDOTc2p*c2aDOTc2a
      -c2pDOTc2p*c2pDOTc2p-c1pDOTc2p*c1pDOTc1p)*(1+a2*S(2,c1p)):
OmegaKAPPA:=getJexpr(0,Expr,0);
```

We obtain

$$\Omega_\kappa = -\sqrt{2\pi}(1+\varepsilon)\left(\frac{49-33\varepsilon}{8} + \frac{19-3\varepsilon}{256}a_2\right) . \tag{97}$$

In linear approximation with respect to a_2 the transport coefficients have been reported in [22]. Substituting (92,97,16) into (89,93,96) we obtain the generalized result for the coefficients η, κ and μ, which accounts for the terms of the next order, $\mathcal{O}\left(a_2^2\right)$.

Transport Coefficients for Granular Gases of Viscoelastic Particles

The same function Ω_η appears in the theory of Granular Gases of viscoelastic particles [18]. We evaluate this function by

```
DefDimension(3);
DefJ();
Expr:=(-1/DIM*c2pDOTc2p*Delta2+c1aDOTc2p*c1aDOTc2p+c2aDOTc2p*c2aDOTc2p
      -c2pDOTc2p*c2pDOTc2p-c1pDOTc2p*c1pDOTc2p)*(1+a2*S(2,c1p)):
OmegaETA:=getJexpr(1,Expr,2);
```

and obtain the corresponding coefficient

$$\Omega_\eta^{\mathrm{vis}} = -\left(w_0 + \delta' w_1 - {\delta'}^2 w_2\right) \tag{98}$$

with

$$\begin{aligned} w_0 &= 4\sqrt{2\pi}\left(1 - \frac{1}{32}a_2\right) \\ w_1 &= \omega_0\left(\frac{1}{15} - \frac{1}{500}a_2\right) \\ w_2 &= \omega_1\left(\frac{97}{165} - \frac{679}{44000}a_2\right) . \end{aligned} \tag{99}$$

Similarly, by

```
DefDimension(3);
DefJ();
Expr:=(c2pDOTc2p-(DIM+2)/2)*(c1aDOTc2p*c1aDOTc1a+c2aDOTc2p*c2aDOTc2a
      -c2pDOTc2p*c2pDOTc2p-c1pDOTc2p*c1pDOTc1p)*(1+a2*S(2,c1p)):
OmegaKAPPA:=getJexpr(1,Expr,2);
```

we obtain the other function

$$\Omega_\kappa^{\text{vis}} = -\left(u_0 + \delta' u_1 - {\delta'}^2 u_2\right) \tag{100}$$

with

$$\begin{aligned} u_0 &= 4\sqrt{2\pi}\left(1 + \frac{1}{32}a_2\right) \\ u_1 &= \omega_0\left(\frac{17}{5} - \frac{9}{500}a_2\right) \\ u_2 &= \omega_1\left(\frac{1817}{440} - \frac{1113}{352000}a_2\right) . \end{aligned} \tag{101}$$

For the gas of viscoelastic particles the viscosity coefficient η is found as the solution of the equation [18]

$$-\zeta T \frac{\partial \eta}{\partial T} = \frac{2}{5}\eta n \sigma^2 \sqrt{\frac{2T}{m}}\,\Omega_\eta^{\text{vis}} + nT\,, \tag{102}$$

where ζ is given by (75). We solve this equation as a perturbative expansion with respect to the dissipative parameter δ',

$$\eta = \eta_0\left(1 + \delta'\tilde{\eta}_1 + {\delta'}^2\tilde{\eta}_2 + \cdots\right) . \tag{103}$$

using the program

```
DefDimension(3):
DefJ():
Expr:=Delta2*(-1/2*(1+a2*(S(2,c1p)+S(2,c2p))+a2^2*S(2,c1p)*S(2,c2p))):
mu2:=getJexpr(1,Expr,2):
mu2:=convert(mu2,polynom):
Expr:=Delta4*(-1/2*(1+a2*(S(2,c1p)+S(2,c2p))+a2^2*S(2,c1p)*S(2,c2p))):
mu4:=getJexpr(1,Expr,2):
tmpA2:={solve(mu4=mu2/DIM*4*DIM/4*(DIM+2)*(1+a2),a2)}:
a2:=tmpA2[1]:
a2:=convert(a2,polynom):
xi0:=2*sigma^(DIM-1)/DIM*n*mu2*sqrt(2*T/m):
ExprETA:=(-1/DIM*c2pDOTc2p*Delta2+c1aDOTc2p*c1aDOTc2p+c2aDOTc2p*c2aDOTc2p
 -c2pDOTc2p*c2pDOTc2p-c1pDOTc2p*c1pDOTc2p)*(1+a2*S(2,c1p)):
OmegaETA:=getJexpr(1,ExprETA,2):
OmegaETA:=convert(OmegaETA,polynom):
EqETA:=xi0*eta0*(1/2+3/5*dprime*eta1+7/10*dprime^2*eta2)
 +4/(DIM-1)/(DIM+2)*eta0*(1+dprime*eta1
 +dprime^2*eta2)*n*sigma^(DIM-1)*sqrt(2*T/m)*OmegaETA+n*T:
EqETA:=taylor(EqETA,dprime,3):
eta0:=solve(coeff(EqETA,dprime,0),eta0);
eta1:=solve(coeff(EqETA,dprime,1),eta1);
eta2:=solve(coeff(EqETA,dprime,2),eta2);
omega1:=9.28569:
```

```
omega0:=6.48562:
evalf(eta1);
evalf(eta2);
etaVIS:=eta0*(1+dprime*eta1+dprime^2*eta2);
plot(etaVIS/eta0,dprime=0..0.5,color=black,linestyle=[1]);
```

The result reads

$$\eta = \frac{5}{16\sigma^2}\sqrt{\frac{mT}{\pi}}\left(1 + 0.483\,\delta' + 0.094\,\delta'^2 + \ldots\right) . \qquad (104)$$

Similarly, the coefficient of thermal conductivity may be obtained from the equation [18],

$$-3\frac{\partial}{\partial T}\zeta\kappa T = \frac{4}{5}\kappa n\sigma^2\sqrt{\frac{2T}{m}}\Omega_\kappa^{\rm vis} + \frac{15}{4}\frac{nT}{m}\left(1 + \frac{21}{10}a_{21}\delta' + \frac{11}{5}a_{22}\delta'^2\right) , \qquad (105)$$

which is also solved perturbatively as an expansion

$$\kappa = \kappa_0\left(1 + \delta'\tilde{\kappa}_1 + \delta'^2\tilde{\kappa}_2 + \cdots\right) . \qquad (106)$$

The result is found by the Maple program

```
DefDimension(3):
DefJ():
ExprKAPPA:=(c2pDOTc2p-(DIM+2)/2)*(c1aDOTc2p*c1aDOTc1a+c2aDOTc2p*c2aDOTc2a
 -c2pDOTc2p*c2pDOTc2p-c1pDOTc2p*c1pDOTc1p)*(1+a2*S(2,c1p)):
OmegaKAPPA:=getJexpr(1,ExprKAPPA,2):
OmegaKAPPA:=convert(OmegaKAPPA,polynom):
Expr:=Delta2*(-1/2*(1+a2*(S(2,c1p)+S(2,c2p))+a2^2*S(2,c1p)*S(2,c2p))):
mu2:=getJexpr(1,Expr,2):
Expr:=Delta4*(-1/2*(1+a2*(S(2,c1p)+S(2,c2p))+a2^2*S(2,c1p)*S(2,c2p))):
mu4:=getJexpr(1,Expr,2):
mu2:=convert(mu2,polynom):
tmpA2:={solve(mu4=mu2/DIM*4*DIM/4*(DIM+2)*(1+a2),a2)}:
a2:=tmpA2[1]:
a2:=convert(a2,polynom):
mu2:=convert(mu2,polynom):
xi0:=2*sigma^(DIM-1)/DIM*n*mu2*sqrt(2*T/m):
xi01:=coeff(xi0,dprime,1)/(2/DIM*sqrt(2*T/m)*sigma^(DIM-1)*n):
xi02:=-coeff(xi0,dprime,2)/(2/DIM*sqrt(2*T/m)*sigma^(DIM-1)*n):
Tdxi0dT:=2/DIM*sqrt(2*T/m)*sigma^(DIM-1)*n*(3/5*xi01*dprime-7/10*xi02*dprime^2):
a21:=coeff(a2,dprime,1):
a22:=coeff(a2,dprime,2):
Tda2dT:=1/10*a21*dprime+1/5*a22*dprime^2:
OmegaKAPPA:=convert(taylor(OmegaKAPPA,dprime,3),polynom):
EqKAPPA:=DIM*(kappa*Tdxi0dT+xi0*TdkappadT+xi0*kappa)
 +4/(DIM+2)*kappa*n*sigma^(DIM-1)*sqrt(2*T/m)*OmegaKAPPA
 +(DIM+2)*DIM/4*n*2*T/m*(1+2*a2+Tda2dT):
kappa:=kappa0*(1+dprime*kappa1+dprime^2*kappa2):
TdkappadT:=kappa0*(1/2+3/5*dprime*kappa1+7/10*dprime^2*kappa2):
EqKAPPA:=taylor(EqKAPPA,dprime,3):
kappa0:=solve(coeff(EqKAPPA,dprime,0),kappa0);
kappa1:=solve(coeff(EqKAPPA,dprime,1),kappa1);
kappa2:=solve(coeff(EqKAPPA,dprime,2),kappa2);
omega1:=9.28569:
omega0:=6.48562:
evalf(kappa1);
evalf(kappa2);
plot(kappa/kappa0,dprime=0..0.1,color=black,linestyle=[1]);
```

Thus we found the coefficient of thermal conductivity

$$\kappa = \frac{75}{64\sigma^2}\sqrt{\frac{T}{\pi m}}\left(1 + 0.393\delta' + 4.904\delta'^2 + \ldots\right) . \tag{107}$$

Completely analogously we obtain the remaining transport coefficient μ:

$$\mu = \frac{75}{64\sigma^2 n}\sqrt{\frac{T^3}{\pi m}}\left(1.229\,\delta' + 1.415\,\delta'^2 + \ldots\right) \tag{108}$$

7 Conclusion

We have reported a method for the analytical evaluation of functions of the collision integral, which we call kinetic integrals, as they are encountered in the kinetic theory of Granular Gases. This method is based on computational symbolic programming, which operates with a set of known basic integrals. To illustrate the application of this method we evaluate various properties of the Granular Gases, ranging from the moments of the velocity distribution function to the transport coefficients. Two different models of the Granular Gases are considered: The simplified model, where particles collide with a constant coefficient of restitution, and the more realistic model where the restitution coefficient depends on the impact velocity, as it follows from the collision law for viscoelastic particles. We demonstrate that evaluation of some kinetic integrals may not in practice be done manually, while the reported method allows to obtain analytical expressions for these quantities. In the examples we generalize some previously known results for the kinetic properties of the dissipative gases.

References

1. S. Chapman and T.G. Cowling. *The Matematical Theory of Non-uniform Gases.* Cambridge University Press, New York, 1970
2. J. Ferziger and H. Kaper. *The Matematical Theory of Transport Processes in Gases.* North-Holland, Amsterdam, 1972
3. T. Pöschel and S. Luding, editors. *Granular Gases*, volume 564 of *Lecture Notes in Physics.* Springer, Berlin, 2001
4. N.V. Brilliantov and T. Pöschel. Deviation from Maxwell distribution in granular gases with constant restitution coefficient. *Phys. Rev. E*, 61:2809, 2000
5. N.V. Brilliantov and T. Pöschel. Velocity distribution of granular gases of viscoelastic particles. *Phys. Rev. E*, 61:5573, 2000
6. S.E. Esipov and T. Pöschel. The granular phase diagram. *J. Stat. Phys.*, 86:1385, 1997
7. T.P.C. van Noije and M. H. Ernst. Velocity distributions in homogeneous granular fluids: the free and the heated case. *Granular Matter*, 1:57, 1998
8. F.G. Bridges, A. Hatzes, and D.N.C. Lin. Structure, stability and evolution of Saturn's rings. *Nature*, 309:333, 1984

9. R. Ramírez, T. Pöschel, N.V. Brilliantov, and T. Schwager. Coefficient of restitution for colliding viscoelastic spheres. *Phys. Rev. E*, 60:4465, 1999
10. H. Hertz. Über die Berührung fester elastischer Körper. *J. f. reine u. angewandte Math.*, 92:156, 1882
11. N.V. Brilliantov, F. Spahn, J.-M. Hertzsch, and T. Pöschel. Model for collisions in granular gases. *Phys. Rev. E*, 53:5382, 1996
12. T. Schwager and T. Pöschel. Coefficient of restitution of viscous particles and cooling rate of granular gases. *Phys. Rev. E*, 57:650, 1998
13. N.V. Brilliantov and T. Pöschel. Hydrodynamics of granular gases of viscoelastic particles. *Phil. Trans. R. Soc. Lond. A*, 360:415, 2001
14. T. Schwager. *Computersimulation granularer Gase und dichter granularer Systeme in Starrkörpernäherung.* PhD thesis, Humboldt-Universität, Berlin, 2002. Logos Verlag
15. T. Schwager. Coefficient of restitution for viscoelastic disks. preprint, 2003
16. P. Resibois and M. de Leener. *Classical Kinetic Theory of Fluids.* Wiley & Sons, New York, 1977
17. The package as well as the examples can be obtained at `http://bioinf.charite.de/kies/GGD/KineticIntegrals/`
18. N.V. Brilliantov and T. Pöschel. Hydrodynamics and transport coefficients for dilute granular gases. *Phys. Rev. E*, in press, 2003
19. N.V. Brilliantov and T. Pöschel. Self-diffusion in granular gases. *Phys. Rev. E*, 61:1716, 2000
20. J.J. Brey, M.J. Ruiz-Montero, D. Cubero, and R. Garcia-Rojo. Self-diffusion in freely evolving granular gases. *Physics of Fluids*, 12:876–883, 2000
21. H. Grad. *Theory of Rarified Gases in Rarified Gas Dynamics.* Pergamon Press, New York, 1960
22. J.J. Brey, J.W. Dufty, C.S. Kim, and A. Santos. Hydrodynamics for granular flow at low density. *Phys. Rev. E*, 58:4638, 1998
23. J.J. Brey and D. Cubero. Hydrodynamic transport coefficients of granular gases. In Pöschel and Luding [3], page 59

Kinetics of Fragmenting Freely Evolving Granular Gases

Ignacio Pagonabarraga[1] and Emmanuel Trizac[2]

[1] Departament de Física Fonamental, Universitat de Barcelona
Carrer Martí i Franqués, 1, 08028-Barcelona, Spain
[2] Laboratoire de Physique Théorique (UMR 8627 du CNRS), Bâtiment 210,
Université de Paris-Sud, 91405 Orsay Cedex, France

Abstract. We investigate the effect of fragmentation on the homogeneous free cooling of inelastic hard spheres, using Boltzmann kinetic theory and Direct Monte Carlo simulations. We analyze in detail a model where dissipative collisions may subsequently lead to a break-up of the grains. With a given probability, two offsprings are then created from one of the two colliding partners, with conservation of mass, momentum and kinetic energy. We observe a scaling regime characterized by a single collisional average, that quantifies the deviations from Gaussian behaviour for the joint size and velocity distribution function. We also discuss the possibility of a catastrophe whereby the number of particles diverges in a finite time. This phenomenon appears correlated to a "shattering" transition marked by a delta singularity at vanishingly small grains for the rescaled size distribution.

1 Introduction

The phenomenon of fragmentation is central to many natural and technological processes, from geology to the pharmaceutical industry (where e.g. the properties of an active principle depend on the size distribution of its constituents). In the realm of granular gases where the fluidization is achieved by a violent excitation or shaking, the fragmentation mechanism is also relevant although experimental studies of these systems usually avoid such a complication.

A theoretical description of fragmentation is nevertheless a difficult task and traditional approaches rely on rate (Smoluchowski-like) equations [1, 2], that describe the time evolution of the size distribution after having integrated out the other degrees of freedom, such as the velocities of the particles. However, when the motion between the collision events is free, the dynamics of a reacting system in general (where, because of collisions, particles may break up, aggregate, annihilate ...) is determined by details of the collision (collisional correlations [3]) which often invalidate the rate equation approach, at least in its usual version. In particular, for dynamics that do not conserve the number of particles, the time evolution of the density crucially depends on the ratio of the kinetic energy dissipated in a "typical" collision, to the mean kinetic energy of the particles [3, 4]. It therefore appears that a more microscopic description than the coarse grained rate equation route is necessary, at the expense of a non negligible complication. When particles disappear upon colliding, it is however possible to prove that the Boltzmann equation becomes exact at late times (in space dimensions $D > 1$ [4]). Although it does not seem possible to show that the validity of Boltzmann's picture (which is less restrictive than the Smoluchowski approach) pertains to a modification of the reaction rule, such an approach captures important collisional correlations missed by the naive rate equation framework. In the following analysis, we will therefore use Boltzmann's kinetic theory (relying on the molecular chaos assumption that the 2 body distribution function of colliding pairs may be factorized in terms of the single particle distribution [5]), to investigate the dynamics of a fragmenting granular gas.

2 Model

We consider a gas of grains that evolve freely in time, from an initial equilibrium distribution, i.e. we assume that up to the initial time, the system is elastic, and hence, that its velocity distribution is Maxwellian.

In the fluidized regime we shall be interested in, grains interact as hard bodies; collisions are always binary, and instantaneous. A colliding pair with velocities $\vec{v}_1$ and $\vec{v}_2$ gives after collision the couple $(\vec{v}_1', \vec{v}_2')$:

$$\begin{aligned} \vec{v}_1' &= \vec{v}_1 - \frac{m_2}{m_1+m_2}\,(1+\alpha)\,\left[\widehat{\varepsilon}\cdot\vec{v}_{12}\right]\widehat{\varepsilon} \\ \vec{v}_2' &= \vec{v}_2 + \frac{m_1}{m_1+m_2}\,(1+\alpha)\,\left[\widehat{\varepsilon}\cdot\vec{v}_{12}\right]\widehat{\varepsilon} \end{aligned} \tag{1}$$

where $\widehat{\varepsilon}$ is a unit vector joining the centers of the colliding grains and $\vec{v}_{12} = \vec{v}_1 - \vec{v}_2$. In a D-dimensional system, at each collision a fraction $\left(1-\alpha^2\right)/(2D)$ of the relative kinetic energy is lost, where α, the restitution coefficient, measures the departure from elastic behaviour ($\alpha = 1$ corresponds to an elastic collision, whereas $\alpha = 0$ describes a completely inelastic collision). After each collision, the stored kinetic energy in internal grain excitations may lead to a break-up of the grains. Such a fragmentation event is stochastic in nature. Previous studies suggest than in a number of materials, the probability that a grain breaks is related to the number of surface defects (or micro-cracks). The more kinetic energy is dissipated, the larger the chance that one of such defects enlarges, leading to the break-up of the grain. Hence, the fracture probability is related both to the energy lost and to the defect distribution (and the related surface stress distribution). There exist different models in the literature that account for such effects. A reasonable fragmentation probability model assumes that the stored energy and the defect distribution are uncorrelated, and that defects are uniformly distributed over a particle's surface; the breaking probability is thus proportional to the particle's surface [7]. Once a grain breaks, we should also specify the number of out-coming particles, and how their sizes and velocities are related to the physical properties of the fragmenting grain. Here the situation is more controversial, and it is highly dependent on whether the breaking process is induced by an energy input from a source which drives the system or whether it is due to the internal dynamics of the system; despite this, the Rosin-Rammler law (and a few variants) is considered to be a reasonable description for the off-spring distribution [1, 7].

Rather than attempting a detailed description of the fragmentation mechanism, in this paper we will assume the simplest break-up process. The idea is to minimize the amount of input at the collision level and to analyze which is the behavior of the system under the simplest possible mechanical rules. We hope to be able to carry out a detailed analysis that will be helpful before investigating more refined approaches.

In our model, after the collision has taken place, with a probability p we break one of the two colliding grains. The grain fragments in 2 off-springs which keep the (spherical) shape of the parent grain. The fragmentation process conserves mass, momentum and kinetic energy. The mass is distributed randomly between the two out-coming grains with a uniform probability density. In this case, the conservation of momentum and kinetic energy imply that both off-springs have the same velocity, equal to that of the fragmenting grain. If we would assume that the two colliding grains may break up due to the collision, there would exist more freedom in the velocity redistribution of the off-springs, even assuming coplanarity of the out-coming grains' trajectories. We will consider that the mass density of the grains is conserved, so that the relation between mass m and radius σ is taken as $m = \sigma^D$.

We will further concentrate on the homogeneous freely evolving state of the fluid under consideration. Such a situation has received much attention in the absence of fragmentation, and the theoretical predictions have been contrasted against molecular dynamics simulations. The homogeneous state is known to be unstable against low wave length fluctuations of the vorticity, and at late times, hydrodynamic flows set up[6]. As a first step, we will concentrate in the regime prior to the appearance of the hydrodynamic instability. In fact, having a hydrodynamics origin, we might also expect an analogous instability in the present model.

In this homogeneous state, the Boltzmann equation, which determines, at a mean-field level, the time evolution of the system through the one-particle distribution function $f(\vec{v}, \sigma, t)$, in a D-dimensional space has the form

$$
\begin{aligned}
\frac{\partial f(\vec{v}, \sigma, t)}{\partial t} &= \frac{1}{2} \int du\, g(u) \int d\vec{v}_1 d\vec{v}_2 d\sigma_1 d\sigma_2 d\widehat{\varepsilon}\ \theta(\vec{v}_{12} \cdot \widehat{\varepsilon}) \left|\vec{v}_{12} \cdot \widehat{\varepsilon}\right| \sigma_{12}^{D-1} \{ \\
& f(\vec{v}_1, \sigma_1, t) f(\vec{v}_2, \sigma_2, t) [-\delta(\vec{v} - \vec{v}_1) \delta(\sigma - \sigma_1) - \delta(\vec{v} - \vec{v}_2) \delta(\sigma - \sigma_2) \\
& + (1-p) \delta(\vec{v} - \vec{v}_1') \delta(\sigma - \sigma_1) + (1-p) \delta(\vec{v} - \vec{v}_2') \delta(\sigma - \sigma_2) \\
& + p\delta(\vec{v} - \vec{v}_1') \left[\delta\left(\sigma - \sigma_1 \cdot u^{1/D}\right) + \delta\left(\sigma - \sigma_1 \cdot (1-u)^{1/D}\right)\right] \\
& + p\delta(\vec{v} - \vec{v}_2') \left[\delta\left(\sigma - \sigma_2 \cdot u^{1/D}\right) + \delta\left(\sigma - \sigma_2 \cdot (1-u)^{1/D}\right)\right]]\}
\end{aligned}
\quad (2)
$$

In the previous expression, $\vec{v}$ stands for the velocity, while σ refers to the particles' radius. The Heaviside function, θ, ensures the appropriate kinematic constraint, that only approaching grains will collide, while the term $|\vec{v}_{12} \cdot \widehat{\varepsilon}|\, \sigma_{12}^{D-1}$ is the collision cross-section; the subscript 12 refers to quantities that involve combinations of the corresponding properties of the colliding pair. For example, $\vec{v}_{12} = \vec{v}_1 - \vec{v}_2$ is the relative velocity of the pair, while $\sigma_{12} = \sigma_1 + \sigma_2$ refers to the distance between their centers. Moreover, $g\,(u)$ is the probability density that one of the off-springs, in case of fragmentation, has a mass u times the parent's mass. Due to mass conservation, $g\,(u) = g\,(1-u)$. In the model analyzed below, we consider the simplest case where this distribution is uniform $[g\,(u) = 1$ and $0 \le u \le 1]$.

For clarity's sake, we have expressed the general evolution equation in terms of the elementary processes affecting the two colliding grains. Obviously, a simplified expression can be derived using the symmetric role that the two grains play during the collision.

The first two terms in the r.h.s. of (2) account for the disappearance of particles of a given velocity due to the collision process itself. For the same reason, new particles of a given species and velocity are produced at the collision. These terms involve the postcollision velocities, $\vec{v}\,'$, because we require that the out-coming velocity of the collision corresponds to the velocity $\vec{v}$. The collision law has been given in (1), which encodes the inelasticity of the process. The two terms in (2) containing the prefactor $1-p$ correspond to the collisions in which, with probability $1-p$, particles do not break. The last four terms account for the fragmentation of grain 1 with mass $m_1\ (= \sigma_1^D)$ (resp. grain 2 with mass m_2), leading to the creation of two off springs with masses um_1 and $(1-u)\,m_1$ [resp. um_2 and $(1-u)\,m_2$]. Mass conservation implies a correlation between the sizes of the newborn particles.

3 Kinetics

A detailed analysis of the dynamics of the system, already at the Boltzmann level, according to (2), is quite involved. However, the key information is already present in the time evolution of the total density and kinetic energy densities, defined respectively as

$$n\,(t) = \int f\,(\vec{v}, \sigma, t)\; d\vec{v} d\sigma \tag{3}$$

$$E_c\,(t) = \int v^2 \sigma^D f\,(\vec{v}, \sigma, t)\; d\vec{v} d\sigma. \tag{4}$$

The total mass density

$$M = \int \sigma^D f\,(\vec{v}, \sigma, t)\; d\vec{v} d\sigma \tag{5}$$

is a conserved quantity and we define the temperature kinetically as

$$k_B T = \frac{2E_c}{Dn} = \frac{\langle mv^2 \rangle}{D}. \tag{6}$$

In the previous equation, the angular brackets of a quantity $\mathcal{Q}$ denote the mean value $\int \mathcal{Q} f(\vec{v}, \sigma, t)\, d\vec{v} d\sigma / n$.

From the Boltzmann equation, the time evolution for the density of particles reads

$$\frac{\mathrm{d}n}{\mathrm{d}t} = p\omega(t)\, n(t) \tag{7}$$

where p is the probability that a particle breaks at a collision, and $\omega(t)$ the collision frequency

$$\omega(t) = \frac{1}{n} \int d\widehat{\varepsilon}\, d1\, d2\; \sigma_{12}^{D-1} \theta(\widehat{\varepsilon} \cdot \vec{v}_{12}) |\widehat{\varepsilon} \cdot \vec{v}_{12}| f(\vec{v}_1, \sigma_1, t) f(\vec{v}_2, \sigma_2, t), \tag{8}$$

where the short-hand notation $d1\, d2$ stands for $d\vec{v}_1 d\sigma_1 d\vec{v}_2 d\sigma_2$. The kinetic energy density is unaffected by the breaking events (unlike the density itself), and only decreases due to the inelastic collisions:

$$\frac{\mathrm{d}E_c(t)}{\mathrm{d}t} = -2b_c(t)\gamma_0 \omega(t)\, E_c(t). \tag{9}$$

In this equation, $\gamma_0 \equiv (1 - \alpha^2)/(2D)$ is related to the energy lost per collision while b_c is a dimensionless collisional average

$$b_c(t) = \frac{D}{2\omega E_c} \int d\widehat{\varepsilon}\, d1\, d2\, \theta(\vec{v}_{12} \cdot \widehat{\varepsilon})\, |\vec{v}_{12} \cdot \widehat{\varepsilon}|^3 \frac{m_1 m_2}{m_1 + m_2} f(\vec{v}_1, \sigma_1, t) f(\vec{v}_2, \sigma_2, t), \tag{10}$$

where it is understood that $m_i = \sigma_i^D$. The parameter b_c is a dissipation parameter: $\gamma_0 b_c$ is the ratio of the kinetic energy dissipated in an "average" collision to the instantaneous mean kinetic energy per particle. Hence, although b_c in a general situation will be a function of time, in the scaling regime it will be a constant, as we will describe and show in the next subsection. Equation (7) expresses the fact that due to the collisions the overall number of particles in the system increases in time. In some cases, this increase is so rapid that the gas exhibits a finite-time singularity, i.e. the number of grains become infinite in a finite amount of time (see next section). The existence of such a singularity, which can be thought of as a limitation of the model, signaling that some of the assumptions are too simple, or that some physical ingredient is missing, will depend on the parameters that characterize the evolution of the system, p, b_c, α and the dimensionality D.

3.1 Scaling Behaviour

We will analyze the evolution of the free evolving gas once the transient effects induced by the particular initial condition chosen have vanished. In particular, we will concentrate in the late time regime, addressing the issue whether the existence of a scaling solution is possible, where there exists a unique typical velocity and size of the particles characterizing the overall evolution of the system.

To this end, a good way to characterize the possible kinetic scenarios, is to introduce an intrinsic time scale, $\mathcal{C}$, which counts the number of collisions [8]. This scale is related to time through, $\mathrm{d}\mathcal{C} = \omega(t)\,\mathrm{d}t$. In terms of this variable, the evolution equations (7) and (9) read

$$\begin{aligned}\frac{\mathrm{d}n}{\mathrm{d}\mathcal{C}} &= pn(t)\\ \frac{\mathrm{d}E_c}{\mathrm{d}\mathcal{C}} &= -2b_c\gamma_0 E_c\end{aligned} \tag{11}$$

with solutions

$$n(t) = n_0 \mathrm{e}^{p\mathcal{C}}\,, \qquad E_c(t) = E_{c0}\exp\left(-2b_c\gamma_0\mathcal{C}\right). \tag{12}$$

Assuming the existence of a scaling regime, the kinetics of the system is determined by one single characteristic velocity $\bar{v}(t)$, and grain size $\bar{r}(t)$. The time dependence of the distribution function $f(\vec{v},\sigma,t)$ consequently occurs through these two quantities [up to a trivial time dependence through the density resulting from the normalization constraint (3)], and the scaling assumption amounts to restricting f to the family of functions of the form:

$$f(\vec{v},\sigma,t) = \frac{n}{\bar{r}\,\bar{v}^D}\,\tilde{f}(\vec{c},\bar{\sigma}), \tag{13}$$

where we have introduced the rescaled size and velocity

$$\vec{c} = \frac{\vec{v}}{\bar{v}(t)} \quad ; \quad \bar{\sigma} = \frac{\sigma}{\bar{r}(t)}. \tag{14}$$

A natural choice is to define $\bar{v}$ as the variance of the velocity distribution and $\bar{r}$ as the mean radius: $\bar{v}^2 = \langle v^2\rangle$, $\bar{r} = \langle\sigma\rangle$.

Within the scaling ansatz, we obtain from (10) that b_c is time independent, and from (8) that ω scales with time as $n\bar{v}\bar{r}^{D-1}$ (i.e. as the ratio of typical velocity over mean free path). If we use the fact that mass density is conserved (which implies that $\bar{r}^D n \propto M \propto t^0$, expressing that mass, being a conserved quantity, does not depend on time), we know that $\bar{r} \propto 1/n^{1/D}$ and $\bar{v}^2 \propto E_c/M \propto E_c$. Combining these relations, we get for the collision frequency

$$\omega \propto \bar{v}n\bar{r}^{D-1} \propto \frac{\bar{v}}{\bar{r}} \propto n^{1/D}E_c^{1/2} \tag{15}$$

from which we can obtain a differential equation for the overall number of collisions as a function of time

$$\frac{\mathrm{d}\mathcal{C}}{\mathrm{d}t} = \omega_0 \left(\frac{n}{n_0}\right)^{\frac{1}{D}} \sqrt{\frac{E_c}{E_{c0}}} = \omega_0 \exp\left[\left(\frac{p}{D} - b_c\gamma_0\right)\mathcal{C}\right] \tag{16}$$

From this relation, we can obtain the explicit time dependence of the number of collisions,

$$\mathcal{C}(t) = \frac{1}{-\frac{p}{D} + b_c\gamma_0} \ln\left[1 + \left(-\frac{p}{D} + b_c\gamma_0\right)\omega_0 t\right] \tag{17}$$

Finally, the explicit time dependence of the mean number of particles and kinetic energy read

$$n(t) = n_0 \left[1 + \left(-\frac{p}{D} + b_c\gamma_0\right)\omega_0 t\right]^{p/(-p/D + b_c\gamma_0)} \tag{18}$$

$$E_c(t) = E_{c0} \left[1 + \left(-\frac{p}{D} + b_c\gamma_0\right)\omega_0 t\right]^{-2b_c\gamma_0/(-p/D + b_c\gamma_0)} \tag{19}$$

These expressions indicate when a finite-time singularity will develop. Due to mass conservation, the divergence in the number of particles implies that the mean size of the particles vanishes. We will see in the following sections that this feature may be associated with a "shattering" transition. In fact, whenever the inequality

$$-\frac{p}{D} + b_c\gamma_0 \leq 0 \tag{20}$$

holds, the number of particles diverges at a time

$$t_f = \frac{1}{\omega_0\left[\frac{p}{D} - b_c\gamma_0\right]} \tag{21}$$

at which the kinetic energy vanishes. This inequality holds at large enough fragmentation probability, $p \geq Db_c\gamma_0$. It therefore appears that the finite-time singularity cannot happen if fragmentation is just a marginal event (low p); alternatively, it will always be present in the elastic limit ($\alpha = 1$ so that $\gamma_0 = 0$). Hence, the kinetics of a gas of freely evolving fragmenting grains is determined by the degree of inelasticity and the collisional average b_c. In principle, this collisional average is itself a function of inelasticity; however, as discussed in the Appendix, such a dependence only appears when the velocity dependence of the distribution function $f(\vec{v}, \sigma, t)$ deviates from a Maxwellian.

In the absence of the above finite-time singularity, both n and E_c show power laws with time, which, from (7) in turn imposes that $\omega \propto 1/t$. The physical meaning of this remarkable feature (also observed in ballistic annihilation [3]) is that there exists a single relevant time scale in the problem (in the sense that the mean collision time that is an *a priori* important time scale, evolves with time as time itself). It is also noteworthy that from (15), the constraint $\omega \propto 1/t$ imposes a relation between the scaling exponents of n and E_c.

The previous results presume the existence of a scaling regime, hypothesis that we have validated numerically, as will be described in Sect. 4. However, note that the existence of a finite-time singularity does not rule out the possibility of a scaling regime. In order to analyze quantitatively such a regime, it is useful to introduce a slower time scale, τ

$$\mathrm{d}\tau = n\,(t)\,\mathrm{d}t \tag{22}$$

In this "fake" time scale, the finite-time singularity is no longer present. This feature favors the numerical analysis of the asymptotic scaling behaviour of the mixture. The internal time scale is related to τ through $\mathrm{d}\mathcal{C} = \omega \mathrm{d}\tau / n$. As a result, in terms of this new time scale, the number of collisions is expressed as

$$\mathcal{C} = \frac{1}{b_c\gamma_0 + \left(1 - \frac{1}{D}\right)p} \ln\left\{1 + \left[b_c\gamma_0 + \left(1 - \frac{1}{D}\right)p\right]\omega_0\tau\right\} \tag{23}$$

Hence, using the exponential growth of both the number of particles and kinetic energy in terms of the number of collisions, we get

$$n\,(\tau) = n_0\left\{1 + \left[b_c\gamma_0 + \left(1 - \frac{1}{D}\right)p\right]\omega_0\tau\right\}^{p/\left[b_c\gamma_0 + \left(1 - \frac{1}{D}\right)p\right]} \tag{24}$$

$$E_c\,(\tau) = E_{c0}\left\{1 + \left[b_c\gamma_0 + \left(1 - \frac{1}{D}\right)p\right]\omega_0\tau\right\}^{-2D/[D + (D-1)p/(b_c\gamma_0)]} \tag{25}$$

When compared with (18) and (19), one can see that, although qualitatively similar, the additional factors that characterize the exponents in this time scale ensure the absence of a finite-time singularity.

The time evolution of the physical quantities of interest in the scaling regime, as depicted in (24) and (25), depend on the restitution coefficient, α, the fragmentation probability, p, the dimensionality, D, and the collisional average b_c. From all of these, only b_c is not known *a priori*. In the Appendix, we show how it is related to the particle distribution. In particular, it is a constant for a Maxwellian distribution of velocities with a single mass-independent temperature. In this limiting case, analytic expressions for the exponents can be obtained. However, as we will see, such a factorization is too simplistic in generic situations. The collisional average will therefore have to be determined numerically, and in all cases it is the only input required to predict the corresponding exponents governing the scaling behaviour of the freely evolving gas.

Incidentally, the existence of a scaling regime can be used at one's advantage in DSMC. As has been explained previously, one of the numerical difficulties in studying the fragmentation of a freely evolving granular gas lies in the fact that the collision frequency decreases with time, making it harder to work with larger systems sizes (which is what inevitably happens in this case due to fragmentation). On the contrary, working with the rescaled velocities $\vec{c}$ instead of the original $\vec{v}$ significantly speeds up the program, and allows to probe longer time scales [9].

3.2 Kinetic Scenarios

As an illustration of the various kinetic scenarios covered by the previous analysis, let us consider in more detail a few limiting cases related to the kinetics of freely evolving gases.

In the absence of fragmentation, $p = 0$, obviously the number of particles remains constant, the system corresponds to that of a mono-disperse granular gas, and the energy decays as

$$E_c = \frac{E_{c0}}{\left[1 + b_c \gamma_0 \omega_0 t\right]^2} \tag{26}$$

recovering Haff's law [10]. Here, the collisional parameter b_c is close to 1, with deviations originating from the non-Gaussian features of the velocity distribution [16]. In this particular case, the two time scales τ and t are just proportional to each other, yielding the same exponent. The assumption of spatial homogeneity prevents us from addressing the late stage deviations from Haff's law that have been described in freely evolving granular gases [10].

For an elastic gas, $\alpha = 1$, the kinetic energy is conserved. If grains can still fragment with a probability p, the number of particles will always diverge at a finite time,

$$n(t) = \frac{n_0}{\left[1 - \frac{p}{D}\omega_0 t\right]^D} \tag{27}$$

In this case it is easier to appreciate the advantage of introducing the slower time scale τ. In terms of this variable, the number of particles will increase as

$$n(\tau) = n_0 \left[1 + \left(1 - \frac{1}{D}\right) p \omega_0 \tau\right]^{D/(D-1)} \tag{28}$$

In order to analyze the existence of an asymptotic regime (hypothesis upon which both (27) and (28) have been derived), it is more convenient to use the latter time scale, because it does not lead to a finite-time divergence. In this way, it is possible to make use of the standard techniques to address the appearance of power-law kinetic regimes, since the indefinite growth ensures the possibility to span several decades. Hence, once the algebraic regime is achieved, accurate values of the exponents can be derived as well. If such exponents are characterized, it is always possible to obtain their counterparts controlling the evolutions in the real time scale t.

The technique of resorting to a slower time scale will always be useful for the present class of fragmentation problems, because the exponent appearing in (24), characterizing the increase in the grain number density, is always positive for any dimension larger than 1. For $D = 1$, the elastic limit is singular. In this case, the scale $\mathcal{C}$ is proportional to τ and the number of particles increases exponentially in τ rather than algebraically. This behavior is recovered, by performing the limit $D \to 1$ in (24), and is due to the fact that in one dimension the collision frequency is proportional to the number of particles for our elastic mixture.

4 Numerical Simulations

4.1 The Method

We have implemented the Direct Simulation Monte Carlo technique (DSMC), which offers a particle method to solve numerically a Boltzmann-type equation, such as (2). This method was originally introduced to study rarefied gases [11], and has been extended to study granular gases [12]. It has been used in a number of different situations and it is easy to adapt to the present system, where the number of grains increases with time. There exist already a number of applications to situations where the number of particles is not a conserved quantity, as it is the case in ballistic annihilation [3].

The assumption of spatial homogeneity simplifies further the implementation of DSMC, because it is enough to sample uniformly the relative orientation of the colliding pair, $\widehat{\varepsilon}$, without the need to keep track of the positions of the particles. Sequentially, a pair of particles is selected at random. Also the vector $\widehat{\varepsilon}$ is chosen randomly with equal probability on the unit circle and the kinematic constraint is checked. In case the particles can collide, the cross section is computed (it corresponds to the factor $|\vec{v}_{12} \cdot \hat{\varepsilon}|\sigma_{12}^{D-1}$ in (8)) and the collision takes place proportionally to its value i.e. the collision is executed if a uniformly chosen random number is smaller than the computed cross section (properly normalized). The decrease in velocity of the particles due to the inelasticity reduces the cross section, hence the simulations become more expensive as time evolves. After the collision is executed (according to (1)), one of the two particles is chosen randomly, and it is broken with probability p.

The collision frequency of each collision i-j can be computed on the basis of the cross-section, and hence the time is advanced accordingly at each collision

$$t \to t + \frac{1}{n^2 \, |v_{ij} \cdot \widehat{\varepsilon}| \, \sigma_{ij}^{D-1}}, \tag{29}$$

where n is the instantaneous number density. It is then possible to keep track of the kinetic evolution of the system, as well as gather statistics on the distribution of sizes and velocities. In the following section we will analyze both the kinetics and the – self-similar in time – polydisperse distributions that the free evolution of the gas gives rise to.

4.2 Numerical Results: Kinetics

We have first analyzed the kinetics in the homogeneous regime to test the scaling hypothesis discussed in Sect. 3. To this end, we have used DSMC simulations to study the evolution of a two-dimensional granular gas of circular disks with constant restitution coefficient α. In all cases we start from an initial condition where all particles have the same size and their velocities are distributed according to a Maxwellian (mono-disperse gas of elastic disks). As it has already been described in Sect. 2, we restrict ourselves to the simplest model where colliding particles fragment with a uniform probability independent of the collision properties, and where only one of the post-colliding disks fragments in two out-coming grains.

In order to validate the existence of a scaling regime, we have computed the total number of particles, $n(t)$, the collision frequency, $\omega(t)$, the mean velocity, $\langle v^2\rangle^{1/2}$, as well as the mean radius, $\langle\sigma\rangle = \bar{r}$. We have also computed different moments of the size, $\langle\sigma^n\rangle(t)$, and velocity, $\langle v^n\rangle(t)$, distribution functions. Within the scaling picture, we expect $\langle\sigma^n\rangle \propto \bar{r}^n$ and $\langle v^p\rangle \propto \bar{v}^p$, as far as the time dependence is concerned. The possibility of a multi-scaling behaviour has been reviewed by Ben-Naim and Krapivsky [13].

In Fig. 1 we display the time evolution of a few relevant quantities in the slower time scale τ. One can clearly see the existence of a scaling regime. The use of this scale allows a detailed analysis of such a regime, extracting the appropriate exponents. An alternative possibility to assess the existence of a scaling regime without introducing τ would be to check if $\langle\sigma^n\rangle$ scales like $\langle\sigma^p\rangle^{n/p}$. For the parameters chosen, the time evolution of the quantities of Fig. 1 shows a finite-time singularity. Such a singularity can be assessed by noticing the steep decrease in the time increment during the simulation. On the contrary, it makes it difficult to analyze numerically the scaling regime; hence the benefit of using τ as the relevant scale.

In Fig. 2a we display different moments of the velocity distribution, as well as the mean collisional velocity $\langle v_{\text{coll}}\rangle$, defined as the mean velocity modulus of colliding partners, i.e restricting to those particles that are in precollisional configurations (since a typical collision involves a particle that is "hotter" than the mean background, this quantity is larger than $\langle v\rangle$, as can be observed on the figure). Due to the inelasticity of the collisions, the typical velocity of the particles decays as a function of time. The figures show that such a decay is algebraic in τ, and also that there is a single exponent characterizing different moments of the velocity. The same holds for the characteristic size of the particles as a function of time, as depicted in Fig. 2b. In this case

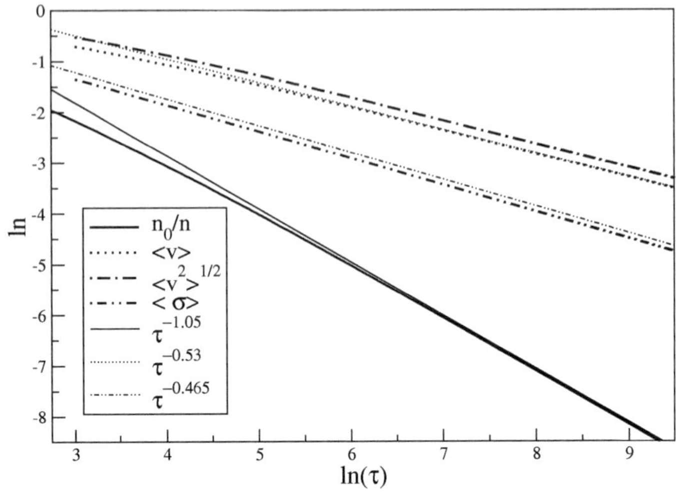

Fig. 1. Time evolution (in terms of the slow time τ) in log-log scale of the total number of particles, mean size and characteristic velocities for $\alpha = 0.5$, and a fragmentation probability $p = 1/2$ in $D = 2$

however, the scaling does not hold for negative powers of the size. We will see in the next section how such a deviation from simple scaling relates to the peculiarities of the size distribution, but it does not imply a breaking of the scaling hypothesis; it only signals the fact that moments of negative order are controlled by the smallest species in the mixture.

We can characterize the asymptotic regime by a set of exponents, that in the terms of τ can be written generically as

$$\begin{aligned} n(\tau) &\sim \tau^{\varepsilon'} \\ E(\tau) &\sim \tau^{-2\gamma'} \\ \sigma(\tau) &\sim \tau^{-\beta'} . \end{aligned} \tag{30}$$

Once these exponents are known, their counterparts associated with the t scale follow. For example, using that $\mathrm{d}\tau = n_0 (1 + a\tau)^{\varepsilon'} \mathrm{d}t$, for $\varepsilon' \neq 1$ we have

$$\frac{1 + a\tau}{1 + a\tau_0} = \left[1 + (1 - \varepsilon') \frac{n_0 a (t - t_0)}{(1 + a\tau_0)^{1-\varepsilon'}} \right]^{1/(1-\varepsilon')} \tag{31}$$

from which we can derive the time evolution of any physical quantity.

In Table 1 we give the values of the exponents fitted numerically for the 2-dimensional gas at different values of the restitution coefficient. We can determine the different exponents independently, but they are related to each other since all of them are eventually functions of the collisional

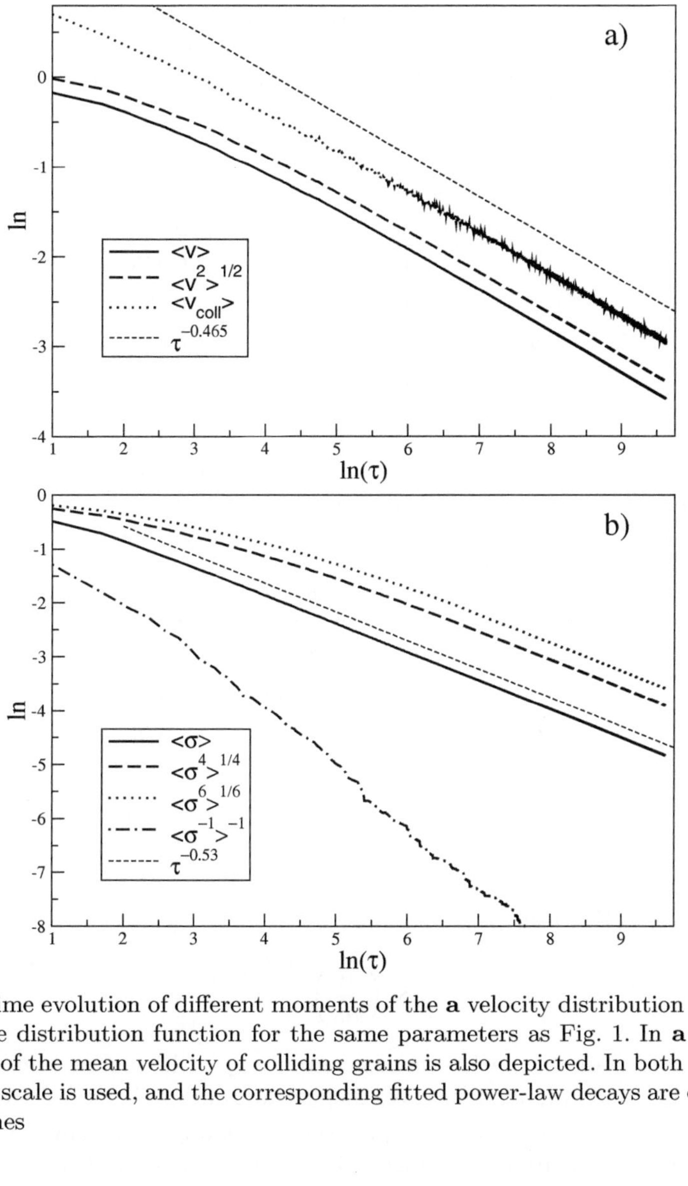

Fig. 2. Time evolution of different moments of the **a** velocity distribution function, and **b** size distribution function for the same parameters as Fig. 1. In **a** the time evolution of the mean velocity of colliding grains is also depicted. In both cases the slow time scale is used, and the corresponding fitted power-law decays are displayed as thin lines

average b_c. It is in fact rather straightforward to check that, once one of the three exponents is known, the remaining ones can be determined using the relationship between them as described in Sect. 3.

In Table 2 we list the values of the collisional average b_c computed from the numerical simulations. We also display the values of the exponent ε' derived using the corresponding computed values of b_c; comparing such values with those measured numerically (as listed in Table 1) shows a good agreement, illustrating the consistency of the analysis. The scaling hypothesis also

Table 1. Exponents characterizing the time evolution of the number and energy densities as well as the mean particle radius (see (30) for definitions) for different values of the restitution parameter α, when the probability of grain fragmentation at a collision is $p = 1/2$

α	ε'	β'	γ'
0.3	0.95	0.45	0.55
0.5	1.05	0.5	0.475
0.7	1.3	0.65	0.35
0.9	1.9	0.95	0.125
0.95	2.3	1.2	0.05

Table 2. Collisional average b_c, and predicted exponent ε' for different values of the restitution parameter α, when the breaking probability is $p = 1/2$. $b_c^{(1)}$ is obtained from (32), plotting the kinetic energy as a function of the number of particles

α	b_c	$b_c^{(1)}$	ε'_{th}
0.3	1.24	1.248	0.94
0.5	1.18	1.173	1.06
0.7	1.07	1.042	1.29
0.9	0.67	0.645	1.77
0.95	0.39	0.2318	1.94

implies, from (12), that

$$E_c \propto \left(\frac{n}{n_0}\right)^{2b_c\gamma_0/p} \tag{32}$$

which provides an alternative way to compute b_c in the scaling regime (where it has to be time independent). We display in Table 2 the collisional average computed following this route [under the symbol $b_c^{(1)}$]. It shows a good agreement with the computed collisional average, giving additional support to the existence of an asymptotic scaling regime.

In the Appendix, we have given a generic expression for b_c, showing that it depends sensitively on both the size and the velocity distribution of the mixture in the scaling regime. If the velocity dependence enters only through a Maxwellian, then the value of b_c is independent of the mass distribution. This surprising result shows that this collisional average is a quantity that will be sensitive to deviations from Maxwell distributions. In this contribution we have only analyzed the behavior of the gas within the molecular chaos assumption, and hence, deviations of b_c from unity signal a deviation from Maxwellian behavior (with the strong assumption that the temperature of a given species does not depend on its size). In a more generic study, deviations from Maxwellian behavior may also result from the breakdown of molecular chaos, so that the use of b_c has to be supplemented by a second quantity that is sensitive to such a breakdown (e.g. the averaged impact parameter [14, 15]).

For a mono-disperse granular gas, deviations from Maxwellian behavior vanish as the elastic limit is approached and can be described quite accurately in terms of a Sonine expansion around the Maxwellian [16]. The values of b_c displayed in Table 2, on the contrary, show that the deviations from a Maxwellian velocity distribution become more pronounced as one approaches the elastic limit. This counter-intuitive trend can be traced back to the peculiar asymptotic size distribution of particles close to elasticity. Such a behavior is described in detail in the next section where it appears that the velocity distribution for nearly elastic gases deviates strongly from a global Maxwellian. At higher inelasticities (which corresponds to small values of the restitution coefficient), such deviations, even if always present, are not so dramatic.

4.3 Numerical Results: Distribution Functions

The existence of a scaling regime implies that the distribution function of a given quantity at two different times has to be invariant after an appropriate rescaling of this quantity [see (13) and (14)]. For instance, the distribution of $\bar{\sigma} = \sigma / \langle\sigma\rangle$ is expected to be time independent. From the DSMC simulation runs, we have computed scaled distribution functions, which do indeed reach a steady state, in agreement with the scenario developed in Sect. 3. We will now discuss in detail the characteristic features of some of the relevant stationary distributions.

Size Distribution. The fragmentation process generates a polydisperse mixture of grains that evolves continuously in time, with a decreasing mean size. Large particles are consequently progressively destroyed, but the shape of the size distribution is preserved, in such a way that the relative amount of particle with respect to the time dependent mean size, $\langle\sigma\rangle(t)$, becomes independent of time.

In this regime, the scaled size distribution exhibits a marked peak for vanishingly small grains, indicating that smaller particles become predominant in the mixture. In Fig. 3 we show such distributions for a number of restitution coefficients, where two regions can be identified. On the extreme of large grains, the decay is exponential while it is algebraic for small particles. The exponent characterizing this algebraic behavior increases with α. For example, for $\alpha = 0.3$ the exponent is consistent with $P(\sigma) \sim \sigma^{-0.1}$, while for $\alpha = 0.7$ we can fit the curve numerically with $P(\sigma) \sim \sigma^{-0.6}$.

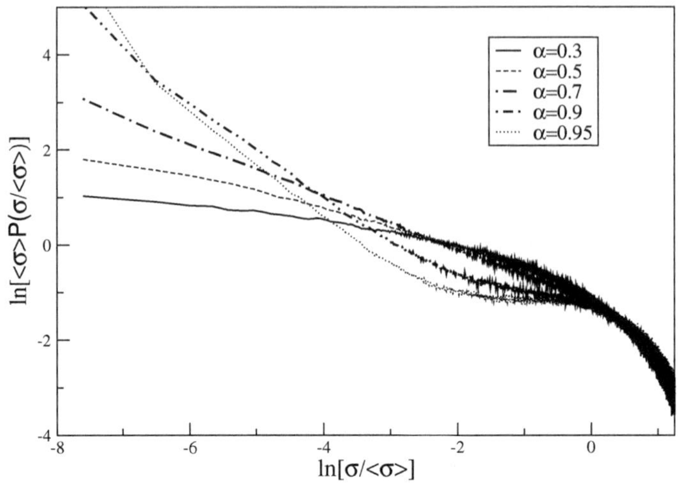

Fig. 3. Log-log plot of the rescaled particle radius distribution, for a uniform random fragmentation probability $p = 1/2$ in $D = 2$

It is interesting to note that at $\alpha = 0.9$, and also at $\alpha = 0.95$, the above exponent becomes close to -1. Such a value is peculiar, because it implies that the probability distribution is not integrable at the origin, incompatible with the requirement that $P(\sigma)$ [or $P(\bar{\sigma})$] has to be normalizable. In fact, a careful inspection of Fig. 3 shows that $P(\sigma)$ at high values of the restitution coefficient develops a jump very close to the origin. Such a jump is consistent with the appearance of a singular contribution, due to a finite fraction of particles with vanishing size. This behaviour is reminiscent of a "shattering" transition [1] and is corroborated by the velocity distributions described in the next section. It indicates the development of an intrinsic size heterogeneity in the system.

The algebraic decay of $P(\sigma)$ implies that, even when the exponent is smaller (in absolute value) than -1, certain averages of the distribution function will not exist. This is consistent with the behavior described in the previous section, and simply indicates that in general, averages of negative powers of the grain size will be controlled by the smallest species available in the mixture, rather than by the overall mixture. However, such a lack of scaling for certain moments of the distribution is not incompatible with the dynamic scaling behavior of the complete distribution function; rather, it shows some of the peculiarities of the asymptotic behavior of the scaling distribution (in this case at small sizes).

The existence of a stationary size distribution suggests the possibility to map the fragmenting granular gas onto an effective polydisperse mixture with a time dependent mean. Although this is indeed possible in principle, we do not have any theoretical framework (except the complicated initial

Boltzmann equation) to predict either the algebraic decay of $P(\sigma)$ or the details of the exponential tail.

Velocity Distribution. We have also studied the velocity distribution to assess how relevant deviations from Maxwellian are, and also to analyze the meaning of a global temperature characterizing the whole system.

In Fig. 4 we display the distribution of rescaled velocities $\vec{c}$ at different inelasticities. Deviations from Maxwellian behavior are observed at all values of α. However, the deviations become more prominent close to the elastic limit, in agreement with the development of a singular contribution coming from vanishingly small grains, explaining the small values for b_c listed in Table 2. In order to gain more insight, we have also computed separately the velocity distribution from those particles with a mass smaller than 5% the mean mass. In Fig. 5a we compare the overall velocity distribution with that of the smaller particles and the contribution coming from the rest of the grains for $\alpha = 0.95$. The low velocity part of the distribution is controlled by the contribution from most of the particles, while the large velocity part is dominated by the small particles. These are characterized by a much larger temperature than the mean, while the rest of the species are much colder (the cross-section changes with size as σ^{d-1} so that small particles suffer less dissipative collisions than the others).

Although less dramatic, the same trend is observed for all values of α. For example, in Fig. 5b we display the equivalent distribution functions at $\alpha = 0.5$. In this case, again the two subsets of grains evolve following separate approximate Maxwellians. According to this interpretation, as α increases, the separation in temperatures between the two sets of particles increases.

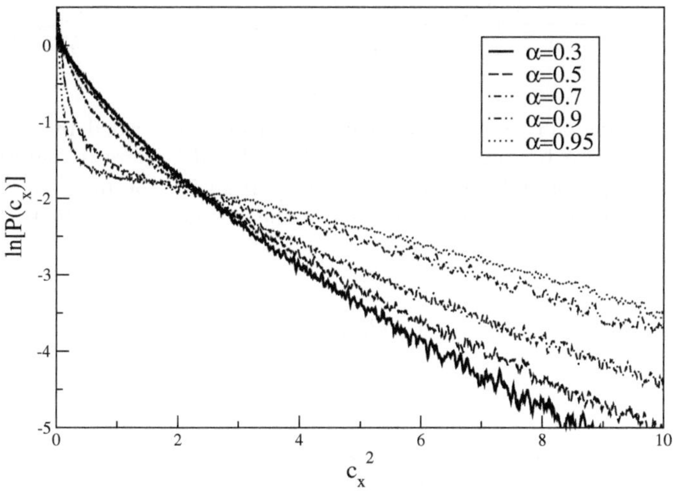

Fig. 4. Linear-log plot of the rescaled velocity distribution $\vec{c} = \vec{v}/\bar{v}$

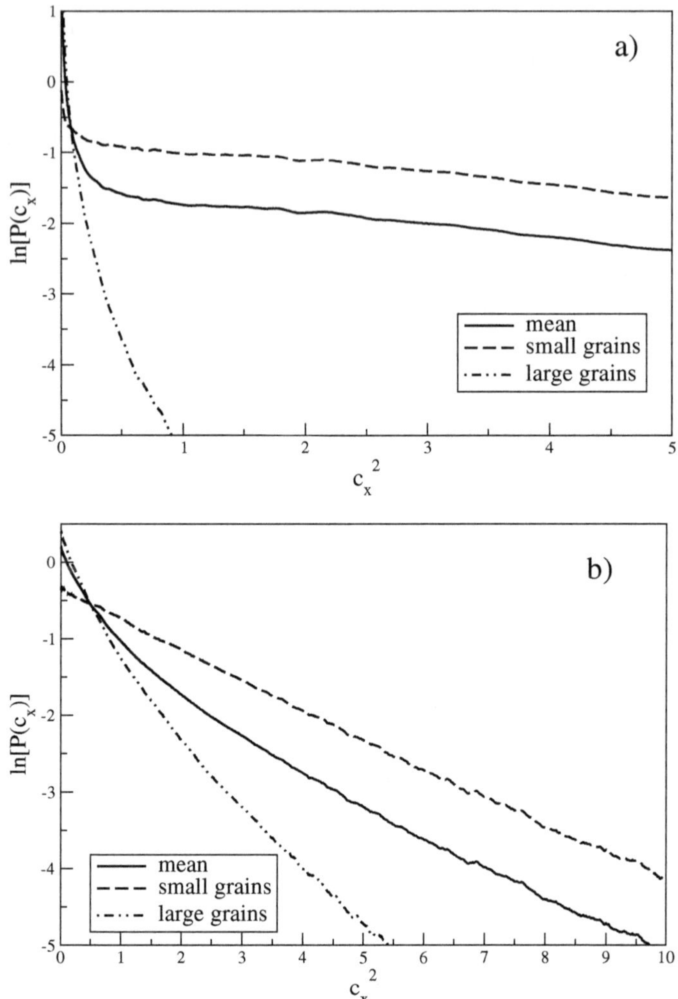

Fig. 5. Linear-log plot of particle velocity distributions, scaled by the mean velocity, for a uniform fragmentation probability. "Small" refers to the distribution of velocities for particles having a mass smaller than 5% the mean mass. "Large" corresponds to the distribution of the rest of the grains. **a** $\alpha = 0.95$; **b** $\alpha = 0.5$

Such an inhomogeneous dependence of the velocity distribution (and in particular its second moment) on grain species, explains the large deviations of b_c from the Maxwellian prediction reported close to the elastic limit. These results also cast doubts on the use of a single temperature to characterize a polydisperse granular mixture. Nonetheless, it is worth noting that such a size heterogeneity does not affect the scaling hypothesis. Even if different temperatures could be defined for different subsets of particles, all of them have the same time evolution, explaining why it is possible to have a simple velocity scaling. Such a behavior has been reported for the free evolution of bidisperse granular mixtures [17]. However, the absence of a common temperature invalidates the use of perturbative calculations, e.g. through a Sonine expansion, to determine the deviations from Gaussian behavior as the we approach the elastic limit, as is customary in the case of monodisperse (and even bidisperse) granular gases.

5 Discussion

We have investigated the evolution of a granular gas in the absence of any driving, taking into account the possibility that the collisions may lead to a fragmentation of the impacting grains.

Rather than performing a detailed analysis of the fragmentation process, we have analyzed the simplest possible model, where the fragmentation is a random event uncorrelated to the details of the collision, except for the requirement that only colliding particles may fragment. This simple fact already leads to a number of interesting results. We have shown the existence of a scaling homogeneous regime characterizing the kinetics of the system. Such a regime is controlled eventually by a single collision average, b_c. The relevance of b_c in the asymptotic regime shows that a theory based on rate equations, usually disregarding collision induced correlations, will be too simplistic to capture the dynamics in these systems.

We have shown how the evolution in the scaling regime may lead to a finite-time singularity where the number of particles diverges. We have derived the conditions leading to this finite-time catastrophe, depending on the different parameters that characterize the system. In the absence of such a singularity, the various moments of the size and velocity distribution exhibit algebraic scaling laws with time, that can be regarded as a generalized Haff's law.

The size distribution exhibits two different limiting kinds of behaviour: an exponential tail at large sizes (compared to the time dependent mean size), and a power-law distribution for small sizes, which may develop into delta singularity close to the elastic limit. The details and character of such a transition, which has been reported in other fragmenting systems [1] deserve a more careful analysis. The existence of a scaling regime is not incompatible with the presence of marked size heterogeneities in the mixture. In particular, the characteristic temperature of small and large grains can be significantly different, in such a way that the mean scaled velocity distribution cannot be regarded as a slightly distorted Maxwellian.

In order to assess to which extent the results discussed are specific of the fragmentation mechanism that has been chosen, we have also carried out numerical studies for a model where the fragmentation probability is correlated to the energy lost in a collision [18]. In that case there exists still a scaling regime characterizing the kinetics. The scaled distribution functions are also similar, except for the fact that size distributions are closer to an exponential. Hence, the results obtained through the analysis of the simplest model provide a reference system, against which other models can be tested, and in this way it should be possible to disentangle the generic aspects involved in the dynamics of fragmenting granular materials. However, it is difficult to assess the robustness of the shattering transition found here, with respect to changes in the modelization. In particular, two important mechanisms have been disregarded here: first, fragmentation is usually related to a certain loss of energy and second the breaking probability increases with the size of the particles (due to the increasing number of micro-cracks) [7]. Due incorporation of these effects is left for future investigations.

Acknowledgments

We would like to thank A. Barrat for useful discussions. I.P. acknowledges the support of CNRS through the attribution of a "poste de chercheur associé", and the hospitality of the "Laboratoire de Physique Théorique" in Orsay where most of this work was performed.

Appendix: On the Collisional Average b_c

The kinetics of the free fragmenting granular gas depends on the value of the collisional average b_c, which in turn is a complicated function of the joint size and velocity distribution. To proceed further, we assume that in the scaling regime, the velocity dependence is Maxwellian. From the distribution function $f(\vec{v}, \sigma, t)$, we change variables and introduce the joint distribution of masses and velocity $f(m, \vec{v}, t)$ that we approximate by

$$f(m_i, \vec{v}_i, t) \simeq \psi(m_i, t)\,\phi\left(m_i v_i^2, t\right) \tag{33}$$

where ψ is the mass distribution, while $\phi\left(m_i v_i^2\right)$ is the Maxwellian,

$$\phi\left(m_i v_i^2, t\right) = \left(\frac{\beta m_i}{2\pi}\right)^{D/2} \exp\left(-m_i v_i^2 \beta/2\right). \tag{34}$$

The time dependence of ϕ is encoded in $\beta = 1/(k_B T)$, see (6). The key assumption here is that $k_B T$ does not depend on the mass m, as would be the case in an equilibrium polydisperse system. We have

$$b_c = \beta \frac{\int \mathrm{d}1\, \mathrm{d}2\, \mathrm{d}\widehat{\varepsilon}\, \theta\left(\vec{v}_{12}\cdot\widehat{\varepsilon}\right) \left|\vec{v}_{12}\cdot\widehat{\varepsilon}\right|^3 \sigma_{12}^{D-1} \mu_{12} \psi(m_1)\,\phi\left(m_1 v_1^2\right) \psi(m_2)\,\phi\left(m_2 v_2^2\right)}{\int \mathrm{d}1\, \mathrm{d}2\, \mathrm{d}\widehat{\varepsilon}\, \theta\left(\vec{v}_{12}\cdot\widehat{\varepsilon}\right) \sigma_{12}^{D-1} \psi(m_1)\,\phi\left(m_1 v_1^2\right) \psi(m_2)\,\phi\left(m_2 v_2^2\right)} \tag{35}$$

where d1 stands for $\mathrm{d}\vec{v}_1\, \mathrm{d}m_1$ and $\mu_{12} = m_1 m_2/(m_1+m_2)$ is the reduced mass. We can express the collisional average b_c in dimensionless quantities, and also transform the velocities to express them in terms of the center-of-mass and relative coordinates. One arrives at

$$b_c = \frac{2\beta}{(D+1)} \frac{\int \mathrm{d}\vec{c}\mathrm{d}m_1\mathrm{d}m_2 c^3 \sigma_{12}^{D-1} \mu_{12} \psi(m_1)\psi(m_2)(m_1 m_2)^{D/2} \mathrm{e}^{-\beta\mu_{12}c^2/2}}{\int \mathrm{d}\vec{c}\mathrm{d}m_1\mathrm{d}m_2 c \sigma_{12}^{D-1} \psi(m_1)\psi(m_2)(m_1 m_2)^{D/2} \mathrm{e}^{-\beta\mu_{12}c^2/2}}$$

$$= \frac{\int \mathrm{d}m_1\mathrm{d}m_2 \left(\frac{1}{\mu_{12}}\right)^{\frac{D+3}{2}} \mu_{12}\, \sigma_{12}^{D-1}\, \psi(m_1)\psi(m_2)(m_1 m_2)^{D/2}}{\int \mathrm{d}m_1\mathrm{d}m_2 \left(\frac{1}{\mu_{12}}\right)^{\frac{D+1}{2}} \sigma_{12}^{D-1}\, \psi(m_1)\psi(m_2)(m_1 m_2)^{D/2}} = 1 \qquad (36)$$

where use has been made of

$$\int \mathrm{d}\widehat{\varepsilon}\theta(\vec{c}\cdot\widehat{\varepsilon})\,|\vec{c}\cdot\widehat{\varepsilon}|^n = \frac{\pi^{\frac{D-1}{2}}\Gamma\left(\frac{n+1}{2}\right)}{\Gamma\left(\frac{n+D}{2}\right)}\, c^n \qquad (37)$$

So, $b_c = 1$ if the velocity distribution follows a Maxwellian. Hence, this collisional average is a good measure of deviations from Maxwellian behavior in polydisperse granular gases.

References

1. S. Redner. *Fragmentation* In H.J Hermann and S. Roux, editors, *Statistical Models for the Fracture of Disordered Media*, 1990, Elsevier Science Pubs. (North Holland)
2. P.L. Krapivsky and E. Ben-Naim. Shattering Transitions in Collision-Induced Fragmentation. cond-mat/0302525
3. E. Trizac. Kinetics and scaling in ballistic annihilation. *Phys. Rev. Lett.* **88**, 160601 (2002)
4. J. Piasecki, E. Trizac and M. Droz. Dynamics of ballistic annihilation. *Phys. Rev. E* **66** 066111 (2002)
5. P. Résibois and M. de Leener, *Classical Kinetic Theory of Fluids*, John Wiley and Sons (1977)
6. I. Goldhirsch and G. Zanetti. Clustering instability in dissipative gases. *Phys. Rev. Lett.* **70**, 1619 (1993)
7. V. Buchholtz, J.A. Freund and T. Pöschel. Molecular dynamics of comminution in ball mills. *Europ. Phys. J. B* **16**, 169 (2000)
8. S. McNamara and W.R. Young. Kinetics of a one-dimensional granular medium in the quasi-elastic limit. *Physics of Fluids* **5**, 34 (1993)
9. J.M. Montanero and A. Santos. Computer simulation of uniformly heated granular fluids. *Granular Matter* **2**, 53 (2000)
10. R. Brito and M.H. Ernst. Extension of Haff's cooling law in granular flows. *Europhys. Lett.* **43**, 497 (1998)
11. G. Bird *Molecular Gas Dynamics and the Direct Simulation of Gas Flows* Clarendon Press, Oxford, 1994
12. J.M. Montanero, V. Garzó, A. Santos and J.J. Brey. Kinetic theory of simple granular shear flows of smooth hard spheres. *J. Fluid Mech.* **389**, 391 (1999)

13. E. Ben-Naim and P.L. Krapivsky. Multiscaling in fragmentation. Physica D **107**, 156 (1997)
14. S. Luding. On the relevance of "molecular chaos" for granular flows. ZAMM **80**, 9 (2000)
15. I. Pagonabarraga, E. Trizac, T.P.C. van Noije and M.H. Ernst. Randomly driven granular fluids: Collisional statistics and short scale structure. *Phys. Rev. E*,**65**, 011303 (2002)
16. T.P.C. van Noije and M.H. Ernst. Velocity distributions in homogeneous granular gases. *Granular Matter* **1**, 57 (1998)
17. V. Garzó and J. Dufty. Homogeneous cooling state for a granular mixture. *Phys. Rev. E* **60**, 5706 (1999)
18. I. Pagonabarraga and E. Trizac. *(In preparation)*

Part II

Granular Hydrodynamics

Shock Waves in Granular Gases

Alexander Goldshtein[1], Alexander Alexeev[2], and Michael Shapiro[3]

[1] MediGuide Ltd., P.O.B. 15003, Haifa 31053, Israel
[2] Laboratory of Aerosol Research, Faculty of Mechanical Engineering, Technion, Haifa 32000, Israel
[3] Laboratory of Transport Processes in Porous Materials, Faculty of Mechanical Engineering, Technion, Haifa 32000, Israel

Abstract. This review is the first attempt to systematize the results on shock waves in granular gases. We present experimental and computational evidences of shock and expansion waves propagating within granular gases. The analysis of model flows with shock and expansion waves shows that even smallest kinetic energy dissipations crucially affects such flows. We discuss the role of these waves for the fluidization of a granular flow by an obstacle, and of a granular layer by a vibrating vessel. The efficacy of hydrodynamic models for the description of shock-expansion wave flows is demonstrated. We also show that these waves play the dominant role in vibrofluidization of highly dissipative granular layers as frequently found in nature and engineering applications.

1 Introduction

When we speak about shock waves the most immediate association is the supersonic motion of a projectile around which a shock wave developes. High pressure and temperature of the surrounding gas and projectile velocities of hundreds of meters per second are inherent for this example, thereby requiring utilization of a special experimental technique to register a shock wave. In contrast, shock waves in granular gases can be easily produced in laboratory conditions with low energy expenses and registered by a home video camera (see Fig. 1). The latter fact can be rationalized by low granular temperature related to chaotic motion of granules and accordingly low speed of sound in granular gases. For typical experimental conditions the speed of sound is about several centimeters per second [1, 2], and any motion of a solid body with a larger velocity is supersonic, thus creating a shock wave.

One of the basic features of granular gases is kinetic energy dissipation associated with inelastic inter-particle collisions, leading to gas cooling. Without external energy pumping, the granular temperature eventually decays to zero, thereby resulting in "solidification" through cluster formation [3] and inelastic collapse [4]. Shock waves serve as a source of such a pumping, as their propagation is accompanied by irreversible transformation of mechanical work into granular kinetic energy. The balance between these two processes is of crucial importance for the existence of the hydrodynamic state of the gas.

To estimate the heat generation by shock waves we consider the following problem: Let an initially resting gas, with (known) initially constant density,

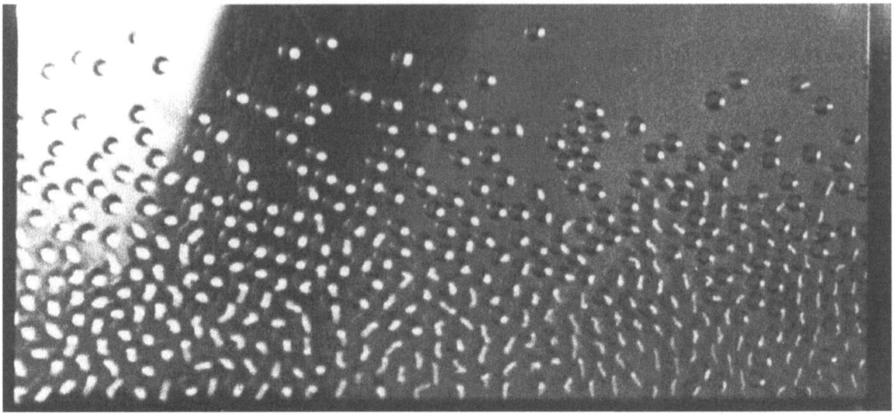

Fig. 1. Compression (shock) wave within the vibrofluidized bed (vessel moves up). Vibrational amplitude, A =1cm, vibrational frequency f = 15 Hz, height of the layer at condensed state, h_m = 10 particle diameters. Ellipsoidal forms of the particle images in the bottom part the layer are caused by their motion with velocities up to 2ms^{-1} in the direction of the long ellipses' axes (from [1], Fig. 7)

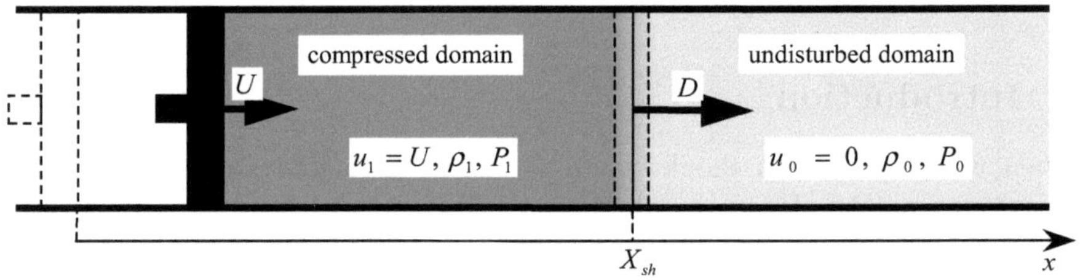

Fig. 2. Schematic of a shock wave caused by a piston steadily moving within an ordinary (molecular) gas

ρ_0, and pressure, P_0, be compressed by a plane piston moving with a constant velocity U, as shown in Fig. 2. The resulting flow can be subdivided into three regions: compressed, undisturbed and transition between them (the latter is indicated by dashed lines in Fig. 2). The simplest description of the problem may be obtained in the approximation of inviscid and non-conducting gas. The problem has no characteristic length, and the transition region connecting compression and undisturbed domains reduces to a discontinuity, which can be considered as a shock wave front. We apply the general laws of conservation of mass, momentum and energy to derive the unknown quantities after the shock front, namely the gas density ρ_1, and the pressure P_1 in the compressed region, and also the velocity D of the shock propagation through the undisturbed region. These conservation laws formulated across the shock wave front result in the so-called Rankine-Hugoniot conditions:

$$[\rho\,(D-u)] = 0 \tag{1}$$

$$\left[\rho\,(D-u)^2 + P\right] = 0 \tag{2}$$

$$\left[\frac{E}{m} + \frac{P}{\rho} + \frac{(D-u)^2}{2}\right] = 0\,, \tag{3}$$

where m and E are the particle mass and internal energy, u is the gas velocity, and the symbol [ ] denotes the jump of any hydrodynamic property across the discontinuity, for example $[P] \equiv P_0 - P_1$.

The general relationships (1) and (2), representing the basic conservation laws of mass and momentum are valid for any flowing medium, including both molecular and granular gases. Equation (3) is valid for molecular gases only; its variant modified to include energy dissipation on the shock front of a granular gas is given in Sect. 2.1 (see (17)).

In order to explain the basic relationships governing changes in the hydrodynamic quantities across the shock wave, consider a perfect molecular gas with constant specific heats. In this simplest case, the gas pressure is proportional to the density and temperature, the latter being proportional to the energy E:

$$P = \rho T \tag{4}$$

$$T = (\gamma - 1)E\,, \tag{5}$$

where γ is the specific heat ratio. For the sake of simplicity we consider the strong shock wave approximation, namely $P_0/P_1 \ll 1$, corresponding to an initially "cold" gas. Substitution of (4,5) into (1,2,3) yields

$$E_1 = \frac{U^2}{2} \tag{6}$$

$$\rho_1 = \frac{\gamma+1}{\gamma-1}\rho_0 \tag{7}$$

$$D = \frac{\gamma+1}{2}U\,. \tag{8}$$

Equation (6) merely states that half of the work performed by the moving piston is converted into the energy of random particle motion. It serves as an upper bound for other, physically more sophisticated models of state. Equation (7) gives an upper bound of ρ for the gas compression by the shock wave, and (8) predicts the shock wave speed. Since $\gamma > 1$ and, correspondingly, $D > U$, the extension of the compressed domain grows with time as $\frac{\gamma-1}{2}U\,t$.

The shock wave, in general, is an example of non-equilibrium phenomena, which are accompanied by irreversible transition of mechanical energy into heat governed by the viscosity and the heat conduction. Hydrodynamic models of these dissipative processes occurring in molecular gases based on the Navier-Stokes and Burnett equations were used for the description of the internal structure and thickness of the shock front [5–7]. It was shown that the linear Burnett theory agrees better with computer simulations than the Navier-Stokes theory except for the special case of a strong shock wave propagating through a dense hard sphere fluid [8], when the gas temperature is anisotropic. Recent achievements of the kinetic theory in predictions of the structure of shock waves are summarized in [9] and references therein.

The above mentioned theoretical and experimental investigations indicate that the shock wave flow is independent of the nature of the gas transition across the shock front having thickness of about several molecular mean free paths. In our analysis of shock waves in granular gases, we consider the simplest Euler-like hydrodynamic model neglecting the effects of gas viscosity and thermal conductivity. This hydrodynamic model is expected to be accurate beyond the shock front, as it occurs in molecular gases.

The review is organized as follows: In Sect. 2.1 we present the simplest hydrodynamic model of a compressible granular gas and discuss how the inelasticity of the particles appears in the hydrodynamic equations and in the conditions at the shock front, and also how the inelasticity affects the speed of sound. In Sects. 2.2 and 2.3 we analyze model problems of shock and expansion wave propagation through a granular gas, and show that even smallest kinetic energy dissipation crucially modifies them. In Sect. 3 we discuss the role of these waves in problems of granular flow interaction with obstacles and compare theses processes in granular and molecular gases. In Sect. 4 we consider the fluidization of granular layers by a vibrating piston and compare the efficacy of vibrofluidization by the shock-expansion waves with other hydrodynamic mechanisms. The applicability of hydrodynamic models for the descriptions of granular flows with shock and expansion waves is demonstrated. We also show that these waves play a dominant role for the vibrofluidization of highly dissipative granular layers.

2 One-Dimensional Waves

In this section, we describe two complementary problems of one-dimensional motion of a granular gas in the absence of gravity. In the first problem, a semi-infinite layer of initially resting granules is compressed by a steadily moving piston that results in the formation of a shock wave. The second problem concerns the expansion of an agitated granular gas into a vacuum. Initially the gas fills a semi-infinite region, the expansion results from the rapid escape of the piston outwards of this region. For these two particular problems, the hydrodynamic equations governing collisional granular motion have analytical solutions.

Generally speaking, the validity of the hydrodynamic approach for the description of granular flows is limited since the granular gas may degrade into a solid state due to energy losses. Therefore, solutions of the piston motion problems are important because they allow us to estimate the validity limits of the hydrodynamic description and to verify its accuracy.

The one-dimensional piston problems are also of practical interest since they may be viewed as limiting cases of oscillations of a deep granular layer, vibrated with large amplitude and low frequency within a vessel (see Sect. 4). The motion of the vessel towards the granular layer corresponds to its compression by the moving piston. The behavior of the layer during the motion of the vessel away from the layer can be viewed as an expansion of heated granular gas into vacuum.

Even for molecular gases, which are more amenable for theoretical treatment, analytical solutions for hydrodynamic equations are limited to very simple cases (see [10, 11]) and, generally, numerical methods should be employed. To this end, Computational Fluid Dynamics (CFD) methods developed for ordinary gases may be adopted for the solution of the hydrodynamic equations governing flows with kinetic energy dissipation of colliding granules, represented by the sink term in the energy equation. The numerical methods have to be verified by comparison with relatively simple model problems having analytical solutions. Such problems are the compression and the expansion of semi-infinite granular gases.

Below, we formulate the simplest hydrodynamic model for a flowing granular gas and discuss solutions of the one-dimensional piston problems in detail. Aiming at comparing the solutions with those obtained for molecular gases [10], we restrict our consideration to the Euler-level hydrodynamic equations.

2.1 Hydrodynamic Model

Consider a dense granular gas consisting of spherical particles of diameter σ, mass m and density ρ_p, characterized by the roughness coefficient β [12] and the restitution coefficient e. It is known that the revised Enskog kinetic theory (RET) provides an accurate description for dense gases composed of elastic spheres [13]. Its generalization for inelastic spheres is straightforward [14–16]. The RET kinetic equation can be solved by the Chapman-Enskog method to derive the Euler [15, 16], or Navier-Stokes [17] type hydrodynamic equations. In one-dimensional representation, the Euler-like hydrodynamic equations for a granular gas consisting of inelastic rough particles are [16]:

$$\frac{\partial \rho}{\partial t} + \frac{\partial \rho u}{\partial x} = 0 \tag{9}$$

$$\frac{\partial \rho u}{\partial t} + \frac{\partial}{\partial x}\left(\rho u^2 + P\right) = -g \tag{10}$$

$$\frac{\rho}{m}\left(\frac{\partial E}{\partial t} + u\frac{\partial E}{\partial x}\right) + P\frac{\partial u}{\partial x} = I\,, \tag{11}$$

where t is the time, x the spatial coordinate, g the gravitational acceleration, and I the volumetric sink term. The granular pressure $P = P_k + P_c$ is composed of the kinetic part P_k (cf. (4,5) for the ideal gas) and the collisional part P_c, given by

$$P_k = \alpha_t \rho \frac{E}{2m} \tag{12}$$

$$P_c = P_k \left[2\left(1+e\right)\nu G\left(\nu\right)\right] , \tag{13}$$

where α_t is the translational constant pressure specific heat, $\rho = \rho_p \nu$ is the granular gas density with ν being the volume solid fraction, and $G\left(\nu\right)$ is the contact value of the pair distribution function. The following approximation was proposed for $G\left(\nu\right)$: [18]

$$G\left(\nu\right) = \left[1 - \left(\frac{\nu}{\nu_m}\right)^{4/3\nu_m}\right]^{-1} , \tag{14}$$

where ν_m is the three-dimensional random close-packed volume fraction. This approximation reproduces asymptotic cases of dilute ($\nu \to 0$) and dense ($\nu \to \nu_m$) gases [19]. The term I accounting for the particle kinetic energy losses during their collisions, may be written in the form [15]

$$I = C_0 \sigma^2 \rho G(\nu) \left(\frac{E}{m}\right)^{3/2} + \frac{\partial u}{\partial x} \left(C_1 P_k + C_2 P_c\right) . \tag{15}$$

The coefficients C_0, C_1 and C_2, and the specific heat α_t are known functions of the particle inelasticity e and the roughness β [16]. The specific heat α_t describes the energy distribution between the translational and rotational modes. Thus, when the particle roughness β changes between -1 and 1, α_t varies between 4/3 and 2/3. These limiting cases model perfectly rough and perfectly smooth spheres, where the rotational energy of the particles is conserved. When $\beta = 1$ or -1 the energy E is equipartitioned between all degrees of freedom or just translational degrees of freedom [20]; for $-1 < \beta < 1$ the system avoids states with energy equipartition [16, 21–23]. Moreover, the coefficient C_0 describes energy losses in incompressible gas; C_1 and C_2 account for the losses due to the gas compressibility. The coefficient C_2 characterizes the "dense gas effect", and C_1 stems from the particle roughness and vanishes for $\beta = \pm 1$. In particular, these coefficients govern the speed of sound in a granular gas via the relation [16]

$$a^2 \frac{m}{E} = \frac{\partial}{\partial n}\left(\frac{P}{E}\right) + \frac{\alpha_t^2 P}{4 P_k}\left[\left(C_2 - C_1\right) + \left(1 - C_2\right)\frac{P}{P_k}\right] , \tag{16}$$

where $n = \rho/m$ is the particle number density. In contrast to molecular gases, the kinetic energy of granular gases is not conserved across the shock wave front and the condition (3) has to be modified to the following form:

$$\left[\left(\frac{E}{m}+\frac{P}{\rho}+\frac{(D-u)^2}{2}\right)\rho\,(D-u)\right]=\{C_1P_k+C_2P_c\}\,[u]\;, \tag{17}$$

where $\{\varphi\}\equiv\frac{1}{2}(\varphi_0+\varphi_1)$ for any hydrodynamic variable φ.

In order get insight into the effect of energy dissipation on shock waves in its simplest form, consider a strong shock wave within a granular gas. Combining conditions at the shock front (1,2,17) with constitutive equations (12,13), one obtains

$$u_1 = D\left(1-\frac{\nu_0}{\nu_1}\right) \tag{18}$$

$$P_1 = \rho_0 D u_1 \tag{19}$$

$$\frac{\nu_1}{\nu_0} = 1+\frac{4}{\alpha_t}\frac{1}{1-C_1+2\,(1-C_2)\,(1+e)\,\nu_1 G\,(\nu_1)} \tag{20}$$

As seen, the kinetic energy dissipation as expressed by e appears only in (20) that relates the particle volume fractions ν_0 and ν_1 respectively before and after the shock front. In the limiting case of a dilute gas, where $\nu \ll \nu_m$, this equation reproduces (7) with the adiabatic exponent $\gamma = 1+\alpha_t\,(1-C_1)\,/2$. For rough and very inelastic granular gases, C_1 may reach 1 [16]. This means that the particle roughness may strongly affect γ even for dilute granular systems.

The majority of problems dealing with wavy motion of granular gases were investigated for a particular collision model of perfectly smooth spherical particles. In this case the parameters appearing in (12,15-17) may be written in the form [16]

$$C_0 = -\sqrt{\frac{32\pi}{27}}\,(1-e^2) \tag{21}$$

$$C_2 = \frac{3}{2}\,(1-e) \tag{22}$$

$$C_1 = 0 \tag{23}$$

$$\alpha_t = \frac{4}{3}\,. \tag{24}$$

Equations (23,24) are exact, whereas (21,22) are approximates; the approximation error of C_0 is less than 1% when $0.6 < e \le 1$ [24, 25]. More accurate approximations of C_0 were derived in [24, 25].

At the end of this section, we discuss the gas-solid transition at the shock front. This transition is impossible in ordinary gases. Indeed, according to (14) when $\nu \to \nu_m$, the equilibrium radial distribution function $G\,(\nu) \to \infty$. For elastic particles $C_2 = 0$ and the right-hand side of (20) turns to unity, when $\nu_1 \to \nu_m$. This means that the solid state $\nu_1 = \nu_m$ may be reached only when $\nu_0 = \nu_m$. However, this is no longer true for strongly dissipative gases, characterized by $C_2 \geq 1$, which implies $e \leq \frac{1}{3}$, correspondingly. For such gases, the hydrodynamic model predicts the gas-solid transition, when the initial volume fraction is about $\nu_0 = \nu_m/4$. Below we consider granular gases consisting of particles with the restitution coefficient $e > \frac{1}{3}$, which allow for thermal energy generation on a shock wave, thereby preventing gas solidification.

2.2 Shock Wave

Above we mentioned that the particle inelasticity modifies the conditions on the shock wave. Now we discuss how the dissipation affects the flow behind the shock front. In contrast to ordinary gases, where the temperature behind the shock is uniform, in granular gases it decreases with distance measured from the front surface. Those particles, which completely lost their kinetic energy, form a dense cluster of the maximal possible volumetric density ν_m (see Fig. 3). This cluster first appears on the piston surface and its thickness grows with time as $(D_\infty - U)\, t$.

It is well known that dense clusters are formed even in freely evolving granular gases [3, 4]. Therefore, the formation of such clusters during the compression by the piston is unavoidable and depends on the rate of the energy dissipation as in freely evolving granular gases [18]. For the homogenous cooling state, the granular thermal energy E is dissipated according to the Haff law [26] as $E = E_0\,(1 + t/t_0)^{-2}$, where E_0 is the initial energy and t_0 is the characteristic time scale of the dissipation. This time scale can be evaluated by substitution of the Haff law into the energy balance equation (11), when the granular volume fraction is constant (say), ν_0 and the other hydrodynamic properties are uniform. This yields $t_0^{-1} = 3C_0/\,(\pi\sigma)\,\nu_0 G(\nu_0)\,(E_0/m)^{1/2}$. In the present problem the volumetric granular density ν and the thermal energy E are non-uniform and vary in the intervals $[\nu_1, \nu_m]$ and $[E_1, 0]$, respectively. Here, ν_1 is the density and E_1 is the thermal energy of a granular gas immediately behind the shock wave as follows from conditions (18–20)

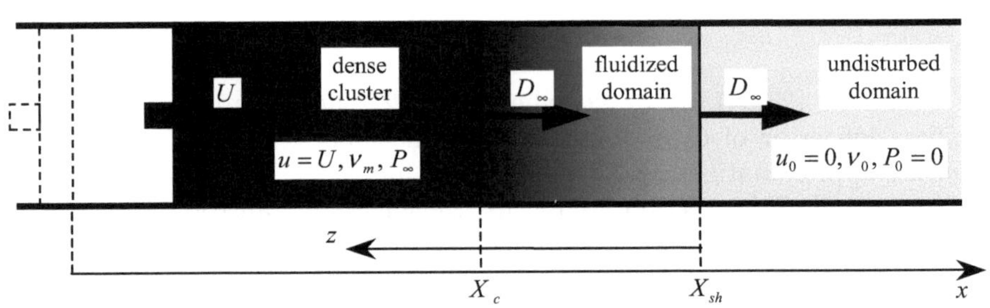

Fig. 3. Schematic of the shock wave propagation within a granular gas

and the constitutive equations (12–14). It was found that the order of magnitude of the dissipation time scale is [27]

$$\frac{1}{t_1} = \frac{3C_0}{\pi\sigma}\nu_1 G(\nu_1)\left(\frac{E_1}{m}\right)^{1/2} . \tag{25}$$

In the long time asymptotic limit $t \gg t_1$ the hydrodynamic equations (9–11) admit a self-similar solution [18]:

$$\psi = \psi(z) , \qquad z = tD_\infty - x + x_0 , \tag{26}$$

where ψ is any hydrodynamic variable, z is the position measured from the shock front (see Fig. 3), D_∞ is the shock wave speed in the limit $t \to \infty$, and x_0 is a constant given by

$$x_0 = \int_0^\infty (D(t) - D_\infty)\, dt . \tag{27}$$

In order to calculate the ultimate limiting values of the shock wave speed and granular pressure on the piston, the mass and momentum conservation conditions (1,2) may be applied to the fluidized domain. Setting the volumetric density, the granular pressure and the velocity at the boundary $x = X_{\text{sh}}$ equal to their values within the undisturbed domain (see Fig. 3), and at the boundary $x = X_c$ to their values within the dense cluster, one obtains

$$D_\infty = \frac{U}{1 - \nu_0/\nu_m} \tag{28}$$

$$P_\infty = \rho_0 D_\infty U \tag{29}$$

It is remarkable that the shock velocity is independent of the isentropic exponent γ (cf. (8)) and the restitution coefficient e. It can be rationalized by noting that the mass and momentum equations underlying (28) do not include these gas properties. The particle inelasticity affects the period during which the shock wave velocity converges to the value given by (28) [27].

Consider a model for the fluidized region separating the densely packed region from the shock front. Here solution (26) should be applied to hydrodynamic equations (9–11). Two of them can be directly integrated by means of conditions (18,19) for the shock wave to yield

$$u(\nu) = D_\infty \left(1 - \frac{\nu_0}{\nu}\right) \tag{30}$$

$$P(\nu) = \rho_0 D_\infty u(\nu) . \tag{31}$$

We see that the velocity and granular pressure depend on the granular volumetric density only. The equation governing the granular gas density immediately follows from (11–13, 15, 23) and can be written in the form

$$z = C_0\sigma^2\rho_p \int_{\nu_1}^{\nu_z} \nu G(\nu) \left(\frac{E}{m}\right)^{3/2} \left[\frac{\rho_0 D_\infty}{m}\frac{dE}{d\nu} - \frac{du}{d\nu}(P - C_2 P_c)\right]^{-1} d\nu\,, \quad (32)$$

where E depends on ν via the relations (12–14)) and (30,31). It follows from (32) that $z = z(\nu, e, \nu_0)$ and it is independent of the piston velocity. The length of the fluidized region is $Z_m = z(\nu_m, e, \nu_0)$.

Expressions (30,32) describe hydrodynamic properties within the fluidized region, and, in particular, allow for the computation of total values of energy $E_{\rm tot}$ and volumetric density $\nu_{\rm tot}$ of granular materials that can be fluidized by a steadily moving piston. The density $\nu_{\rm tot}$ is [18]

$$\nu_{\rm tot} = \nu_m N(\nu_0, e) \quad (33)$$

$$N(\nu_0, e) = \int_0^{Z_m} \nu(z, \nu_0, e)\, dz\,, \quad (34)$$

where $\nu(z, \nu_0, e)$ satisfies (32), and $N(\nu_0, e)$ is equal to the number of rows in the random dense packing of the granular gas. The function $N(\nu_0, e)$ was shown to monotonically increase with decreasing particle inelasticity and increasing density ν_0 [18]. This can be rationalized by noting that with decreasing ν_0 the pressure on the piston P_∞ also decreases (see (29)), which results in a smaller fluidized mass. Below we estimate the maximal value of the number of rows, which can be fluidized by a steadily moving piston:

$$N_{\max} = \sqrt{\frac{3\pi}{8}}\,(1-e)^{-1}\,, \quad (35)$$

where $N_{\max}$ corresponds to the limit $N(\nu_0, e)$, when $\nu_0 \to \nu_m$ and $e \to 1$.

It follows from the analytical solution that the gas pressure on the piston approaches a constant value. The granular kinetic energy tends to zero, and the granular gas density reaches the maximal value ν_m. Such a behavior is caused by a singularity, existing in the radial distribution function $G(\nu)$ in the limit $\nu \to \nu_m$ (see (14)). This singularity is a manifestation of the gas-solid phase transition. In fact, such a transition cannot be modeled numerically in the framework of the hard particle model. The basic deficiency of this model is its inability to account for the energy of elastic particle deformations. Such deformations in the granular layer create pressure P_∞ within the solid cluster (see Fig. 3). To account for the energy of particle deformations a viscoelastic [28] or cut-off collisional models [29–31] can be used. According to the cut-off model, the constant coefficient of restitution e should be replaced by

$$e_{\rm cut} = \begin{cases} e, & \text{if} \quad \zeta < \zeta_{\rm cut} \\ 1, & \text{if} \quad \zeta > \zeta_{\rm cut}\,, \end{cases} \quad (36)$$

where ζ is a variable, which stands for the duration of particle collisions or for the distance between them or for their relative velocity prior to collision, and ζ_{cut} is a cut-off value of this variable. Below this threshold, the condition (36) yields an inelastic hard particle, whereas above it, an elastic collision is assumed.

Two variants of this model have been shown to be useful for the description of the long time asymptotic of the "homogenous cooling state" problem in the presence of clusters [29–31]. A hydrodynamic variant of this model, formulated in terms of cut-off granular density, was suggested in [27].

Kamenetsky et al. [27] solved numerically the Euler-like equations describing granular layer compression by a steadily moving piston using the cut-off model by a CFD algorithm. They found that the results of simulations practically do not depend on ν_{cut} when it is close to ν_m. The CFD calculations reproduce the long time asymptotic solution given by (28–32) with high accuracy, as seen in Figs. 4 and 5. This provides a certain justification for the cut off density model (36).

Figure 4 shows volumetric granular density distributions within the disturbed domain at different moments. In this figure the distance is measured from the piston. The granular density is maximal at the piston and decreases with distance measured from it. The density jump caused by the shock wave indicates the shock front position. In the interval between $t = 32$ and $t = 40$, the granular density reaches $\nu_m = 0.64$ thereby indicating the appearance of the condense region within the layer. At $t = 75$, the numerical and analytical solutions practically coincide. At $t = 40$ the length of the fluidized region calculated numerically by CFD is larger than that predicted by the asymptotic solution. It means that the mass as well as the total kinetic energy in the fluidized region change non-monotonically with time. Therefore, at some time t_1 (see (25) and the discussion above) the energy reaches its maximum.

Figure 5 shows the temporal evolution of the total kinetic energy corresponding to different initial densities. Here, time is rescaled as $t_{\text{res}} = t/t_1$. For relatively advanced times, i.e. $t_{\text{res}} \approx 1 - 2$, we see good agreement between the numerical data and the asymptotic values. It follows from the numerical calculations that the condense region is formed at $t_{\text{res}} \approx 1$ independently of the initial density. Moreover, it is notable that for all initial densities, the fluidized energy reaches a maximum at $t_{\text{res}} \approx 0.5$. It means that at $t_{\text{res}} \approx 0.5$ the gain in granular layer's chaotic kinetic energy is maximal. Hence, this time estimate may serve as a criterion for optimal vibrofluidization of granular layers.

2.3 Expansion Wave

Consider a dilute ($\nu \ll \nu_m$) granular gas filling a semi-infinite spatial domain $x > 0$ separated from vacuum by a partition. At time $t = 0^+$ the partition is withdrawn and the gas expands into vacuum along the negative x-axis. To describe this process we use the Lagrangian coordinate $h = \int_{x_1}^{x} \nu \, dx$ where

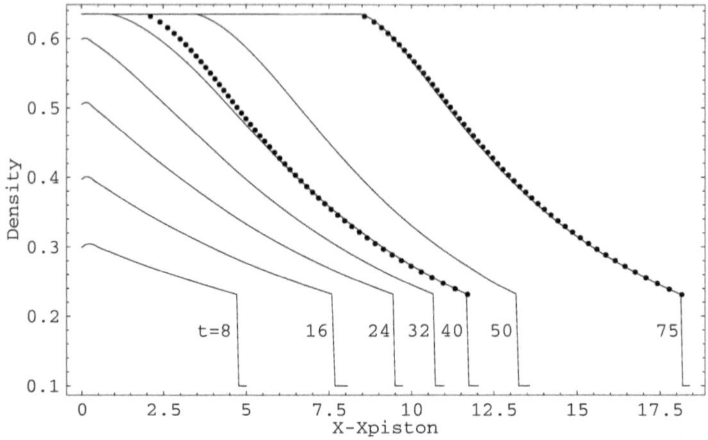

Fig. 4. Evolution of the density distribution for $e = 0.9$, $\nu_0 = 0.1$, $\nu_{\mathrm{cut}} = 0.635$, and $\nu_m = 0.64$. Time and length are scaled by the piston velocity U and the particle radius $\sigma/2$. Solid lines show numerical results; points indicate the asymptotic analytical solution (from [27], Fig. 3)

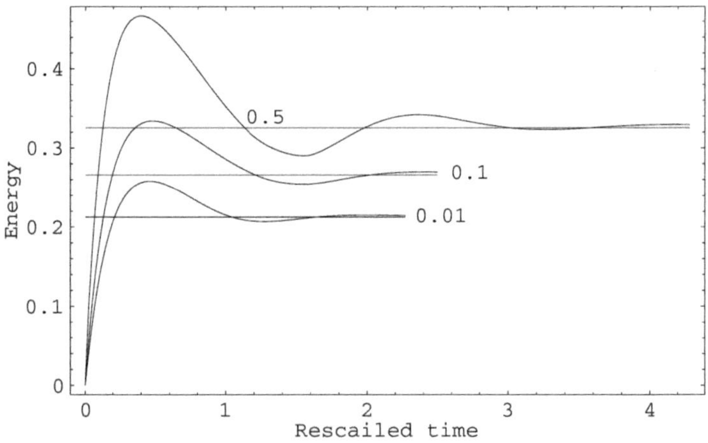

Fig. 5. Dependencies of the fluidized kinetic energy on time for different initial densities $\rho_0 = 0.01$, 0.1 and 0.5, and $e = 0.9$ (from [27], Fig. 18)

x and x_1 denote the Eulerian coordinates of a test and a reference particle, respectively [10]. As reference particle, we choose one near the gas-vacuum interface i.e., $P(t, h = 0) = 0$. Taking the initial values of the gas density ν_0, granular pressure P_0, particle mass m and diameter σ as reference values, one can rewrite the hydrodynamic equations (9–11) in the non-dimensional forms [32]:

$$\frac{\partial}{\partial t}\left(\frac{1}{\nu}\right) = \frac{\partial u}{\partial h} \tag{37}$$

$$\frac{\partial u}{\partial t} = -\frac{1}{\gamma}\frac{\partial P}{\partial h} \tag{38}$$

$$\nu\frac{\partial P}{\partial t} - \gamma P\frac{\partial \nu}{\partial t} = -2\delta\left(\nu P\right)^{3/2}, \tag{39}$$

where the pressure P is expressed via the speed of sound as $P = \rho a^2/\gamma$, $\gamma = 5/3$ (see (16)), and $\delta = \left(4\pi/15\right)^{1/2}\left(1-e^2\right)$. The density and the (zero) bulk velocity within the undisturbed domain do not change. The granular pressure and the speed of sound in the undisturbed region diminish due to granular energy dissipation:

$$\nu_u = 1 \tag{40}$$

$$u_u = 0 \tag{41}$$

$$P_u = \left(1+\delta\, t\right)^{-2} \tag{42}$$

$$a_u = \left(1+\delta\, t\right)^{-1}, \tag{43}$$

where the subscript u denotes the undisturbed quantities. It follows from (43) that the trajectory of the rarefaction wave front is

$$H\left(t\right) = \int_0^t a_u\left(\tau\right) d\tau = \frac{\ln\left(1+\delta\, t\right)}{\delta}. \tag{44}$$

The flow resulting from the expansion of the granular gas into vacuum is thus confined to the region $0 < h < H\left(t\right)$. Since this region is adjacent to the undisturbed domain, the solution at the wave head $h = H\left(t\right)$ should obey conditions (40–43). On the other hand, at the wave tail, $h = 0$, the granular pressure vanishes, i.e. $P = 0$.

This problem was investigated by three different methods: (i) analytical, (ii) computational (CFD) treatment using the hydrodynamic model and (iii) Discrete Element Method (DEM) simulations [33]. We first discuss the key features of the analytical solution. It was shown in [33] that the hydrodynamic velocity u may be constructed in the following self-similar form:

$$u\left(\tau, x\right) = \frac{\gamma-1}{\gamma+1}\frac{U\left(x\right)}{f\left(x\right)}\left(1-\eta^{f\left(x\right)}\right) \tag{45}$$

$$\eta = \frac{\exp\left(x\right)-1}{\tau} \tag{46}$$

$$\tau = \delta\, t \tag{47}$$

$$x = \delta\, h, \tag{48}$$

wherein the functions $f(x)$ and $U(x)$ satisfy the following requirements:

$$U(0) = U_{\text{esc}} = 2/(\gamma - 1) \tag{49}$$

$$f(0) = (\gamma - 1)/(\gamma + 1)\,. \tag{50}$$

The chosen form of the solution provides a transition to the known solution for conservative gases, u_c (see [10]) in the limit $\delta \to 0$:

$$u_c = -U_{\text{esc}} \left[1 - \left(\frac{h}{t} \right)^{\frac{\gamma-1}{\gamma+1}} \right], \tag{51}$$

where U_{esc} is the escape speed - the ultimate upper limit velocity of each particle at large times. Moreover, solution (45–50) reproduces (51) at the wave tail ($x \ll 1$), where the collision frequency is low and, hence, the energy dissipation is weak. The dissipation modifies solution in several aspects: First, the ratio h/t is replaced by η given by (46) to provide for the boundary condition (41) at the wave head $h = H(t)$. Second, the escape speed depends on the coordinate x in the following manner:

$$U(x) = \frac{\exp(x) - 1}{\gamma(\gamma - 1)\exp(2x)\left(1 - \exp(x/2\gamma)\right)}, \tag{52}$$

indicating exponential decay of U for large x. Third, the function $f(x)$ recovers the escape momentum of the granular gas. This momentum may be directly calculated from (38) by mean of integration subject to the boundary conditions imposed on the granular pressure at the wave tail, $h = 0$ and the wave head, $h = H(t)$ (see (44)):

$$M(\tau) \equiv \int_0^\infty u(t,h)\,dh = \frac{1}{\gamma}\int_0^t P_u(t_1)\,dt_1 = \frac{\tau}{\gamma\delta(1+\tau)}, \tag{53}$$

where t_1 is a dummy variable. The gas pressure P_u within the undisturbed domain governs the escape momentum (53). It is valid for any gas inelasticity δ. In particular, for conservative gases (in the limit $\delta \to 0$) expression (53) yields the linear time dependence $M_c(t) = t/\gamma$. In contrast to conservative gases, for granular gases the escape momentum is finite for all times. For large times the escape momentum approaches the limiting value

$$M^\infty = 1/\gamma\delta\,, \quad \tau \to \infty\,. \tag{54}$$

The function f was chosen to satisfy the global momentum conservation condition (53) and condition (50). This was done by choosing the functional form

$$f(x) = \frac{ax + b(\gamma - 1)/(\gamma + 1)}{x + b}, \tag{55}$$

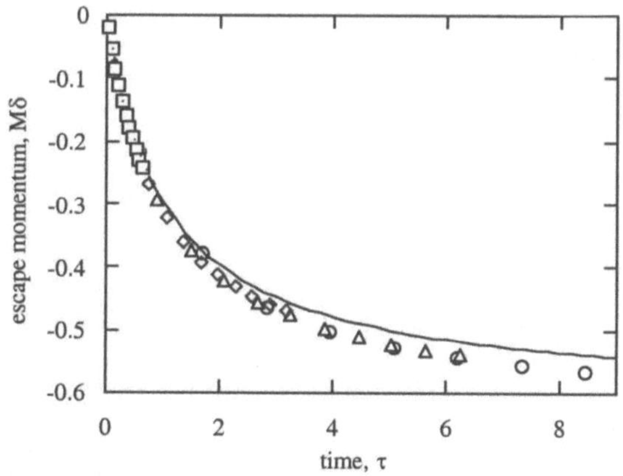

Fig. 6. Dimensionless escape momentum $\delta M\ (\tau)$ versus dimensionless time, τ: solid line corresponds to the analytical (see (53)) and CFD results; DEM calculations (number particles $N = 12000$) *squares:* $e = 0.99$, *diamonds:* $e = 0.95$, *triangles:* $e = 0.9$, *circles:* $e = 0.8$ (from [33], Fig. 3)

which was shown to be an excellent approximation with the parameters $a = 2.0$ and $b = 0.68$.

The approximative analytical solution given by (45–49), (52) and (55) has been compared with the numerical solution of problem (37–44) by CFD and DEM methods. The escape momentum calculated by the three methods for different e is plotted in Fig. 6. Analytical and CFD models predict the self-similar law (53), where $M\delta$ depends on the inelasticity e only via time τ. One can see that DEM simulations virtually reproduce the above universal law with maximal deviation about 5%. The escape energy calculated by the three methods is plotted in Fig. 7. As seen, the analytical model slightly overestimates the CFD data, though both methods predict the self-similar form of the energy evolution. The disagreement between the analytical and the CFD solutions may be attributed to the asymptotic nature of the analytical solution (45–48), which might be improved by including more terms of the expansion [32, 33].

As seen in Fig. 7, the escape energy obtained in the DEM simulations depends weakly on the particle inelasticity $(1 - e)$. The discrepancy between CFD and DEM data grows with both increasing time and inelasticity. Two possible reasons for this discrepancy are inefficient statistics and cluster formation. In order to distinguish between these reasons the DEM calculations were repeated for $e = 0.8$ using a larger number of particles [33]. It was found that increase in particle number results in better agreement with CFD data. Even at large time $\tau \approx 20$ the discrepancy between CFD and DEM calculations did not exceed 4% [33]. This means that cluster formation occurring in the homogeneous cooling state [3] exerts only a minor effect on the hydrodynamic quantities in the problem considered here.

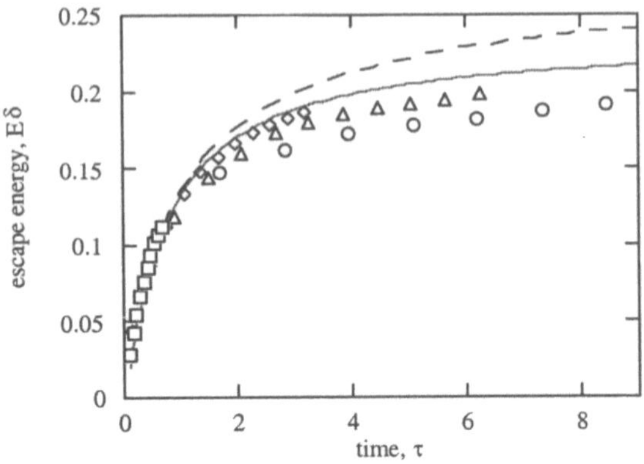

Fig. 7. Dimensionless escape energy, $\delta E(\tau) = \frac{1}{2}\delta \int_0^\infty u^2(t,h)\,dh$ versus dimensionless time, τ: *the solid line* corresponds to CFD; the *dashed line* to the analytical methods; symbols correspond to DEM (number particles $N = 12000$), *squares:* $e = 0.99$, *diamonds:* $e = 0.95$, *triangles:* $e = 0.9$, *circles:* $e = 0.8$ (from [33], Fig. 4)

For conservative gases the ultimate escape energy, E_∞ and momentum, M_∞ (i.e., retrieved for long times by the expanding part of the layer) are finite quantities in contrast with the comparable classical results. They characterize the conversion of the granular thermal energy into the energy of the gas hydrodynamic motion. Having analytical approximations for the escape energy and momentum, one can calculate for slightly inelastic spheres ($e \to 1$) the effective particle volume escaping into vacuum as:

$$N_{\max} \approx M_\infty^2 / E_\infty \approx 0.89\,(1-e)^{-1} \,. \tag{56}$$

This characterizes the maximal number of layers, which may retrieve their hydrodynamic velocity from thermal energy due to the expansion wave. It is remarkable that the estimate (56) of the penetration depth of an expansion wave into a gas is rather close to the maximal depth of a granular layer, which can be fluidized by a steadily moving piston given by (35). These estimations mean that layers deeper than $N_{\max} \approx (1-e)^{-1}$ cannot participate in the energy conversion processes between the two modes of granular motions: chaotic and directed.

Another important similarity between the two model problems is the rate of energy conversion. As can be seen in Fig. 7, after a period of order $\tau \approx 5$ this conversion process is practically completed. The latter conclusion qualitatively agrees with our estimate of the formation of solid clusters expressed in terms of the rescaled time.

3 Two-Dimensional Waves

Now we turn to two-dimensional granular flows. We consider a parallel flow of granules having equal initial velocities. In such a flow no relative granular motion exists. Therefore, both speed of sound and granular temperature vanish. Interactions of this flow with any obstacle are always supersonic and result in the formation of an oblique shock wave. In ordinary gases oblique shock waves can be subdivided into detached and attached shock waves depending on the shape of the obstacle. When a supersonic flow approaches a blunt body a detached shock wave is formed, whereas the interaction of such a flow with a sharp body results in an attached shock wave. Below, we discuss some numerical and experimental results obtained for two-dimensional granular flows around a disk (a blunt body) and a wedge (a sharp body).

No analytical solution of granular hydrodynamic equations is available for two-dimensional granular flows. Moreover, even for molecular gases exact solutions for compressible flow problems are limited to the case of inviscid supersonic flow near a semi-infinite wedge [10]. Therefore, only this flow geometry may possibly be treated analytically for granular gases in view of complexity of their governing equations with respect to those of molecular gases. Generally, numerical methods are required to solve two-dimensional granular gas problems.

The problems considered in this section have many practical applications. The interaction of dilute particle systems with obstacles is met in various two-phase flow situations [34]. Another extreme, namely flows of dense granular gases around obstacles is important for modeling of mixers of granular materials, where the granules interact with mixer blades, impellers etc.

3.1 Flow around a Disk

Buchholtz and Pöschel [35] investigated numerically the flow of rough, inelastic granules interacting with a resting disk obstacle of diameter much larger than that of the granules. They implemented a soft-particle molecular dynamics model assuming some particle deformation during their collisions. The granules are initially uniformly distributed and move with equal velocities. Buchholtz and Pöschel calculated the averaged (hydrodynamic-level) fields of granular density and temperature. They found that in front of the obstacle a "corona" appears in the flow field that separates the undisturbed flow from that affected by the obstacle, as seen in Fig. 8. They pointed out that in the region behind the "corona," granular temperature and density strongly increase and achieve their maxima. We therefore recognize this "corona" as a shock wave front (shown in Fig. 8 by the dashed line), across which the flow properties change abruptly.

We rationalize the interaction between the granular flow and the disk as follows. The incoming flow has a constant hydrodynamic velocity and zero granular temperature resulting in nil sound velocity. The boundary condition on the solid surface requires that the velocity vector be close to the vertical in the vicinity of the front stagnation point. Thus, the flow is deflected by the disk. This causes chaotic particle motion resulting in an increase in granular temperature and speed of sound, a. On the other hand, the gas hydrodyna-

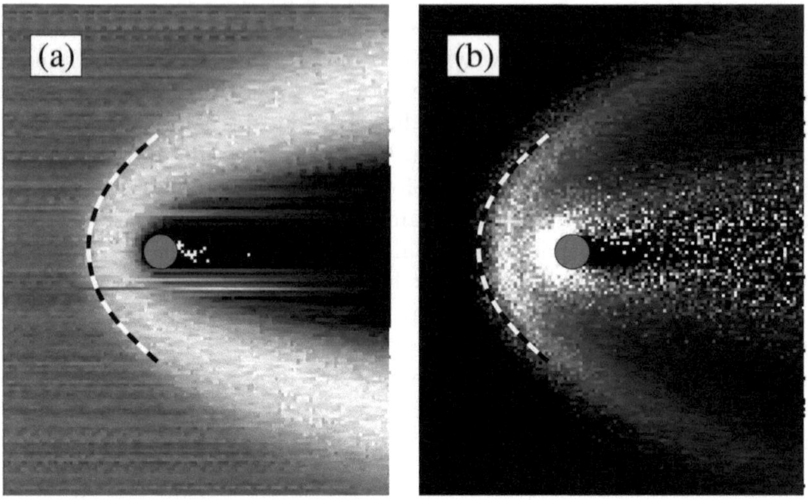

Fig. 8. The fields of granular density **a** and temperature **b** of the flow around a disk. The plots are shaded according to the values of granular density and temperature, so that high values are light and low values are dark. The *dashed lines* show the shock wave front position (from [35], Fig. 12)

mic velocity, u, decreases near the stagnation point due to the flow-obstacle interaction and eventually becomes smaller than the local speed of sound. A region of subsonic flow, where the Mach number $\mathrm{M} = u/a < 1$, appears in the front of the stagnation point, thereby allowing propagation of the disturbances in this region against the flow as seen in Fig. 8, whereas the other parts of the flow remain supersonic with $\mathrm{M} > 1$.

All these granular flow features are typical of ordinary gases as well. In particular, the density distribution presented in Fig. 8b resembles the well-known shadow photographs of a supersonic flow of a molecular gas around a sphere (see Fig. 9). When a sphere is immersed into a supersonic flow field, a detached shock wave appears in front of it as follows from the boundary condition. Moreover, the flow is subsonic near the stagnation point ($\mathrm{M} < 1$) and the rest of the flow field is supersonic ($\mathrm{M} > 1$).

However, there are several features distinguishing the supersonic granular flow from those of molecular gases. Buchholtz and Pöschel [35] found a region of very high density and low temperature in the vicinity of the obstacle nose, for large enough radius of the disk obstacle, R. The granular temperature was shown to decrease when the particles move towards the nose. Such a behavior of the flow field was discussed in Sect. 2.2 in the context of compression wave problem albeit for the limiting situation $R \to \infty$. In particular, the density at the solid boundary was shown to be equal to the maximal packing density ν_m (see Fig. 4).

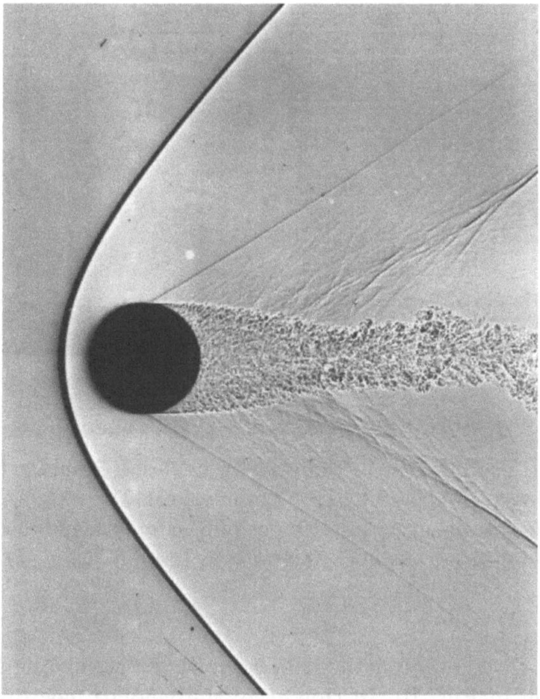

Fig. 9. A sphere at M = 1.53. The shadowgraph catches a 1/2-inch sphere in free flight through air. The flow is subsonic behind the part of the bow wave that is ahead of the sphere, and over its surface back to 45°. At about 90° the laminar boundary layer separates through an oblique shock wave, and quickly becomes turbulent. The fluctuating wake generates a system of weak disturbances that merge into the second shock wave. *Photograph by A.C. Charters* (from [36], Fig. 266)

Another feature untypical for ordinary gases is the appearance of a region of very low density and granular temperature (see Fig. 8). In this region, the granular density is close to zero (vacuum) with very few particles reaching this region. Calculations show that the granular gas steadily expands into this vacuum region with distance away from the obstacle. Several properties of such an expansion wave are discussed in Sect. 2.3.

3.2 Flow around a Wedge

The flow of a granular stream around a wedge was investigated experimentally and numerically by Rericha et al. [37]. In their experiments, stainless steel particles were used falling under gravity past a wedge sandwiched between two glass plates. High speed digital photography was utilized for measurements of the particle positions and velocities during their motion. The authors found that the Mach number of the granular flow in the vicinity of the wedge top was about 12. As in ordinary gases, when a supersonic granular

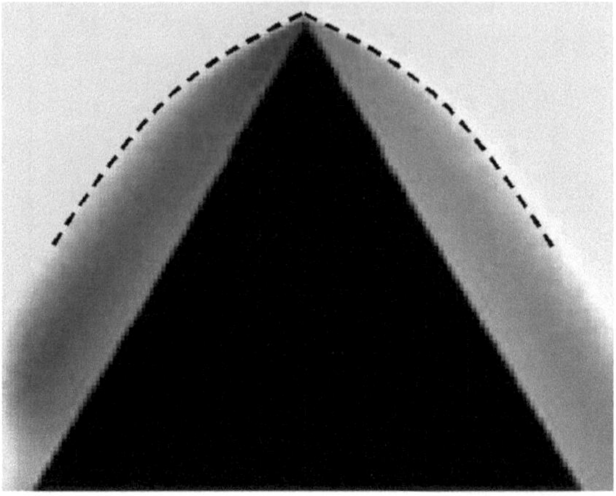

Fig. 10. Horizontal component of the velocity field of a granular flow incident downward on a wedge as predicted by DEM simulations. The plot is shaded according to values of horizontal velocity, so that high values are dark and low values are light. *Dashed line* shows position of shock wave front (from [37], Fig. 2)

flow encounters a wedge, an attached shock wave originating from the wedge tip arises as seen in Fig. 10. In this case, the velocity vector is deflected by the shock wave through just the correct angle to satisfy the boundary condition that the wedge surface be a streamline. Hence, a non-zero horizontal component in the flow velocity vector appears in the granular flow after the shock. The flow downstream of the shock wave has a higher density and temperature, and lower mean velocity, which is similar to shocks in molecular gas flows (see Fig. 11). Moreover, in the case of an attached shock wave the flow remains supersonic everywhere behind the shock front. This is in contrast to the flows around a disk, where in front of the disk, flow is subsonic. The authors also found that when the granular flow reaches the bottom of the wedge an expansion wave is formed leading to granular flow expansion into the vacuum region behind the wedge.

For the numerical analysis of this granular flow Rericha et al. used an event-driven molecular dynamics algorithm in three dimensions. In this algorithm, the granules are inelastic, rough, hard spheres, following parabolic motion between their collisions. The dissipative properties of granular materials are characterized by the coefficients of restitution e and friction μ, which were chosen to fit their experimental data. By means of this fitting, excellent agreement between experimental data and the results of molecular dynamics simulations was obtained. It was found also that the material parameters affect the flow behind the shock front only, whereas the qualitative behavior of the flow remains the same.

Moreover, Rericha et al. solved numerically granular hydrodynamic equations for a flow around a wedge by means of a finite difference algorithm.

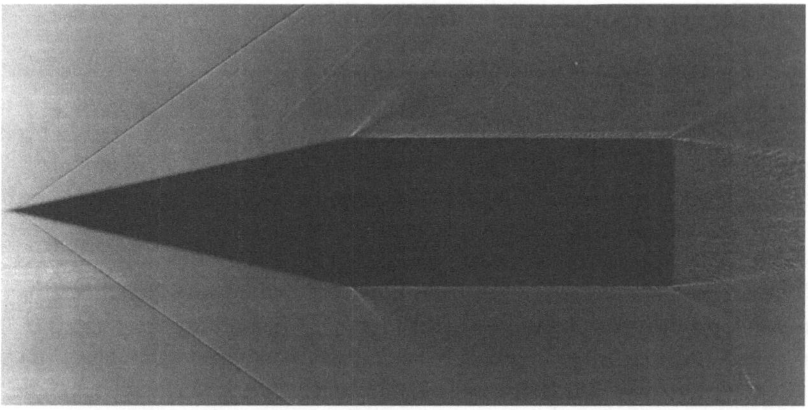

Fig. 11. Cone-cylinder in supersonic free flight. A cone-cylinder of 12.5^o semi-vertex angle is shot through air at M = 1.84. The boundary layer becomes turbulent shortly behind the vertex, and generates Mach waves that are visible in this shadowgraph. *Photographs by A.C. Charters* (from [36], Fig. 259)

They used a two-dimensional problem formulation disregarding the friction between the flow and the side walls. As a result, they obtained some quantitative disagreement between the continuum modeling and experimental results; yet the flow behavior predictions were qualitatively correct. Moreover, their molecular dynamic calculations without wall friction agreed well with solutions of the hydrodynamic equations. This justifies the applicability of hydrodynamic models to two-dimensional problems, where viscous friction on the boundaries can be neglected. The disagreement between the viscous hydrodynamic model and the experimental data probably may be explained by a deficiency of the viscous boundary conditions, as formulated in [38].

4 Waves in Vibrated Granular Media

Vibrated beds of granular materials are frequently encountered in various technological operations. The purpose of vibrations is to intensify the heat and/or mass transfer to or from the granules for the benefit of different industrial processes, such as milling, drying, polishing, etc. Experiments show that vibroagitated granular materials exhibit fluid-like properties. For example, they fill vessels of arbitrary shapes. Vibrated granular materials can be characterized by self-diffusivity, granular temperature [39, 40] and dynamic viscosity [41–43]. Other phenomena characteristic of vibrofluidized granular layers are convective rolls [43–45], transverse waves [41, 46], Faraday waves (see [43] and references therein) and shock waves [1, 2, 47]. In what follows, we discuss basic mechanisms of thermal energy injection from the vibrating vessel into the granular material sustaining the fluid-like state.

4.1 Basic Mechanisms of Vibrofluidization

We identify three major mechanisms whereby thermal-like energy is pumped into a granular system: (i) shear work, (ii) heat flux and (iii) shock waves. The relevance of these mechanisms for granular gases were highlighted by Bagnold [48], Haff [26] and Goldshtein [14], respectively. In vibrated beds several of these mechanisms act simultaneously. The vibrating base produces the work on the layer by means of either mechanism (ii) or (iii), whereas the lateral wall produces the work by means of shear (i).

The work of shear forces had been most extensively studied as a source of granular temperature production [12, 49–55]. It is well known that the effect of a tangentially vibrated wall on viscous gas is restricted to the "penetration depth" [56]. Within this boundary layer the work performed by the wall transforms into the heat through viscous dissipation. For a granular gas the basic mechanism of the viscous energy dissipation is the same, though the resulting state of the gas is apparently governed by the dissipation through inelastic granular collisions. Below we estimate the penetration depth in granular gases.

Following the Stokes solution for the flow near an oscillating plate [56], one estimates the boundary layer thickness as $\delta \approx \sqrt{\nu/\omega}$, where ν is the kinematic viscosity of the fluid and ω is the frequency of the wall oscillations. The work performed by the shear force per unit time within the boundary layer can be approximated as

$$\nu \left(\frac{\partial v}{\partial x}\right)^2 \approx \nu \left(\frac{V_w}{\delta}\right)^2 = \omega V_w^2 , \tag{57}$$

where $V_w = A\omega$ and A is the oscillation amplitude. In the steady state [33] this work is balanced by the energy dissipation characterized by the rate $T^{3/2}(1-e)/\lambda$ [26], where λ is the mean free path, leading to the following expression for the granular temperature within the boundary layer:

$$T = \left(\frac{\lambda\omega}{(1-e)V_w}\right)^{2/3} V_w^2 . \tag{58}$$

Taking into account the known relation between the kinematic viscosity, mean free path and temperature $\nu \approx \lambda\sqrt{T}$, we obtain that the boundary layer thickness is

$$\delta = A\left[\frac{\lambda}{(1-e)A}\right]^{1/3} . \tag{59}$$

In [39], $\lambda \approx 300^{-1}$ cm was measured within granular beds consisting of particles characterized by $e = 0.95$ and vibrated with $7\times10^{-3} < A < 1.6\times10^{-2}$ cm. For their experimental data one can estimate the penetration depth of the shearing wall motion as $\delta \approx 2\times10^{-2}$ cm, which is less than the particle diameter $\sigma \approx 0.1$ cm. This means that the shear work is inefficient as a

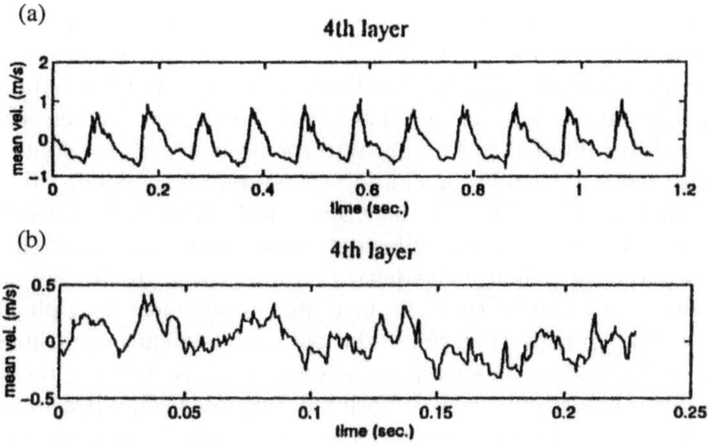

Fig. 12. Mean vertical velocity in the 4th layer at the same velocity amplitude $V \approx 0.88\,\mathrm{ms}^{-1}$ of the piston and the same granular bed ($e = 0.9$, $h_m \approx 4\sigma$), but different vibrational parameters: (a) $A = 0.141\,\mathrm{ms}^{-1}$, $f = 10$ Hz, $\Gamma = 5.68$, (b) $A = 0.0141\,\mathrm{ms}^{-1}$, $f = 100$ Hz, $\Gamma = 56.8$, where $\Gamma = A\omega^2/g$ is the dimensionless acceleration amplitude (from [57], Fig. 15)

mechanism for generating granular temperature for vibrated beds under the above (rather typical) experimental conditions.

Consider a granular gas moving vertically above a vibrating base. Computer simulations evidence [57, 58] that the base can thermally agitate the gas through heat flux or shock wave mechanism. The illustrations presented in Figs. 12a and b depict mean vertical velocities of particles subjected to vertical vibrations in the gravity field. The velocity amplitude was kept constant, but the vibrational amplitude differed by the factor 10. As seen in Fig. 12a, when the vibrational amplitude is small, particle velocities change with time rather chaotically without any correlations. Such a behavior may be viewed macroscopically as stationary. This has been referred to as a heat flux mechanism [26], implying that the granular temperature on the wall is prescribed by the wall kinetics [59]. However, for larger vibrational amplitude as in Fig. 12b, particle velocities are well correlated with that of the oscillating base. The mean particle velocity experiences jumps (see Fig. 12a) which is typical for flows containing shocks. Other experimental and computational evidence of shock wave in vibrated granular beds was presented in [1, 2, 47]. As shown below, significance of each of these two mechanisms is governed by the dissipative properties of the granular layer, namely its depth and granular restitution coefficient.

A criterion governing the depth of the region that may be vibrofluidized includes a parameter $D = N\,(1 - e)$, where N is the number of particle layers [60]. Note that this parameter does not include any information on the granular gas density. Therefore, this parameter may be used to characterize those granular layers where clusters cannot be destroyed via external vibrations [61]. For granular systems vibrated in the absence of gravity, clusters were observed for one-dimensional [60], two-dimensional [62] and three-dimensional [63] geometries. It had been shown that for the latter case, clusters are formed when D exceeds about 0.3. The clusters themselves perform some (chaotic or periodic) motion, thereby precluding the system from coming to a steady state, which is necessary for the heat flux mechanism to be operative [26]. Experiments with granular beds vibrated in the gravity field show that steady state ceases to exist, when inelasticity D approaches a limit (say) D_H [64]. It is reasonable to expect that the heat flux mechanism is applicable only when $D \leq D_H$. Outside that range the only mechanism responsible for the expansion in vertically vibrated layers is shock-expansion waves propagation.

Below we compare efficiency of both mechanisms for fluidization of vibrated beds for microgravity conditions. The problem is of practical importance in space-related applications [63] or situations where the dimensionless piston acceleration $\Gamma = A\omega^2/g \gg 1$, such as shaker ball mills [65].

4.2 Granular Gases Vibrated in Microgravity Field

We start from a brief description of the resonance phenomenon [66, 67], which is observed in microgravity (acceleration g is dropped from (10)). Consider a tube of length L filled with air closed at one end by a rigid plug, and at the other end by a harmonically oscillating piston. Assuming one dimensional gas motion, the following boundary conditions should be imposed on the gas velocity:

$$\begin{aligned} u\left(L, t\right) &= 0 \\ u\left(x_p\left(t\right), t\right) &= \dot{x}_p \\ x_p\left(t\right) &= A \sin\left(2\pi f t\right) , \end{aligned} \tag{60}$$

where x_p is the piston coordinate, and A and f are the oscillation amplitude and frequency, respectively. If f is far from a certain resonance frequency f_r the piston produces standing waves, as shown in Fig. 13, with amplitude of the axial gas velocity of $O\left(\delta^2 a_0\right)$, where $\delta = \sqrt{A/L}$ is the small parameter and a_0 is the speed of sound. When f is close to f_r shock waves develop and the gas oscillations are amplified (Fig. 13), so that the gas velocity is $O\left(\delta\, a_0\right)$. The shock waves travel periodically along the tube between the piston and the resting plug with the resonance frequency f_r given by

$$f_r = \frac{a_0}{2L} \, . \tag{61}$$

A time-space diagram of the gas density at the first resonance is presented in Fig. 14. One can see that at any moment a single shock wave propagates along the tube and separates the region of high gas density behind the shock front from that of low density in the front of the shock. This wave travels

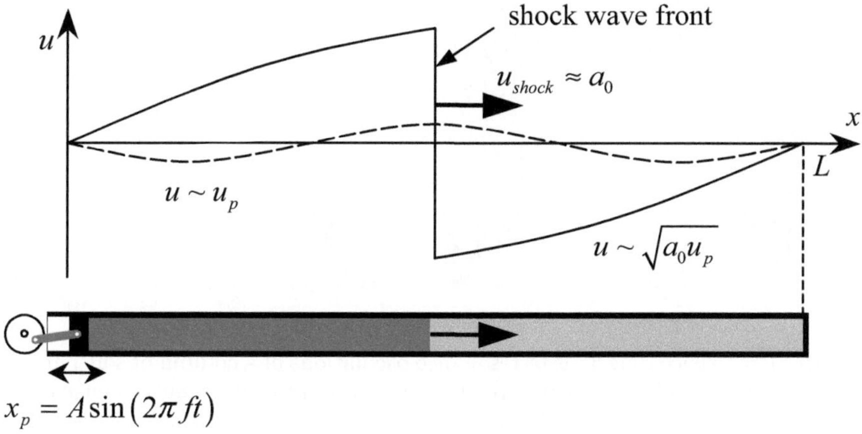

Fig. 13. Schematic of molecular gas oscillations within a closed tube driven by an oscillating piston. The *dashed line* shows velocity distribution with the tube at a frequency f far from the resonance frequency $f_r = a_0/2L$. The *solid line* shows velocity distribution at resonance, when $f = f_r$. Here $u_p = 2\pi A f$ is the maximal piston velocity. Amplification in amplitude of gas oscillations caused by resonance is $\sqrt{a_0/u_p} \approx \sqrt{L/A}$

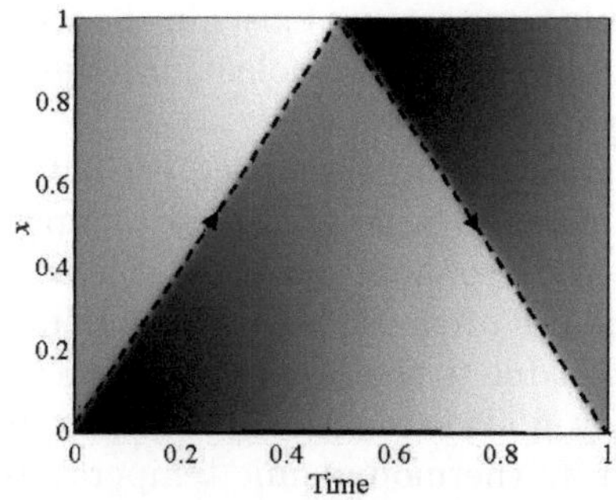

Fig. 14. Time-space diagram of density distribution within a closed tube filled by an ordinary gas at resonance. The plot is shaded according to gas density, so that high density is dark and low density is light. The *dashed line* shows the shock wave front trajectory

back and forth with a speed close to a_0 and strikes the plug and piston once per vibrational period. We note also that time-space diagrams for other thermodynamic variables have qualitatively similar features.

In [58, 68] resonance oscillations in granular gases were investigated by means of DEM simulations in which soft and hard particle models were utilized. Time-space diagrams of granular density and temperature calculated for a one-dimensional column of point-size inelastic soft particles are shown

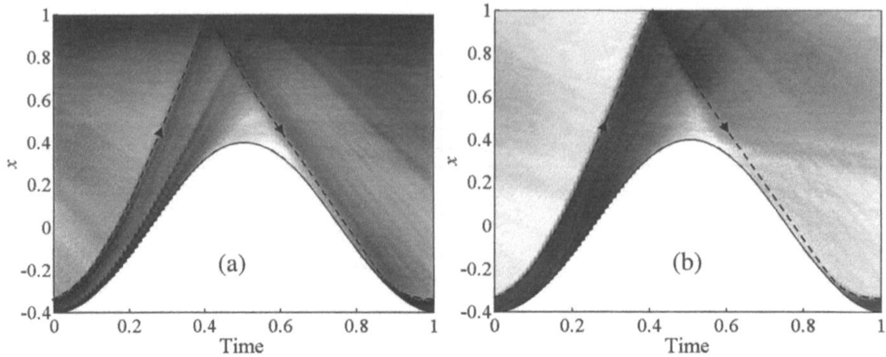

Fig. 15. Time-space diagrams of resonance oscillations of a column of soft particles ($D = 0.6$, $N = 100$). The plots are shaded according to **a** granular density or **b** granular temperature, so that high density is dark, while low density is light. The *dashed line* shows shock wave front trajectory

in Figs. 15a and b, respectively. One can see that the shock travels to and fro along the tube with frequency equal to that of the oscillating piston. As in molecular gases, the shock wave propagates from higher to lower temperature regions within the granular gas. But the granular density distribution differs from that for ordinary gases. As seen in Fig. 15a, a high-density region with granules having negligible temperature exists in the vicinity of the plug. Such a situation is practically impossible in molecular gases due to their relatively high thermal diffusivity.

Another feature distinguishing resonance oscillations of ordinary gases from those of granular gases is large vibrational amplitudes (relative to the tube length) needed to get resonance patterns. It is because of different physical origins of the resonance oscillations in the molecular and granular gases. In the former resonance is due to the proximity of the piston frequency to the natural system frequency f_r. It is practically constant and determined by the gas speed of sound and its thermodynamic temperature. In contrast, in the granular gas the speed of sound is determined by the kinetic energy transferred by the oscillating piston to the granules via shock waves. Thus, resonance granular gas oscillations are a manifestation of the balance existing between the kinetic energy inelastic dissipation and its generation by the oscillating piston.

To obtain vibration parameters providing time-periodic resonance gas oscillations, we consider energy balance condition. Assuming the particle density, ρ, and energy, E, to be close to their initial constant values ρ_0 and E_0 [58, 68] respectively, and for a dilute gas limit ($v \ll v_m$), one can express from (11) and (15) the leading order term of the energy dissipation rate per unit area of the tube as $C_0\,(e)\,\sigma^2 \rho_0\,(E_0/m)^{3/2}\,(Ln_0)$. When gas oscillations are time-periodic these losses are balanced by the energy transmitted to the gas by the piston. For dilute particle systems, inelasticity affects neither the

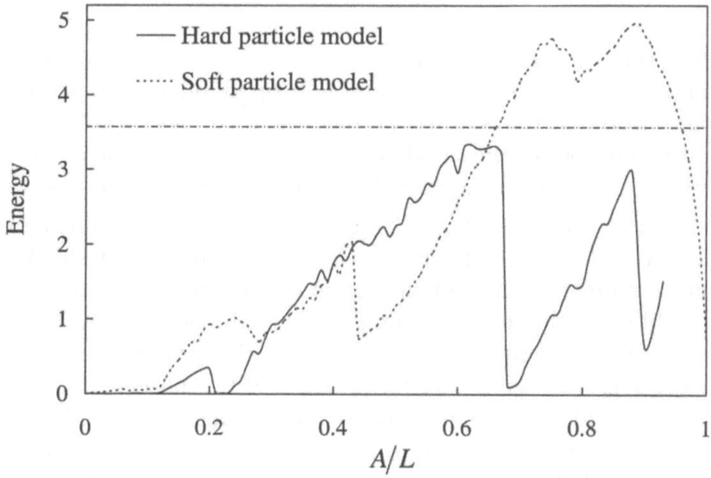

Fig. 16. Kinetic energy versus vibrational amplitude for a column of hard and soft particles ($D = 0.6$, $N = 100$). The *dash-dotted lines* correspond to (65)

speed of sound (16) nor condition (17) at the shock front. Thus, the time-averaged power per unit area generated by the piston may be calculated as for ordinary gases [66]:

$$E_p = -f \int_0^{1/f} Pu|_{x=L}\, dt = \frac{16}{3\sqrt{\gamma+1}} \rho_0 a_0^3 \left(\frac{A}{L}\right)^{3/2} . \tag{62}$$

Equating the expressions for energy generation and dissipation, we get the necessary condition for resonance oscillations of a granular gas:

$$\delta^3 = (A_r/L)^{3/2} = 0.284\,(1-e^2)\,N\,, \tag{63}$$

where A_r is the resonance amplitude and $N = L\rho_0\sigma^2$ is the number of particles in the volume $L\sigma^2$. For e close to unity, the right hand side of (63) is close to the parameter $D = N(1-e)$ used in [60] as a characteristic of the dissipative properties of granular systems. According to (63) the total energy generation is of order δ^3, which is consistent with the weak shock wave approximation [66] implicitly employed to derive this resonance condition. Using condition (63), we obtain that for oscillations in the condition specified in Fig. 15, $A/L = \delta^2 \approx 0.49$. This is rather close to the results calculated numerically with a soft particle model, namely, $A/L \approx 0.4$, and a hard particle model, for which $A/L \approx 0.66$ [58, 68] (see also Fig. 16).

The work produced by the piston is dissipated by the shock waves, which act to heat the gas and to increase the granular temperature. This means that relation (62) estimates the rate of energy pumping by the shock wave mechanism. It was hypothesized in [69] that acoustic waves also can dissipate enough energy to compensate losses due to granular inelasticity. Below we compare the rates of energy dissipation by acoustic and shock waves. Taking into account that acoustic wave energy is proportional to the square of the velocity amplitude and this wave propagates with the speed of sound, one can estimate the power generated by a piston producing acoustic waves as $\frac{1}{2}\rho_0 a_0 (2\pi f A)^2$. This estimation divided by relation (62) under condition (61) yields the power ratio $\sim (A/L)^{1/2}$. The latter result, obtained under the assumption $A \ll L$, accepted for acoustic waves, imply that shock waves in the resonance regime dissipate much more energy than acoustic waves.

It is notable that the resonance amplitude is independent of the oscillation frequency; it is completely defined by the dissipative properties of the granular system. Therefore, any vibration frequency will be the resonance one, since any initial state will eventually evolve to the resonance state [58], where the dissipated and generated amounts of energy are balanced. In this state the speed of sound and the kinetic energy in the system are given by [58]

$$a_r = 2Lf \,, \qquad E_r = a_r^2/\gamma\,(\gamma - 1) \,, \tag{64}$$

where $\gamma = 5/3$. Thus, the kinetic energy per particle at resonance is

$$E_r = 3.6\,(Lf)^2 \,. \tag{65}$$

Apparently for other amplitudes the oscillations are out of resonance and, therefore, the kinetic energy is below that given by (65), which is the upper limit of energy that can be sustained within a granular layer by means of external vibrations.

Figure 16 shows the kinetic energy of an oscillating granular column with $D = 0.6$ versus piston amplitude as calculated by DEM for soft and hard particle collision models. Both curves have similar qualitative behavior. They have several global maxima corresponding to different patterns of resonance oscillations. Thus, the maxima at $A/L = 0.4$ and $A/L = 0.66$ correspond to the first resonance oscillations according to the soft and hard particle models, respectively. The energy maximum at $A/L = 0.2$ as predicted by the hard particle model is caused by the second resonance, when the shock interacts with the plug once per two piston periods. At large amplitudes, the maxima are associated with the half resonance, when the shock is reflected by the plug twice per piston period [58].

When $A/L < 0.1$ the kinetic energy is small compared with its maximal value and practically indistinguishable in Fig. 16. This state is associated with the formation of a cluster, which contains all the particles in the column except very few (one to three particles). The cluster moves near the plug and only few particles oscillate non-periodically between the piston and the cluster as shown in [60]. It was found that for non-resonant oscillations the number of fast particles is practically independent of D, when $D > 0.2$ [58]. Such energy distribution between the particles cannot be treated by known hydrodynamic models. However, for layers with $D < D_H \approx 0.2$, the number of chaotically moving fast particles increases, which indicates partial fluidization of the column by the heat flux mechanism. Under resonance conditions maximal kinetic energy is injected by the piston to the inelastic column, yielding maximal number of rapidly moving particles, thus providing maximal fluidization.

As seen in Fig. 16, the energy calculated by hard particle DEM at the first resonance for $A/L = 0.66$ is rather close to that predicted by the hydrodynamic model of (65) shown in this figure by the dash-dotted line. The kinetic energy never exceeds the limiting value given by (65) calculated for the hard particle model. This confirms the hydrodynamic nature of the column oscillations. On the other hand, when the soft particle model is used, the kinetic energy at the first resonance is about half of E_r, whereas for $A/L > 0.7$ it significantly overestimates this limiting value. This discrepancy with the hydrodynamic prediction is related to the so-called "detachment" effect [70, 71], which prevails for soft particles. Namely, when a layer consisting of soft particles collides with a wall, a solid particle cluster may be formed within the layer due to its compression. Such a cluster conserves part of the kinetic energy associated with the elastic deformation of the particles. After the rebound from the wall the elastic deformation energy of the cluster is retrieved from the particles and converted to their relative motion resulting in a gain in the granular temperature of the layer. One possible way to describe the detachment effect by means of hydrodynamic models is to utilize the cut-off collision models [29]. As was mentioned above, this model was found efficient for the description of the formation of solid granular clusters within granular gases [27]. Hence, this model is expected to be also efficient for the description of the inverse effect of the detachment. For such a goal the phenomenological hydrodynamic model [27] should be replaced by a more rigorous one [72, 73].

4.3 Granular Gases Vibrated in Gravity Field

The dynamics of granular beds vibrated in the gravity field is of applied [74] and also of fundamental interest. We believe that shock waves play a major (though hitherto unrecognized) role, especially for beds that are not too shallow (few layers in depth). It seems that in thick layers "granular heat" flux cannot reach very deep into the layer bulk. In this case, shock waves' propagation is to assume the leading position among the energy injection mechanisms. This claim clearly needs substantiation by further research. Nevertheless, we wish to mention several findings, supporting it. In [63] it was discovered that steady states of beds deeper than four layers and vibrated in the gravity field are unstable to the emergence of bouncing clusters. In addition, [1, 2, 47] present experimental and computational evidences of shock wave propagation through granular beds.

Figure 17 shows the evolution of the granular density of a layer consisting of nine monolayers of particles with $e = 0.9$ and vertically vibrated with

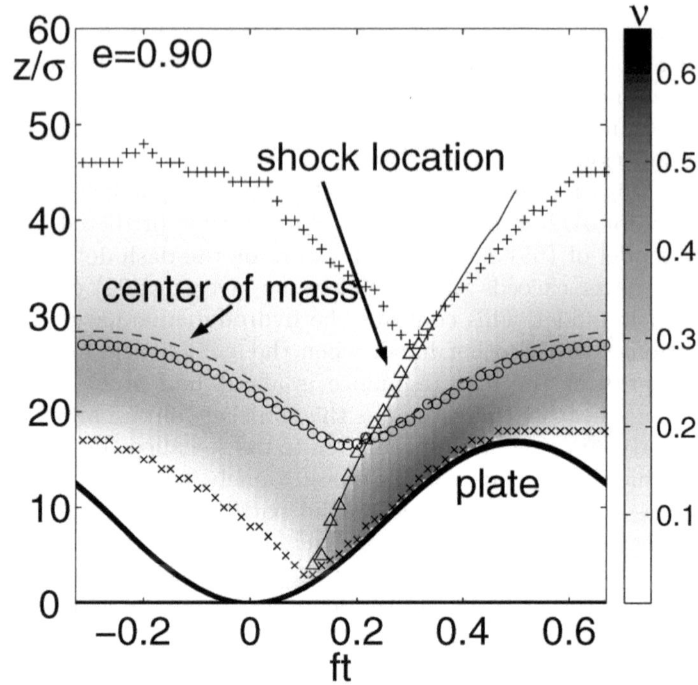

Fig. 17. Location of the shock (solid lines for continuum, triangles for MD) and the center of mass of the layer (dashed lines for continuum, circles for MD) as a function of the oscillation period ft of the plate (thick solid line) for particles with $e = 0.9$. The plot is shaded according to the volume fraction from the continuum simulation, so that high volume fraction is dark and low volume fraction is light. The "top" and the "bottom" of the layer from MD (when the volume fraction drops to less than 4% of random close packed) are shown as + and × symbols, respectively. The material below the shock is visibly compressed as compared to the region above the shock, as can be seen from the shading (from [2], Fig. 2)

dimensionless vibrational acceleration $\Gamma = 3$ and frequency $f \approx 30$ Hz as calculated by Bougie et al. [2]. We now describe the motion of this layer on the basis of a simplistic hydrodynamic concept proposed in [1]. This concept underlies the model of a deep granular bed vibrated within a vessel in the gravity field, which will be discussed later. It is assumed that the hydrodynamic motion of the granular layer is periodic with period $1/f$. We start the description from the moment when the layer having low granular temperature falls under the action of gravity towards the vibrating bottom. Short time after the layer collides with the vessel, particles on the layer bottom attain hydrodynamic velocity close to that of the vessel and a shock wave is formed. In front of (above) the shock wave, the undisturbed particles still move downwards in free fall. As a result, the center of mass of the layer continues its motion downwards as seen in Fig. 17. The shock wave propagates from the lowermost to the uppermost part of the granular layer, compressing the granular gas and generating random motion of the particles behind the front. The physics of shock wave propagation in granular gases is discussed in Sect. 2.2. When the shock wave reaches the layer top, where eventually the granular pressure vanishes, $P = 0$, an expansion wave is formed, which propagates in the opposite direction from the uppermost to the lowermost part of the layer (see Sect. 2.3). The propagation of the expansion wave results in a transformation of the chaotic kinetic energy of the particles (granular temperature) into the energy of average (hydrodynamic) motion. Owing to this energy transformation, the particles involved in the expansion process accelerate upwards, while the granular temperature decreases. The expansion wave propagating downwards eventually reaches the bottom of the layer causing its detachment from the vessel. This creates another expansion wave propagating upwards. Both expansion waves expand the layer during its free flight, recovering its density to the value prevailing before the shock wave passage (see Fig. 17).

This description explains several experimental observations pertaining to vibrated systems [43]: (i) slow density variation and expansion of the granular layer during its free flight, and (ii) rapid (over a small fraction of an oscillation cycle) compression of the granular layer when it hits the floor. As follows from the above description, the viborofluidized state is governed by a shock-expansion wave mechanism. The shock wave heats the granular gas compensating energy losses due to inelastic collisions, whereas the expansion wave transforms granular temperature into the energy of average hydrodynamic motion, precluding granular layer collapse to a compacted state.

It is instructive to get a simple analytic model, which grasps the overall behavior and integral properties of a fluidized layer, and gives insight into the shock wave dynamics. Such a model was devised in [1] on the basis of several simplifying assumptions: (i) particle volume concentration ν have to be close to the maximal packing density, ν_m, i.e. $\Delta \equiv 1 - \nu/\nu_m \ll 1$; (ii) the period t_c during which the layer is in contact with the vessel, is much shorter than the period of layer free-flight t_f, i.e., $t_c/t_f \ll 1$; (iii) the layer detaches from the vessel bottom when its acceleration is about $-g$ (like for a solid body). Equations (9–11) were solved analytically [1] under these assumptions to give the average bed porosity, granular kinetic energy and time of propagation of the shock wave through the layer in the form:

$$\overline{\Delta} = \frac{3}{2} k V^2 \tag{66}$$

$$\frac{\overline{E}}{mgh_m} = \frac{k}{4} V^2 \tag{67}$$

$$t_{\rm sh} f = \frac{k}{4\pi} \Gamma \,, \tag{68}$$

where $V = (2\pi Af)/\sqrt{gh_m}$ and $\Gamma = A(2\pi f)^2/g$ are the dimensionless amplitude of vessel velocity and acceleration, respectively, and h_m is the height of the granular layer in its condensed state. The coefficient $k(H)$ is given by the following relations

$$k(H) = \frac{f(H)}{0.5 + q(H)} \tag{69}$$

$$f(H) = \frac{1}{\sqrt{3}H} \ln\left(1 + \sqrt{3} - \sqrt{3}\exp(-H)\right) \tag{70}$$

$$\begin{aligned} q(H) = & \frac{1}{H}\left[\left(1 + \frac{1}{\sqrt{3}}\right)(\exp(H) - 1) - H\right] \\ & - \frac{1}{H^2}\left[\left(1 + \frac{1}{\sqrt{3}}\right)(\exp(H) - 1 - H) - \frac{H^2}{2}\right] . \end{aligned} \tag{71}$$

Here, $H = \left(1 - e^2\right) h_m / \left(\sqrt{\pi}\sigma\right)$ characterizes the kinetic energy losses within the layer. The scaling relations (66,67) may be used to underline the advantageous nature of high amplitude and low frequency vessel oscillations for layer fluidization by means of vertical vibrations. For a given constant acceleration of the vibrated vessel such vibrational regimes yield low layer porosity and high granular kinetic energy, thereby sustaining the gaseous state of the granular bed. On the other hand, these regimes allow easy observation of shock waves with simple equipment such as a home video camera (see Fig. 1). Moreover, these regimes were shown efficient for mixing of different granular materials [75].

The scaling relations (66–68) have been verified by processing data obtained from photographs of the particle distribution within vibrated beds consisting of glass beads. Theoretical prediction of the particle average porosity was shown to be in a good qualitative agreement with the experimental data [1]. In [75], relation (68) was used for the estimation of t_c. The authors have reported qualitative agreement between the experimental data and hydrodynamic model predictions.

A still better understanding of the wave dynamics of vibrofluidized granular layer may be achieved on the basis of a combined numerical analysis based on simulations by both DEM and CFD methods. Comparison of these methods allows to determine the range of applicability of the hydrodynamic description of vibrated granular media. The first step in this direction was done by Bougie et al. [2], who calculated the motion of a 3D granular layer vertically vibrated within a vessel open from the top. In their calculations, the authors implemented a hard sphere event-driven algorithm for DEM simulations. In parallel continuous hydrodynamic equations were solved by a finite-difference algorithm. The authors reported that after a transient during several oscillation cycles, periodic layer motion was reached with a period equal to that of the external vibrations. They found that during each oscillation period, a shock wave is formed due to the collision between the layer and the oscillating vessel bottom. This shock propagates through the granular layer up to its uppermost part with the Mach number being about 10, thus indicating that this flow is strongly supersonic. A sharp increase in granular temperature was found behind the shock front. This implies that the shock transfers the kinetic energy from the oscillating piston deep into the granular layer.

The density and temperature distributions behind the shock front qualitatively agree with those obtained for the problem of piston motion in a cold granular gas, as described in Sect. 2.2. Therefore, this emphasizes the relevance of the piston problem to understand the physics of the processes governing vibrated granular gas. Moreover, the prediction of the shock wave propagation time given by (68) agrees with the value computed in simulations by Bogie et al. [2]. Indeed, substitution particle inelasticity $e = 0.9$ and $h_m \approx 10\sigma$, used in simulations in [2], into (69–71) yields $H \approx 1.07$ and $k(H) \approx 0.25$. Introducing the latter value into (68) and $\Gamma = 3$ yields $t_{\mathrm{sh}} f \approx 1/16$. According to [1], the propagation time of an expansion wave is of the same order as the shock wave, which yields the contact time of about $\approx 1/8$. This time is only two times larger than that reported in [2]. The discrepancy can be attributed to the asymptotic nature of the model (66–71) and the underlying hypotheses (i–iii).

Bogie et al. [2] also reported a good agreement between their DEM and CFD simulations even for strong dissipations, where the validity of the continuum equations is in question since the volume fraction approaches the close-packing value. In particular, they observed shocks in granular layers with the dissipative parameter D ranging from 0.4 to 1.5. When D exceeded 1.5, they stopped their hard sphere DEM simulations because of clusters' formation.

The limiting value $D \approx 1.5$ obtained in [2] for a vibrating piston is rather close to the theoretical limit (35) calculated for a steadily moving piston when $t \to \infty$. It was shown numerically [27] that D becomes maximal when the time of the piston motion is equal to that given by (25). This maximum is very close to 1.5. However, this limit observed in simulations in [2], was obtained for fixed vibrational parameters. Experiments [1] evidence that an appropriate choice of vibrational parameters allows vibrofluidization of layers with $D \approx 2$. Does an upper limit of D exist for which cluster formation can be avoided? In numerical and theoretical studies [61, 76] dealing with vertically vibrated one-dimensional columns this limit was found to be $D \approx 3$. Granular layers characterized by $D > 3$ cannot be investigated in the framework of the hard sphere model due to the cluster formation, and the soft particle model allowing particle deformations should be utilized. Highly dissipative vertically vibrated systems with $D > 3$ were investigated experimentally [75] and numerically [71], and some fluid-like layer properties were registered: mixing, waves, convection. This unexpected behavior was explained by the fact that highly dissipative layers prevail interchangeably in either condensed or fluidized state. The transition from the condensed to the fluidized state was found to result from the "detachment effect" [71], discussed in Sect. 4.2.

Bogie et al. [2] found that granular inelasticity does not affect qualitatively the motion of the granular layer up to the layer dissipation parameter $D >$ 0.4. For layers with D as small as 0.4, the authors observed gaseous steady states without a coherent layer motion typical for more dissipative layers. One may hypothesize that this steady state is governed by the heat flux mechanism, although the critical D_H value defining the transition between the shock-wave and heat-flux mechanisms was not calculated. In [57] the possibility of vibrofluidization of layers with $D \approx 0.4$ by means of the heat-flux mechanism was reported. However, experimental verification of their simulations is difficult since the granular layer should be vibrated with a huge acceleration $\Gamma = 56.8$. This value exceeds the limit imposed by mechanical strength of typical vibrational equipment.

The observations in [2] support the dominant role of shock waves in vibrofluidization of highly dissipative granular layers that mostly met in nature and industry. Finally, the results demonstrate that continuum equations can describe rather well the complicated time-depended behavior of granular layers. Further DEM and CFD simulations verified by laboratory measurements will advance our understanding of the role of shock waves as a mechanism providing energy transport from solid obstacles or vibrated surfaces into granular materials.

5 Summary

The dissipative nature of grain collisions is the main source of the principal difference between "granular" and molecular gases. Due to inelastic collisions, granular gases can rapidly lose their temperature and collapse into clusters even in the absence of attractive inter-particle forces. In order to prevent cluster formation a permanent external source is needed to perform work and pump energy into the granular system. We identified three major mechanisms responsible for this energy pumping: (i) shear induced work, (ii) heat flux and (iii) shock waves. The present review is the first attempt to systematize results on the shock wave mechanism and compare it with other mechanisms.

We presented experimental and computational evidences of shock and expansion waves propagating within granular gases. The role of these waves was discussed in granular flow interactions with obstacles and in vibrated granular materials. We showed that a shock wave propagating through a granular gas transfers the energy of directed (hydrodynamic) motion of the granular bulk into granular temperature and thereby causing gas compression. On the other hand, an expansion wave transfers granular temperature into energy of directed motion of the granular gas, expanding it into regions of low granular pressure. Interactions of these waves govern the resulting hydrodynamic state of granular gases in these flows.

In Sect. 2 we discussed the simplest model of the wavy granular motion based on the Euler-type hydrodynamic equations written for a system of hard, inelastic, rough spheres. We also studied model problems of shock and expansion wave propagation in a semi-infinite granular gas, namely, a shock wave formed by a steadily moving piston and expansion of a granular gas into vacuum. It was shown that only a limited part of piston's work may be converted into the granular temperature and, similarly, only a limited part of the semi-infinite layer may be fluidized. In the complementary expansion wave process, only a limited part of the kinetic energy of a semi-infinite layer may be transformed into the energy of directed granular motion. The limits of energy transformation resulting from granular inelasticity were estimated in terms of the dissipative parameter $D = N(1-e)$. It was shown that for both model problems the maximal number of granular layers, $N_{\max}$ participating into the process of energy transformation is proportional to $(1-e)^{-1}$ (see (35,56)), thereby resulting in $D \approx 1$ In Sect. 4 we show that this estimation agrees qualitatively and even quantitatively with the estimation of the maximal depth of a vibrated granular layer, sustainable in a hydrodynamic state by means of vibrations.

We compared the mechanisms (i)–(iii) with respect to their ability to vibrofluidize granular layers. We showed that for typical experimental conditions of a vertically vibrated layer the effect of shear is negligible. Experiments and computer simulations show that the heat flux mechanism is restricted to $D_H \approx 0.4$ in the gravity field and $D_H \approx 0.2$ in microgravity conditions, respectively. For $D > 0.4$, the granular layer may be fluidized by means of the shock/expansion mechanism only.

Scaling relations for parameters governing optimal vibrational regimes providing maximal energy pumping into granular layers vibrated in microgravity were presented. Namely, the maximal value of energy of relative granular motion can be achieved (see (65)) if the vibrational amplitude is sufficiently large (as given by (63)). It is important to stress that the hydrodynamic models of the shock-expansion wave problems considered in Sects. 2–4 were favorably compared with computer simulations and a good agreement was observed. This provides a basis for concluding that hydrodynamic models are efficient tools for investigation of compressible flows of granular gases.

Acknowledgment

AG gratefully acknowledges CECAM for funding his participation in the Workshop on Granular Gases. Participation in this workshop inspired writing this review. The authors are indebted to T. Pöschel for his energetic assistance in preparation of the paper. One of us (AG) would like to acknowledge partial support by the Israel Science Foundation, grant no. 53/01. Fruitful discussions with S. Luding, E. Ben-Naim, T. Pöschel and N. Brilliantov are acknowledged.

References

1. A. Goldshtein, M. Shapiro, L. Moldavsky, and M. Fichman. Mechanics of collisional motion of granular materials. Part II. Wave propagation through vibrofluidized granular layers. *J. Fluid Mech.*, 287:349, 1995
2. J. Bougie, S.J. Moon, J.B. Swift, and H.L. Swinney. Shocks in vertically oscillated granular layers. *Phys. Rev. E*, 66:051301, 2002
3. I. Goldhirsch and G. Zanetti. Clustering instability in dissipative gases. *Phys. Rev. Lett.*, 70:1619, 1993
4. S. McNamara and W.R. Young. Inelastic collapse and clumping in a one-dimensional granular medium. *Phys. Fluids A*, 4:496, 1992.
5. E. Salomon and M. Mareschal. Usefulness of the Burnett description of strong shock waves. *Phys. Rev. Lett.*, 69:269, 1992
6. F.J. Uribe, R.M. Velasco, and L.S. Garcia-Colin. Burnett description of strong shock waves. *Phys. Rev. Lett.*, 81:2044, 1998
7. J.M. Montanero, M.L. de Haro, V. Garzo, and A. Santos. Strong shock waves in a dense gas: Burnett theory versus Monte Carlo simulation. *Phys. Rev E*, 58:7319, 1998
8. J.M. Montanero, M.L. de Haro, A. Santos, and V. Garzo. Simple and accurate theory for strong shock waves in a dense hard-sphere fluid. *Phys. Rev E*, 60:7592, 1999
9. C. Cercignagni, A. Frezzoti, and P. Grosfits. The structure of an infinitely strong shock wave. *Phys. Fluids*, 11:2775, 1999
10. R. Courant and K.O. Friedrichs. *Supersonic Flow and Shock Waves.* Interscience, New York, 1948.
11. Y.B. Zeldovich and Y.P. Raizer. *Physics of shock waves and high-temperature hydrodynamic phenomena.* Academic Press, New York, 1966
12. C.K.K. Lun and S.B. Savage. A simple kinetic theory for granular flow of rough, inelastic, spherical particles. *J. Appl. Mech. - T. ASME*, 54:47, 1987
13. H. van Beijeren and M.H. Ernst. Kinetic theory of hard spheres. *J. Stat. Phys.*, 21:125, 1979
14. A.I. Goldshtein. *Determination of structure of hydrodynamic equations and investigation of self-similar motion of gas consisting of inelastic smooth spheres.* PhD thesis, Institute of Theoretical and Applied Mechanics, USSR Acad. Sci., Novosibirsk, 1989
15. A.I. Goldshtein, V.N. Poturaev, and I.A. Shulyak. Structure of the equations of hydrodynamics for a medium consisting of inelastic rough spheres. *Fluid Dynamics*, 25:305, 1990
16. A. Goldshtein and M. Shapiro. Mechanics of collisional motion of granular materials. Part I. General hydrodynamic equations. *J. Fluid Mech.*, 282:75, 1995
17. V. Garzo and J.W. Dufty. Dense fluid transport for inelastic hard spheres. *Phys. Rev. E*, 59:5895, 1999
18. A. Goldshtein, M. Shapiro, and C. Gutfinger. Mechanics of collisional motion of granular materials. Part III. Self-similar shock-wave motion. *J. Fluid Mech.*, 316:29, 1996
19. B.J. Alder and W.G. Hoover. Numerical statistical mechanics. In H.N.V. Temperley, J.S. Rowlingson, and G.S. Rushbrooke, editors, *Physics of Simple Liquids*, page 79, Amsterdam, 1968. North-Holland

20. S. Chapman and T.G. Cowling. *The Mathematical Theory of Non-Uniform Gases.* Cambridge University Press, Cambridge, 3rd edition, 1970
21. M. Huthmann and A. Zippelius. Dynamics of inelastically colliding rough spheres: Relaxation of translational and rotational energy. *Phys. Rev. E*, 56:R6275, 1997
22. S. Luding, M. Huthmann, S. McNamara, and A. Zippelius. Homogeneous cooling of rough, dissipative particles: Theory and simulations. *Phys. Rev. E*, 58:3416, 1998
23. S. McNamara and S. Luding. Energy nonequipartition in systems of inelastic rough spheres. *Phys. Rev. E*, 58:2247, 1998
24. T.P.C. van Noije and M.H. Ernst. Velocity distribution in homogeneous granular fluids: the free and the heated case. *Granular Matter*, 1:57, 1998
25. T.P.C. van Noije and M.H. Ernst. Kinetic theory of granular gases. In Pöschel and Luding [77], page 3
26. P.K. Haff. Grain flow as a fluid-mechanical phenomenon. *J. Fluid Mech.*, 134:401, 1983
27. V. Kamenetsky, A. Goldshtein, M. Shapiro, and D. Degani. Evolution of shock wave in a granular gas. *Phys. Fluids*, 12:3036, 2000
28. N.V. Brilliantov, F. Spahn, J.-M. Hertzsch, and T. Pöschel. Model for collisions in granular gases. *Phys. Rev. E*, 53:5382, 1996
29. S. Luding and S. McNamara. How to handle the inelastic collapse of a dissipative hard-sphere gas with the TC model. *Granular Matter*, 1:113, 1998
30. E. Ben-Naim, S.Y. Chen, G.D. Doolen, and S. Redner. Shocklike dynamics of inelastic gases. *Phys. Rev. Lett.*, 83:4070, 1999
31. X. Nie, E. Ben-Naim, and S.Y. Chen. Dynamics of freely cooling granular gases. *Phys. Rev. Lett.*, 89:204301, 2002
32. A. Goldshtein, M. Shapiro, and C. Gutfinger. Mechanics of collisional motion of granular materials, Part IV. Expansion wave. *J. Fluid Mech.*, 327:117, 1996
33. A. Goldshtein, V. Kamenetsky, A. Potapov, M. Shapiro, C. Campbell, and D. Degani. Hydrodynamics of rapid granular flow of inelastic particles into vacuum. *Granular Matter*, 4:115, 2002
34. B.A. Balanin. Influence of reflected particles on the ablation of a body in a two-phase flow. *Fluid Dynamics*, 19:841, 1984
35. V. Buchholtz and T. Pöschel. Interaction of a granular stream with an obstacle. *Granular Matter*, 1:33, 1998
36. M. Van Dyke. *An album of fluid motion.* The Parabolic Press, Stanford, 1988
37. E.C. Rericha, C. Bizon, M.D. Shattuck, and H.L. Swinney. Shocks in supersonic sand. *Phys. Rev. Lett.*, 88:014302, 2002
38. I. Goldhirsch. Probing the boundaries of hydrodynamics. In Pöschel and Luding [77], page 79
39. O. Zik and J. Stavans. Self-diffusion in granular flows. *Europhys. Lett.*, 16:255, 1991
40. R.D. Wildman, J.M. Huntley, and J.-P. Hansen. Self-diffusion of grains in a two-dimensional vibrofluidized bed. *Phys. Rev. E*, 60:7066, 1999
41. E. Clement, L. Vanel, J. Rajchenbach, and J. Duran. Pattern formation in a vibrated granular layer. *Phys. Rev. E*, 53:2972, 1996
42. O. Zik, J. Stavans, and Y. Rabin. Mobility of a sphere in vibrated granular media. *Europhys. Lett.*, 17:315, 1992
43. C. Bizon, M.D. Shattuck, and J.B. Swift. Linear stability of a vertically oscillated granular layer. *Phys. Rev. E*, 60:3664, 1999

44. E.E. Ehrichs, H.M. Jaeger, G.S. Karczmar, J.B. Knight, V.Y. Kuperman, and S.R. Nagel. Granular convection connection observed by magnetic resonance imaging. *Science*, 267:1632, 1995
45. Y. Lan and A.D. Rosato. Convection related phenomena in granular dynamics simulations of vibrated beds. *Phys. Fluids*, 9:3615, 1997
46. S. Doudy, S. Fauve, and C. Laroshe. Subharmonic instabilities and defects in a granular layer under vertical vibrations. *Europhys. Lett.*, 8:621, 1889
47. A.V. Potapov and C. Campbell. Propagation of inelastic waves in deep vertically shaken particle beds. *Phys. Rev. Lett.*, 77:4760, 1996
48. R.A. Bagnold. Experiments on a gravity-free dispersion of large solid particles in a Newtonian fluid under shear. *Proc. Roy. Soc. London Ser. A*, 225:49, 1954
49. C.K.K. Lun, S.B. Savage, D.J. Jeffery, and N. Chepurniy. Kinetic theories for granular flow: inelastic particles in Couette flow and slightly inelastic particles in a general flow field. *J. Fluid Mech.*, 140:223, 1984
50. J.T. Jenkins and M.W. Richman. Grad's 13-moment system for a dense gas of inelastic spheres. *Arch. Rat. Mech. Anal.*, 87:355, 1985
51. J.T. Jenkins and M.W. Richman. Kinetic theory for plane flows of a dense gas of identical, rough, inelastic, circular disks. *Phys. Fluids*, 28:3485, 1985
52. J.T. Jenkins and M.W. Richman. Plane simple shear of smooth inelastic circular disks: the anisotropy of the second moment in the dilute and dense limits. *J. Fluid Mech.*, 192:313, 1988
53. C.K.K. Lun. Kinetic theory for granular flow of dense, slightly inelastic, slightly rough spheres. *J. Fluid Mech.*, 233:539, 1991
54. M. Alam and P.R. Nott. Stability of plane Couette flow of a granular material. *J. Fluid Mech.*, 377:99, 1998
55. M. Babic. Unsteady Couette granular flows. *Phys. Fluids*, 9:2486, 1997
56. H. Schlichting. *Boundary-layer theory.* McGraw-Hill, New York, 1979
57. Y. Lan and A.D. Rosato. Behavior of vibrating beds of smooth inelastic spheres. *Phys. Fluids*, 7:1818, 1995
58. A. Goldshtein, A. Alexeev, and M. Shapiro. Hydrodynamics of resonance oscillations of column of inelastic particles. *Phys. Rev. E*, 59:6967, 1999
59. J.J. Brey, M.J. Ruiz-Montero, and F. Moreno. Boundary conditions and normal state for a vibrated granular fluid. *Phys. Rev. E*, 62:5339, 2000
60. Y. Du, H. Li, and L.P. Kadanoff. Breakdown of hydrodynamics in a one-dimensional system of inelastic particles. *Phys. Rev. Lett.*, 74:1268, 1995
61. S. Luding, E. Clement, A. Blumen, J. Rajchenbach, and J. Duran. Studies of columns of beads under external vibrations. *Phys. Rev. E*, 49:1634, 1994
62. A. Kudrolli, M. Volpert, and J.P. Gollub. Cluster formation due to collision in granular materials. *Phys. Rev. Lett.*, 78:1383, 1996
63. E. Falcon, R. Wunenburger, P. Evesque, S. Fauve, C. Chabot, Y. Garrabos, and D. Beysens. Cluster formation in a granular medium fluidized by vibrations in low gravity. *Phys. Rev. Lett.*, 83:440, 1999
64. E. Falcon, S. Fauve, and C. Laroche. Cluster formation, pressure and density measurements in a granular medium fluidized by vibrations. *Eur. Phys. J. B*, 9:183, 1999
65. D. Gavrilov, O. Vinogradov, and W.J.D. Shaw. Simulation of grinding in a shaker ball mill. *Powder Tech.*, 101:63, 1999
66. B. Betchov. Non-linear oscillations of the column of a gas. *Phys. Fluids*, 1:205, 1958

67. A. Goldshtein, P. Vainshtein, M. Fichman, and C. Gutfinger. Resonance gas oscillations in closed tubes. *J. Fluid Mech.*, 322:147, 1996
68. A. Goldshtein, A. Alexeev, and M. Shapiro. Resonance oscillations in granular gases in granular gases. In Pöschel and Luding [77], page 266
69. S. McNamara and J.L. Barrat. Energy flux into a fluidized granular medium at a vibrating wall. *Phys. Rev. E*, 55:7767, 1997
70. S. Luding, E. Clement, A. Blumen, J. Rajchenbach, and J. Duran. Anomalous energy dissipation in molecular-dynamics simulations of grains: The "detachment" effect. *Phys. Rev. E*, 50:4113, 1994
71. A. Alexeev, A. Goldshtein, and M. Shapiro. The liquid and solid states of highly dissipative vibrated granular columns: one-dimensional computer simulations. *Powder Technol.*, 123:83, 2002
72. S. Luding and A. Goldshtein. Hydrodynamic theory for dissipative hard spheres with multi-particle interactions. *Granular Matter*, in press, 2003
73. T. Pöschel, N.V. Brilliantov, and T. Schwager. Long-time behavior of granular gases with impact-velocity dependent coefficient of restitution. *Physica A*, 325:274, 2003
74. V.A. Chlenov and N.D. Mikhailov. *Vibrofluidized Beds.* Nauka, Moscow, 1972
75. A. Alexeev, V. Royzen, V. Dudko, A. Goldshtein, and M. Shapiro. Dynamics of vertically vibrated two-dimensional granular layers. *Phys. Rev. E*, 59:3231, 1999
76. B. Bernu, F. Delyon, and R. Mazighi. Steady-states of a column of shaken inelastic beads. *Phys. Rev. E*, 50:4551, 1994
77. T. Pöschel and S. Luding, editors. *Granular Gases*, volume 564 of *Lecture Notes in Physics.* Springer, Berlin, Heidelberg, New York, 2001

Linearized Boltzmann Equation and Hydrodynamics for Granular Gases

J. Javier Brey[1], James W. Dufty[2], and María J. Ruiz-Montero[1]

[1] Física Teórica, Facultad de Física, Universidad de Sevilla, 41080 Sevilla, Spain. brey@us.es, majose@us.es

[2] Department of Physics, University of Florida, Gainesville, Florida 32611, USA. dufty@phys.efl.edu

Abstract. The linearized Boltzmann equation is considered to describe small spatial perturbations of the homogeneous cooling state. The corresponding macroscopic balance equations for the density, temperature, and flow velocity are derived from it as the basis for a hydrodynamic description. Hydrodynamics is defined in terms of the spectrum of the generator for the dynamics of the linearized Boltzmann equation. The hydrodynamic eigenfunctions and eigenvalues are calculated in the long wavelength limit. The results allow identification of the hydrodynamic part of the solution to the linearized Boltzmann equation. This contribution is used to calculate the fluxes in the macroscopic balance equations, leading to the Navier-Stokes equations and associated transport coefficients. The results agree with those obtained earlier by the Chapman-Enskog method. The implications of this analysis for application of methods of linear response to granular fluids and derivation of Green-Kubo expressions for transport coefficients are discussed.

1 Introduction

The derivation of hydrodynamic equations for granular gases and identification of expressions for the associated transport coefficients has been a problem of theoretical and practical interest for the last two decades [1–7]. Early direct extensions of the Chapman-Enskog method to inelastic systems have been refined recently so that the transport coefficients for the Boltzmann gas are now known accurately [5, 8]. These refinements have clarified the role of the time dependent reference state, the local homogeneous cooling state (HCS), and its effects on the qualitative and quantitative characterization of the conditions for the validity of a hydrodynamic description. The extension of these results to higher densities also has been accomplished in the context of the phenomenological Boltzmann-Enskog equation [6]. However, more general conclusions regarding transport in dense granular fluids are quite limited. On the other hand, for elastic collisions the analysis of transport in fluids is quite sophisticated using, for instance, methods of linear response theory [9]. Such methods provide formally exact expressions for transport properties in terms of time dependent fluctuations in the equilibrium reference state. These representations have proven to be particularly useful due to the possibility of their direct evaluation by molecular dynamics simulation. In this context, it is natural to ask to what extent the methods of linear response theory can be applied to granular fluids.

In applications of linear response for fluids with elastic collisions, the local conserved densities of mass, energy, and momentum play a special role. They represent the slow variables as the system approaches uniformity. More significantly, they are the hydrodynamic modes in the long wavelength limit, i.e the formal eigenfunctions of the Liouville operator corresponding to the hydrodynamic eigenvalues. Standard linear response methods consider perturbations of the homogeneous state (the Gibbs ensemble) that are linear combinations of the conserved densities. In this way, hydrodynamic excitations are selected and, by formal manipulation of the response functions, Green-Kubo expressions for the transport coefficients are derived directly. A quite reasonable extension of this method to granular systems, using linear combinations of the local conserved densities as initial perturbations, has been described in [10]. The resulting Green-Kubo expressions are similar to those for systems with elastic collisions, but they do not agree at low density with those from the Chapman-Enskog method noted above. This difference has been made more precise by reformulating the Chapman-Enskog results in a Green-Kubo representation [11]. The resulting time correlation functions are not simply those for the fluxes of the conserved densities, as it is the case for elastic collisions. The reason is that the hydrodynamic modes for granular systems are no longer simply related to the conserved densities, even in the long wavelength limit. Instead, the hydrodynamic modes and conserved densities are conjugate sets in the sense that the former generate the hydrodynamic excitation while the latter measure the response. An objective here is to clarify this subtle difference in the context of the linearized Boltzmann equation for inelastic hard particles, and to discuss the consequences for more general applications.

Clearly, an important first issue is the unambiguous definition of hydrodynamic modes. This is accomplished by considering the eigenvalue problem for the operator generating the dynamics for the Boltzmann equation, linearized around the reference homogeneous cooling state (see Sect. 3 below). The hydrodynamic part of its spectrum is identified first in the long wavelength limit. Hydrodynamics refers to systematic approximations to the macroscopic balance equations for the local relevant densities (here our choice is the deviations of number density, temperature, and flow velocity from their values in the homogeneous state). Consequently, we first derive these equations and find they are closed without approximation in the long wavelength limit. These asymptotic equations determine $d+2$ hydrodynamic eigenvalues, where d is the dimensionality of the system. For normal fluids these are all zero, while for granular systems there are three distinct eigenvalues, one of which is d-fold degenerate. The corresponding eigenvalues for the linearized Boltzmann equation then identify its asymptotic hydrodynamic spectrum. The hydrodynamic spectrum more generally is defined to be the eigenvalues that are continuously connected to these asymptotic values in the long wavelength limit. The expression "hydrodynamic mode" means the hydrodynamic eigenfunction with an exponential time dependence, in the appropriate time units, characterized by the corresponding hydrodynamic eigenvalue. A primary new contribution here is the calculation of these hydrodynamic eigenfunctions in the long wavelength limit, showing their difference from the local conserved densities.

With the hydrodynamic eigenfunctions identified, the hydrodynamic part of the solution to the linearized Boltzmann equation can be extracted. This is the dominant contribution for large times and long wavelengths. More specifically, this hydrodynamic part is calculated to first order in the gradients of the hydrodynamic fields and the result is used to calculate the average fluxes in the macroscopic balance equations at finite wavelengths. This provides the Navier-Stokes hydrodynamics for granular gases, and allows identification of the transport coefficients. The results agree with those from the Chapman-Enskog method [5, 8] and the Green-Kubo representation of [11]. However, their derivation here exposes more clearly the hydrodynamic eigenfunctions and their role as the origin of key differences from gases with elastic collisions.

In the next section, the nonlinear Boltzmann equation for a granular gas is defined, and the exact balance equations for the density, temperature, and flow velocity are derived from it. The homogeneous state of these equations corresponds to global cooling, and the corresponding homogeneous cooling solution (HCS) to the nonlinear Boltzmann equation is described. This solution represents the homogeneous state about which small spatial perturbations are considered. The corresponding linear Boltzmann equation and linear balance equations for these small perturbations are then described. Next, the eigenvalue problem for the generator of the time dependence in the linearized Boltzmann equation is defined in Sect. 4, and the hydrodynamic eigenfunctions and eigenvalues are calculated in the long wavelength limit. These results are used in Sect. 5 to derive the Navier-Stokes hydrodynamic equations, with expressions for the transport coefficients in terms of the linearized Boltzmann collision operator. Their equivalence with previous results from the Chapman-Enskog method is demonstrated, and these expressions are also given a Green-Kubo representation. Finally, the analysis is recapitulated, and implications for a more general treatment of transport in rapid granular fluids beyond the low density context are discussed.

2 Nonlinear Boltzmann Equation and the Homogeneous Cooling State

The Boltzmann equation for a system of inelastic smooth hard disks ($d = 2$) or spheres ($d = 3$) of mass m and diameter σ has the form [3, 12]

$$\left(\frac{\partial}{\partial t} + \vec{v} \cdot \nabla\right) f(\vec{r}, \vec{r}, t) = J[f|f], \tag{1}$$

where $J[f|f]$ is the nonlinear inelastic Boltzmann collision operator,

$$J[f|f] = \int d\vec{v}\, \bar{T}(\vec{v},\vec{v}_1) f(\vec{r},\vec{v},t) f(\vec{r},\vec{v}_1,t). \tag{2}$$

Here $\bar{T}(\vec{v},\vec{v}_1)$ is the inelastic binary collision operator

$$\bar{T}(\vec{v},\vec{v}_1) = \sigma^{d-1} \int d\hat{\vec{\sigma}}\, \Theta\left(\vec{g}\cdot\hat{\vec{\sigma}}\right) \vec{g}\cdot\hat{\vec{\sigma}} \left(\alpha^{-2} b_{\vec{\sigma}}^{-1} - 1\right), \tag{3}$$

with $\vec{g} = \vec{v} - \vec{v}_1$, Θ the Heaviside step function, $\hat{\vec{\sigma}}$ a unit vector joining the center of the two particles at contact, α the coefficient of normal restitution, and $b_{\vec{\sigma}}^{-1}$ an operator replacing all the velocities $\vec{v}$ and $\vec{v}_1$ appearing to its right by their precollisional values $\vec{v}^*$ and $\vec{v}_1^*$, given by

$$b_{\vec{\sigma}}^{-1}\vec{v} = \vec{v}^* = \vec{v} - \frac{1+\alpha}{2\alpha}\left(\hat{\vec{\sigma}}\cdot\vec{g}\right)\hat{\vec{\sigma}}, \quad b_{\vec{\sigma}}^{-1}\vec{v}_1 = \vec{v}_1^* = \vec{v}_1 + \frac{1+\alpha}{2\alpha}\left(\hat{\vec{\sigma}}\cdot\vec{g}\right)\hat{\vec{\sigma}}. \tag{4}$$

For arbitrary functions $g(\vec{v},\vec{v}_1)$ and $h(\vec{v},\vec{v}_1)$, it is

$$\int d\vec{v} \int d\vec{v}_1 g(\vec{v},\vec{v}_1)\bar{T}(\vec{v},\vec{v}_1)h(\vec{v},\vec{v}_1) = \int d\vec{v} \int d\vec{v}_1 h(\vec{v},\vec{v}_1)T(\vec{v},\vec{v}_1)g(\vec{v},\vec{v}_1), \tag{5}$$

where

$$T(\vec{v},\vec{v}_1) = \sigma^{d-1} \int d\hat{\vec{\sigma}}\, \Theta\left(\vec{g}\cdot\hat{\vec{\sigma}}\right) \vec{g}\cdot\hat{\vec{\sigma}} \left(b_{\vec{\sigma}} - 1\right). \tag{6}$$

The operator $b_{\vec{\sigma}}$ changes the velocities $\vec{v}$ and $\vec{v}_1$ to its right into their post-collisional values $\vec{v}'$ and $\vec{v}_1'$,

$$b_{\vec{\sigma}}\vec{v} = \vec{v}' = \vec{v} - \frac{1+\alpha}{2}\left(\hat{\vec{\sigma}}\cdot\vec{g}\right)\hat{\vec{\sigma}}, \quad b_{\vec{\sigma}}\vec{v}_1 = \vec{v}_1' = \vec{v}_1 + \frac{1+\alpha}{2}\left(\hat{\vec{\sigma}}\cdot\vec{g}\right)\hat{\vec{\sigma}}. \tag{7}$$

By using (5), the equalities

$$\int d\vec{v} \begin{pmatrix} 1 \\ m\vec{v} \\ \frac{mv^2}{2} \end{pmatrix} J[f|f] = \begin{pmatrix} 0 \\ \vec{0} \\ -\frac{pd}{2}\zeta[f] \end{pmatrix}. \tag{8}$$

are easily obtained. In the above expressions, $p = nT$ is the pressure (Boltzmann's constant has been set equal to unity as usual), n being the number density and T the granular temperature, and ζ is a a nonlinear functional of the distribution function,

$$\zeta[f] = \frac{(1-\alpha^2)\pi^{\frac{d-1}{2}} m\sigma^{d-1}}{4\Gamma\left(\frac{d+3}{2}\right) pd} \int d\vec{v} \int d\vec{v}_1\, f(\vec{r},\vec{v},t) f(\vec{r},\vec{v}_1,t) g^3. \tag{9}$$

The hydrodynamic fields are defined in the standard way in terms of velocity moments, i.e.

$$n(\vec{r},t) = \int d\vec{v}\, f(\vec{r},\vec{v},t), \tag{10}$$

$$n(\vec{r},t)\vec{u}(\vec{r},t) = \int d\vec{v}\, \vec{v} f(\vec{r},\vec{v},t), \tag{11}$$

$$\frac{d}{2} n(\vec{r},t)T(\vec{r},t) = \int d\vec{v}\, \frac{m}{2} V^2 f(\vec{r},\vec{v},t), \tag{12}$$

where $\vec{V} = \vec{v} - \vec{u}$ is the velocity of the particle relative to the local velocity flow $\vec{u}$. Their balance equations follow directly from (8),

$$\frac{\partial n}{\partial t} + \vec{\nabla} \cdot (n\vec{u}) = 0, \tag{13}$$

$$\frac{\partial \vec{u}}{\partial t} + \vec{u} \cdot \vec{\nabla}\vec{u} + (mn)^{-1} \vec{\nabla} \cdot \mathsf{P} = 0, \tag{14}$$

$$\frac{\partial T}{\partial t} + \vec{u} \cdot \vec{\nabla} T + \frac{2}{nd} \left(\mathsf{P} : \vec{\nabla}\vec{u} + \vec{\nabla} \cdot \vec{q} \right) + T\zeta = 0. \tag{15}$$

The functionals giving the pressure tensor P and the heat flux $\vec{q}$ are:

$$\mathsf{P}(\vec{r},t) = m \int d\vec{v}\, \vec{V}\vec{V} f(\vec{r},\vec{v},t) = p(\vec{r},t)\mathsf{I} + \int d\vec{v}\, \mathsf{D}(\vec{V}) f(\vec{r},\vec{v},t), \tag{16}$$

$$\vec{q}(\vec{r},t) = \frac{m}{2} \int d\vec{v}\, V^2 \vec{V} f(\vec{r},\vec{v},t) = \int d\vec{v}\, \vec{S}(\vec{V}) f(\vec{r},\vec{v},t). \tag{17}$$

Here I is the unit tensor, and

$$\mathsf{D}(\vec{V}) = m \left(\vec{V}\vec{V} - \frac{1}{d} V^2 \mathsf{I} \right), \quad \vec{S}(\vec{V}) = \left(\frac{m}{2} V^2 - \frac{d+2}{2} T \right) \vec{V}. \tag{18}$$

Equations (13)–(15) become a closed set of equations for the $d+2$ hydrodynamic fields once the pressure tensor and heat flux have been expressed in terms of these fields (e.g. via Newton's law and Fourier's law for a normal fluid). For inelastic systems, also a similar expression for the cooling rate ζ is needed. The $d+2$ independent solutions to these equations are the macroscopic hydrodynamic excitations. To illustrate this, it is useful to record the corresponding "Euler level" equations, obtained by neglecting all spatial gradients beyond first order. Then the pressure tensor reduces to the local pressure, since the second term on the right hand side of (16) must be traceless, and the heat flux vanishes by similar symmetry considerations. Furthermore, the cooling rate as a function of the density and the temperature becomes, on dimensional grounds, $\zeta \to \bar{\zeta} T^{1/2} n$, where $\bar{\zeta}$ is a constant. There is no first order gradient contribution to the cooling rate, again due to symmetry. Therefore, the Euler equations for a granular gas are

$$\frac{\partial n}{\partial t} + \vec{\nabla} \cdot (n\vec{u}) = 0, \tag{19}$$

$$\frac{\partial \vec{u}}{\partial t} + \vec{u} \cdot \vec{\nabla} \vec{u} + (mn)^{-1} \vec{\nabla} p = 0, \tag{20}$$

$$\left(\frac{\partial}{\partial t} + \bar{\zeta} T^{1/2} n \right) T + \vec{u} \cdot \vec{\nabla} T + \frac{2p}{nd} \vec{\nabla} \cdot \vec{u} = 0. \tag{21}$$

These are a set of closed nonlinear hydrodynamic equations for n, $\vec{u}$, and T.

In the remainder of this work, the hydrodynamic excitations are studied for small perturbations about a special homogeneous state. In that case, the equations become linear and the hydrodynamic excitations are associated with a $d+2$ dimensional eigenvalue problem. The spatially homogeneous reference state is called the homogeneous cooling state (HCS). It corresponds to a solution f_{HCS} of the Boltzmann equation (1) having the scaling form [3]

$$f_{\text{HCS}}(\vec{v}, t) = n_H v_0^{-d}(t) \chi_{\text{HCS}}(\vec{c}), \quad \vec{c} = \frac{\vec{v}}{v_0(t)}, \tag{22}$$

where

$$v_0(t) = \left[\frac{2T_{\text{HCS}}(t)}{m} \right]^{1/2} \tag{23}$$

is the "thermal velocity" and $\chi_{\text{HCS}}(\vec{c})$ is an isotropic function of $\vec{c}$. As indicated in (22), all the time dependence of f_{HCS} occurs through the homogeneous temperature $T_{\text{HCS}}(t)$, which obeys the equation

$$\frac{\partial T_{\text{HCS}}(t)}{\partial t} + T_{\text{HCS}}(t) \zeta_{\text{HCS}}(t) = 0, \tag{24}$$

with

$$\zeta_{\text{HCS}}(t) = \frac{(1-\alpha^2) \pi^{\frac{d-1}{2}} \sigma^{d-1} n_H v_0(t)}{2\Gamma\left(\frac{d+3}{2}\right) d} \int d\vec{c} \int d\vec{c}_1 \chi_{\text{HCS}}(\vec{c}) \chi_{\text{HCS}}(\vec{c}_1) |\vec{c} - \vec{c}_1|^3. \tag{25}$$

Substitution of the scaling form, (22), into the Boltzmann equation leads to

$$\frac{1}{2} \zeta_{\text{HCS}} \frac{\partial}{\partial \vec{v}} \cdot (\vec{v} f_{\text{HCS}}) = J[f_{\text{HCS}} | f_{\text{HCS}}]. \tag{26}$$

The explicit time dependence of the HCS can be eliminated by introducing a dimensionless time scale s by

$$s = \int_0^t dt' \, \frac{v_0(t')}{\ell}, \tag{27}$$

where $\ell \equiv \left(n_H \sigma^{d-1}\right)^{-1}$ is proportional to the mean free path of the particles in the HCS. In terms of the s variable, (24) becomes

$$\frac{\partial}{\partial s} T_{\mathrm{HCS}}(s) = -\zeta_0 T_{\mathrm{HCS}}(s), \tag{28}$$

with

$$\zeta_0 = \frac{\ell \zeta_{\mathrm{HCS}}}{v_0(t)}, \tag{29}$$

that does not depend on time. Equation (28) is the differential form of the Haff's homogeneous cooling law [13]. Similarly, (26) takes the form

$$\frac{1}{2}\zeta_0 \frac{\partial}{\partial \vec{c}} \cdot (\vec{c} \chi_{\mathrm{HCS}}) = J'[\chi_{\mathrm{HCS}}|\chi_{\mathrm{HCS}}], \tag{30}$$

$$J'\left[\chi_{\mathrm{HCS}}|\chi_{\mathrm{HCS}}\right] = \int d\vec{c}_1 \, \bar{T}(\vec{c}, \vec{c}_1) \chi_{\mathrm{HCS}}(\vec{c}) \chi_{\mathrm{HCS}}(\vec{c}_1), \tag{31}$$

$$\bar{T}(\vec{c}, \vec{c}_1) = \int d\hat{\vec{\sigma}} \Theta \left[(\vec{c} - \vec{c}_1) \cdot \hat{\vec{\sigma}} \right] (\vec{c} - \vec{c}_1) \cdot \hat{\vec{\sigma}} (\alpha^{-2} b_{\vec{\sigma}}^{-1} - 1). \tag{32}$$

The operator $b_{\vec{\sigma}}^{-1}$ in the last equation is defined again by (4), but substituting $(\vec{v}, \vec{v}_1)$ by $(\vec{c}, \vec{c}_1)$. The form of the distribution function χ_{HCS} given by the solution of (30) has been extensively studied. A summary of the results is given in [14].

3 Linearized Boltzmann Equation

Let us now consider a situation where the inelastic gas is very close to the HCS, so that we can write

$$f(\vec{r}, \vec{v}, t) = f_{\mathrm{HCS}}(\vec{v}, t) + \delta f(\vec{r}, \vec{v}, t), \quad |\delta f(\vec{r}, \vec{v}, t)| \ll f_{\mathrm{HCS}}(\vec{v}, t). \tag{33}$$

Substitution of (33) into the Boltzmann equation (1), keeping only terms up to first order in δf yields

$$\frac{\partial}{\partial t} \delta f(\vec{r}, \vec{v}, t) + \vec{v} \cdot \nabla \delta f(\vec{r}, \vec{v}, t) = K \delta f(\vec{r}, \vec{v}, t). \tag{34}$$

The linearized Boltzmann collision operator is defined by

$$K\delta f(\vec{r},\vec{v},t) = J[f_{\mathrm{HCS}}|\delta f] + J[\delta f|f_{\mathrm{HCS}}]. \tag{35}$$

Equation (34) can be written in dimensionless form, using the scaled velocity $\vec{c}$ defined in (22), the reduced time s introduced in (27), and the dimensionless length scale

$$\vec{l} = \frac{\vec{r}}{\ell}. \tag{36}$$

Then, the linearized inelastic Boltzmann equation becomes

$$\frac{\partial}{\partial s}\delta\chi + \vec{c}\cdot\frac{\partial}{\partial\vec{l}}\delta\chi = \Lambda\delta\chi, \tag{37}$$

where

$$\delta\chi(\vec{l},\vec{c},s) = n_H^{-1} v_0^d(t)\delta f(\vec{r},\vec{v},t) \tag{38}$$

and

$$\Lambda\delta\chi = J'[\chi_{\mathrm{HCS}}|\delta\chi] + J'[\delta\chi|\chi_{\mathrm{HCS}}] - \frac{\zeta_0}{2}\frac{\partial}{\partial\vec{c}}\cdot(\vec{c}\delta\chi). \tag{39}$$

In this representation, there is not any explicit dependence on time due to the reference state, and all the stationary methods known from the case of elastic collisions [9] can now be applied to the analysis of (37).

We define relative deviations of the hydrodynamic fields from their values in the HCS by

$$\rho\left(\vec{l},s\right) \equiv \frac{\delta n(\vec{r},t)}{n_H} = \int d\vec{c}\,\delta\chi\left(\vec{l},\vec{c},s\right), \tag{40}$$

$$\vec{\omega}\left(\vec{l},s\right) \equiv \frac{\delta\vec{u}(\vec{r},t)}{v_0(t)} = \int d\vec{c}\,\vec{c}\delta\chi\left(\vec{l},\vec{c},s\right), \tag{41}$$

$$\theta\left(\vec{l},s\right) \equiv \frac{\delta T(\vec{r},t)}{T_{\mathrm{HCS}}(t)} = \int d\vec{c}\left(\frac{2c^2}{d}-1\right)\delta\chi\left(\vec{l},\vec{c},s\right), \tag{42}$$

where $\delta y(\vec{r},t) \equiv y(\vec{r},t) - y_{\mathrm{HCS}}(t)$ denotes the deviation of the local value of the macroscopic variable y from its value in the HCS. Taking velocity moments in the Boltzmann equation (37), the linearized balance equations for the hydrodynamic fields are obtained,

$$\frac{\partial}{\partial s}\rho\left(\vec{l},s\right) + \frac{\partial}{\partial\vec{l}}\cdot\vec{\omega}\left(\vec{l},s\right) = 0, \tag{43}$$

$$\left(\frac{\partial}{\partial s}-\frac{\zeta_0}{2}\right)\vec{\omega}\left(\vec{l},s\right)+\frac{\partial}{\partial\vec{l}}\cdot\Pi\left(\vec{l},s\right)=0, \tag{44}$$

$$\frac{\partial}{\partial s}\theta\left(\vec{l},s\right)+\frac{2}{d}\frac{\partial}{\partial\vec{l}}\cdot\vec{\omega}\left(\vec{l},s\right)+\frac{2}{d}\frac{\partial}{\partial\vec{l}}\cdot\vec{\phi}\left(\vec{l},s\right)+\delta\zeta_0\left(\vec{l},s\right) \\ -\zeta_0\left[\theta\left(\vec{l},s\right)+\rho\left(\vec{l},s\right)\right]=0. \tag{45}$$

The reduced linear pressure tensor Π and heat flux $\vec{\phi}$ are defined as

$$\Pi=\frac{1}{2}\left[\theta\left(\vec{l},s\right)+\rho\left(\vec{l},s\right)\right]\mathsf{I}+\int d\vec{c}\,\Delta\left(\vec{c}\right)\delta\chi\left(\vec{l},\vec{c},s\right), \tag{46}$$

$$\vec{\phi}\left(\vec{l},s\right)=\int d\vec{c}\,\vec{\Sigma}\left(\vec{c}\right)\delta\chi\left(\vec{l},\vec{c},s\right), \tag{47}$$

where Δ and $\vec{\phi}$ are the dimensionless forms of D and $\vec{S}$,

$$\Delta(\vec{c})=\vec{c}\vec{c}-\frac{c^2}{d}\mathsf{I},\quad \vec{\Sigma}(\vec{c})=\left(c^2-\frac{d+2}{2}\right)\vec{c}, \tag{48}$$

The quantities Π and $\vec{\phi}$ are related with the linearization of the fluxes given in (16) and (17) by

$$\Pi\left(\vec{l},s\right)=\frac{\mathsf{P}(\vec{r},t)}{2n_H T_{\mathrm{HCS}}(t)}-\frac{\mathsf{I}}{2},\quad \vec{\phi}(\vec{l},s)=\frac{m^{1/2}\vec{q}(\vec{r},t)}{2^{1/2}n_H T_{\mathrm{HCS}}^{3/2}(t)}. \tag{49}$$

Finally, the term $\delta\zeta_0(\vec{l},s)$ is the dimensionless linear deviation of the cooling term from its value in the HCS,

$$\delta\zeta_0\left(\vec{l},s\right)=\frac{\left(1-\alpha^2\right)\pi^{\frac{d-1}{2}}}{\Gamma\left(\frac{d+3}{3}\right)d}\int d\vec{c}\int d\vec{c}_1\,\chi_{\mathrm{HCS}}(\vec{c})\delta\chi\left(\vec{l},\vec{c},s\right)|\vec{c}-\vec{c}_1|^3. \tag{50}$$

Equations (43)–(45) are the linearized transport equations corresponding to the more general forms (13)-(15). They are the starting point for identifying the hydrodynamic modes of the system, once the pressure tensor and the heat flux have been expressed as linear functions of the hydrodynamic fields. Then, (43)–(45) become a closed set of equations for these fields whose solutions are the hydrodynamic modes. For example, at very long wavelengths the spatial gradients can be neglected and these equations reduce to

$$\frac{\partial}{\partial s}\rho\left(\vec{l},s\right)=0,\quad \left(\frac{\partial}{\partial s}-\frac{\zeta_0}{2}\right)\vec{\omega}\left(\vec{l},s\right)=0,$$

$$\left(\frac{\partial}{\partial s}+\frac{\zeta_0}{2}\right)\theta\left(\vec{l},s\right)+\zeta_0\rho(\vec{l},s)=0. \tag{51}$$

In the last equation, we have used that $\delta\zeta_0 = \zeta_0(2\rho + 3\theta/2)$, following from (9), particularized for a homogenous system, and the property $\zeta \to \bar{\zeta}T^{1/2}n$, noticed above (19). In fact, these long wavelength equations can be obtained directly from linearization of the Euler equations (19)-(21). The solutions of (51) have the time dependence $\exp(\lambda_i s)$, with $\lambda_i = 0, \pm\zeta_0/2$. The zero eigenvalue is d-fold degenerate.

4 Eigenvalue Problem

In this section we explore solutions to the linearized Boltzmann equation and characterize the hydrodynamic modes. This is done through a formal analysis of the eigenvalue problem for the associated linear operator. Since the equation is linear and the collision operator Λ does not change the space variable, it is sufficient to consider a single Fourier mode, i.e. we take

$$\delta\chi\left(\vec{l},\vec{c},s\right) = e^{i\vec{k}\cdot\vec{l}}\delta\widetilde{\chi}\left(\vec{k},\vec{c},s\right). \tag{52}$$

The linearized Boltzmann equation, (37), then becomes

$$\frac{\partial}{\partial s}\delta\widetilde{\chi} = \left(\Lambda - i\vec{k}\cdot\vec{c}\right)\delta\widetilde{\chi}, \tag{53}$$

whose formal solution is

$$\delta\widetilde{\chi}\left(\vec{k},\vec{c},s\right) = e^{s\left(\Lambda - i\vec{k}\cdot\vec{c}\right)}\delta\widetilde{\chi}\left(\vec{k},\vec{c},0\right). \tag{54}$$

The above expression shows most clearly that the possible excitations for the gas under these conditions of small initial perturbations are determined by the spectrum of the linear operator $\Lambda - i\vec{k}\cdot\vec{c}$. This suggests a study of the eigenvalue problem

$$\left(\Lambda - i\vec{k}\cdot\vec{c}\right)\xi_i\left(\vec{k},\vec{c}\right) = \lambda_i\left(\vec{k}\right)\xi_i(\vec{k},\vec{c}). \tag{55}$$

The problem is posed in a Hilbert space of functions of $\vec{c}$ with a scalar product given by

$$\langle g|h\rangle = \int d\vec{c}\,\chi_{\text{HCS}}^{-1}(\vec{c})g^*(\vec{c})h(\vec{c}), \tag{56}$$

where g^* denotes the complex conjugate of g. Of particular interest here are the eigenvalues λ_β^h and eigenfunctions $\chi_\beta^h(\vec{c})$ associated with linear hydrodynamics. These can be identified as follows. First, at $\vec{k} = 0$ they correspond to those eigenvalues of Λ alone that coincide with the zeroth order in the gradients eigenvalues of the linearized balance equations (51), $\lambda_i(0) = 0, \pm\zeta_0/2$, i.e.

$$\Lambda\xi_\beta^h(\vec{c}) = \lambda_\beta^h\xi_\beta^h(\vec{c}), \quad \lambda_\beta^h = 0, \pm\frac{\zeta_0}{2}, \tag{57}$$

where $\xi_\beta^h(\vec{c})$ stands for $\xi_\beta^h(\vec{k}=0,\vec{c})$. More generally, at finite $\vec{k}$ the hydrodynamic modes are defined to be the solutions of (55) that are continuously connected to those of (57) as $k \to 0$. The eigenvalue problem (57) is solved below, so the existence of the hydrodynamic modes is assured. The formal solution (54) then can be decomposed in the form

$$\delta\widetilde{\chi}\left(\vec{k},\vec{c},s\right) = \sum_\beta \widetilde{a}_\beta^h\left(\vec{k},s\right)\xi_\beta^h\left(\vec{k},\vec{c}\right) + \delta\widetilde{\chi}^m\left(\vec{k},\vec{c},s\right). \tag{58}$$

The first term arises from the hydrodynamic modes while the second one represents all the other "microscopic" excitations. The hydrodynamic coefficients $\widetilde{a}_\beta^h(\vec{k},s)$ can be expressed as

$$\begin{aligned}\widetilde{a}_\beta^h\left(\vec{k},s\right) &= \exp\left[s\lambda_\beta^h\left(\vec{k}\right)\right]\left\langle\bar{\xi}_\beta^h\left(\vec{k},\vec{c}\right)\middle|\,\delta\chi\left(\vec{k},\vec{c},0\right)\right\rangle \\ &= \left\langle\bar{\xi}_\beta^h\left(\vec{k},\vec{c}\right)\middle|\,\delta\chi\left(\vec{k},\vec{c},s\right)\right\rangle. \end{aligned}\tag{59}$$

Since $\Lambda - i\vec{k}\cdot\vec{c}$ is not self-adjoint, the functions $\bar{\xi}_\beta^h(\vec{k},\vec{c})$ differ from $\xi_\beta^h(\vec{k},\vec{c})$ but together they must form a biorthogonal set. It will appear below that these coefficients are closely related to the hydrodynamic fields of the linearized balance equations.

The expected context for a hydrodynamic description is now clear. For given initial conditions both hydrodynamic and microscopic excitations contribute to (58). However, for long wavelengths (small k) it is expected that the microscopic excitations decay on time scales short compared to those for the hydrodynamic modes, leaving only the latter on long time scales. The relevant separation of time scales is expected to occur after a few collisions, so there is a large and interesting time scale for the hydrodynamic description. Preliminary analysis based on kinetic models for the Boltzmann equation [15–17] and also on low order matrix representations of the linearized Boltzmann collision operator [18], suggest that the separation of modes assumed here does occur, and hydrodynamics does dominate for wavelengths and times long compared with the mean free path and the mean free time, respectively. Furthermore, it appears from these models that the cooling rate ζ_0 remains smaller than the collision frequency even for strong inelasticity, so that inclusion of the non-conserved energy among the hydrodynamic variables seems justified.

Hydrodynamic Modes for $k = 0$

We consider now the eigenproblem associated with the linearized homogeneous Boltzmann operator,

$$\Lambda\xi_i(\vec{c}) = \lambda_i\xi_i(\vec{c}). \tag{60}$$

Of course, finding all the solutions of this equation is an impossible task, even in the elastic limit. Nevertheless, it is quite easy to obtain some particular solutions, which will turn out to be the relevant ones in the hydrodynamic regime. We begin by constructing the function

$$F(\vec{c},\rho,\vec{\omega},\gamma) \equiv \rho\chi_{\mathrm{HCS}}\left(\gamma\vec{C}\right), \tag{61}$$

where $\vec{C} = \vec{c} - \vec{\omega}$. Although F has the functional form of a "local" HCS distribution, the parameters ρ, $\vec{\omega}$, and γ do not need to have any particular meaning here, playing just the role of auxiliary parameters. From (30) it is obtained that F verifies

$$\frac{\rho}{2}\zeta_0\frac{\partial}{\partial\vec{c}}\cdot\left(\vec{C}F\right) = \gamma^{d+1}J'\left[F|F\right]. \tag{62}$$

Then, taking derivatives with respect to ρ, $\vec{\omega}$, and γ in the above equation and, afterwards, the limit $\rho = \gamma = 1$, $\vec{\omega} = 0$, leads to the set of equations

$$\Lambda\psi_1(\vec{c}) = -\frac{1}{2}\zeta_0\psi_3(\vec{c}),\quad \Lambda\vec{\psi}_2(\vec{c}) = \frac{1}{2}\zeta_0\vec{\psi}_2(\vec{c}),\quad \Lambda\psi_3(\vec{c}) = -\frac{1}{2}\zeta_0\psi_3(\vec{c}), \tag{63}$$

where

$$\psi_1(\vec{c}) \equiv \left(\frac{\partial F}{\partial\rho}\right)_0 = \chi_{\mathrm{HCS}}(\vec{c}), \tag{64}$$

$$\vec{\psi}_2(\vec{c}) \equiv \left(\frac{\partial F}{\partial\vec{\omega}}\right)_0 = -\frac{\partial\chi_{\mathrm{HCS}}(\vec{c})}{\partial\vec{c}}, \tag{65}$$

$$\psi_3(\vec{c}) \equiv -\left(\frac{\partial F}{\partial\gamma}\right)_0 - \chi_{\mathrm{HCS}}(\vec{c})d = -\frac{\partial}{\partial\vec{c}}\cdot\left[\vec{c}\chi_{\mathrm{HCS}}(\vec{c})\right]. \tag{66}$$

Here the zero subindex indicates that the derivatives are evaluated at $\rho = \gamma = 1, \vec{\omega} = 0$. From (63) it is easily seen that the set of functions

$$\{\xi_\beta(\vec{c})\} \equiv \left\{\psi_1(\vec{c}) - \psi_3(\vec{c}), \vec{\psi}_2(\vec{c}), \psi_3(\vec{c})\right\} \tag{67}$$

are solutions of (60) corresponding to the eigenvalues

$$\lambda_1 = 0,\quad \lambda_2 = \frac{\zeta_0}{2},\quad \lambda_3 = -\frac{\zeta_0}{2}, \tag{68}$$

respectively. The eigenvalue λ_2 is, therefore, d-fold degenerate. This confirms (57) above, indicating points in the spectrum at $k = 0$ corresponding to the linear balance equations. Hence, these are the hydrodynamic modes at longest wavelengths, $\xi_\beta^h(\vec{c})$. For elastic collisions, the eigenvalues are all zero, with $d+2$ degeneracy. The eigenfunctions in that case are constructed from linear combinations of the collisional invariants $1, \vec{c}$, and c^2. For inelastic collisions, the degeneracy is partially removed by collisional cooling, and the velocity dependence of the eigenfunctions is more complex, since χ_{HCS} is not longer Maxwellian. The identification of the hydrodynamic eigenfunctions at $\vec{k}=0$ is a primary new result of this analysis,

The eigenfunctions $\xi_\beta^h(\vec{c})$ can be interpreted as follows. Consider a local form of the HCS distribution, $f_{\text{HCS}}^{(l)}$, whereby the hydrodynamics fields are replaced by their actual space and time dependent values in a general nonequilibrium state,

$$f_{\text{HCS}}^{(l)}(\vec{r},\vec{v},t) = n(\vec{r},t) v_0^{(l)-d}(t) \chi_{\text{HCS}}(\vec{C}\,), \tag{69}$$

where

$$\vec{C} = \frac{\vec{v}-\delta\vec{u}(\vec{r},t)}{v_0^{(l)}(t)}, \quad v_0^{(l)}(t) = \left[\frac{2T(\vec{r},t)}{m}\right]^{1/2}. \tag{70}$$

This is the analogous to the local equilibrium distribution for elastic collisions. It represents a state whose averages values of 1, $\vec{v}$, and v^2 are those of the nonequilibrium state. This seems a reasonable choice to describe an experimentally prepared state. The linearization of such an state around the HCS gives

$$f_{\text{HCS}}^{(l)}(\vec{r},\vec{v},t) \rightarrow f_{\text{HCS}}(\vec{v},t) + n_H v_0^{-d}(t) \delta\chi^{(l)}(\vec{c},\vec{l},s), \tag{71}$$

$$\delta\chi^{(l)}(\vec{c},\vec{l},s) = \left[\xi_1^h(\vec{c}) + \xi_3^h(\vec{c})\right] \rho(\vec{l},s) + \vec{\xi}_2^h(\vec{c}) \cdot \vec{\omega}(\vec{l},s) + \frac{1}{2}\xi_3^h(\vec{c})\theta(\vec{l},s). \tag{72}$$

Thus, the eigenfunctions $\xi_\beta^h(\vec{c})$ describe the linear deviation of the local HCS, appearing then as the natural perturbations for a macroscopic preparation of the initial nonequilibrium state. This will be discussed further below.

It is easy to show that the functions $\{\xi_\beta^h(\vec{c})\}$ are linearly independent so that they define a $d+2$ dimensional subspace of the Hilbert space. Although the functions $\psi_\beta(\vec{c})$ are orthogonal with the definition of scalar product in (56), the eigenfunctions $\xi_\beta^h(\vec{c})$ are not, as a consequence of the operator Λ being non-Hermitian. This leads, in principle, to consider the left eigenvalue problem and, therefore, to study the adjoint operator Λ^+ defined by

$$\langle g|\Lambda h\rangle^* = \left\langle h|\Lambda^+ g\right\rangle, \tag{73}$$

for arbitrary $g(\vec{c})$ and $h(\vec{c})$. Using the property given in (5) it is found

$$\begin{aligned}\Lambda^+ g(\vec{c}) = & \int d\vec{c}_1\, \chi_{\text{HCS}}(\vec{c})\chi_{\text{HCS}}(\vec{c}_1) T(\vec{c},\vec{c}_1) \left[\chi_{\text{HCS}}^{-1}(\vec{c})g(\vec{c}) + \chi_{\text{HCS}}^{-1}(\vec{c}_1)g(\vec{c}_1)\right] \\ & + \frac{\zeta_0}{2}\chi_{\text{HCS}}(\vec{c})\vec{c}\cdot\frac{\partial}{\partial\vec{c}}\left[\chi_{\text{HCS}}^{-1}(\vec{c})g(\vec{c})\right].\end{aligned} \tag{74}$$

By an analysis similar to that for the right eigenfunctions, it is possible to show that $\chi_{HCS}(\vec{c})$ is an eigenfunction of Λ^+ with eigenvalue zero, and $\vec{c}\chi_{HCS}(\vec{c})$ are d eigenfunctions with eigenvalue $\zeta_0/2$. We have not been able to find the remaining eigenfunction of Λ^+ with eigenvalue $-\zeta_0/2$. Nevertheless, for our purposes here, and also for many other applications, it will suffice to identify a set of functions $\left\{\bar{\xi}^h_\beta(\vec{c}), \beta = 1,2,3\right\}$ verifying the biorthonormality condition

$$\langle \bar{\xi}^h_\beta | \xi^h_{\beta'} \rangle = \delta_{\beta,\beta'}, \tag{75}$$

$\beta, \beta' = 1,2,3$. A convenient choice is given by

$$\bar{\xi}^h_1(\vec{c}) = \chi_{\text{HCS}}(\vec{c}), \quad \bar{\xi}^h_2(\vec{c}) = \vec{c}\chi_{\text{HCS}}(\vec{c}), \quad \bar{\xi}^h_3 = \left(\frac{c^2}{d} + \frac{1}{2}\right)\chi_{\text{HCS}}(\vec{c}). \tag{76}$$

The above functions span a dual subspace of that spanned by the right eigenfunctions $\{\xi^h_\beta(\vec{c})\}$. The fact that they are linear combinations of 1, $\vec{c}$, and c^2 is closely related with the existence and approach to a hydrodynamic description, as it will be shown in the next section. Although $\bar{\xi}^h_3(\vec{c})$ is not an eigenfunction of Λ^+, as it can be verified by direct substitution into (74), it nevertheless completes the dual space such that for any linear combination of the form

$$g(\vec{c}) = \sum_{\beta=1}^{3} a_\beta \xi^h_\beta(\vec{c}), \tag{77}$$

the coefficients a_β are given by

$$a_\beta = \langle \bar{\xi}^h_\beta | \, g \rangle = \int d\vec{c}\, \chi^{-1}_{\text{HCS}}(\vec{c}) \bar{\xi}^h_\beta(\vec{c}) g(\vec{c}). \tag{78}$$

It is instructive to consider what happens in the elastic limit $\alpha \to 1$. Then $\chi_{\text{HCS}}(\vec{c})$ becomes the Maxwellian $\chi_{\text{MB}}(\vec{c}) = \pi^{-d/2}\exp(-c^2)$ and the right eigenfunctions are given by

$$\left\{\left[2c^2 - (d+1)\right]\chi_{\text{MB}}(\vec{c}), 2\vec{c}\chi_{\text{MB}}(\vec{c}), (2c^2 - d)\chi_{\text{MB}}(\vec{c})\right\}. \tag{79}$$

All of these eigenfunctions now correspond to the same eigenvalue zero, as it is the case for any linear combination of $\chi_{HCS}(\vec{c})$ times $1, \vec{c}$ and c^2. Usually, for elastic collisions an orthonormal set is constructed to span the space, and the specific linear combinations for the eigenfunctions are obtained following perturbation theory to remove the degeneracy (e.g., at finite k) [19]. The non-orthogonal linear combinations in (79) follow in a similar way, with the small inelasticity breaking the degeneracy to determine the right linear combinations. Actually, the specific combinations obtained for $\alpha = 1$, $k = 0$ depend on how the degenerate limit is obtained ($\alpha = 1, k \to 0$ is different from $k = 0$, $\alpha \to 1$). This is expected since the degenerate homogeneous elastic state is being recovered from physically different perturbations.

5 Navier-Stokes and Green-Kubo Expressions

In the last section the hydrodynamic modes were defined and calculated at asymptotically long wavelengths. It remains to identify the hydrodynamic coefficients in (58) as well as the hydrodynamic modes at finite k. Consider first the coefficients defined by (59) and evaluate the biorthogonal function $\bar{\xi}_\beta^h(\vec{k},\vec{c})$ to leading order in k, so that

$$\widetilde{a}_\beta^h\left(\vec{k},s\right) \to \left\langle \bar{\xi}_\beta^h(\vec{c})\right| \delta\chi\left(\vec{k},\vec{c},s\right)\right\rangle . \tag{80}$$

Use of the expressions given in (76) shows that, in coordinate representation, these are just the hydrodynamic fields of the linear balance equations

$$\left\{a_\beta^h\left(\vec{l},s\right)\right\} \to \left\{\rho\left(\vec{l},s\right),\vec{\omega}\left(\vec{l},s\right),\frac{1}{2}\theta\left(\vec{l},s\right)+\rho\left(\vec{l},s\right)\right\}. \tag{81}$$

This completes the connection between the hydrodynamic modes and the dynamics determined from the linear balance equations. They are the same for long wavelengths and small perturbations, as expected.

5.1 Navier-Stokes Approximation

To obtain the hydrodynamic eigenvalues $\lambda_\beta^h\left(\vec{k}\right)$ to Navier-Stokes (k^2) order, the eigenvalue problem (55) can be solved treating $i\vec{c}\cdot\vec{k}$ as a small perturbation of the operator Λ [20]. This method implicitly restricts k to be small relative to the deviation of the restitution coefficient α from unity, the elastic limit. A more uniform description, without restrictions on k relative to α, is obtained by calculating the fluxes in the balance equations as functions of $\widetilde{a}_\beta^h(\vec{k},s)$, to first order in k. The resulting Navier-Stokes hydrodynamic equations can then be used to determine $\lambda_i^h\left(\vec{k}\right)$. This is the approach used here. The first step is accomplished by using the identity

$$e^{(A+B)s} = e^{As} + \int_0^s ds'\, e^{As'}Be^{(A+B)(s-s')}, \tag{82}$$

valid for arbitrary operators A and B, in the formal solution (54) to get

$$\delta\widetilde{\chi}\left(\vec{k},\vec{c},s\right) = e^{s\Lambda}\delta\widetilde{\chi}\left(\vec{k},\vec{c},0\right) - \int_0^s ds'\, e^{s'\Lambda} i\vec{k}\cdot\vec{c}e^{-s'\left(\Lambda - i\vec{k}\cdot\vec{c}\right)}\delta\widetilde{\chi}\left(\vec{k},\vec{c},s\right). \tag{83}$$

Next, substitute the representation (58) to identify the dependence on $\widetilde{a}_\beta^h(\vec{k},s)$,

$$\begin{aligned}\delta\widetilde{\chi}\left(\vec{k},\vec{c},s\right) &= e^{s\Lambda}\delta\widetilde{\chi}\left(\vec{k},\vec{c},0\right) - \int_0^s ds'\, e^{s'\Lambda} i\vec{k}\cdot\vec{c}e^{-s'(\Lambda - i\vec{k}\cdot\vec{c})} \\ &\quad\times\left[\sum_\beta \widetilde{a}_\beta^h\left(\vec{k},s\right)\xi_\beta^h\left(\vec{k},\vec{c}\right) + \delta\widetilde{\chi}^m\left(\vec{k},\vec{c},s\right)\right] \\ &\rightarrow e^{s\Lambda}\delta\widetilde{\chi}\left(\vec{k},\vec{c},0\right) - \sum_\beta\left[\int_0^s ds'\, e^{s'\Lambda}\vec{c}e^{-s'\Lambda}\xi_\beta^h(\vec{c})\right]\cdot i\vec{k}\widetilde{a}_\beta^h\left(\vec{k},s\right) \\ &= e^{s\Lambda}\delta\widetilde{\chi}\left(\vec{k},\vec{c},0\right) - \sum_\beta\left[\int_0^s ds'\, e^{s'(\Lambda-\lambda_\beta^h)}\vec{c}\xi_\beta^h(\vec{c})\right]\cdot i\vec{k}\widetilde{a}_\beta^h\left(\vec{k},s\right).\end{aligned} \tag{84}$$

In the second transformation, it has been assumed that k is sufficiently small so that only leading order terms need to be retained, and that s is sufficiently large to neglect the microscopic excitations relative to the hydrodynamic modes contributions. Moreover, in the last equality use has been made of the fact that ξ_β^h is an eigenfunction of Λ. In coordinate representation this result becomes more explicitly

$$\delta\chi\left(\vec{l},\vec{c},s\right) \rightarrow e^{s\Lambda}\delta\chi\left(\vec{l},\vec{c},0\right) - \sum_{\beta=1}^{3}\vec{F}_\beta\left(\vec{c},s\right)\cdot\frac{\partial}{\partial\vec{l}}\nu_\beta\left(\vec{l},s\right), \tag{85}$$

where

$$\left\{\nu_\beta\left(\vec{l},s\right)\right\} \equiv \left\{\rho\left(\vec{l},s\right), \vec{\omega}\left(\vec{l},s\right), \theta\left(\vec{l},s\right)\right\} \tag{86}$$

and

$$\vec{F}_1(\vec{c},s) = \int_0^s ds' e^{s'\Lambda}\xi_1(\vec{c})\vec{c} + 2\vec{F}_3(\vec{c},s), \tag{87}$$

$$F_{2,ij}(\vec{c},s) = \int_0^s ds' e^{s'\left(\Lambda - \frac{\zeta_0}{2}\right)}\xi_{2,i}(\vec{c})c_j, \tag{88}$$

$$\vec{F}_3(\vec{c},s) = \frac{1}{2}\int_0^s ds'\, e^{s'\left(\Lambda + \frac{\zeta_0}{2}\right)}\xi_3(\vec{c})\vec{c}. \tag{89}$$

The pressure tensor and the heat flux up to first order in the gradients of the fields are now obtained by substituting (85) into (46) and (47). The term involving the initial condition $\delta\xi(\vec{l},\vec{c},0)$ does not contribute to these quantities since, in the long time limit, it is given by a linear combination of the functions $\chi_\beta^h(\vec{c})$, that are all orthogonal to both $\Delta(\vec{c})\chi_{\mathrm{HCS}}(\vec{c})$ and

$\vec{\Sigma}(\vec{c})\chi_{\text{HCS}}(\vec{c})$. Taking into account the symmetry properties of the system, the resulting expressions can be written in the form

$$\Pi_{ij}\left(\vec{l},s\right) = \frac{1}{2}\left[\theta\left(\vec{l},s\right) + \rho\left(\vec{l},s\right)\right]\delta_{ij}$$
$$-\tilde{\eta}\left[\frac{\partial}{\partial l_i}\omega_j\left(\vec{l},s\right) + \frac{\partial}{\partial l_j}\omega_i\left(\vec{l},s\right) - \frac{2}{d}\frac{\partial}{\partial \vec{l}}\cdot\vec{\omega}\left(\vec{l},s\right)\delta_{ij}\right], \quad (90)$$

$$\vec{\phi}\left(\vec{l},s\right) = -\tilde{\kappa}\frac{\partial}{\partial \vec{l}}\theta\left(\vec{l},s\right) - \tilde{\mu}\frac{\partial}{\partial \vec{l}}\rho\left(\vec{l},s\right). \quad (91)$$

Equation (90) is the expected Navier-Stokes expression for the pressure tensor, involving the shear viscosity coefficient $\tilde{\eta}$, but (91) contains, besides the usual Fourier law characterized by the heat conductivity $\tilde{\kappa}$, an additional contribution proportional to the density gradient and with an associated transport coefficient $\tilde{\mu}$. This latter term has no analogue in elastic gases.

The expressions of the (time-dependent) transport coefficients are

$$\tilde{\eta}(s) = \frac{1}{d^2+d-2}\sum_{i,j}^{d}\int d\vec{c}\,\Delta_{ij}(\vec{c})F_{2,ij}(\vec{c},s), \quad (92)$$

$$\tilde{\kappa}(s) = \frac{1}{d}\int d\vec{c}\,\vec{\Sigma}(\vec{c})\cdot\vec{F}_3(\vec{c},s), \quad (93)$$

$$\tilde{\mu} = \frac{1}{d}\int d\vec{c}\vec{\Sigma}(\vec{c})\cdot\vec{F}_1(\vec{c},s). \quad (94)$$

The hydrodynamic eigenfunctions $\xi_\beta^h(\vec{c})$ correspond to the smallest eigenvalues and hence to the dynamics on hydrodynamic time scales (see the discussion following (59)). As already mentioned, the functions $\Delta_{ij}(\vec{c})\chi_{\text{HCS}}(\vec{c})$ and $\Sigma_i(\vec{c})\chi_{\text{HCS}}(\vec{c})$ are orthogonal to the subspace spanned by the right eigenfunctions $\xi_\beta(\vec{c})$. Therefore, the decay of the "correlation functions" in (92)–(94) lies outside the spectrum of the slowest modes, and the s-time integral in the expression of the functions F_β converges. Consequently, it is possible to set $s\to\infty$ for times scales long compared with the microscopic times. In this way, the reduced transport coefficients become time-independent. Returning to the original variables, the above expressions for the hydrodynamic fluxes are equivalent to

$$P_{ij}(\vec{r},t) = n_H T_{\text{HCS}}(t)\left[1 + \frac{\delta T(\vec{r},t)}{T_{\text{HCS}}(t)} + \frac{\delta n(\vec{r},t)}{n_H}\right]\delta_{ij}$$
$$-\eta\left[\frac{\partial}{\partial r_i}\delta u_j + \frac{\partial}{\partial r_j}\delta u_i - \frac{2}{d}\left(\vec{\nabla}\cdot\delta\vec{u}\right)\delta_{ij}\right], \quad (95)$$

$$\vec{q}(\vec{r},t) = -\kappa \vec{\nabla} \delta T(\vec{r},t) - \mu \vec{\nabla} \delta n(\vec{r},t), \tag{96}$$

with the transport coefficients given by

$$\eta = n_H m \ell v_0(t) \tilde{\eta}, \tag{97}$$

$$\kappa = n_H \ell v_0(t) \tilde{\kappa}, \tag{98}$$

$$\mu = \frac{m \ell v_0^3(t)}{2} \tilde{\mu}. \tag{99}$$

5.2 Green-Kubo Relations

It is possible to write (92)–(94) in the form of Green-Kubo relations. Define an operator $\tilde{\Lambda}$ by

$$\left[\int d\vec{c}\, g^*(\vec{c}) \Lambda h(\vec{c}) \right]^* = \int d\vec{c}\, h^*(\vec{c}) \tilde{\Lambda} g(\vec{c}), \tag{100}$$

for arbitrary h and g. Comparison with (73) yields

$$\begin{aligned} \tilde{\Lambda} g(\vec{c}) &= \chi_{\text{HCS}}^{-1}(\vec{c}) \Lambda^+ \left[g(\vec{c}) \chi_{\text{HCS}}(\vec{c}) \right] \\ &= \int d\vec{c}_1 \chi_{\text{HCS}}(\vec{c}_1) T(\vec{c}, \vec{c}_1) [g(\vec{c}) + g(\vec{c}_1)] + \frac{\zeta_0}{2} \vec{c} \cdot \frac{\partial}{\partial \vec{c}} g(\vec{c}). \end{aligned} \tag{101}$$

Also define "correlation functions" in the HCS by

$$\langle gh \rangle \equiv \int d\vec{c}\, \chi_{\text{HCS}}(\vec{c}) g(\vec{c}) h(\vec{c}). \tag{102}$$

Then, the expressions for the transport coefficients can be rewritten as

$$\tilde{\eta} = \frac{1}{d^2 + d - 2} \sum_{i,j}^{d} \int_0^\infty ds \, \langle \Delta_{ij}(s) \Phi_{2,ij} \rangle \, e^{-s\zeta_0/2}, \tag{103}$$

$$\tilde{\kappa} = \frac{1}{d} \int_0^\infty ds \, \left\langle \vec{\Sigma}(s) \cdot \vec{\Phi}_3 \right\rangle e^{s\zeta_0/2}, \tag{104}$$

$$\tilde{\mu} = 2\tilde{\kappa} + \frac{1}{d} \int_0^\infty ds \, \left\langle \vec{\Sigma}(s) \cdot \vec{\Phi}_1 \right\rangle, \tag{105}$$

with the definitions

$$\vec{\Phi}_1 \equiv \vec{c} - 2\vec{\Phi}_3(\vec{c}), \tag{106}$$

$$\Phi_{2,ij}(\vec{c}) = -c_j \frac{\partial \ln \chi_{\text{HCS}}(\vec{c})}{\partial c_i}, \tag{107}$$

$$\vec{\Phi}_3(\vec{c}) = -\frac{1}{2}\vec{c}\left[d + \vec{c} \cdot \frac{\partial \ln \chi_{\text{HCS}}(\vec{c})}{\partial \vec{c}}\right]. \tag{108}$$

The time dependence of the dynamical variables is given by

$$g(\vec{c}, s) = e^{s\tilde{\Lambda}} g(\vec{c}). \tag{109}$$

Equations (103)–(105) are the Green-Kubo formulae for a dilute granular gas described by the Boltzmann equation. They express the transport coefficients as integrals of time-correlations functions in the HCS. Although they are not identical to those derived in [11] by a different method, it is easy to show they are equivalent, since they differ in terms giving vanishing contributions.

Let us also compare them with the expressions derived from the nonlinear Boltzmann equation by the Chapmann-Enskog method [5, 8]. Consider the reduced shear viscosity coefficient. It is given by (92) with $F_{2,ij}$ defined in (88). In the limit of large s, we have seen that the integral in the latter expression can be extended up to infinity. Performing then the integral, it is obtained

$$\tilde{\eta} = -\frac{1}{d^2 + d - 2} \sum_{i,j}^{d} \int d\vec{c}\, \Delta_{ij}(\vec{c}) \left(\Lambda - \frac{\zeta_0}{2}\right)^{-1} \xi_{2,i}(\vec{c}) c_j, \tag{110}$$

or, equivalently,

$$\tilde{\eta} = \frac{1}{d^2 + d - 2} \sum_{i,j}^{d} \int d\vec{c}\, \Delta_{ij}(\vec{c}) \mathcal{C}_{ij}(\vec{c}), \tag{111}$$

where $\mathcal{C}_{ij}$ is the solution of the integral equation

$$\left(\Lambda - \frac{\zeta_0}{2}\right) \mathcal{C}_{ij}(\vec{c}) = -\xi_{2,i}(\vec{c}) c_j, \tag{112}$$

being orthogonal to the subspace spanned by the lowest order right eigenfunctions of Λ. Equations (111) and (112) are equivalent to those derived by the Chapman-Enskog procedure in [5, 8], where the explicit expressions of the transport coefficients in the first Sonine approximation is obtained. A similar analysis leading to the same conclusion can be carried out for the other transport coefficients. The only conceptually relevant difference is in the reference state. In the present linearized analysis the transport coefficients are defined in the HCS, whereas in the Chapmann-Enskog method they are defined in a *local* HCS, characterized by the parameters $n(\vec{r}, t)$, $\vec{u}(\vec{r}, t)$ and $T(\vec{r}, t)$. As a consequence, these are the fields appearing in the scaling in (97)–(99).

It is also instructive to consider the elastic limit $\alpha \to 1$ of the Green-Kubo expressions derived here. In that case, we already know that the HCS distribution becomes the Maxwellian χ_{MB}, and $\zeta_0 \to 0$. Then, the functions defined in (106)–(108) become

$$\vec{\Phi}_1(\vec{c}) \to -2\vec{\Sigma}(\vec{c}), \quad \Phi_{2,ij}(\vec{c}) \to 2c_i c_j, \quad \vec{\Phi}_3(\vec{c}) \to \left(c^2 - \frac{d}{2}\right)\vec{c}. \tag{113}$$

Therefore, the transport coefficient $\tilde{\mu}$, given by (105) vanishes in the elastic limit, as expected, and

$$\begin{aligned} \eta \to \eta_0 &= \frac{2nm\ell v_0(t)}{d^2+d-2} \sum_{i,j}^{d} \int_0^\infty ds \, \langle \Delta_{ij}(s) c_i c_j \rangle \\ &= \frac{2nm\ell v_0(t)}{d^2+d-2} \sum_{i,j}^{d} \int_0^\infty ds \, \langle \Delta_{ij}(s) \Delta_{ij} \rangle . \end{aligned} \tag{114}$$

$$\begin{aligned} \kappa \to \kappa_0 &= \frac{n\ell v_0(t)}{d} \int_0^\infty ds \left\langle \vec{\Sigma}(s) \cdot \left(c^2 - \frac{d}{2}\right)\vec{c} \right\rangle \\ &= \frac{n\ell v_0(t)}{d} \int_0^\infty ds \left\langle \vec{\Sigma}(s) \cdot \vec{\Sigma} \right\rangle . \end{aligned} \tag{115}$$

The last equalities in the above equations follow from the symmetry of the time correlation functions. For elastic systems, the temperature is constant to lowest order in the gradients of the hydrodynamic fields and, therefore, in the integrands appearing in the above expressions the definition of the scale s, given by (27), reduces to $s = v_0 t/\ell$, i.e. it is simply proportional to the actual time. Then, (114) and (115) are the low density limit of the usual Green-Kubo expressions for the transport coefficients in terms of the autocorrelation functions of the microscopic fluxes of the relevant densities [9].

6 Discussion

The primary results of the above analysis have been the identification of hydrodynamic modes in the spectrum of the linearized Boltzmann operator, their leading order calculation in Sect. 4, and their use to derive the Navier-Stokes approximations to the heat and momentum fluxes. Of particular interest are the long wavelength forms of the hydrodynamic eigenfunctions ξ_β^h given in (67), which are not simply linear combinations of the conserved densities $1, \vec{v}, v^2$. It is clear from (58) that perturbations of the homogeneous state due to $\xi_\beta^h(\vec{c})$ generate the hydrodynamic excitations, while the response to these excitations is measured by the fields $\tilde{a}_\beta^h(s)$. These fields are averages of the biorthogonal set of functions $\bar{\xi}_\beta$, which are linear combinations of the conserved densities $1, \vec{v}, v^2$, as required by the macroscopic balance equations. As anticipated in the Introduction, this "conjugate" relationship between the sets $\left\{\xi_\beta^h\right\}$ and $\left\{\bar{\xi}_\beta\right\}$ is a key difference between normal and granular gases. For elastic collisions, the two sets of fields are the same. This has important consequences for the derivation of transport properties as it has been illustrated here. For example, the reduction leading to the Navier-Stokes approximation in (84), depends critically on the fact that the functions ξ_β^h are eigenfunctions of Λ. Any other choice for the initial perturbation would lead to a different time dependence, mixing microscopic and hydrodynamic excitations, and precluding direct identification of transport properties.

The identification of the hydrodynamic modes discussed here is important for the application of linear response methods [21] to granular systems. In order to illustrate it, let us consider again the formal solution to the linear Boltzmann equation given in (54) and choose for the initial condition the linearization of a *local* HCS as given by (71),

$$\begin{aligned}\delta\widetilde{\chi}\left(\vec{k}, \vec{c}, s\right) &= e^{s\left(\Lambda - i\vec{k}\cdot\vec{c}\right)} \sum_\beta \langle \bar{\xi}_\beta^h | \delta\widetilde{\chi}(0)\rangle \, \xi_\beta^h\left(\vec{c}\right) \\ &= e^{s\left(\Lambda - i\vec{k}\cdot\vec{c}\right)} \sum_\beta a_\beta\left(\vec{k}, s=0\right) \xi_\beta^h\left(\vec{c}\right). \end{aligned} \tag{116}$$

This initial state clearly represents a hydrodynamic perturbation. The response $a_\beta(\vec{k}, s)$ is then given by

$$a_\beta\left(\vec{k}, s\right) = \sum_{\beta'} R_{\beta\beta'}\left(\vec{k}, s\right) a_{\beta'}\left(\vec{k}, s=0\right), \quad R_{\beta\beta'}\left(\vec{k}, s\right) = \left\langle \bar{\xi}_\beta^h | e^{s\left(\Lambda - i\vec{k}\cdot\vec{c}\right)} \xi_{\beta'}^h \right\rangle. \tag{117}$$

The response functions $R_{\beta\beta'}(\vec{k}, s)$ are time correlation functions comprised of the pair of the conjugate densities. It can be shown that application of standard linear response methods to these response functions leads to the correct Green-Kubo relations with the proper convergence properties for the time integrals. Conversely, the same procedure applied to correlation functions comprised of the $\left\{\bar{\xi}_\beta\right\}$ alone does not lead directly to well-posed Green-Kubo expressions. Let us also mention that linear perturbations of the kind considered in (116) have been used [22] in direct Monte Carlo simulations of the Boltzmann equation [23], to verify the validity of the linear hydrodynamic description and the explicit expressions of the transport coefficients.

Although the analysis here has been limited to the low density linearized Boltzmann equation, these considerations point the way for a proper application of linear response methods in the more general context of nonequilibrium statistical mechanics.

Acknowledgements

The research of J.W.D. reported here was supported in part by the Department of Energy Grants DE-FG03-98DP00218 and DE-FG02ER54677. The research of J.J.B. and M.J.R. was partially supported by the Ministerio de Ciencia y Tecnología (Spain) through Grant No. BFM2002-00307.

References

1. J.T. Jenkins and M.W. Richman. Grad's 13-moment system for a dense gas of inelastic spheres. *Arch. Ration.Mech. Anal.*, 87:355, 1985
2. C.K.W. Lun, S.B. Savage, D.J. Jeffrey, and N. Chepurnuy. Kinetic theories for granular flow: Inelastic particles in Couette flow and slightly inelastic particles in a general flowfield. *J. Fluid Mech.*, 140:223, 1984
3. A. Goldsthein and M. Shapiro. Mechanics of collisional motion of granular materials. Part I. General hydrodynamic equations. *J. Fluid Mech.*, 282:75, 1995
4. N. Sela and I. Goldhirsch. Hydrodynamic equations for rapid flows of smooth inelastic spheres, to Burnett order. *J. Fluid Mech.*, 361:41, 1998
5. J.J. Brey, J.W. Dufty, C.S. Kim, and A. Santos. Hydrodynamics for granular flow at low density. *Phys. Rev. E*, 58:4638, 1998
6. V. Garzó and J.W. Dufty. Dense fluid transport for inelastic hard spheres. *Phys. Rev. E*, 59:5895, 1999
7. V. Garzó and J.W. Dufty. Hydrodynamics for a granular binary mixture at low density. *Phys. of Fluids*, 14:1476, 2002
8. J.J. Brey and D. Cubero. Hydrodynamic transport coefficients in granular gases. In T. Pöschel and S. Luding, editors, *Granular Gases*, volume 564 of *Lecture Notes in Physics*, pages 59–78, Berlin, 2001. Springer
9. J.A. McLennan. *Introduction to Nonequilibrium Statistical Mechanics.* Prentice-Hall, New Jersey, 1989
10. I. Goldhirsch and T.P.C. van Noije. Green-Kubo relations for granular fluids. *Phys. Rev. E*, 61:3241, 2000
11. J.W. Dufty and J.J. Brey. Green-Kubo expressions for a granular gas. *J. Stat. Phys.*, 109:433, 2002
12. J.J. Brey, J.W. Dufty, and A. Santos. Dissipative dynamics for hard spheres. *J. Stat. Phys.*, 87:1051, 1997
13. P.K. Haff. Grain flow as a fluid-mechanical phenomenon. *J. Fluid Mech.*, 134:401, 1983
14. T.P.C. van Noije and M.H. Ernst. Kinetic theory of granular gases. In T. Pöschel and S. Luding, editors, *Granular Gases*, volume 564 of *Lectures Notes in Physics*, pages 3–30, Berlin, 2001. Springer
15. J.J. Brey, F. Moreno, and J.W. Dufty. Model kinetic equation for low-density granular flow. *Phys. Rev. E*, 54:445, 1996
16. J.J. Brey, W.W. Dufty, and A. Santos. Kinetic models for granular flow. *J. Stat. Phys.*, 97:281, 1999
17. J.J. Brey, F. Moreno, and M.J. Ruiz-Montero. Spatial correlations in dilute granular flows: A kinetic model study. *Phys. of Fluids*, 10:2965, 1998

18. J.W. Dufty, J.J. Brey, and J.F. Lutsko. Diffusion in a granular fluid. I. Theory. *Phys. Rev. E*, 65:051303, 2002
19. P. Résibois and M. de Leener. *Classical Kinetic Theory of Fluids*. John Wiley, New-York, 1977
20. J.W. Dufty, J.J. Brey, and M.J. Ruiz-Montero. unpublished
21. K. Kubo, M. Toda, and N. Hashitsume. *Statistical Physics II*. Springer-Verlag, Berlin, 1991
22. J.J. Brey, M.J. Ruiz-Montero, and D. Cubero. On the validity of linear hydrodynamics for low-density granular flows described by the boltzmann equation. *Europhys. Lett*, 48:359, 1999
23. G.A. Bird. *Molecular Gas Dynamics and the Direct Simulation of Gas Flows*. Clarendon, Oxford, 1994

Development of a Density Inversion in Driven Granular Gases

Yaron Bromberg[1], Eli Livne[2], and Baruch Meerson[2]

[1] School of Physics and Astronomy, Tel Aviv University, Tel Aviv 69978, Israel
[2] Racah Institute of Physics, Hebrew University of Jerusalem, Jerusalem 91904, Israel

Abstract. Granular materials fluidized by a rapidly vibrating bottom plate often develop a fascinating density inversion: a heavier layer of granulate supported by a lower-density region. We employ the Navier-Stokes granular hydrodynamics to follow a density inversion as it develops in time. Assuming a dilute low-Mach-number flow, we derive a reduced time-dependent model of the late stage of the dynamics. The model looks especially simple in the Lagrangian coordinates. The time-dependent solution describes the growth of a density peak at an intermediate height. A transient temperature minimum is predicted to develop in the region of the density peak. The temperature minimum disappears at later times, as the system approaches the steady state. At late times, the predictions of the low-Mach-number model are in good agreement with a numerical solution of the full hydrodynamic equations. At an early stage of the dynamics, pressure oscillations are predicted.

1 Introduction

An ensemble of hard spheres in motion, that collide inelastically and are characterized by a constant coefficient of normal restitution, represents the simplest model of granular gas [1]. It also provides an excellent example of complexity. Indeed, the properties of the granular gas as a whole, such as clustering [2–5], come out as non-trivial consequences of the simple algebraic laws that govern the velocities of the binary collisions of individual particles. Furthermore, being intrinsically far from thermal equilibrium, granular gas defies statistical mechanics, the cornerstone of the equilibrium science. As the result, formulating a universally applicable continuum theory of granular gas, even in the dilute limit, is not a simple task [6, 7]. The use of the Boltzmann (or Enskog) equation, the starting point of a systematic derivation of the constitutive relations, is based on the *Molecular Chaos* hypothesis. This hypothesis is justified, in the dilute limit, for an ensemble of *elastic* hard spheres. Its use for *inelastic* hard spheres is, rigorously speaking, an uncontrolled assumption. Indeed, inelasticity of the particle collisions introduces inter-particle correlations which can invalidate the *Molecular Chaos* hypothesis. The correlations become more pronounced as the inelasticity of the collision increases. On the contrary, for *nearly elastic* collisions, $1 - r^2 \ll 1$ (where r is the coefficient of normal restitution) the correlations are small, and the Boltzmann equation can be used safely (again, in the dilute limit).

An additional crucial assumption, made in the process of the derivation of the hydrodynamics from the Boltzmann equation, is scale separation. The hydrodynamics is valid if the mean free path of the particles is much less than any characteristic length scale, and the mean time between two consecutive collisions is much less than any characteristic time scale, described hydrodynamically. This assumption (as well as the dilute gas assumption, if used) should be verified, in every specific system, after the hydrodynamic problem is solved and the characteristic length and time scales determined. One more criterion for the validity of granular hydrodynamics is the smallness of fluctuations, driven by the noise caused by the discrete nature of the particles. The accuracy of hydrodynamics improves with an increase of the total number of particles in the system N [8].

Once hydrodynamics is valid, it is highly rewarding using it. Hydrodynamics has a great predictive power and helps to identify important collective phenomena (shear flows and vortices, shocks, different modes of clustering flows, phase separation etc.) that are impossible to perceive in the language of individual particles. Once identified, these collective phenomena can then be investigated in experiment and simulations in more general (not necessarily hydrodynamic) formulations. A recent example: employing a variant of granular hydrodynamics, Livne *et al.* [9] predicted a novel phase separation instability in a system of inelastic hard spheres, driven by a rapidly vibrating plate at zero gravity. This system had been investigated earlier by many workers. However, the phase separation instability, and a plethora of interesting effects accompanying it, were overlooked. Undoubtedly, the reason for that was that hydrodynamics had not been fully appreciated and exploited [10].

In this paper we employ the Navier-Stokes granular hydrodynamics for a theoretical analysis of the *process* of formation of a density inversion in a dilute granular gas driven by a rapidly vibrating bottom plate in a gravity field. The phenomenon of a density inversion in the *steady state* of this system is well known in the granular community. On the contrary, the *dynamics* of the formation of the density inversion has never been investigated previously. As we will see, this dynamics is quite instructive. The hydrodynamic theory yields predictions of time-dependent quantities that can be checked in molecular dynamics simulations and experiment. Our additional motivation here is pedagogical. While dealing with this problem, we will present two useful hydrodynamic techniques. The first is reduction of the order of the hydrodynamic equations, based on a hierarchy of the length/time scales in the problem. The second technique is the use of the Lagrangian coordinates.

Here is a layout of the remainder of this paper. In Sect. 2 we will write down a set of hydrodynamic equations and boundary conditions that describe the dynamics of formation of a density inversion. By introducing scaled variables, we will delineate the two scaled governing parameters of the problem. In Sect. 3 we briefly review the properties of the steady state and the condition for a density inversion. In Sect. 4 we assume a low-Mach-number flow, derive a reduced hydrodynamic model and use it for an investigation of the late-time dynamics of the formation of the density inversion. Section 5 reports a numerical solution of the full hydrodynamic equations and compares the numerical results with the predictions of the low-Mach-number model. A discussion of the results is presented in Sect. 6.

2 The Model Problem and Hydrodynamic Equations

We will adopt the simplest possible model, in two dimensions, that exhibits a density inversion. Let $N \gg 1$ identical nearly elastic hard disks of diameter d and mass $m = 1$ move without friction in a two-dimensional box of width L_x and infinite height. The driving base is at $y = 0$, the side walls are at $x = 0$ and $x = L_x$. The gravity acceleration g is in the negative y direction. The rapid vibrations of the base can be modeled, in a simplified way, by prescribing a constant granular temperature $T = T_0$ at $y = 0$ (we measure the granular temperature in the units of the velocity squared). This means that, upon collision with the base, the particle velocity is drawn from a Maxwell's distribution with temperature T_0. The kinetic energy of the particles is being lost by inelastic hard-core collisions parameterized by a constant coefficient of normal restitution r. Particle collisions with the side walls are assumed elastic. Alternatively, one can specify periodic boundary conditions in the x-direction.

Introducing a hydrodynamic description, we deal with three coarse-grained fields: the particle number density $n(\vec{r}, t)$, mean flow velocity $\vec{v}(\vec{r}, t)$ and granular temperature $T(\vec{r}, t)$. In addition, we make two important assumptions. First, we assume that that the collisions are nearly elastic: $1 - r^2 \ll 1$. Second, we assume that the granular gas is dilute everywhere (including the density peak region, see below), that is n is small enough compared to the close-packing density n_c. These assumptions enable us to use the classic version of the Navier-Stokes granular hydrodynamics [11], and to limit ourselves to the leading order terms in n/n_c in the constitutive relations. The dilute gas condition will be checked *a posteriori*.

It has been shown recently that, when this system is large enough in the lateral direction, thermal convection may occur [12–14]. In this work we will assume that L_x is sufficiently small, and the convection is suppressed by the viscosity and lateral heat conduction. Therefore, we assume flow solely in the vertical direction. Therefore, the governing equations are:

$$\frac{\partial n}{\partial t} + \frac{\partial}{\partial y}(nv) = 0, \tag{1}$$

$$n\left(\frac{\partial v}{\partial t} + v\frac{\partial v}{\partial y}\right) = \frac{\partial P}{\partial y} - ng, \tag{2}$$

$$n\left(\frac{\partial T}{\partial t}+v\frac{\partial T}{\partial y}\right)=-\frac{\partial q}{\partial y}+P\frac{\partial v}{\partial y}-I. \tag{3}$$

Here $P = -nT + \eta\partial v/\partial y$ is the vertical component of the stress tensor, $q = -\kappa\partial T/\partial y$ is the heat flux, and I is the heat loss rate by the inelastic collisions. In the nearly elastic limit, the leading order contributions to the viscosity η and heat conductivity κ are given by their classical values for for *elastic* hard disks:

$$\eta=\frac{1}{2d}\left(\frac{T}{\pi}\right)^{1/2} \quad \text{and} \quad \kappa=\frac{2}{d}\left(\frac{T}{\pi}\right)^{1/2}.$$

The heat loss rate is the following:

$$I=\pi^{1/2}(1-r^2)\,dn^2T^{3/2}.$$

The boundary conditions for the problem include the fixed temperature

$$T(y=0,t)=T_0 \tag{4}$$

and zero velocity

$$v(y=0,t)=0 \tag{5}$$

at the base. Now, as the total number of grains N is constant:

$$\int_0^\infty n(y,t)\,dy=N/L_x=const\,, \tag{6}$$

the mass flux should vanish at infinity:

$$nv=0 \ \text{ at } \ y=\infty. \tag{7}$$

Similarly, one should assume that the momentum flux at infinity is zero:

$$nT+nv^2-\eta\frac{\partial v}{\partial y}=0 \ \text{ at } \ y=\infty, \tag{8}$$

as well as the energy flux:

$$nv\left(\frac{v^2}{2}+2T\right)-\eta v\frac{\partial v}{\partial y}-\kappa\frac{\partial T}{\partial y}=0 \ \text{ at } \ y=\infty. \tag{9}$$

We refer the reader to the book of Landau and Lifshitz [15] for a detailed discussion of the divergent forms of the hydrodynamic equations, and for the expressions for the momentum and energy flux density used in (8) and (9).

What are the characteristic length/time scales of the problem? The fixed base temperature T_0 and gravity acceleration g define a characteristic macroscopic length scale $\lambda = T_0/g$. The characteristic number density of the gas is therefore $n_0 = N/(\lambda L_x)$. Correspondingly, there are three characteristic time scales in the problem. The fast time scale of the problem, $\tau_f = T_0^{1/2}/g$, is independent of the particle size. The heat conduction time scale is $\tau = (\pi\lambda/g)^{1/2}(Nd/L_x)$. The inelastic heat loss time scale τ_l can be defined as $\tau_l^{-1} = (1-r^2)d\,n_0 T_0^{1/2}$.

How do we know that the time scale τ_f is indeed fast? Before we answer this question, let us first make a preliminary evaluation of the validity of hydrodynamics. We should demand that the characteristic mean free path of the particles $\sim (n_0 d)^{-1}$ be much less than the macroscopic length scale λ. This immediately follows $L_x/(Nd) \ll 1$. Notice that the quantity Nd/L_x is of the order of the number of monolayers at rest (that is, when the system is not fluidized). Therefore, the validity of the hydrodynamics requires that the number of monolayers be much larger than 1. Now we compare the characteristic time scales τ_f and τ introduced above. We see that the ratio τ_f/τ is of the order of $L_x/(Nd)$, which *should* be much less than unity if we want the hydrodynamics to be valid.

To make a full use of hydrodynamics, we introduce scaled variables. Let us measure time in the units of the heat conduction time τ, the coordinate in the units of λ, the temperature in the units of T_0, the density in the units of n_0 and the mean flow velocity in the units of λ/τ. In the scaled variables, the governing equations (1)-(3) become

$$\frac{\partial n}{\partial t} + \frac{\partial}{\partial y}(nv) = 0, \tag{10}$$

$$\varepsilon^2 n\left(\frac{\partial v}{\partial t} + v\frac{\partial v}{\partial y}\right) = -\frac{\partial}{\partial y}(nT) - n + \frac{\varepsilon^2}{2}\frac{\partial}{\partial y}\left(T^{1/2}\frac{\partial v}{\partial y}\right), \tag{11}$$

$$n\left(\frac{\partial T}{\partial t} + v\frac{\partial T}{\partial y}\right) + nT\frac{\partial v}{\partial y} = \frac{\varepsilon^2}{2}T^{1/2}\left(\frac{\partial v}{\partial y}\right)^2 + \frac{4}{3}\frac{\partial^2}{\partial y^2}T^{3/2} - 4\Lambda^2 n^2 T^{3/2}, \tag{12}$$

where

$$\varepsilon = \frac{L_x}{\pi^{1/2}Nd} \ll 1 \quad \text{and} \quad \Lambda = \frac{(1-r^2)^{1/2}}{2\,\varepsilon}$$

are the two governing scaled parameters of the problem. Parameter Λ is of the order of the square root of the ratio between the heat conduction time scale and inelastic heat loss time scale.

The reduction of the four-dimensional space of governing parameters d, N, r and L_x to the (ε, Λ) - plane is an immediate benefit from using hydrodynamics. Indeed, series of experiments or simulations with different d, N, r and L_x, but with the same ε and Λ, should produce the same dynamics in the scaled variables. We will see in the following that, in the late stage of the dynamics (and in the steady state), the only relevant parameter is Λ.

In the scaled variables, the boundary conditions (4), (5), (7), (8) and (9) become $T = 1$ and $v = 0$ at $y = 0$; and

$$nv = 0, \tag{13}$$

$$nT + \varepsilon^2 nv^2 - \frac{\varepsilon^2}{2} T^{1/2} \frac{\partial v}{\partial y} = 0 \tag{14}$$

and

$$nv\left(\frac{\varepsilon^2}{4} v^2 + T\right) - \frac{\varepsilon^2}{4} T^{1/2} v \frac{\partial v}{\partial y} - T^{1/2} \frac{\partial T}{\partial y} = 0 \tag{15}$$

at $y = \infty$.

The simplest way to satisfy the boundary conditions (14) and (15) is to assume a zero heat flux at infinity, which implies $\partial T/\partial y = 0$ at $y = \infty$. As the total amount of material is finite, the density n should vanish at $y = \infty$. Therefore, the nT term in (14) should also vanish there. Now, using what remains of (14) together with (13), we see that (15) can be reduced to the simple condition $\partial v/\partial y = 0$ at $y = \infty$. Therefore, we are left with simple boundary conditions: $T = 1$ and $v = 0$ at $y = 0$, and $\partial T/\partial y = \partial v/\partial y = 0$ at $y = \infty$ [16]. A full formulation of the time-dependent problem requires that we specify the initial conditions $n(y, t = 0), T(y, t = 0)$ and $v(y, t = 0)$. We postpone this issue until later.

Like in many other one-dimensional gasdynamic problems [17], it is convenient to make a transformation of variables from the Eulerian coordinate y to the Lagrangian mass coordinate $m(y, t) = \int_0^y n(y', t)\, dy'$. The physical meaning of the Lagrangian mass coordinate is very simple: it is the mass content of the gas on the Eulerian interval $(0, y)$ per unit length in the x-direction. Notice that the infinite interval $0 \le y < \infty$ in the Eulerian coordinates is mapped, by this coordinate transformation, into a finite interval $0 \le m < 1$, at all times. Going over to the Lagrangian coordinate m, we can rewrite (10)–(12) as follows:

$$\frac{\partial}{\partial t}\left(\frac{1}{n}\right) = \frac{\partial v}{\partial m}, \tag{16}$$

$$\varepsilon^2 \frac{\partial v}{\partial t} = -\frac{\partial}{\partial m}(nT) - 1 + \frac{\varepsilon^2}{2} \frac{\partial}{\partial m}\left(T^{1/2} n \frac{\partial v}{\partial m}\right) \tag{17}$$

$$\frac{\partial T}{\partial t} + nT \frac{\partial v}{\partial m} = \frac{\varepsilon^2}{2} T^{1/2} n \left(\frac{\partial v}{\partial m}\right)^2 + \frac{4}{3} \frac{\partial}{\partial m}\left(n \frac{\partial}{\partial m} T^{3/2}\right) - 4\Lambda^2 n T^{3/2}. \tag{18}$$

The boundary conditions become $T = 1$ and $v = 0$ at $m = 0$, and $n\, \partial T/\partial m = n\, \partial v/\partial m = 0$ at $m = 1$.

Notice that the transformation from the Eulerian to Lagrangian coordinate involves the density $n(y,t)$ that is unknown *a priori*. We should not worry about it. After having determined the density $n(m,t)$ in the Lagrangian coordinates, one can go back, at any time t, to the Eulerian coordinate by using the relation

$$y(m,t) = \int_0^m \frac{dm'}{n(m',t)} . \tag{19}$$

3 Steady State Profiles and Density Inversion

Let us briefly review the steady state solution of the problem [13, 14, 18]. To obtain the steady state profiles $n_s(m)$ and $T_s(m)$ we simply put $\partial/\partial t = v = 0$ in (16)–(18). Then (16) is obeyed automatically, while (17) and (18) become

$$\frac{d}{dm}(n_s T_s) + 1 = 0 \tag{20}$$

and

$$\frac{d}{dm}\left(n_s \frac{d}{dm} T_s^{3/2}\right) - 3\Lambda^2 n_s T_s^{3/2} = 0, \tag{21}$$

respectively. As we see, the steady state is described by a single scaled parameter Λ. Integrating (20) with respect to m and using the boundary condition at $m = 1$ (that becomes simply $nT = 0$), we obtain $n_s T_s = 1 - m$. Now we express the density through the temperature: $n_s(m) = (1-m)/T_s(m)$. Using this relation in (21), we arrive at the following *linear* equation [13, 14, 18] for $\omega_s(m) = T_s^{1/2}(m)$:

$$(1-m)\,\omega_s'' - \omega_s' - \Lambda^2(1-m)\,\omega_s = 0, \tag{22}$$

where primes stand for the m-derivatives. This is a Bessel equation whose general solution is a linear combination of the modified Bessel functions of the first kind $I_0\left[\Lambda(1-m)\right]$ and $K_0\left[\Lambda(1-m)\right]$. Demanding that the heat flux vanish at $m = 1$ and returning to the temperature variable, we obtain:

$$T_s(m) = \frac{I_0^2\left(\Lambda(1-m)\right)}{I_0^2(\Lambda)} \tag{23}$$

Respectively,

$$n_s(m) = \frac{1-m}{T_s(m)} = \frac{(1-m)I_0^2(\Lambda)}{I_0^2\left(\Lambda(1-m)\right)} . \tag{24}$$

To transform back to the Eulerian coordinate y, we should evaluate the integral

$$y(m) = \int_0^m \frac{dm'}{n_s(m')} \tag{25}$$

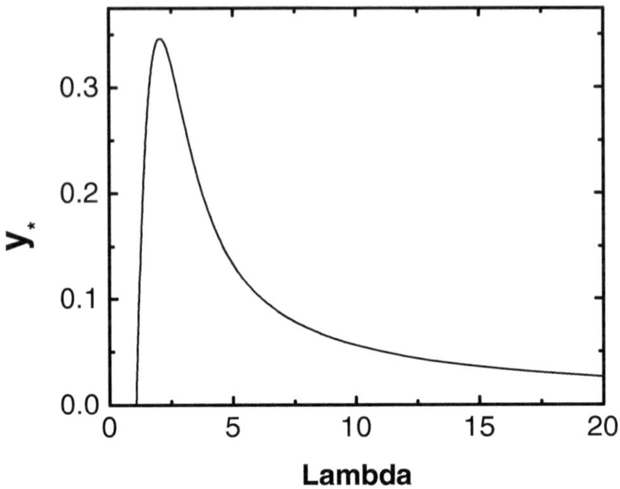

Fig. 1. The density peak height in the (scaled) Eulerian coordinate versus Λ

numerically. Figures 2 and 3 show the steady state temperature and density profiles in the Lagrangian and Eulerian coordinates for $\Lambda = \sqrt{10}$.

Equation (24) enables one to find the critical (minimum) value of Λ for the density inversion. At small Λ the density decreases monotonically with m (and therefore with y). The density inversion is born, at $m = 0$, at $\Lambda = \Lambda_c$. Therefore, to find Λ_c, we demand that the derivative $n_s'(m)$ vanish at $m = 0$. Using (24), we can reduce this condition to the equation $I_0(\Lambda) = 2\Lambda I_1(\Lambda)$, first obtained in [18]. Its solution yields $\Lambda_c \simeq 1.06569\ldots$. At $\Lambda > \Lambda_c$ the density reaches its maximum in the Lagrangian point $m_* = 1 - \Lambda_c/\Lambda > 0$. Therefore, the position of the density peak in the Lagrangian coordinate grows with Λ monotonically. The respective (scaled) density maximum value is

$$n_{max} = \frac{\Lambda_c \, I_0^2(\Lambda)}{\Lambda \, I_0^2(\Lambda_c)} \,. \tag{26}$$

Using (25), we can compute the scaled Eulerian coordinate y_* at which the maximum density maximum occurs:

$$y_* = \frac{1}{I_0^2(\Lambda)} \int_{\Lambda_c}^{\Lambda} \frac{I_0^2(\xi)}{\xi} \, d\xi \,. \tag{27}$$

The dependence of y_* on Λ is non-monotonic, see Fig. 1. Very close to the density inversion birth (at $\Lambda > \Lambda_c$) this dependence is linear: $y_* \simeq \Lambda/\Lambda_c - 1$, while at $\Lambda \gg \Lambda_c$ one obtains $y_* \simeq (2\Lambda)^{-1}$. Actually, the integral in (27) can be calculated analytically, in terms of a hypergeometric function.

4 Formation of a Density Inversion: Low-Mach-Number Flow

Now let us return to the time-dependent problem. Can we exploit the presence of a small parameter ε^2 in (17) and (18) to simplify the equations? One can expect that the respective terms *cannot* be neglected during the rapid initial stage of the dynamics. The duration of this stage is of the order of the acoustic time or, in our scaled units, of the order of ε. A comparison of different terms in the momentum equation (17) shows that the typical scaling of the velocity during this stage is $\varepsilon^{-1} \gg 1$. In the physical units, the flow velocity here is comparable to, or larger than, the speed of sound. Correspondingly, at early times the ε^2-terms are comparable to, or larger than, the other terms and therefore should be kept.

The situation changes at later times, determined by the inelastic heat losses and heat conduction. Here the characteristic Mach number of the flow becomes small, and we can neglect, in the leading order, the ε^2-terms in (17) and (18). As the result, the momentum equation (17) is reduced to the hydrostatic condition (20), while in the energy equation (18) one can drop the viscous heating term. The continuity equation (16) remains the same as before. The boundary conditions at $m = 0$ do not change, but those at $m = 1$ are reduced to $n\,\partial T/\partial m = 0$. The boundary condition $n\,\partial v/\partial m = 0$ should be dropped altogether.

What is the physical meaning of the reduced model? A low-Mach-number flow proceeds on a time scale that is much longer than the time needed for establishing the (approximate) hydrostatic equilibrium, so the flow does not violate the hydrostatics. Similar reductions of nonlinear time-dependent gas-dynamic problems, that employ a separation of time/length scales, have been successfully used in the context of the dynamics of an optically thin plasma that is cooled by radiation [19]. The presence of a heat loss term in the energy equation makes the latter problem similar to problems of granular dynamics.

The hydrostatic condition immediately yields $p(m,t) \equiv n(m,t)T(m,t) = 1 - m$, like in the steady state solution! Using it together with the continuity equation (16), we can rewrite the energy equation as an evolution equation for $\omega(m,t) = T^{1/2}(m,t)$:

$$\omega \frac{\partial \omega}{\partial t} = (1-m)\frac{\partial^2 \omega}{\partial m^2} - \frac{\partial \omega}{\partial m} - \Lambda^2 (1-m)\,\omega\,. \tag{28}$$

This nonlinear parabolic equation should be solved with two boundary conditions: $\omega(0,t) = 1$ and $(1-m)\,\partial\omega/\partial m \to 0$ as $m \to 1$. Notice that the only relevant scaled parameter in the low-Mach-number problem is Λ.

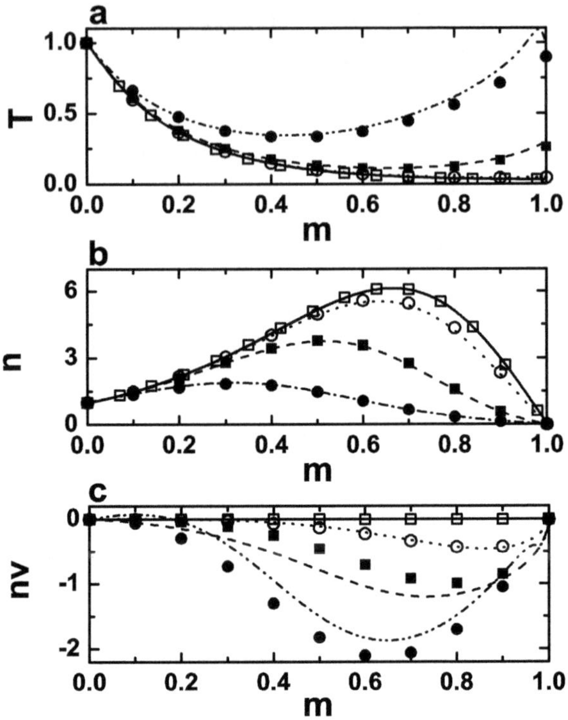

Fig. 2. The temperature **a**, number density **b** and mass flux nv **c** versus the Lagrangian mass coordinate m at late times, as computed from the low-Mach-number model (*symbols*) and full hydrodynamics (*lines*). The scaled parameters are $\Lambda = \sqrt{10}$ and $\varepsilon = \sqrt{5} \cdot 10^{-2}$. The scaled times are $t = 0.1$ (*solid circle, dash-dot-dot line*), 0.3 (*solid square, dashed line*), 0.5 (*empty circle, dotted line*) and 0.85 (*empty square, solid line*). The analytic steady state solutions $T_s(m)$ and $n_s(m)$ are indistinguishable from the respective *solid lines*

Once (28) is solved, we can find $n(m,t)$ and $v(m,t)$:

$$v(m,t) = \frac{\partial}{\partial t} \int_0^m \frac{dm'}{n(m',t)} \,. \tag{29}$$

To transform back to the Eulerian coordinates we use (19).

Equation (28) was solved numerically with a MATLAB PDE solver. Figure 2 shows an example of numerical solution, at different times, for $\Lambda = \sqrt{10}$. The initial condition for the temperature, $T(m, t = 0) = 1$, corresponds to a barometric density profile, which looks especially simple in the Lagrangian coordinate: $n(m, t = 0) = 1 - m$. The solution shows cooling of the granular gas because of the inelastic heat losses (see Fig. 2a). Simultaneously, a density peak develops at an intermediate height (Fig. 2b). The material flux (Fig. 2c) is everywhere negative: the material slowly falls from above

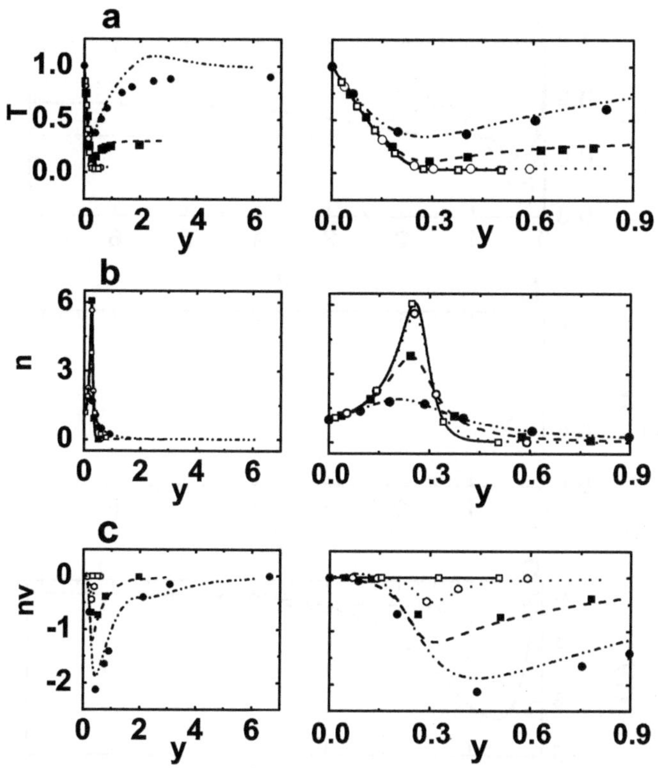

Fig. 3. The temperature **a**, number density **b** and mass flux nv **c** versus the Eulerian coordinate y at late times, as computed from the low-Mach-number model (*symbols*) and full hydrodynamics (*lines*). The scaled parameters are $\Lambda = \sqrt{10}$ and $\varepsilon = \sqrt{5} \cdot 10^{-2}$. The figures on the left and on the right show the same profiles on different scales of height. The scaled times are $t = 0.1$ (*solid circle, dash-dot-dot line*), 0.3 (*solid square, dashed line*), 0.5 (*empty circle, dotted line*) and 0.85 (*empty square, solid line*). The steady state solutions $T_s(y)$ and $n_s(y)$ are indistinguishable from the respective *solid lines*

until the steady-state with a zero mean velocity is reached. An important feature of the time-dependent solution is a transient temperature minimum that develops at intermediate times. It is caused by the fact that the inelastic heat losses are very small at large heights, where the gas density is small. Therefore, the gas cools down relatively fast at intermediate heights, which causes a temperature minimum there. Then a downward heat flux develops that removes the heat from above, finally creating a monotonically decreasing steady state temperature profile. One can see from Fig. 2 that the steady state temperature and density profiles coincide with the analytic solutions. The respective solutions in the Eulerian coordinates are shown in Fig. 3.

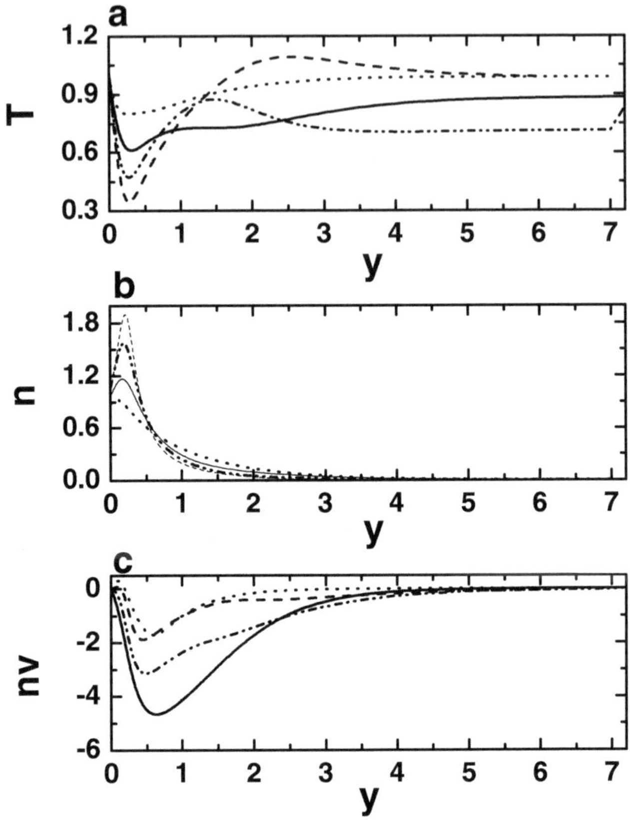

Fig. 4. The temperature **a**, number density **b** and mass flux nv **c** versus the Eulerian coordinate y at early times, as computed from the full hydrodynamics. The scaled parameters are $\Lambda = \sqrt{10}$ and $\varepsilon = \sqrt{5} \cdot 10^{-2}$. The scaled times are $t = 0.01$ (*dotted line*), 0.04 (*solid line*), 0.07 (*dash-dot-dot line*) and 0.1 (*dashed line*)

5 Formation of a Density Inversion: Early Times

The low-Mach-number theory is invalid at early times, when the hydrostatics has not settled down yet, and the mean flow velocitiy can reach, or even exceed, the local speed of sound. To investigate these early times, and to check the validity of the low-Mach-number theory at later times, we solved numerically the full set of hydrodynamic equations (16)-(18) in the Lagrangian form. The parameter ε was taken to be $\sqrt{5} \cdot 10^{-2} \simeq 0.02236$. Equations (16) and (17) were solved with an explicit numerical scheme, based on the standard finite-difference method described in [20]. The viscous term in (17) was integrated explicitly, using a two-steps method, which is both stable and accurate. The energy equation (18) was treated implicitly, with a conservative scheme.

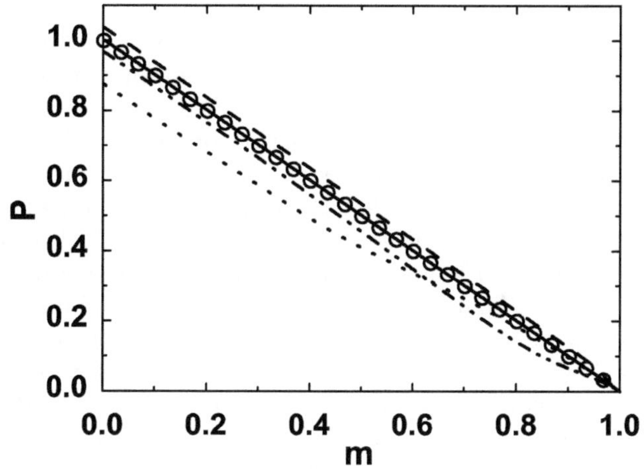

Fig. 5. The pressure $p = nT$ versus the Lagrangian mass coordinate m at different times, as computed from the full hydrodynamics. The scaled parameters are $\Lambda = \sqrt{10}$ and $\varepsilon = \sqrt{5} \cdot 10^{-2}$. The scaled times are $t = 0.01$ (*dashed line*), 0.04 (*dash-dot-dot line*), 0.5 (*solid line*) and 0.85 (*empty circles*)

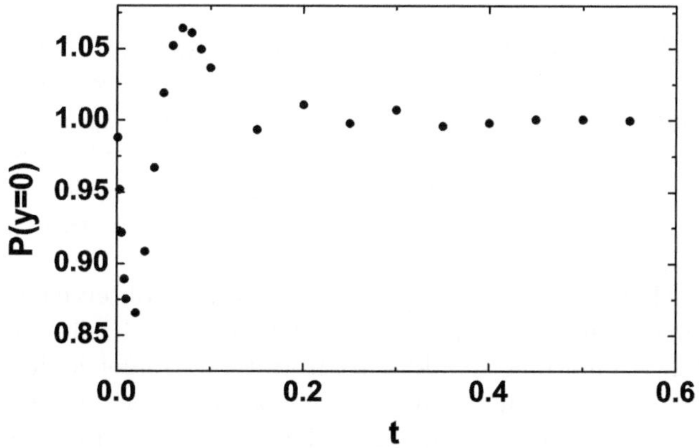

Fig. 6. The pressure $p = nT$ at the bottom plate $y = 0$, versus time, as computed from the full hydrodynamics. The scaled parameters are $\Lambda = \sqrt{10}$ and $\varepsilon = \sqrt{5} \cdot 10^{-2}$

We started with simple initial conditions: the constant temperature $T(m, t = 0) = 1$, the barometric density profile $n(m, t = 0) = 1 - m$ and a zero mean velocity. As expected, the late-time results show good agreement with the predictions of the low-Mach-number theory. Figure 2 shows that, at latest times, the full-hydrodynamics solutions coincide with those obtained with the low-Mach-number theory. At an intermediate time one can see a local increase of the temperature at high altitudes, which is predicted by the full hydrodynamics, but missed by the low-Mach-number theory. The reason for this transient temperature maximum is the relatively large viscous heating that occurs at high altitudes because of a steep velocity gradient there.

The early-time results of the full numerical solution, transformed back to the Eulerian coordinate, are shown in Fig. 4. One can see that the development of the density peak, and the formation of the transient temperature minimum, start already at early times. A more detailed analysis does show differences between the fast and slow theories. Indeed, a necessary condition for the validity of the low-Mach-number theory is the hydrostatic balance condition. This condition looks especially simple in the Lagrangian coordinate: $p(m,t) = 1 - m$. That is, when the hydrostatics sets in, the plots of the pressure versus m, at different times, should collapse into a single straight line. Figure 5 shows the pressure plots at different times. One can see that, at late times, the hydrostatic condition sets in as expected. At early times pressure oscillations are observed, and hydrostatics does not hold. Figure 6 shows the pressure history at the bottom plate $m = 0$, or $y = 0$. One can see that the pressure undergoes oscillations on the fast time scale of the system $\mathcal{O}(\varepsilon)$. The oscillations rapidly decay with time and, starting from $t \simeq 0.4$, the pressure stays almost constant which implies a low-Mach-number flow.

6 Discussion

We have employed granular hydrodynamics to describe a time-dependent granular flow that gives rise to a density inversion in a vibrofluidized granular system. We have predicted the formation of a transient temperature minimum at intermediate times. The temperature minimum ultimately disappears as the system approaches the steady state with a density inversion. We have also predicted pressure oscillations at early times, when the hydrostatic condition is not yet satisfied.

We hope that the predicted phenomena will be observed in molecular dynamics (MD) simulations and in experiments with vibrofluidized granular beds. The MD simulations should start from the barometric density profile, corresponding to the temperature of the bottom plate. A natural way to initiate the initial state is to first run the simulation with *elastic* particle collisions. After the barometric profile sets in, the collision rule should be switched into inelastic regime.

The hydrodynamic theory considered in this work is expected to be quantitatively accurate for nearly elastic collisions, $1 - r^2 \ll 1$, and in the dilute limit. The latter condition demands that the maximum number density of the particles be less than, say, $0.2\, n_c$, where $n_c = 2/(\sqrt{3}d^2)$ is the hexagonal close packing density. The maximum density peak is achieved at the steady state. Therefore, using (26) and going back to the "physical" density, we can write the dilute regime criterion as

$$\frac{\Lambda_c}{I_0^2(\Lambda_c)} \frac{\sqrt{3}Nd^2}{2\lambda L_x} \frac{I_0^2(\Lambda)}{\Lambda} < 0.2\,. \tag{30}$$

An alternative form of the same criterion is

$$\frac{I_0^2(\Lambda)}{\varepsilon\Lambda}\,\frac{d}{\lambda} < 0.65\,. \tag{31}$$

We expect that the theory is still valid qualitatively when this criterion is *not* satisfied, but the maximum density is still less than about $0.5\,n_c$. To achieve a better accuracy for such moderately dense flows, one can use the constitutive relations of Jenkins and Richman [11] that account for excluded volume effects.

Acknowledgements

This work was supported by a grant No. 180/02 from the Israel Science Foundation, administered by the Israel Academy of Sciences and Humanities.

References

1. L. P. Kadanoff. Built upon sand. *Rev. Mod. Phys.*, 71:435, 1999.
2. M. A. Hopkins and M. Y. Louge. Inelastic microstructure in rapid granular flows of smooth disks. *Phys. Fluids A*, 3:47, 1991.
3. I. Goldhirsch and G. Zanetti. Clustering instability in dissipative gases. *Phys. Rev. Lett.*, 70:1619, 1993.
4. S. McNamara and W. R. Young. Dynamics of a freely evolving, two-dimensional granular medium. *Phys. Rev. E*, 53:5089, 1996.
5. J. S. Olafsen and J. S. Urbach. Clustering, order, and collapse in a driven granular monolayer. *Phys. Rev. Lett.*, 81:4369, 1998.
6. T. C. P van Noije and Ernst. M. Kinetic theory of granular gases. In Pöschel and Luding [21], page 3.
7. I. Goldhirsch. Granular gases: Probing the boundaries of hydrodynamics. In Pöschel and Luding [21], page 79.
8. It has been found recently that fluctuations are *anomalously* large [22, 23], and their strength decreases very slowly with N [22], in the parameter region corresponding to phase separation in driven granular gases. As the result, hydrodynamics can still be invalid in a system of as many as $4 \cdot 10^4$ particles [22].
9. E. Livne, B. Meerson, and P. V. Sasorov. Symmetry-breaking instability and strongly peaked periodic clustering states in a driven granular gas. *Phys. Rev. E*, 65:021302, 2002.
10. The phase separation instability has been recently observed in molecular dynamics simulations [22, 24, 25], and the subject is rapidly developing [26, 27].
11. J. T. Jenkins and M. W. Richman. Kinetic theory for plane flows of a dense gas of identical, rough inelastic circular disks. *Phys. Fluids*, 28:3485, 1985.
12. R. Ramìrez, D. Risso, and P. Cordero. Thermal convection in fluidized granular systems. *Phys. Rev. Lett.*, 85:1230, 2000.
13. X. He, B. Meerson, and G. Doolen. Hydrodynamics of thermal granular convection. *Phys. Rev. E*, 65:030301(R), 2002.

14. E. Khain and B. Meerson. Onset of thermal convection in a horizontal layer of granular gas. *Phys. Rev. E*, 67:021306, 2003.
15. L. D. Landau and E. M. Lifshitz. *Course of Theoretical Physics*, volume 6 Fluid Mechanics. Pergamon, Oxford, 1987.
16. It was observed, in some experiments [28, 29] and simulations [18, 29, 30] with vibrofluidized granular materials that, at large heights, the temperature starts *increasing* with the height. As the particle density at these heights is very small, hydrodynamics might be inapplicable. In any case, most of the density profile is apparently insensitive to the presence of such a temperature inversion, if any.
17. Y. B. Zeldovich and Y. P. Raizer. *The Physics of Shock Waves and High Temperature Hydrodynamic Phenomena.* Academic, New York, 1967.
18. J. J. Brey, M. J. Ruiz-Montero, and F. Moreno. Hydrodynamics of an open vibrated granular system. *Phys. Rev. E*, 63:061305, 2001.
19. B. Meerson. Nonlinear dynamics of radiative condensations in optically thin plasmas. *Rev. Mod. Phys.*, 68:215, 1996.
20. R. D. Richtmyer and K. W. Morton. *Difference Methods for Initial Value Problems.* Interscience, New York, 1967.
21. T. Pöschel and S. Luding, editors. *Granular Gases.* Lecture Notes in Physics. Springer, Berlin, 2001.
22. B. Meerson, T. Pöschel, P. V. Sasorov, and T. Schwager. Breakdown of granular hydrodynamics at a phase separation threshold. cond-mat/020826.
23. A. Barrat and E. Trizac. A molecular dynamics "Maxwell demon" experiment for granular mixtures. *Molecular Physics (to appear)*, 2003. cond-mat/0212054.
24. J. J. Brey, M. J. Ruiz-Montero, F. Moreno, and R. García-Rojo. Transversal inhomogeneities in dilute vibrofluidized granular fluids. *Phys. Rev. E*, 65:061302, 2002.
25. M. Argentina, M. G. Clerc, and R. Soto. Van der Waals-like transition in fluidized granular matter. *Phys. Rev. Lett.*, 89:044301, 2002.
26. E. Khain and B. Meerson. Symmetry-breaking instability in a prototypical driven granular gas. *Phys. Rev. E*, 66:021306, 2002.
27. E. Livne, B. Meerson, and P. V. Sasorov. Symmetry breaking and coarsening of clusters in a prototypical driven granular gas. *Phys. Rev. E*, 66:050301(R), 2002.
28. E. Clement and J. Rajchenbach. Fluidization of a bidimensional powder. *Europhys. Lett.*, 16:133, 1991.
29. K. Helal, T. Biben, and J. P. Hansen. Local fluctuations in a fluidized granular media. *Physica A*, 240:361, 1997.
30. R. Ramìrez and R. Soto. Temperature inversion in granular fluids under gravity. cond-mat/0210471.

Kinetic Theory for Inertia Flows of Dilute Turbulent Gas-Solids Mixtures

Cliff K.K. Lun[1] and Stuart B. Savage[2]

[1] Center for Multiphase Flows Research, 55 Cairnside Crescent, Toronto, Canada
[2] McGill University, Department of Civil Engineering and Applied Mechanics, Montreal, Canada

Abstract. In this work, we develop a new kinetic theory for the inertia flows of dilute turbulent gas-solids two-phase suspensions. The gas-solids interactions in momentum and energy are incorporated in the conservation equations by using the Fokker-Planck technique. On the other hand, the effects of the solid particles' kinetic motions and the solid-solid collisional interactions are formulated in the model by utilizing the approach of the kinetic theory for rapid granular flows of identical, smooth, slightly inelastic hard spheres. A set of solids-phase conservation equations, transport coefficients, and constitutive relations such as solids-phase stresses and energy fluxes are derived systematically by using a moment method. The solids-phase transport coefficients are found to depend on the gas-friction coefficient which is in essence the inverse of the mean drag relaxation time for the solids-phase. We employ a two-equation k–ε closure together with a set of time-averaged turbulence conservation equations to characterize the gas-phase medium. The important effect of turbulence modulation due to the presence of solid particles is also accounted for in the energy equations. We apply the present theory to study the case of an axisymmetric round jet of turbulent gas and solid particles. There is favorable agreement between the theoretical predictions and the experimental measurements for a number of macroscopic flow properties.

1 Introduction

Gas-solids two-phase flows occur widely in natural phenomena and industrial processes; some examples are snow and rock avalanches, desert sandstorms, dispersions of pollutants, pollen and allergens in air, cyclone separators, circulating fluidized beds, gas-solids fuel injector, chemical reactors, gas-solids pollutants emitted from internal combustion engines.

During the past four decades there has been a large amount of work aimed at modelling gas-solids two-phase flows by using the continuum mechanical approach in which each constituent is regarded as a continuum occupying the same region in space and interacting with one another. Even a brief description of each model would present a formidable task (see for example [1–3]). The main difficulty with the continuum models is that the solids-phase is treated simply as another fluid constituent. Such a coarse representation neglects the microscopic flow properties of the solids-phase which, in fact, can influence immensely the macroscopic flow behaviour of the two-phase system. For this reason, there is still no general agreement for any basic transport coefficients and continuum constitutive relations such as the solids-phase viscosity, stresses and kinetic energy fluxes.

As an alternative to the continuum approach, there have been several attempts to model gas-solids two-phase flows by employing the Fokker-Planck method in the kinetic theory of gases [4–7]. These theories mainly focused stochastically on the gas-solids interactions, and ignored the effects of the individual particle motion and the interparticle collisions. As a result, they were unable to yield any transport coefficients and constitutive relationships.

Hyland et al. [8] developed a kinetic equation of the probability distribution function to describe the motion of particle dispersion in an unbounded Gaussian random flow field in which the mean flow was simple shear and the turbulence was homogeneous and stationary. They were able to determine the time variation of various macroscopic flow properties such as the particle concentration, mean particle velocity, and the particle stresses for this particular flow system.

The development of the kinetic theory of granular flows in the 1980s [9–13] not only improved our understanding of rapid flows of granular materials, but also has shed much light on the roles played by the solids-phase in the complex system of gas-solids two-phase suspensions [14–20].

The kinetic theory of rapid granular flows of Lun et al. [11] has been incorporated and in some cases extended to the studies of gas-solids two-phase flows by a number of researchers [21–26]. The general idea was to combine the kinetic theory for the solids-phase with a continuum model for the gas-phase. Interaction forces such as the drag were included to account for the transfer of momentum between the phases. For example, by following an approach similar to the kinetic theory, Sangani et al. [27] developed a theory for rapidly sheared granular flows of inelastic grains which includes the viscous effects of the interstitial gas for large Stokes number. Turbulence modulations and couplings in dissipation of fluctuation kinetic energies between the phases are often neglected in these models.

In the present study, we extend the kinetic theory developed by Lun et al. [11] for smooth, slightly inelastic spherical particles to incorporate the effects of turbulent interstitial gas in the inertia flow regime. The interfacial couplings in momentum and energy are formulated in the transport equations by means of the Fokker-Plank method. Through the utilization of time averaging for the gas-phase flow properties and ensemble averaging for the solids-phase ones, we derive a set of conservation equations for mass, momentum and energy for the suspension in a systematic way. The gas-solids momentum and energy interaction terms are found to arise naturally in the balance equations. The placement of these coupling terms in the equations turns out to be crucial. Constitutive relations for the solids-phase stresses and energy fluxes, and transport coefficients such as the solids-phase viscosity are determined explicitly. A two-equation k–ε model is used to close the set of gas-phase turbulence equations.

2 Gas-Solids Interactions

In a dilute gas-solids suspension, momentum and energy are transferred among the flow constituents in various ways. Unlike gas molecules, the motion of the granular particles is not self-sustaining. Instead, it must be driven by other means; for example, gravity, a mechanical device such as a shear cell, or the interaction forces generated by the carrier gas.

In the present study, we consider the case where the carrier gas is intrinsically turbulent, independent of whether the solids-phase is present or not. Let L represent a characteristic length of the flow system, then the corresponding Reynolds number $Re_{\mathrm{L}} = u_{\mathrm{fm}} L/\nu_{\mathrm{f}}$ would be high (i.e. $Re_{\mathrm{L}} \gg 1$); u_{fm} and ν_{f} are the gas mean speed and kinematic viscosity, respectively. The mass density ratio of solid to gas ($\rho_{\mathrm{s}}/\rho_{\mathrm{f}}$) under consideration is of the order 10^3. The major gas-solids interaction force is the drag. The lift force and the rotational motion of the solid particles will be neglected in this treatment.

The variation of the specific drag force with particle Reynolds number for a single sphere may be written as

$$\vec{f} = \tau_{\mathrm{d}}^{-1} \left(\vec{u}_{\mathrm{f}}^{*} - \vec{c} \right) , \tag{1}$$

where $\vec{u}_{\mathrm{f}}^{*}$ is the gas velocity approaching the particle from a far distance, and $\vec{c}$ is the particle velocity. The drag relaxation time τ_{d}, which in general varies with the difference in velocities between the phases, is expressed as

$$\tau_{\mathrm{d}} = \tau_{\mathrm{v}} / (a_{\mathrm{o}} Re_{\mathrm{p}}^{\gamma}) . \tag{2}$$

The viscous relaxation time for a single sphere is $\tau_{\mathrm{v}} = \rho_{\mathrm{s}} d^2 / 18\mu_{\mathrm{f}}$, where d is the particle diameter, and μ_{f} is the usual dynamic gas viscosity. The particle Reynolds number is defined as $Re_{\mathrm{p}} = \rho_{\mathrm{f}} d |\vec{u}_{\mathrm{f}}^{*} - \vec{c}| / \mu_{\mathrm{f}}$, where ρ_{f} is the gas mass density. From the Stokes' regime to the Newton's, the coefficients a_{o} and γ in (2) may be correlated in the following manner

$$\begin{array}{lll} a_{\mathrm{o}} = 1 & \gamma = 0 & Re_{\mathrm{p}} < 1 \\ a_{\mathrm{o}} = 1 & \gamma = 0.27 & 1 \le Re_{\mathrm{p}} < 30 \\ a_{\mathrm{o}} = 0.4 & \gamma = 0.54 & 30 \le Re_{\mathrm{p}} < 10^3 \\ a_{\mathrm{o}} = 0.0175 & \gamma = 1 & 10^3 \le Re_{\mathrm{p}} < 2 \times 10^5 . \end{array} \tag{3}$$

Similar expressions for the coefficients, which were less accurate in comparison with the standard drag experimental result than those in (3), had been proposed some time ago by Lapple [28]. The drag law in (1)–(3), which is applicable to a wide range of Re_{p} from the laminar to the 'turbulent' flow condition, is comparable to some of those previously proposed [17, 29].

It is worthwhile to note that the above equations for the drag force on a single sphere are based on experimental results [30, 31] which were obtained mostly by testing spheres of various sizes in quiescent fluids. In the case of spheres moving in a turbulent gas field, the exact form for the drag is somewhat unclear [17, 29]. Often, formulations like the ones in (3) which are essentially inferred from results of spheres moving in a laminar flow are assumed to apply in turbulent flows.

In a single particle system, the drag relaxation time (2) is inversely proportional to the magnitude of the relative velocity between the particle and the gas far upstream raised to a certain power (as embedded in the particle Reynolds number Re_{p}). However, in the case of a turbulent suspension undergoing rapid shearing, it is not clear what the most appropriate way is for defining the relative velocity between the gas medium and a single particle in the midst of identical ones. Here, we assume that the particle Reynolds number Re_{p} in (2) and (3) may be replaced by the local mean particle Reynolds number, i.e.,

$$Re_{\mathrm{m}} = \rho_{\mathrm{f}} d |\vec{u}_{\mathrm{f}} - \vec{u}_{\mathrm{s}}| / \mu_{\mathrm{f}} \ , \tag{4}$$

where $\vec{u}_{\mathrm{f}}$ and $\vec{u}_{\mathrm{s}}$ are the local mean gas and particle velocities, respectively. In other words, we approximate the drag relaxation time for individual particle by the local mean value. This simplifying assumption permits the drag relationship in (1) to be readily broken down into a mean component and a fluctuating component, i.e.

$$\vec{f}_{\mathrm{m}} = \tau_{\mathrm{d}}^{-1} (\vec{u}_{\mathrm{f}} - \vec{u}_{\mathrm{s}}) \ , \tag{5}$$

$$\vec{f}' = \tau_{\mathrm{d}}^{-1} \left(\vec{u}'_{\mathrm{f}} - \vec{C} \right) \ , \tag{6}$$

where now

$$\tau_{\mathrm{d}} = \tau_{\mathrm{v}} / (a_{\mathrm{o}} Re_{\mathrm{m}}^{\gamma}) \ , \tag{7}$$

and the correlation in (3) becomes

$$\begin{array}{lll} a_{\mathrm{o}} = 1 & \gamma = 0 & Re_{\mathrm{m}} < 1 \\ a_{\mathrm{o}} = 1 & \gamma = 0.27 & 1 \leq Re_{\mathrm{m}} < 30 \\ a_{\mathrm{o}} = 0.4 & \gamma = 0.54 & 30 \leq Re_{\mathrm{m}} < 10^3 \\ a_{\mathrm{o}} = 0.0175 & \gamma = 1 & 10^3 \leq Re_{\mathrm{m}} < 2 \times 10^5 \ . \end{array} \tag{8}$$

In (5)–(6), subscripts f and s denote the gas and solids phases respectively. Primed gas-phase quantities denote fluctuating ones, and quantities with subscript m represent mean values, unless otherwise specified. For the notations of the solids-phase quantities, we primarily employ the set normally used in the kinetic theory of granular flows [11]. For example, the instantaneous particle velocity is given as $\vec{c} = \vec{u}_{\mathrm{s}} + \vec{C}$, where $\vec{C}$ is the fluctuating part.

If we estimate the eddy turnover time by expressing $\tau_{\mathrm{e}} = \ell_{\mathrm{e}}/\sqrt{2k}$, where ℓ_{e} is the integral length scale of the turbulence and k is the turbulence intensity, then the relevant Stokes number would be the ratio of the particle drag relaxation time to the eddy turnover time, i.e. $St = \tau_{\mathrm{d}}/\tau_{\mathrm{e}}$. Using the local mean drag relaxation time (7), the Stoke's number may be expressed as

$$St = \frac{1}{9\sqrt{2}} \frac{\rho_{\mathrm{s}}}{\rho_{\mathrm{f}}} \left(\frac{d}{\ell_{\mathrm{e}}}\right)^2 \frac{Re_{\mathrm{T}}}{a_{\mathrm{o}} Re_{\mathrm{m}}^{\gamma}} \,. \tag{9}$$

In the above equation the turbulence Reynolds number is $Re_{\mathrm{T}} = k^{1/2}\ell_{\mathrm{e}}/\nu_{\mathrm{f}}$, where ν_{f} is the gas kinematic viscosity. The diameter of the solid particles, d, under consideration here, ranges typically from 0.1 to 10 mm. For the inertia flow regime, we have the conditions that $\rho_{\mathrm{s}}/\rho_{\mathrm{f}} > 10^3$, $d/\ell_{\mathrm{e}} \simeq 1$ and $Re_{\mathrm{T}}/9\sqrt{2}a_{\mathrm{o}}Re_{\mathrm{m}}^{\gamma} > 1$. From (9), it is clear that $St \gg 1$. Consequently, the fluctuations of the flow properties of the solids-phase and the gas-phase may be considered to be uncorrelated, simply because the influence of the gas turbulence on the particles' motion would be relatively small.

3 Granular Transport Theory

We follow the approach of the kinetic theory of granular flows [11, 32] to derive a set of conservation equations for the solids phase. The ensemble average of a single-particle quantity ψ is defined as

$$\langle \psi \rangle = \frac{1}{n} \int \psi f^{(1)}(\vec{r}, \vec{c}; t)\, \mathrm{d}\vec{c} \,, \tag{10}$$

where $f^{(1)}(\vec{r}, \vec{c}; t)$ is the single-particle velocity distribution function of the solids-phase, and n is the particle number density.

The evolution equation of the single-particle velocity distribution function $f^{(1)}$ in the phase-space may be derived from the Liouville's equation [33]. The equation so obtained is a Boltzmann-type of equation

$$\frac{\partial f^{(1)}}{\partial t} + \vec{c} \cdot \nabla f^{(1)} + \vec{b} \cdot \frac{\partial f^{(1)}}{\partial \vec{c}} = \left(\frac{\partial f^{(1)}}{\partial t}\right)_{\mathrm{fs}} + \left(\frac{\partial f^{(1)}}{\partial t}\right)_{\mathrm{ss}} , \tag{11}$$

where $\vec{b}$ is the external body force per unit mass. The two terms on the right hand side of (11) account for the rates of change of $f^{(1)}$ due to the gas-solid interactions (with subscript fs) and the solid-solid interactions (with subscript ss).

We may model the turbulent gas-solid interaction term in (11) in the following manner. The solid particle is relatively massive, i.e. $\rho_{\mathrm{s}}/\rho_{\mathrm{f}} > 10^3$. Since $St \gg 1$, as far as the grain is concerned, it experiences small changes of direction and momentum within a time interval τ due mainly to the interaction with the gas in its vicinity. The local turbulent motion of the gas-phase is assumed to be completely random and possess no history effect. Consequently, the interaction process between the solid particle and the turbulent gas may be regarded as essentially Markovian. Such a process can be described by the Chapman-Kolmogorov equation [34–36]

$$f^{(1)}(\vec{c},t) = \int f^{(1)}(\vec{c}-\triangle\vec{c}, t-\tau)\Pi_{\rm fs}(\vec{c}-\triangle\vec{c},\triangle\vec{c})\mathrm{d}(\triangle\vec{c}) \,, \tag{12}$$

where $\Pi_{\rm fs}(\vec{c}-\triangle\vec{c},\triangle\vec{c})$ denotes the transition probability density that particle P experiences the change in velocity $\triangle\vec{c}$ in τ. Expanding (12) in a Taylor series to first order in τ and to second-order in $\triangle\vec{c}$ and after some algebraic manipulation, we obtain the well-known Fokker-Planck equation

$$\left(\frac{\partial f^{(1)}}{\partial t}\right)_{\rm fs} = -\frac{\partial}{\partial\vec{c}}\cdot\left(\frac{\langle\triangle\vec{c}\rangle_{\rm f}}{\tau}f^{(1)}\right) + \frac{1}{2}\frac{\partial^2}{\partial\vec{c}\partial\vec{c}}:\left(\frac{\langle\triangle\vec{c}\triangle\vec{c}\rangle_{\rm f}}{\tau}f^{(1)}\right) \,, \tag{13}$$

where $\langle\triangle\vec{c}\rangle_{\rm f}$ and $\langle\triangle\vec{c}\triangle\vec{c}\rangle_{\rm f}$ denote respectively the gas-phase ensemble averages of interfacial momentum and energy exchanges.

To determine the two Fokker-Planck terms in (13), it seems reasonable to assume that the ergodic hypothesis applies in the present situation, and we may equate the gas-phase ensemble average to that of the time average

$$\frac{\langle\triangle\vec{c}\rangle_{\rm f}}{\tau} = \frac{1}{\tau}\int_{t-\tau}^{t}\frac{\mathrm{d}\vec{c}}{\mathrm{d}t}\,\mathrm{d}t = \overline{\frac{\mathrm{d}\vec{c}}{\mathrm{d}t}} = \vec{f}_{\rm m} + \overline{\vec{f}'} \,, \tag{14}$$

where an overbar signifies gas-phase time averaging. From (6), the time averaged fluctuating force is simply

$$\overline{\vec{f}'} = -\varphi_{\rm d}\vec{C} \,, \tag{15}$$

where

$$\varphi_{\rm d} = \tau_{\rm d}^{-1} \,, \tag{16}$$

is referred to as the gas-friction coefficient here. In the process of obtaining (15), we have taken the time τ to be long compared to the drag relaxation time and the eddy turnover time $\tau_{\rm e}$, i.e. $\tau_{\rm e} < \tau_{\rm d} < \tau$, such that

$$\frac{1}{\tau}\int_{t-\tau}^{t} u_{\rm f}'\,\mathrm{d}t = 0 \,. \tag{17}$$

Equations (15) and (17) simply affirm the condition that the fluctuations between the gas and the solids phases are uncorrelated as discussed earlier. Furthermore, the solids volume fraction $\alpha_{\rm s}$, which is a volumetric mean quantity, is treated as independent of the turbulence averaging time-scale of the gas-phase.

The mean ($\vec{C}$-independent) force $\vec{f}_{\rm m}$ in (14) and subsequently in (13) may be regarded as *external* to the solids-phase, and it can be included in the external body force term, $\vec{b}$, on the right hand side of (11). Thus from here onward, we write

$$\vec{b} = \vec{g} + \vec{f}_{\mathrm{m}} \,, \tag{18}$$

where $\vec{g}$ is the gravitational acceleration.

On the basis of ensuring the entropy source strength being non-negative in the Fokker-Planck model for gases, Woods [35] showed that the second Fokker-Planck term in (13) can be related to the fluctuating component of the first one. Letting $\vec{A} = \overline{\vec{f}'}$ and $\vec{B} = \langle \triangle\vec{c}\triangle\vec{c} \rangle_{\mathrm{f}}/\tau$, we have

$$2\vec{A} = \frac{\partial}{\partial\vec{c}} \cdot \vec{B} - \frac{1}{T}\vec{B} \cdot \vec{C} \,, \tag{19}$$

From (15) and (19), one finds readily

$$\vec{B} = 2\varphi_{\mathrm{d}} T \vec{\delta} \,, \tag{20}$$

where $\vec{\delta}$ is the unit tensor, $T = \left\langle C^2 \right\rangle /3$ is the granular temperature, and $\vec{B}/2 = \langle \triangle\vec{c}\triangle\vec{c} \rangle_{\mathrm{f}}/2\tau$ is in essence the tensorial rate of kinetic energy transfer to the solid-phase from the gas-phase through the fluctuating interfacial forces.

At high solids concentration, the hydrodynamic interactions between solid particles become important [37–40]. In the present study, we limit our investigation to dilute gas-solids suspensions. From the numerical simulation work of Kim et al. [38], it is evident that in a laminar flow field if two identical spheres are separated center-to-center by more than about four particle diameters apart perpendicular to the streamwise direction, then there is little mutual influence between the individual drag forces. The lift and torque mutual interactions on the spheres are relatively small compared to the drag. The range of Re_{p} studied was between 50 to 150. In flow systems with intrinsic gas turbulence, one would expect the mutual hydrodynamic interactions between the solid particles to be less pronounced than in the case of the laminar flow, because the turbulent gas would be able to diffuse the hydrodynamic disturbances due to the proximity of a neighboring particle much more effectively. Nevertheless, using Bagnold's formula [41], $\alpha_{\mathrm{s}} = 0.74(d/s)^3$, a separation of four particle diameters (i.e., $s = 4d$) translates into roughly a solids volume fraction of about 0.01. In other words, if the solids concentration $\alpha_{\mathrm{s}} < 0.01$, the suspension may be regarded as dilute since the interparticle hydrodynamic interactions may be neglected. In conjunction with this condition, the particle drag relaxation time would be much less than the interparticle collision time, i.e. $\tau_{\mathrm{d}} < \tau_{\mathrm{c}}$, which serves as an alternate indicator for dilute suspension flows.

Multiplying the evolution equation (11) for $f^{(1)}$ by ψ and integrating over the entire velocity space yields

$$\frac{\partial\langle n\psi\rangle}{\partial t}+\nabla\cdot\langle n\vec{c}\psi\rangle-\langle n\mathrm{D}\psi\rangle=\left\langle\overline{\vec{f'}}\cdot\frac{\partial\psi}{\partial\vec{c}}\right\rangle+\frac{1}{2}\left\langle\vec{B}:\frac{\partial^2\psi}{\partial\vec{c}\partial\vec{c}}\right\rangle-\nabla\cdot\vec{\theta}(\psi)+\chi(\psi)\,, \tag{21}$$

where

$$\mathrm{D}\psi=\frac{\partial\psi}{\partial t}+\vec{c}\cdot\nabla\psi+\vec{b}\cdot\frac{\partial\psi}{\partial\vec{c}}\,.$$

The last two terms in (21) are derived from the solid-solid collisional interaction term in (11). Their formulations were dealt with by Lun et al. [11] and will not be repeated here. One can express

$$\int\psi\left(\frac{\partial f^{(1)}}{\partial t}\right)_{\mathrm{ss}}\mathrm{d}\vec{c}_1=-\nabla\cdot\vec{\theta}(\psi)+\chi(\psi)\,. \tag{22}$$

The collisional flux term is

$$\vec{\theta}(\psi)\approx-\frac{d^3}{2}\int_{\vec{c}_{12}\cdot\vec{k}}(\psi_1'-\psi_1)\,(\vec{c}_{12}\cdot\vec{k})\vec{k}f^{(2)}\left(\vec{r}-\frac{d}{2}\vec{k},\vec{c}_1;\vec{r}+\frac{d}{2}\vec{k},\vec{c}_2;t\right)\mathrm{d}\vec{k}\,\mathrm{d}\vec{c}_1\,\mathrm{d}\vec{c}_2\,, \tag{23}$$

and the collisional source or sink term is

$$\chi(\psi)=\frac{d^2}{2}\int_{\vec{c}_{12}\cdot\vec{k}}(\psi_2'+\psi_1'-\psi_2'-\psi_1')\left(\vec{c}_{12}\cdot\vec{k}\right)f^{(2)}\left(\vec{r}-d\vec{k},\vec{c}_1;\vec{r},\vec{c}_2;t\right)\mathrm{d}\vec{k}\,\mathrm{d}\vec{c}_1\,\mathrm{d}\vec{c}_2\,, \tag{24}$$

where $\vec{c}_{12}=\vec{c}_1-\vec{c}_2$ is the relative velocity of particles 1 and 2, $\vec{k}$ is the unit vector from the center of particle 1 to that of particle 2. The primed solid-phase quantities represent post-collisional values. Since we are dealing with flows in the inertia regime, the influence of the interstitial gas and the interparticle hydrodynamic interactions during the relatively high momentum particle-particle collisions may be neglected.

Taking ψ to be m, $m\vec{c}$ and $mc^2/2$ in (21) we obtain a set of conservation equations for mass, momentum and fluctuating kinetic energy

$$\frac{\partial}{\partial t}(\alpha_\mathrm{s}\rho_\mathrm{s})+\nabla\cdot(\alpha_\mathrm{s}\rho_\mathrm{s}\vec{u}_\mathrm{s})=0\,, \tag{25}$$

$$\frac{\partial}{\partial t}(\alpha_\mathrm{s}\rho_\mathrm{s}\vec{u}_\mathrm{s})+\nabla\cdot(\alpha_\mathrm{s}\rho_\mathrm{s}\vec{u}_\mathrm{s}\vec{u}_\mathrm{s})=\alpha_\mathrm{s}\rho_\mathrm{s}\vec{b}-\nabla\cdot\vec{P}_\mathrm{s}\,, \tag{26}$$

$$\frac{3}{2}\frac{\partial}{\partial t}(\alpha_\mathrm{s}\rho_\mathrm{s}T)+\frac{3}{2}\nabla\cdot(\alpha_\mathrm{s}\rho_\mathrm{s}\vec{u}_\mathrm{s}T)=-\vec{P}_\mathrm{s}:\nabla\vec{u}_\mathrm{s}-\nabla\cdot\vec{q}_\mathrm{s}-\gamma_\mathrm{s}$$
$$+\alpha_\mathrm{s}\rho_\mathrm{s}\left\langle\overline{\vec{f'}}\cdot\vec{C}\right\rangle+\frac{1}{2}\alpha_\mathrm{s}\rho_\mathrm{s}\left\langle\vec{B}:\vec{\delta}\right\rangle\,. \tag{27}$$

The mean solids-phase velocity is $\vec{u}_\mathrm{s} = \langle \vec{c} \rangle$. The total grain stress tensor is $\vec{P}_\mathrm{s} = \vec{P}_\mathrm{sk} + \vec{P}_\mathrm{sc}$, where the kinetic and collisional contributions are

$$\vec{P}_\mathrm{sk} = \alpha_\mathrm{s} \rho_\mathrm{s} \left\langle \vec{C}\vec{C} \right\rangle , \tag{28a}$$

$$\vec{P}_\mathrm{sc} = \vec{\theta} \left(m\vec{C} \right) . \tag{28b}$$

Similarly the total translational energy flux is $\vec{q}_\mathrm{s} = \vec{q}_\mathrm{sk} + \vec{q}_\mathrm{sc}$, where

$$\vec{q}_\mathrm{sk} = \alpha_\mathrm{s} \rho_\mathrm{s} \left\langle C^2 \vec{C} \right\rangle /2 , \tag{29a}$$

$$\vec{q}_\mathrm{sc} = \vec{\theta}(mC^2/2) . \tag{29b}$$

The rate of collisional energy dissipation per unit volume is

$$\gamma_\mathrm{s} = -\chi \left(mC^2/2 \right) . \tag{30}$$

4 Moment Method

Following the approach of Lun [32], and Dahler and Theodosopulu [42], the single-particle velocity distribution function $f^{(1)}$ for smooth, inelastic spheres may be written as

$$f^{(1)} = f^{(\mathrm{o})}(1 + \phi) , \tag{31}$$

where $\phi < 1$, and

$$f^{(\mathrm{o})} = \frac{n}{(2\pi T)^{3/2}} \exp \left(-\frac{(\vec{c} - \vec{u}_\mathrm{s})^2}{2T} \right) , \tag{32}$$

is the local Maxwellian equilibrium distribution function. The perturbation ϕ may be approximated by the trial function

$$\phi = \frac{1}{2\alpha_\mathrm{s}\rho_\mathrm{s}T^2} \vec{P}^*_\mathrm{sk} : \vec{C}\vec{C} + \frac{2}{5\alpha_\mathrm{s}\rho_\mathrm{s}T^2} \vec{q}_\mathrm{sk} \cdot \vec{C} \left(\frac{C^2}{2T} - \frac{5}{2} \right) , \tag{33}$$

where $\vec{P}^*_\mathrm{sk}$ is the kinetic stress deviator defined as $\vec{P}^*_\mathrm{sk} = \alpha_\mathrm{s}\rho_\mathrm{s} \left\langle \left[\vec{C}\vec{C} \right]^* \right\rangle$, and $\left[\vec{C}\vec{C} \right]^* = \vec{C}\vec{C} - (C^2/3)\vec{\delta}$. The terms in the perturbation ϕ corresponds essentially to small gradients of mean solids velocity and granular temperature.

The pair velocity distribution function $f^{(2)}$ may be approximated by

$$f^{(2)} = g_\mathrm{o}(\alpha_\mathrm{s}) f_1^{(1)} f_2^{(1)} , \tag{34}$$

where $g_\mathrm{o}(\alpha_\mathrm{s})$ is the radial distribution function at contact and may be expressed as [43]

$$g_\mathrm{o}(\alpha_\mathrm{s}) = \frac{2 - \alpha_\mathrm{s}}{2(1 - \alpha_\mathrm{s})^3} . \tag{35}$$

The collisional model for smooth, slightly inelastic particles used by Lun et al. [11] is adopted in the present study.

Although the flow system considered here is dilute, due to the facts that the granular particles are relative large and massive compared to the fluid molecules and that they are not negligibly small compared to the size of the flow system typically, these finite size solids cannot simply be regarded as point masses as in the case of rarefied gases. Furthermore, Lun [44, 45] found that in the gas-solids suspension flows in vertical pipes and in horizontal channels with α_s of the order 10^{-3}, the interparticle collisional frequency is of the same order of magnitude as the particle-wall one. In other words, particle collisions still play an important role even in these so-called dilute systems. Thus, we must of necessity account for the transfer of dynamic properties by means of collisions through a finite distance of the order of a particle diameter. Consequently, a full derivation of the kind found in the dense kinetic theory of granular flow is required.

The quantities $\vec{P}^*_{\mathrm{sk}}$ and $\vec{q}_{\mathrm{sk}}$ in (33) can be determined by satisfying the steady state moment equations which are generated by taking ψ in (21) to be the higher moments of particle velocities, namely letting $\psi = m\,[\vec{c}\,\vec{c}\,]^*$ with the condition for the temperature gradient as $\nabla T = 0$, and $\psi = \frac{1}{2}c^2\vec{c}$ with the shear rates as $\vec{S}_{\mathrm{s}} = 0$, respectively. Such a method of solution is basically a simplified version of Grad's moment method [46]. As a result, we obtain the constitutive relationships

$$\vec{P}^*_{\mathrm{sk}} = -\frac{2\mu_{\mathrm{s}}}{g_{\mathrm{o}}}\left[1 + \frac{8}{5}\eta(3\eta - 2)\alpha_{\mathrm{s}}g_{\mathrm{o}}\right]\vec{S}_{\mathrm{s}}\,, \tag{36}$$

and

$$\vec{q}_{\mathrm{sk}} = -\frac{\lambda_{\mathrm{s}}}{g_{\mathrm{o}}}\left[1 + \frac{12}{5}\eta^2(4\eta - 3)\alpha_{\mathrm{s}}g_{\mathrm{o}}\right]\nabla T\,, \tag{37}$$

where

$$\mu_{\mathrm{s}} = \mu_{\mathrm{i}}\left[1 + \frac{2\varphi_{\mathrm{d}}\mu_{\mathrm{i}}}{\alpha_{\mathrm{s}}\rho_{\mathrm{s}}T}\right]^{-1}\,, \tag{38}$$

$$\mu_{\mathrm{i}} = \frac{5m}{16d^2\eta(2-\eta)}\left(\frac{T}{\pi}\right)^{1/2}\,, \tag{39}$$

$$\lambda_{\mathrm{s}} = \lambda_{\mathrm{i}}\left[1 + \frac{6\varphi_{\mathrm{d}}\lambda_{\mathrm{i}}}{5\alpha_{\mathrm{s}}\rho_{\mathrm{s}}T}\right]^{-1}\,, \tag{40}$$

$$\lambda_{\mathrm{i}} = \frac{75m}{8d^2\eta(41-33\eta)}\left(\frac{T}{\pi}\right)^{1/2}\,. \tag{41}$$

The rate of shear tensor is $\vec{S}_{\mathrm{s}} = (u_{s_{i,j}} + u_{s_{j,i}})/2 - (u_{s_{k,k}}/3)\delta_{ij}$. The coefficient of restitution e, which is taken to be constant, characterizes the inelastic collisional property of the solid particles, and it is cast in terms of the parameter

$\eta = (1+e)/2$. The present result for transport coefficients expressed through (38)–(41) are similar to those obtained by Rice and Allnatt [36] for a simple liquid. In (38) and (40), one finds that the solids-phase viscosity and kinetic energy fluctuation (or pseudo-thermal) conductivity both decrease with increasing friction coefficient φ_{d} (15). This is consistent with what one might expect for the physical behaviour of the solids-phase. As the particles move about, part of their energy would be spent on overcoming interfacial friction. If the gas-friction increases, the granular temperature will decrease, as will the solids-phase viscosity (38) and pseudo-thermal conductivity (40).

The kinetic grain stress tensor defined in (28a) is

$$\vec{P}_{\mathrm{sk}} = \alpha_{\mathrm{s}}\rho_{\mathrm{s}}T\vec{\delta} - \frac{2\mu_{\mathrm{s}}}{g_{\mathrm{o}}}\left[1 + \frac{8}{5}\eta(3\eta-2)\alpha_{\mathrm{s}}g_{\mathrm{o}}\right]\vec{S}_{\mathrm{s}}\,. \tag{42}$$

From (28b) and (29b), the collisional stress and energy flux contributions are respectively

$$\begin{aligned}\vec{P}_{\mathrm{sc}} &= 4\eta\alpha_{\mathrm{s}}^2\rho_{\mathrm{s}}g_{\mathrm{o}}T\vec{\delta} - \frac{16\eta\alpha_{\mathrm{s}}\mu_{\mathrm{s}}}{5}\left[1 + \frac{8}{5}\eta(3\eta-2)\alpha_{\mathrm{s}}g_{\mathrm{o}}\right]\vec{S}_{\mathrm{s}}\\ &\quad -\eta\mu_{\mathrm{b}}\left[\frac{6}{5}\vec{S}_{\mathrm{s}} + (\nabla\cdot\vec{u}_{\mathrm{s}}\vec{\delta})\right]\,, \qquad (43)\\ \vec{q}_{\mathrm{sc}} &= -\frac{12\eta\alpha_{\mathrm{s}}\lambda_{\mathrm{s}}}{g_{\mathrm{o}}}\left[1 + \frac{12}{5}\eta^2(4\eta-3)\alpha_{\mathrm{s}}g_{\mathrm{o}}\right]\nabla T - \frac{64}{25}\lambda_{\mathrm{i}}\eta^2(41-33\eta)\nabla T\,. \qquad (44)\end{aligned}$$

The total grain stress tensor is

$$\begin{aligned}\vec{P}_{\mathrm{s}} &= \alpha_{\mathrm{s}}\rho_{\mathrm{s}}\left[(1+4\eta\alpha_{\mathrm{s}}g_{\mathrm{o}})T - \eta\mu_{\mathrm{b}}\nabla\cdot\vec{u}_{\mathrm{s}}\right]\vec{\delta}\\ &\quad -\left\{\frac{2\mu_{\mathrm{s}}}{g_{\mathrm{o}}}(1+\frac{8}{5}\eta\alpha_{\mathrm{s}}g_{\mathrm{o}})\left[1+\frac{8}{5}\eta(3\eta-2)\alpha_{\mathrm{s}}g_{\mathrm{o}}\right] + \frac{6}{5}\eta\mu_{\mathrm{b}}\right\}\vec{S}_{\mathrm{s}}\,. \qquad (45)\end{aligned}$$

where

$$\mu_{\mathrm{b}} = \frac{16m\alpha_{\mathrm{s}}^2 g_{\mathrm{o}}}{\pi d^2}\left(\frac{T}{\pi}\right)^{1/2}, \tag{46}$$

is the bulk viscosity for perfectly elastic particles. Similarly, the total granular energy flux is

$$\begin{aligned}\vec{q}_{\mathrm{s}} = -\frac{\lambda_{\mathrm{s}}}{g_{\mathrm{o}}}\left(1+\frac{12}{5}\eta\alpha_{\mathrm{s}}g_{\mathrm{o}}\right)&\left[1+\frac{12}{5}\eta^2(4\eta-3)\alpha_{\mathrm{s}}g_{\mathrm{o}}\right]\\ &\nabla T - \frac{64}{25}\lambda_{\mathrm{i}}\eta^2(41-33\eta)\nabla T\,. \qquad (47)\end{aligned}$$

The expression for the rate of collisional energy dissipation per unit volume in (30) carried out to the same order of approximation as the expressions for stresses and energy flux is

$$\gamma_{\mathrm{s}} = \frac{48\eta(1-\eta)}{\pi^{1/2}d}\alpha_{\mathrm{s}}^2\rho_{\mathrm{s}}g_{\mathrm{o}}T^{3/2}\,. \tag{48}$$

Other methods of solution to a generalized Boltzmann equation for the evolution of the single solids-phase velocity distribution function which is equivalent to (21) had been proposed by a number of researchers [47–49]. However, due to the complexity of the two-phase flow problem being treated here, we prefer the simplicity of the present moment method employed above at the present time.

5 Mixture Theory

We may derive a set of conservation equations for the gas-solids mixture by following the traditional control volume approach. The instantaneous differential balance equations for mass, momentum and kinetic energy are

$$\frac{\partial}{\partial t}(\alpha_{\mathrm{f}}\rho_{\mathrm{f}} + \alpha_{\mathrm{s}}\rho_{\mathrm{s}}) + \nabla\cdot(\alpha_{\mathrm{f}}\rho_{\mathrm{f}}\vec{u}_{\mathrm{f}}^{*} + \alpha_{\mathrm{s}}\rho_{\mathrm{s}}\vec{c}) = 0\;, \tag{49}$$

$$\begin{aligned}\frac{\partial}{\partial t}(\alpha_{\mathrm{f}}\rho_{\mathrm{f}}\vec{u}_{\mathrm{f}}^{*} + \alpha_{\mathrm{s}}\rho_{\mathrm{s}}\vec{c}) &+ \nabla\cdot(\alpha_{\mathrm{f}}\rho_{\mathrm{f}}\vec{u}_{\mathrm{f}}^{*}\vec{u}_{\mathrm{f}}^{*} + \alpha_{\mathrm{s}}\rho_{\mathrm{s}}\vec{c}\,\vec{c})\\ &= (\alpha_{\mathrm{f}}\rho_{\mathrm{f}} + \alpha_{\mathrm{s}}\rho_{\mathrm{s}})\vec{g} - \nabla\cdot\left(\alpha_{\mathrm{f}}\vec{P}_{\mathrm{f}}^{*}\right) - \Psi_c(m\vec{c})\;,\end{aligned} \tag{50}$$

$$\begin{aligned}\frac{1}{2}\frac{\partial}{\partial t}&\left(\alpha_{\mathrm{f}}\rho_{\mathrm{f}}u_{\mathrm{f}}^{*2} + \alpha_{\mathrm{s}}\rho_{\mathrm{s}}c^{2}\right) + \frac{1}{2}\nabla\cdot\left(\alpha_{\mathrm{f}}\rho_{\mathrm{f}}u_{\mathrm{f}}^{*2}\vec{u}_{\mathrm{f}}^{*} + \alpha_{\mathrm{s}}\rho_{\mathrm{s}}c^{2}\vec{c}\right)\\ &= \alpha_{\mathrm{f}}\rho_{\mathrm{f}}\vec{g}\cdot\vec{u}_{\mathrm{f}}^{*} + \alpha_{\mathrm{s}}\rho_{\mathrm{s}}\vec{g}\cdot\vec{c} - \nabla\cdot\left(\alpha_{\mathrm{f}}\vec{u}_{\mathrm{f}}^{*}\cdot\vec{P}_{\mathrm{f}}^{*}\right) + \alpha_{\mathrm{f}}\vec{P}_{\mathrm{f}}^{*}:\nabla\vec{u}_{\mathrm{f}}^{*}\\ &\quad + \Psi_c\left(\frac{1}{2}mc^{2}\right) + \alpha_{\mathrm{s}}\rho_{\mathrm{s}}\vec{f}\cdot(\vec{c}-\vec{u}_{\mathrm{f}}^{*}) + \dot{E}_{\mathrm{w}}\;.\end{aligned} \tag{51}$$

In (50) and (51), $\vec{P}_{\mathrm{f}}^{*}$ is the instantaneous fluid molecular stresses. The term $\Psi_c(\psi)$ accounts for the instantaneous solid-solid collisional contributions of ψ. The second to last term on the right hand side of (51), $\alpha_{\mathrm{s}}\rho_{\mathrm{s}}\vec{f}\cdot(\vec{c}-\vec{u}_{\mathrm{f}}^{*})$, is the rate of work done by the interfacial friction. The last term $\dot{E}_{\mathrm{w}}$ represents the rate of turbulence kinetic energy production due to the presence of the solid particles [44].

The time average for an instantaneous gas-phase flow property, $\xi_{\mathrm{f}}^{*}(\vec{r},t) = \xi_{\mathrm{f}}(\vec{r},t) + \xi_{\mathrm{f}}'(\vec{r},t)$, is defined as

$$\xi_{\mathrm{f}}(\vec{r},t) = \frac{1}{\tau_{\mathrm{a}}}\int_{t}^{t+\tau_{\mathrm{a}}} \xi_{\mathrm{f}}^{*}(\vec{r},t')\mathrm{d}t'\;, \tag{52}$$

for

$$\frac{1}{\tau_{\mathrm{a}}}\int_{t}^{t+\tau_{\mathrm{a}}} \xi_{\mathrm{f}}'(\vec{r},t')\mathrm{d}t' = 0\;, \tag{53}$$

where τ_{a} is the averaging time scale with $\tau_{\mathrm{d}} < \tau_{\mathrm{a}} < \tau_{\mathrm{m}}$; τ_{m} is the time scale characteristic of the slow variations in the gas-phase. We should point out that the time averaging defined in (52) and (53) is consistent with the one utilized earlier in the Fokker-Planck method for characterizing the gas-solids interaction in (14) and (17) for $\tau_\mathrm{a} \approx \tau$.

After we perform both gas-phase time averaging and solids-phase ensemble averaging in the above equations (49)–(51), we obtain a set of mean conservation laws for the mixture

$$\frac{\partial}{\partial t}(\alpha_\mathrm{f}\rho_\mathrm{f} + \alpha_\mathrm{s}\rho_\mathrm{s}) + \nabla\cdot(\alpha_\mathrm{f}\rho_\mathrm{f}\vec{u}_\mathrm{f} + \alpha_\mathrm{s}\rho_\mathrm{s}\vec{u}_\mathrm{s}) = 0\,, \tag{54}$$

$$\begin{aligned}\frac{\partial}{\partial t}&(\alpha_\mathrm{f}\rho_\mathrm{f}\vec{u}_\mathrm{f} + \alpha_\mathrm{s}\rho_\mathrm{s}\vec{u}_\mathrm{s}) + \nabla\cdot(\alpha_\mathrm{f}\rho_\mathrm{f}\vec{u}_\mathrm{f}\vec{u}_\mathrm{f} + \alpha_\mathrm{s}\rho_\mathrm{s}\vec{u}_\mathrm{s}\vec{u}_\mathrm{s})\\ &= (\alpha_\mathrm{f}\rho_\mathrm{f} + \alpha_\mathrm{s}\rho_\mathrm{s})\vec{g} - \nabla\cdot(\alpha_\mathrm{f}\vec{P}_\mathrm{f}) - \nabla\cdot(\alpha_\mathrm{f}\vec{\sigma}_\mathrm{f}) - \nabla\cdot\vec{P}_\mathrm{s}\,,\end{aligned} \tag{55}$$

$$\begin{aligned}\frac{\partial}{\partial t}&\left(\alpha_\mathrm{f}\rho_\mathrm{f}k + \frac{3}{2}\alpha_\mathrm{s}\rho_\mathrm{s}T\right) + \nabla\cdot\left(\alpha_\mathrm{f}\rho_\mathrm{f}\vec{u}_\mathrm{f}k + \frac{3}{2}\alpha_\mathrm{s}\rho_\mathrm{s}\vec{u}_\mathrm{s}T\right)\\ &= -\alpha_\mathrm{f}\vec{\sigma}_\mathrm{f} : \nabla\vec{u}_\mathrm{f} - \nabla\cdot(\alpha_\mathrm{f}\vec{q}_\mathrm{f}) - \overline{\vec{u}'_\mathrm{f}\cdot\nabla\cdot(\alpha_\mathrm{f}\vec{P}'_\mathrm{f})} - \vec{P}_\mathrm{s} : \nabla\vec{u}_\mathrm{s} - \nabla\cdot\vec{q}_\mathrm{s}\\ &\quad - \gamma_\mathrm{s} + \alpha_\mathrm{s}\rho_\mathrm{s}\overline{\left\langle\vec{f}'\cdot(\vec{C}-\vec{u}'_\mathrm{f})\right\rangle} + \left\langle\overline{\dot{E}_\mathrm{w}}\right\rangle\,.\end{aligned} \tag{56}$$

The specific turbulence kinetic energy is defined as $k = \overline{u'^2_\mathrm{f}}/2$. If we subtract the set of solids-phase conservation equations (25)–(27) from the above mixture one, we obtain the set for the gas-phase only

$$\frac{\partial}{\partial t}(\alpha_\mathrm{f}\rho_\mathrm{f}) + \nabla\cdot(\alpha_\mathrm{f}\rho_\mathrm{f}\vec{u}_\mathrm{f}) = 0\,, \tag{57}$$

$$\frac{\partial}{\partial t}(\alpha_\mathrm{f}\rho_\mathrm{f}\vec{u}_\mathrm{f}) + \nabla\cdot(\alpha_\mathrm{f}\rho_\mathrm{f}\vec{u}_\mathrm{f}\vec{u}_\mathrm{f}) = \alpha_\mathrm{f}\rho_\mathrm{f}\vec{g} - \alpha_\mathrm{s}\rho_\mathrm{s}\vec{f}_\mathrm{m} - \nabla\cdot(\alpha_\mathrm{f}\vec{P}_\mathrm{f}) + \nabla\cdot(\alpha_\mathrm{f}\vec{\sigma}_\mathrm{f})\,, \tag{58}$$

$$\begin{aligned}\frac{\partial}{\partial t}(\alpha_\mathrm{f}\rho_\mathrm{f}k) + \nabla\cdot(\alpha_\mathrm{f}\rho_\mathrm{f}\vec{u}_\mathrm{f}k) &= -\alpha_\mathrm{f}\vec{\sigma}_\mathrm{f} : \nabla\vec{u}_\mathrm{f} - \nabla\cdot(\alpha_\mathrm{f}\vec{q}_\mathrm{f}) - \overline{\vec{u}'_\mathrm{f}\cdot\nabla\cdot(\alpha_\mathrm{f}\vec{P}'_\mathrm{f})}\\ &\quad - \frac{1}{2}\alpha_\mathrm{s}\rho_\mathrm{s}\left\langle\vec{B}:\vec{\delta}\right\rangle - \alpha_\mathrm{s}\rho_\mathrm{s}\overline{\left\langle\vec{f}'\cdot\vec{u}'_\mathrm{f}\right\rangle} + \left\langle\overline{\dot{E}_\mathrm{w}}\right\rangle\,.\end{aligned} \tag{59}$$

In the above equations, the mean fluid molecular stress tensor is

$$\vec{P}_\mathrm{f} = p_\mathrm{f}\vec{\delta} - 2\mu_\mathrm{f}\vec{S}_\mathrm{f}\,, \tag{60}$$

where $\vec{S}_\mathrm{f} = (u_{\mathrm{f}_{i,j}} + u_{\mathrm{f}_{j,i}})/2$ is the mean fluid strain-rate tensor. The Reynolds turbulence stress tensor is defined here as

$$\begin{aligned}\vec{\sigma}_\mathrm{f} &= \rho_\mathrm{f}\overline{\vec{u}'_\mathrm{f}\vec{u}'_\mathrm{f}}\\ &= \frac{2}{3}k\vec{\delta} - 2\mu_\mathrm{T}\vec{S}_\mathrm{f}\,,\end{aligned} \tag{61}$$

where μ_T is the gas-phase eddy viscosity. The turbulence kinetic energy flux is $\vec{q}_\mathrm{f} = \rho_\mathrm{f}\overline{u'^2_\mathrm{f}\vec{u}'_\mathrm{f}}/2$.

In order to close the set of gas-phase conservation equations, we extend the standard k–ε turbulence model for a single gas-phase [50, 51] to treat the case of a two-phase mixture. In view of the complexity of the analytical formulation at hand, it seems sensible that a first-order two-equation turbulence model would suffice. The gas turbulence kinetic energy is regarded as isotropic. This is consistent with the assumption made earlier for the granular temperature in the kinetic theory, i.e. the granular temperature is isotropic as well. The modified turbulence kinetic energy equation for the two-phase flows can be expressed as

$$\frac{\partial}{\partial t}(\alpha_{\mathrm{f}}\rho_{\mathrm{f}}k) + \nabla\cdot(\alpha_{\mathrm{f}}\rho_{\mathrm{f}}\vec{u}_{\mathrm{f}}k) = -\alpha_{\mathrm{f}}\vec{\sigma}_{\mathrm{f}} : \nabla\vec{u}_{\mathrm{f}} - \alpha_{\mathrm{f}}\rho_{\mathrm{f}}\varepsilon + \nabla\cdot[\alpha_{\mathrm{f}}(\mu_{\mathrm{f}} + \mu_{\mathrm{T}}/\sigma_k)\nabla k] \\ -\frac{1}{2}\alpha_{\mathrm{s}}\rho_{\mathrm{s}}\left\langle\vec{B}:\vec{\delta}\right\rangle - \alpha_{\mathrm{s}}\rho_{\mathrm{s}}\left\langle\overline{\vec{f}'\cdot\vec{u}'_{\mathrm{f}}}\right\rangle + \left\langle\dot{\bar{E}}_{\mathrm{w}}\right\rangle \,, \quad (62)$$

where $\varepsilon = \nu_{\mathrm{f}}\overline{u'_{\mathrm{f}_{i,j}}u'_{\mathrm{f}_{i,j}}}$ is the dissipation per unit mass, and the gas-phase eddy viscosity is

$$\mu_{\mathrm{T}} = \rho_{\mathrm{f}}C_{\mu}k^2/\varepsilon \,. \quad (63)$$

The dissipation rate equation is

$$\frac{\partial}{\partial t}(\alpha_{\mathrm{f}}\rho_{\mathrm{f}}\varepsilon) + \nabla\cdot(\alpha_{\mathrm{f}}\rho_{\mathrm{f}}\vec{u}_{\mathrm{f}}\varepsilon) \\ = -\alpha_{\mathrm{f}}C_{\varepsilon 1}\frac{\varepsilon}{k}\vec{\sigma}_{\mathrm{f}} : \nabla\vec{u}_{\mathrm{f}} - \alpha_{\mathrm{f}}\rho_{\mathrm{f}}C_{\varepsilon 2}\frac{\varepsilon^2}{k} + \nabla\cdot[\alpha_{\mathrm{f}}(\mu_{\mathrm{f}} + \mu_{\mathrm{T}}/\sigma_{\varepsilon})\nabla\varepsilon] \\ + \left[\frac{1}{2}\alpha_{\mathrm{s}}\rho_{\mathrm{s}}\left\langle\vec{B}:\vec{\delta}\right\rangle + \alpha_{\mathrm{s}}\rho_{\mathrm{s}}\left\langle\overline{\vec{f}'\cdot\vec{u}'_{\mathrm{f}}}\right\rangle - \left\langle\dot{\bar{E}}_{\mathrm{w}}\right\rangle\right]\frac{\varepsilon}{k} \,, \quad (64)$$

The closure coefficients in the above equations are $C_{\varepsilon 1} = 1.44$, $C_{\varepsilon 2} = 1.92$, $C_{\mu} = 0.09$, $\sigma_k = 1$, and $\sigma_{\varepsilon} = 1.3$ [50, 51].

6 Turbulence Modulation

The last three terms in (62) and (64) are explicitly responsible for the turbulence modulations in the gas-phase caused by the solid particles.

The term $\alpha_{\mathrm{s}}\rho_{\mathrm{s}}\left\langle\vec{B}:\vec{\delta}\right\rangle/2$ is the mean rate of energy transfer per unit volume from the carrier gas-phase to the solids-phase. From (20), we obtain

$$\frac{1}{2}\alpha_{\mathrm{s}}\rho_{\mathrm{s}}\left\langle\vec{B}:\vec{\delta}\right\rangle = 3\alpha_{\mathrm{s}}\rho_{\mathrm{s}}\varphi_{\mathrm{d}}T \,. \quad (65)$$

As far as the gas-phase is concerned, this term is an energy dissipation term since it carries a negative sign in front as shown in (62). However, if we focus our attention at the solids-phase energy equation (27) for a moment, the above term basically cancels the mean rate of energy dissipation due to the fluctuating interfacial forces acting upon the particles in fluctuating motion, i.e., from (15)

$$\alpha_s \rho_s \left\langle \vec{f'} \cdot \vec{C} \right\rangle = -3\alpha_s \rho_s \varphi_d T \; . \tag{66}$$

In other words, the energy induced by the carrier gas-phase into the solids-phase *via* the fluctuating interfacial forces supports the fluctuating motion of the solid particles. Any additional energy transfered into the solids-phase from for example rapid shearing of the bulk mixture, would likely augment the granular temperature accordingly.

By using (6), the second to last term in (62) becomes

$$-\alpha_s \rho_s \left\langle \overline{\vec{f'} \cdot \vec{u}'_f} \right\rangle = -2\alpha_s \rho_s \varphi_d k \; , \tag{67}$$

which represents the mean rate of turbulence kinetic energy dissipation per unit volume due to the fluctuating interfacial forces acting upon the fluid elements.

The last term $\left\langle \overline{\dot{E}}_w \right\rangle$ in (62) is the mean rate of turbulence kinetic energy production per unit volume due to the finite size effect of the solid particles which includes, for example, the effects of shedding wakes and vortices, deflection and distortion of large energetic eddies. Lun [44] obtained the following formulation for $\left\langle \overline{\dot{E}}_w \right\rangle$ by inferring from the experimental result of Wu and Faeth [52] for a single sphere,

$$\left\langle \overline{\dot{E}}_w \right\rangle = 2\pi C_w n d \rho_f \nu_t k \; . \tag{68}$$

The phenomenological coefficient C_w may be expressed as

$$\begin{aligned} C_w &= 0 && Re_m \le 1 \\ C_w &= \frac{1}{3}\delta_w^2 && 1 < Re_m < 310 \\ C_w &= \frac{17}{24}\delta_w^2 && 310 \le Re_m \le 1560 \; , \end{aligned} \tag{69}$$

where δ_w is a phenomenological parameter to be determined *via* experiments. The local turbulence kinematic viscosity ν_t is given by

$$\begin{aligned} \nu_t/\nu_f &= 1 && 0 \le Re_m < 1 \\ \nu_t/\nu_f &= 1 + 0.00966(Re_m - 1) && 1 \le Re_m < 135 \\ \nu_t/\nu_f &= 0.017 Re_m && 135 \le Re_m < 310 \\ \nu_t/\nu_f &= 1.2 + 0.000042 Re_m^2 && 310 \le Re_m < 610 \\ \nu_t/\nu_f &= 0.029 Re_m && 610 \le Re_m < 1560 \; . \end{aligned} \tag{70}$$

In (69) and (70), we have replaced the instantaneous particle Reynolds number Re_p given in Lun [44] by the local averaged one, i.e., Re_m. According to the experimental result [52], the parameter δ_{w} ranges roughly from 4 to 6, whereas Lun had taken it to be a constant at 4. Furthermore, the second expression in (70) differs from what was proposed originally. Lun assumed that the effect of turbulence modulation started when the particle Reynolds number reached a 'critical' value (Re_{c}) of about 18 at which flow separation on a sphere was initiated. However, modification of the turbulence intensity such as stretching and deflection of energetic eddies in the gas by the sphere should actually start at a much lower value of Re_{c}. In the present study, we have taken Re_{c} to be unity.

7 Application

We apply the above kinetic theory to simulate numerically the experiment of Mostafa et al. [53] that dealt with an unconfined non-swirl axisymmetric round jet aimed vertically downward. Air was used to propel glass beads having a mean diameter of about 0.105 mm and a mass density of 2500 kg/m^3 through a jet nozzle (with diameter $D = 2r_{\mathrm{o}} = 25.3$ mm) into a large chamber. The jet Reynolds number, $Re_{jet} = 4\dot{m}_{\mathrm{f}}/\pi\mu_{\mathrm{f}}D$, was kept constant at 5712; $\dot{m}_{\mathrm{f}}$ is the bulk mass flow rate of air. The solids fraction tested in the experiment was of the order of 10^{-4} to 10^{-3}. The radial distribution function $g_o(\alpha_{\mathrm{s}})$ in (35) is taken as unity. The coefficient of restitution used in the present computation for the glass beads is 0.95, and the parameter δ_{w} in (69) is 4.55.

After eliminating the higher order terms and assuming a steady state, the set of conservation equations [(25)-(27), (57), (58), (62), (64)] for the gas and the solids phases in cylindrical coordinates (r, θ, x) reduce to the following. The mass conservation equations are

$$\frac{\partial}{\partial x}(\alpha_{\mathrm{s}}u_{\mathrm{s}}) + \frac{1}{r}\frac{\partial}{\partial r}(r\alpha_{\mathrm{s}}v_{\mathrm{s}}) = 0\ , \tag{71}$$

$$\frac{\partial}{\partial x}(\alpha_{\mathrm{f}}u_{\mathrm{f}}) + \frac{1}{r}\frac{\partial}{\partial r}(r\alpha_{\mathrm{f}}v_{\mathrm{f}}) = 0\ . \tag{72}$$

The momentum equations are

$$\frac{\partial}{\partial x}\left(\alpha_{\mathrm{s}}u_{\mathrm{s}}^2\right) + \frac{1}{r}\frac{\partial}{\partial r}\left(r\alpha_{\mathrm{s}}v_{\mathrm{s}}u_{\mathrm{s}}\right) = \alpha_{\mathrm{s}}g + \alpha_{\mathrm{s}}f_{\mathrm{mx}} - \frac{1}{\rho_{\mathrm{s}}}\frac{\partial p_{\mathrm{s}}}{\partial x} + \frac{1}{r}\frac{\partial}{\partial r}\left(r\nu_{\mathrm{s}}\frac{\partial u_{\mathrm{s}}}{\partial r}\right)\ , \tag{73}$$

$$\frac{\partial}{\partial x}(\alpha_{\mathrm{s}}u_{\mathrm{s}}v_{\mathrm{s}}) + \frac{1}{r}\frac{\partial}{\partial r}\left(r\alpha_{\mathrm{s}}v_{\mathrm{s}}^2\right) = \alpha_{\mathrm{s}}f_{\mathrm{mr}} - \frac{1}{\rho_{\mathrm{s}}}\frac{\partial p_{\mathrm{s}}}{\partial r} + \frac{2}{r}\frac{\partial}{\partial r}\left(r\nu_{\mathrm{s}}\frac{\partial v_{\mathrm{s}}}{\partial r}\right) - 2\nu_{\mathrm{s}}\frac{v_{\mathrm{s}}}{r^2}\ , \tag{74}$$

$$\frac{\partial}{\partial x}\left(\alpha_{\mathrm{f}}u_{\mathrm{f}}^2\right) + \frac{1}{r}\frac{\partial}{\partial r}\left(r\alpha_{\mathrm{f}}v_{\mathrm{f}}u_{\mathrm{f}}\right) = -\frac{\alpha_{\mathrm{s}}\rho_{\mathrm{s}}}{\rho_{\mathrm{f}}}f_{\mathrm{mx}} - \frac{1}{\rho_{\mathrm{f}}}\frac{\partial(\alpha_{\mathrm{f}}p_{\mathrm{f}})}{\partial x} + \frac{1}{r}\frac{\partial}{\partial r}\left[r\alpha_{\mathrm{f}}\nu_{\mathrm{eff}}\frac{\partial u_{\mathrm{f}}}{\partial r}\right], \tag{75}$$

$$\frac{\partial}{\partial x}\left(\alpha_{\mathrm{f}}u_{\mathrm{f}}v_{\mathrm{f}}\right) + \frac{1}{r}\frac{\partial}{\partial r}\left(r\alpha_{\mathrm{f}}v_{\mathrm{f}}^2\right) = -\frac{\alpha_{\mathrm{s}}\rho_{\mathrm{s}}}{\rho_{\mathrm{f}}}f_{\mathrm{mr}} - \frac{1}{\rho_{\mathrm{f}}}\frac{\partial\left(\alpha_{\mathrm{f}}p_{\mathrm{f}}\right)}{\partial r}\ . \tag{76}$$

In (73)–(74), $p_{\mathrm{s}} = \alpha_{\mathrm{s}}\rho_{\mathrm{s}}(1 + 4\eta\alpha_{\mathrm{s}}g_{\mathrm{o}})T$ is the solids normal pressure, and ν_{s} is the solids kinematic viscosity which can be obtained from (45). Similarly in

(75), p_f is the fluid pressure, the effective gas viscosity is $\nu_\mathrm{eff} = \nu_\mathrm{f} + \nu_\mathrm{T}$, and $\nu_\mathrm{T} = \mu_\mathrm{T}/\rho_\mathrm{f}$ is the eddy kinematic viscosity. Assuming T to be a function of x–direction only, the solids-phase kinetic energy equation (27) becomes

$$\frac{\partial}{\partial x}(\alpha_\mathrm{s} u_\mathrm{s} T) + \frac{T}{r}\frac{\partial}{\partial r}(r\alpha_\mathrm{s} v_\mathrm{s}) = \frac{2}{3}\nu_\mathrm{s}\left(\frac{\partial u_\mathrm{s}}{\partial r}\right)^2 - \frac{2\gamma_\mathrm{s}}{3\rho_\mathrm{s}} \,. \tag{77}$$

The k–ε equations (62) and (64) are now

$$\frac{\partial}{\partial x}(\alpha_\mathrm{f} u_\mathrm{f} k)+\frac{1}{r}\frac{\partial}{\partial r}(r\alpha_\mathrm{f} v_\mathrm{f} k) = \alpha_\mathrm{f}\nu_\mathrm{T}\left(\frac{\partial u_\mathrm{f}}{\partial r}\right)^2 - \alpha_\mathrm{f}\varepsilon + \frac{1}{r}\frac{\partial}{\partial r}\left[r\alpha_\mathrm{f}\left(\nu_\mathrm{f} + \frac{\nu_\mathrm{T}}{\sigma_\mathrm{k}}\right)\frac{\partial k}{\partial r}\right] - \frac{\alpha_\mathrm{s}\rho_\mathrm{s}\varphi_\mathrm{d}}{\rho_\mathrm{f}}(3T + 2k) + \frac{\left\langle \dot{E}_\mathrm{w}\right\rangle}{\rho_\mathrm{f}} \,, \tag{78}$$

$$\frac{\partial}{\partial x}(\alpha_\mathrm{f} u_\mathrm{f}\varepsilon) + \frac{1}{r}\frac{\partial}{\partial r}(r\alpha_\mathrm{f} v_\mathrm{f}\varepsilon) = \alpha_\mathrm{f} C_{\varepsilon 1}\nu_\mathrm{T}\left(\frac{\partial u_\mathrm{f}}{\partial r}\right)^2\frac{\varepsilon}{k} - \alpha_\mathrm{f} C_{\varepsilon 2}\frac{\varepsilon^2}{k} + \frac{1}{r}\frac{\partial}{\partial r}\left[r\alpha_\mathrm{f}(\nu_\mathrm{f} + \frac{\nu_\mathrm{T}}{\sigma_\varepsilon})\frac{\partial\varepsilon}{\partial r}\right] + \frac{\alpha_\mathrm{s}\rho_\mathrm{s}\varphi_\mathrm{d}\varepsilon}{\rho_\mathrm{f} k}(3T + 2k) - \frac{\left\langle \dot{E}_\mathrm{w}\right\rangle\varepsilon}{\rho_\mathrm{f} k} \,. \tag{79}$$

The global mass conservation relationship for the mixture is

$$\alpha_\mathrm{s} + \alpha_\mathrm{f} = 1 \,. \tag{80}$$

With appropriate boundary conditions, the above ten equations (71)-(80) for the unknowns $\alpha_\mathrm{s}, \alpha_\mathrm{f}, p_\mathrm{f}, u_\mathrm{s}, v_\mathrm{s}, u_\mathrm{f}, v_\mathrm{f}, T, k$, and ε were integrated numerically by using a pressure-correction method with a 2*nd* order staggered central difference scheme [54, 55]. The numerical residuals for mass, momentum and energy were kept to the order of about 10^{-7}.

8 Comparisons between Predictions and Experiments

Figure 1 shows the comparisons between the present theoretical predictions and the experimental measurements for the dimensionless (a) gas mean velocity $u_\mathrm{f}/u_\mathrm{fo}$, (b) turbulence intensity $\sqrt{2k/3}/u_\mathrm{fc}$ and (c) Reynolds shear stress $\overline{u'_\mathrm{f} v'_\mathrm{f}}/u^2_\mathrm{fc}$ for the jet at the downstream location of $x/D = 12.45$ for various mass loading ratios (Lr). The mass loading ratio is defined as the ratio of the mass flux of the solids to that of the air. The gas mean centerline velocities u_fo measured at $x/D = 0.04$ from the nozzle exit for the cases of $Lr = 0, 0.2$ and 1.0 were 4.74, 4.70 and 4.20 m/s respectively, whereas the particle mean centerline velocities u_so were 4.41 and 3.39 m/s for $Lr = 0.2$ and 1.0, respectively. The quantity u_fc represents the *local* gas mean centerline velocity.

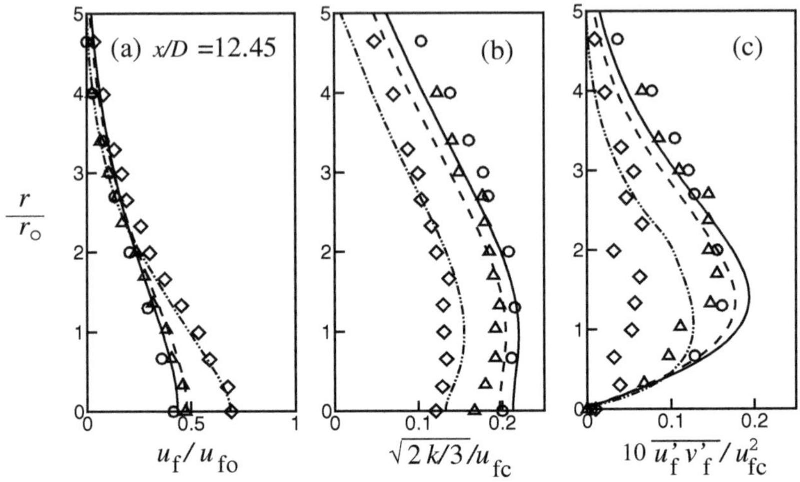

Fig. 1. Variations of dimensionless **a** gas mean velocity, **b** turbulence intensity and **c** Reynolds shear stress. Open symbols represent experimental data (Mostafa et al. [53]) and curves are the present theoretical predictions. Circles and solid lines are for $Lr = 0$; triangles and dashed lines are for $Lr = 0.2$; diamonds and dotted are lines for $Lr = 1$

In general, there is good agreement between the experimental measurements and the present theoretical predictions.

Typically, the spreading rate of the air decreased with increasing loading ratio (see Fig. 1a). As the jet spread radially going downstream, the air velocity slowed down at the core region. At a certain location ($x/D < 12.45$), the particle mean centerline velocity became larger than that of the air. From this point onward the solids-phase became the carrier. In other words, the solids-phase was now transferring momentum and energy to the gas-phase. As a result, the spreading of the air jet was retarded to an extent that could be quite substantial depending on the mass loading. As shown in Figs. 1b and 1c, there was obvious turbulence attenuation in the jet with increasing mass loading. Even though the solids-phase fed back part of its energy to the gas-phase, the amount was still insufficient to compensate for the overall energy loss in the air due mainly to eddy viscosity and interfacial friction.

In Fig. 2, we have plotted the dimensionless profiles of (a) mean solids volume fraction, α_s/α_{sc}, (b) particle mean velocity u_s/u_{sc}, and (c) granular temperature $\sqrt{T}/u_{sc}$, where α_{sc} and u_{sc} are the local solids mean centerline volume fraction and velocity, respectively. Overall, there is reasonable agreement between the predictions and the measurements. The solids concentration decreased radially as one might expect. The particle mean velocity

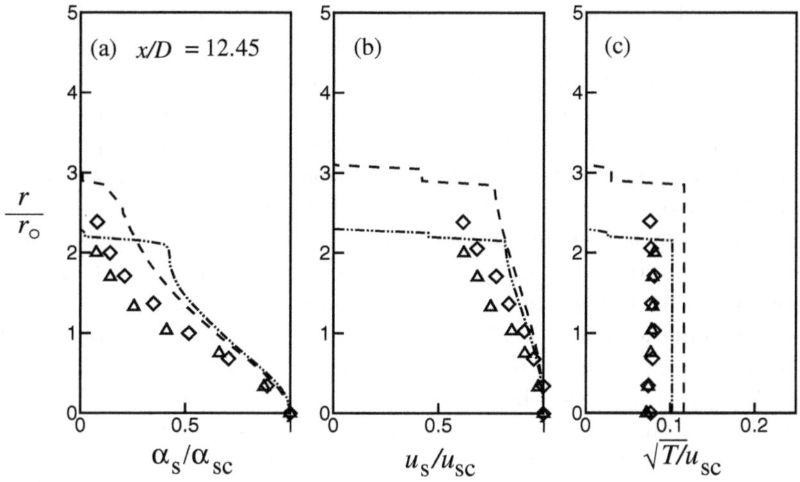

Fig. 2. Variations of dimensionless **a** mean solids volume fraction, **b** particle mean velocity and **c** granular temperature. Triangles and dashed lines are for $Lr = 0.2$; diamonds and dashed-dotted lines are for $Lr = 1$

profiles in Fig. 2b showed a finite degree of shear gradient in the radial direction, while the granular temperature profiles in Fig. 2c remained to be almost uniform across the cross section. One interesting point about the comparisons was that the present theory predicted a smaller spreading rate of the solids with increasing mass loading, whereas the experiments seemed to show otherwise. However, the exact position of the free boundary of the solids-phase is often difficult to locate experimentally. Since the solids concentration near the free boundary is supposed to be very small, one would need to take very large samples for long periods of time in close proximity there. As far as this possible discrepancy is concerned, further study is required.

9 Conclusion

We have developed a new kinetic theory for the inertia flow regime of dilute turbulent gas-solids two-phase mixtures where the Stokes number $St \gg 1$. A complete set of conservation equations is derived systematically for the flow mixture. Closure for the gas-phase conservation equations is achieved by extending the standard k–ε model for a single gas-phase to the case of two-phase flow taking into account the couplings in mass, momentum and energy between the phases. In this study, we propose a drag law which is applicable for a wide range of particle Reynolds number. The gas-solids interactions in momentum and energy are characterized in the present theory by using the Fokker-Plank technique. Furthermore, the kinetic and collisional contributions to momentum and energy transfers in the solids-phase are incorporated in the model by following the approach of the kinetic theory of granular flows. Turbulence modulation of the gas-phase due to the presence of the solid particles is modelled based on both theory and experimental data. We have determined the transport coefficients and constitutive relationships for the solids-phase by solving a generalized Maxwell-Boltzmann-type transport equation using a moment method. In addition to the usual flow properties such as the solids fraction and the granular temperature, the transport coefficients are found to depend on the gas-friction coefficient φ_{d}, which for the present case is simply the inverse of the mean drag relaxation time for the solid particles. In contrast to the traditional continuum mechanical approach, the present kinetic theory is able to yield a set of constitutive relationships for the solids-phase, which is obviously crucial in the study of the complex turbulent flow of any two-phase gas-solids mixture.

The kinetic theory is applied to simulate the case of a turbulent axisymmetric round jet of a mixture of air and glass-beads. In general, there is favorable agreement between the theoretical predictions and the experimental result for a number of macroscopic flow properties such as gas and solids mean velocities, turbulence intensity, Reynolds shear stress, solids concentration and granular temperature. With increasing mass loading of the solids, the spreading rate of the air jet decreases while the turbulence intensity in the air is attenuated.

The present theory can be extended to the case of a moderately dense gas-solids flow system. In that case the hydrodynamic interactions between solid particles may not be neglected. At the same time, the interparticle collision processes will become important. Another pertinent and important subject is the modelling of the interactions between the gas-solids mixture and the solid-wall boundaries.

References

1. A. Bedford and D.S. Drumheller. Theory of immiscible and structured mixtures. *Int. J. Eng. Sc.*, 21:863–960, 1983
2. D.A. Drew. Mathematical modeling of two-phase flow. *Ann. Rev. Fluid Mech.*, 15:261–291, 1983
3. Y.A. Buyevich and I.N. Shchelchkova. Flow of dense suspensions. *Prog. Aerosp. Sci.*, 18:121–150, 1978
4. M.W. Reeks. On a kinetic equation for the transport of particles in turbulent flows. *Phys. Fluids A*, 3:446–456, 1991
5. R. Calflisch and G.C. Papanicolaou. Dynamic theory of suspensions with brownian effects. *SIAM J. Appl. Math.*, 43:885–906, 1983
6. Y.A. Buyevich. Statistical hydrodynamics of disperse systems. part i. physical background and general equations. *J. Fluid Mech.*, 49:489–507, 1971
7. G. Houghton. Particles and fluid diffusion in homogeneous fluidization. *I. & E. C. Fundam.*, 5:153–164, 1966

8. K.E. Hyland, S. McKee, and M.W. Reeks. Exact analytic solutions to turbulent particle flow equations. *Phys. Fluids A*, 11:1249–1261, 1999
9. S.B. Savage and D.J. Jeffrey. The stress tensor in a granular flow at high shear rates. *J. Fluid Mech.*, 110:255–272, 1981
10. J.T. Jenkins and S.B. Savage. A theory for the rapid flow of identical, smooth, nearly elastic particles. *J. Fluid Mech.*, 130:187–202, 1983
11. C.K.K. Lun, S.B. Savage, D.J. Jeffrey, and N. Chepurniy. Kinetic theories for granular flow: inelastic particles in couette flow and slightly inelstic particles in a general flow field. *J. Fluid Mech.*, 140:223–256, 1984
12. J.T. Jenkins and M.W. Richmond. Grad's 13-moment system for a dense gas of inelastic spheres. *Arch. Rat. Mech. and Anal.*, 87:355–377, 1985
13. C.K.K. Lun and S. B. Savage. A simple kinetic theory for granular flow of rough, inelastic, spherical particles. *J. Appl. Mech.*, 54:47–53, 1987
14. B.G.M. van Wachem, J.C. Schouten, C.M. van den Bleek, R. Krishna, and J.L. Sinclair. Comparative analysis of cfd models of dense gas-solid systems. *AIChE J.*, 47:1035–1051, 2001
15. S. Sundaresan. Modeling the hydrodynamics of multiphase flow reactors: Current status and challenges. *AIChE J.*, 46:1102–1105, 2000
16. R. Jackson. *The Dynamics of Fluidized Particles.* Cambridge University Press, Cambridge, 2000
17. L.S. Fan and C. Zhu. *Principles of Gas-Solid Flows.* Cambridge University Press, Cambridge, 1998
18. E. Periano and B. Leckner. Fundamental of turbulent gas-solid flows applied to circulating fluidized bed combustion. *Prog. Energy Combust. Sci.*, 24:259–296, 1998
19. C.T. Crowe, T.R. Troutt, and J. N. Chung. Numerical models for two-phase turbulent flows. *Ann. Rev. Fluid Mech.*, 28:11–43, 1996
20. D. Gidaspow. *Multiphase Flow and Fluidization – Continuum and Kinetic Theory Descriptions.* Academic Press, San Diego, 1994
21. A. Neri and D. Gidaspow. Riser hydrodynamics: simulation using kinetic theory. *AIChE J.*, 46:52–67, 2000
22. D.C. Swailes, Y.A. Sergeev, and A. Parker. Chapman-enskog closure approximation in the kinetic theory of dilute turbulent gas-particulate suspensions. *Physica A*, 254:517–547, 1998
23. C.M. Hrenya and J.L. Sinclair. Effects of particle-phase turbulence in gas-solid flows. *AIChE J.*, 43:853–869, 1997
24. J.L. Sinclair and R. Jackson. Gas-particle flow in a vertical pipe with particle-particle interactions. *AIChE J.*, 35:1473–1486, 1989
25. J.A. Pita and S. Sundaresan. Gas-solid flow in vertical tubes. *AIChE J.*, 37:1009–1018, 1991
26. J. Ding and D. Gidaspow. A bubbling fluidization model using kinetic theory of granular flow. *AIChE J.*, 36:523–538, 1990
27. A.S. Sangani, G. Mo, H.K. Tsao, and D.L. Koch. Simple shear flows of dense gas-solid suspensions at finite stokes numbers. *J. Fluid Mech.*, 313:309–341, 1996
28. C.E. Lapple. Dust and mist collection. In J. H. Perry, editor, *Chemical Engineer's Handbook*, 3rd edn, pages 50–63. McGraw-Hill, 1950
29. C. Crowe, M. Sommerfled, and Y. Tsuji. *Multiphase Flows with Droplets and Particles.* CRC Press, Boca Raton, 1997

30. R.B. Bird, W.E. Stewart, and E.N. Lightfoot. *Transport Phenomena.* John Wiley, New York, 1960
31. H. Schlichting. *Boundary Layer Theory.* McGraw Hill, New York, 1979
32. C.K.K. Lun. Kinetic theory for granular flow of dense, slightly inelastic, slightly rough spheres. *J. Fluid Mech.*, 233:539–559, 1991
33. J.O. Hirshfelder, C.F. Curtiss, and R.B. Bird. *Molecular Theory of Gases and Liquids.* John Wiley, New York, 1967
34. R.L. Liboff. *Kinetic Theory - Classical, Quantum, and Relativistic Descriptions.* Prentice Hall, New Jersey, 1990
35. L.C. Woods. *Thermodynamics of Fluid Systems.* Clarendon Press, New York, 1975
36. A.S. Rice and P. Gray. *The Statistical Mechanics of Simple Liquids. Monographs in Statistical Physics and Thermodynamics.* John Wiley, New York, 1965
37. R. Folkersma, H.N. Stein, and F.N. van de Vosse. Hydrodynamic interactions between two identical spheres held fixed side by side against a uniform stream directed perpendicular to the line connecting the spheres' centres. *Int. J. Mult. Flow*, 26:877–887, 2000
38. I. Kim, S. Elghobashi, and W.A. Sirignano. Three-dimensional flow over two spheres placed side by side. *J. Fluid Mech.*, 246:465–488, 1993
39. M.L. Allen. The effect of particle proximity on the drag force generated at high reynolds numbers. *Trans. I. Chem. Engrs.*, 58:187–194, 1980
40. P.N. Rowe and G.A. Henwood. Drag forces in a hydraulic model of a fluidized bed - part 1. *Trans. I. Chem. Engrs.*, 39:43–54, 1961
41. R.A. Bagnold. Experiments on a gravity free dispersion of large solid spheres in a newtonian fluid under shear. *Proc. Roy. Soc. Lond.*, 225:49–63, 1954
42. J.S. Dahler and M. Theodosopulu. The kinetic theory of dense polyatomic fluids. *Adv. Chem. Phys.*, 31:155–229, 1975
43. N.F. Carnahan and K.E. Starling. Equations of state for non-attracting rigid spheres. *J. Chem. Phys.*, 51:635–636, 1969
44. C.K.K. Lun. Numerical simulation of dilute turbulent gas-solid flow. *Int. J. Mult. Flow*, 26:1707–1736, 2000
45. C.K.K. Lun and H. S. Liu. Numerical simulation of dilute turbulent gas-solid flows in horizontal channels. *Int. J. Mult. Flow*, 23:575–605, 1997
46. H. Grad. On the kinetic theory of rarefied gases. *Comm. Pure and Appl. Math.*, 2:331–407, 1949
47. N. Sela and I. Goldhirsch. Hydrodynamic equations for rapid flows of smooth inelastic spheres, to burnett order. *J. Fluid Mech.*, 361:41–74, 1998
48. J.M. Montanero, V. Garzo, A. Santos, and J.J. Brey. Kinetic theory of simple granular shear flows of smooth hard spheres. *J. Fluid Mech.*, 389:391–411, 1999
49. A. Goldshtein and M. Shapiro. Mechanics of collisional motion of granular materials. part 1. gerneral hydrodynamic equations. *J. Fluid Mech.*, 282:75–114, 1995
50. W.P. Jones and B.E. Launder. The prediction of laminarization with a two-equation model of turbulence. *Int. J. Heat and Mass Trans.*, 15:301–314, 1972
51. B.E. Launder and B.I. Sharma. Application of the energy dissipation model of turbulence to the calculation of flow near a spinning disc. *Let. in Heat and Mass Trans.*, 1:131–138, 1974
52. J.S. Wu and G.M. Faeth. Spheres wakes at moderate reynolds numbers in turbulent environment. *AIAA J.*, 32:535–541, 1994

53. A.A. Mostafa, H.C. Mongia, V.G. McDonell, and G.S. Samuelsen. Evolution of particle-laden jet flows: a theoretical and experimental study. *AIAA J.*, 27:167–183, 1989
54. S.V. Patankar. *Numerical Heat Transfer and Fluid Flow.* Hemisphere, New York, 1980
55. J.P. Van Doormaal and G.D. Raithby. Enhancements of the simple method for predicting incompressible fluid flows. *AIAA J.*, 7:147–163, 1984

Part III

Driven Gases and Structure Formation

Driven Granular Gases

Stefan Luding[1,2], Raffaele Cafiero[3], and Hans J. Herrmann[2]

[1] Particle Technology, DelftChemTech, TU Delft, Julianalaan 136, 2628 BL Delft, The Netherlands, s.luding@tnw.tudelft.nl
[2] Institute for Computer Applications 1, University of Stuttgart, Pfaffenwaldring 27, 70569 Stuttgart, Germany, lui@ica1.uni-stuttgart.de
[3] P.M.M.H., École Supérieure de Physique et de Chimie Industrielles (ESPCI), 10, rue Vauquelin, 75251 Paris CEDEX 05, France

Abstract. In this paper, a two-dimensional granular gas of inelastic, rough spheres subject to driving is examined. Either the translational degrees of freedom are agitated proportional to a power $|v(\vec{x})|^\delta$ of the local particle velocity $v(\vec{x})$, or the rotational degrees are agitated randomly with respect to the angular velocity $\omega(\vec{x})$.

The steady state properties of the model, with respect to energy, partition of energy, and velocity distributions, are examined for different values of δ, and compared with the homogeneous driving case $\delta = 0$. A driving linearly proportional to $v(\vec{x})$ seems to reproduce some experimental observations which could not be reproduced by a homogeneous driving. Furthermore, we obtain that the system can be homogenized even for strong dissipation, if a driving inversely proportional to the velocity is used ($\delta < 0$). In the case of rotational driving, the system is well randomized and clusters are hindered. Even though rotational driving may be difficult to realize experimentally, this is an opportunity to avoid or delay the often unwanted effect of clustering.

1 Introduction

Granular materials belong to the fascinating world of non-linear, dissipative, non-equilibrium systems [1–3]. Hard spheres, as a special case, are used also as basic model for gases, liquids, and e.g. glasses [4]. Adding dissipation and rotation to hard spheres, one has the simplest yet realistic model granulate. Granular media are (more complicated) collections of macroscopic particles with rough surfaces and dissipative, frictional interactions, but also involve eigen-modes of vibration and possibly plastic deformation or fracture. Disregarding the latter two, one has Molecular dynamics (MD) simulations as an established tool to complement advanced theoretical approaches and difficult experimental studies.

Granular gases are the more dynamic limit of granular media as opposed to static systems. One of the outstanding effects in granular gases is the so called clustering, a self-stabilized density instability due to dissipation [5–7], where large, dense collections of particles co-exist with almost empty areas. Clustering occurs in initially homogeneous systems [5–10] and should not be confused with the so-called "inelastic collapse" [5, 11], the divergence of the collision rate, which is inherent to the frequently used hard (rigid) sphere model [5, 6, 12, 13]. Freely cooling systems – mainly examined numerically and theoretically – are almost impossible to realize experimentally. Only in the last years, laboratory experiments were performed, where clustering could be examined in driven systems [14–20], where also kinetic theory approaches [21–32] complemented by numerical simulations have proven to be successful [12, 25, 33–39].

The driving of a granular material can be realized by moving walls [2, 40] which lead to local heating and possibly shock-waves [41], or the system can alternatively be driven by a global homogeneous, random energy source in different variations [12, 25, 29, 36, 42–45]. The latter type of energy input does not exactly resemble the experimental situation, where a two-dimensional (2D) horizontal layer of spheres is agitated by vertical vibrations of the bottom surface and the horizontal degrees of freedom are indirectly agitated due to the different vertical jump heights of the colliding particles [15, 19, 20, 38]. Translational energy input (due to vibrations) was also applied for other boundary conditions and a variety of interesting experimental results were obtained just recently [14–17, 19, 20, 46].

Generic seems to be that one can obtain a gas and a liquid state, together with dense, solid-like clusters which form due to dissipation. Thus, the choice of the driving term to put into a theory for dissipative systems is an open problem and we expect that it depends on the nature of the driving (vibrating wall, airflow, Brownian noise, etc.). In this paper we present a more detailed and complete study as in [12, 13] and also combine both approaches in the framework of a mean field theory for the evolution to the steady state.

In [12], in order to find possible candidates for a realistic energy input, the inelastic hard sphere (IHS) model with an inhomogeneous, *multiplicative driving* was examined. The driving is proportional to the local velocity $|v(\vec{x})|^\delta$, with a given power δ which can take both positive and negative values. The classical homogeneous driving with $\delta = 0$ is contained in our approach as a special case. A positive power leads to weak energy input for slow particles, e.g. those moving collectively inside a cluster [15, 19, 38]. We present a series of results from numerical simulations performed for different sizes N of the system, different values of r and different exponents δ of the driving. Then we compare our results with an analytical MF approach.

A driving linearly proportional to $v(\vec{x})$ ($\delta = 1$) seems to reproduce some experimental observations which could not be reproduced by a homogeneous driving, like an anticipated onset of clustering with respect to the case of homogeneous driving, stable clustering in equilibrium with a low density (gas) phase, anomalous velocity distribution in the clustering regime with exponents different from the value 3/2 obtained analytically and numerically for the case with homogeneous driving [47]. Furthermore, we obtain that the system can be homogenized even for strong dissipation, if a driving inversely proportional to the velocity is used ($\delta < 0$).

Depending on the experimental setup, energy can be given to the translational degrees of freedom or, alternatively, to the rotational ones, or to both. Since the first case catched most of the attention, we change the focus and feed also rotational energy instead of translational. Driving of the rotational degrees of freedom is scarcely realized so far [M. Markus, private communication, A. Kudrolli, private communication, D. Wolf, private communication], and requires either electro- or magneto-dynamic forces in order to couple the rotational degrees of freedom. Such a situation could correspond, for example, to a gas of rough magnetic particles subject to a rapidly varying, homogeneous, magnetic field. Besides a possible experimental application, we believe that this study is interesting in itself. We examine the case of isolated rotational driving, since the correct modeling of the driving mechanism is of great importance for a theory of granular gases to describe realistic experimental situations. In the future, translational and rotational driving can also be combined, however, we focus only on the isolated cases here because this allows at least partially for analytical solutions.

The paper is organized as follows. Section 2 introduces and describes the model, and in Sect. 3 we the mean field (MF) equations for the model with multiplicative translational driving and homogeneous rotational driving are provided. In Sect. 4 we present numerical simulations for translationally driven smooth particles, and compare numerical results with analytical predictions. Section 5 is devoted to the comparison of simulations of rotationally driven particles with theory. A study of the velocity distributions, and in particular of its high energy tail follows in Sect. 6, and in Sect. 7 we draw the conclusions of our study and outline future research directions.

As a remark to our nomenclature. When we write steady state in the following, this has to be understood as non-equilibrium steady state (NESS) [36]. Furthermore, the word equilibrium means a state of the system, where energy-input and dissipation cancel each other so that energy-loss and energy-input rates are in "equilibrium"; we do not have an equilibrium in the classical sense here.

2 The Model

In this section the numerical model is introduced and a few typical examples of the simulation are presented in Fig. 1.

A system of N three-dimensional spheres with radius a and mass m is considered, interacting via a hard-core potential and confined to a 2D plane of linear extension L, with periodic boundary conditions. The degrees of freedom are the positions $\vec{r}_i(t)$, the translational velocities $\vec{v}_i(t)$, and the rotational velocities $\vec{\omega}_i(t)$ for each sphere numbered by $i = 1, \dots, N$.

dissipation in tangential direction is quantified by the (constant) tangential restitution r_t, such that $\vec{v}'^{(t)}_{\rm c} = -r_t \vec{v}^{(t)}_{\rm c}$.

The magnitude of dissipation is proportional to $1 - r^2$ (normal) and $1 - r_t^2$ (tangential), while the strength of the coupling between rotational and translational motion is connected to $1 + r_t$, where the (constant) normal restitution r can vary between 1 (elastic) and 0 (inelastic) and the tangential restitution r_t varies between -1 (smooth) and $+1$ (rough), corresponding to zero and maximum coupling, respectively [6, 9, 22, 48]. The details of more realistic contact models are discussed in other chapters in this book [31, 49].

2.2 Translational, Multiplicative Driving

The system is agitated each time interval $\Delta t = f_{\rm dr}^{-1}$, with a driving rate $f_{\rm dr}$ (in the experiment this is according to the frequency of the vibrating bottom). For homogeneous driving, it is costumary to assume $f_{\rm dr} \gg t_n^{-1}$, where t_n^{-1} is the collision frequency of the granular gas. This, however, does not correspond to the experiments, were $f_{\rm dr}$ can be rather small [15, 19, 20]. In our simulations we will use driving frequencies around $100\,{\rm s}^{-1}$ comparable to those used experimentally [15, 19]. We did numerical checks with strongly different values of $f_{\rm dr}$ and found a similar behavior of the system even for driving frequencies lower than, but of the same order as t_n^{-1}, provided that a stationary state is reached.

Every time interval Δt, the velocity of particle i is changed:

$$\begin{aligned} v'^x_i(t) &= v^x_i(t) + r^x_i |\vec{v}_i(t)|^\delta v_r^{1-\delta} \\ v'^y_i(t) &= v^y_i(t) + r^y_i |\vec{v}_i(t)|^\delta v_r^{1-\delta} \ , \end{aligned} \tag{1}$$

where the driving occurs at time t and the prime on the left hand side indicates the value after the driving event. v_r is a reference velocity (in this study we use $v_r = 1\,{\rm m\,s}^{-1}$) which allows to define the dimensionless translational particle temperature $T = E/(NT_r)$, with $E = (m/2)\sum_{i=1}^N \boldsymbol{v}_i^2$ and the reference temperature $T_r = mv_r^2$. The variance of the uncorrelated Gaussian random numbers r_i^x and r_i^y (with zero mean) can now be interpreted as a dimensionless driving temperature $T_{\rm dr}$.

The power δ is a parameter that takes into account the type of driving. In the special case of $\delta = 0$, one has random driving, as typically used in the literature. The random numbers are dimensionless and v_r provides the dimension and scale of the driving velocity. For $\delta > 0$, particles with large velocities obtain overproportional energy input, whereas in the case $\delta < 0$ slow particles obtain stronger energy input as compared to fast particles. The former case resembles the experimental situation in so far that particles with slow relative velocities presumably move in phase and thus obtain less energy input than particles that move out of phase.

The stochastic driving rule in (1) leads thus to an average rate of change of temperature

$$\Delta T/\Delta t = H_{\rm dr} T^{\delta} \quad \text{with} \quad H_{\rm dr} = f_{\rm dr} T_{\rm dr} \ . \tag{2}$$

In the next section we introduce this driving rate in the MF equation for the evolution of T [6], but first the case of rotational driving is discussed.

2.3 Rotational Driving

In the case of a rotational driving event, with frequency $f_{\rm dr}$, see above, the translational velocity remains unchanged, but the angular velocity ω_i of particle i is modified at each time of agitation t so that

$$\omega_i'(t) = \omega_i(t) + r_i^{\omega}\, \omega_0, \tag{3}$$

where the prime on the left hand side indicates the value after the driving event. Due to the two-dimensionality of the system, we apply the driving force only to the z-direction, so that the scalar ω is to be understood as the z-component of $\vec{\omega}$.

The reference angular velocity, ω_0, allows to define the dimensionless translational and rotational particle temperatures

$$T_{\rm tr} = E_{\rm tr}/(N T_0) \ , \tag{4}$$

and

$$T_{\rm rot} = 2 E_{\rm rot}/(N T_0) \ , \tag{5}$$

with the translational energy $E_{\rm tr} = (m/2) \sum_{i=1}^{N} \vec{v}_i^{\,2}$, the rotational energy $E_{\rm rot} = (q m a^2/2) \sum_{i=1}^{N} \vec{\omega}_i^2$ (with the moment of inertia prefactor $q = 2/5$ for $3D$ spheres), and the reference temperature $T_0 = m v_0^2$, with $v_0 = a\omega_0$. The variance of the uncorrelated Gaussian random numbers r_i^{ω} (with zero mean) can now be interpreted as a dimensionless driving temperature $T_{\rm dr}^{\omega}$ [13]. The stochastic driving leads thus to an average rate of change of temperature

$$\Delta T_{\rm rot}/\Delta t = J_{\rm dr} \ , \quad \text{with} \quad J_{\rm dr} = f_{\rm dr} T_{\rm dr}^{\omega} \ . \tag{6}$$

In the case of driving of the translational degrees of freedom, the reference temperature T_r will be used, whereas in the case of rotational driving, the reference T_0 will be used for scaling.

3 Mean Field Evolution Equations

The starting point for our mean-field analysis is the theory of Huthmann and Zippelius [9], for a freely cooling gas of infinitely rough particles, which was recently complemented by event driven (ED) simulations in 2D and 3D [6] and by studies of driven systems as well [12, 13]. The main outcome of this approach is a set of coupled evolution equations for the translational and rotational MF temperatures $T_{\rm tr}$ and $T_{\rm rot}$ [6, 9], which can be extended to describe arbitrary energy input (driving) [12, 13]. The equations were derived assuming uncorrelated collisions and a Maxwellian velocity distribution, together with the contact model assumptions as described above.

In the present study, given the random driving temperatures $T_{\rm dr}$ or $T_{\rm dr}^{\omega}$ and an energy input rate $f_{\rm dr}$, as defined above, one just has to add the positive rate of change of translational energy $H_{\rm dr}$ and rotational energy $J_{\rm dr}$ to the system of equations:

$$\frac{d}{dt}T_{\rm tr}(t) = G\left[-AT_{\rm tr}^{3/2} + BT_{\rm tr}^{1/2}T_{\rm rot}\right] + H_{\rm dr}T^{\delta} \ , \tag{7}$$

$$\frac{d}{dt}T_{\rm rot}(t) = 2G\left[BT_{\rm tr}^{3/2} - CT_{\rm tr}^{1/2}T_{\rm rot}\right] + J_{\rm dr} \ , \tag{8}$$

with

$$G = \frac{8\nu g(\nu)}{a\sqrt{\pi m}} \ , \tag{9}$$

and the pair correlation function at contact $g(\nu) = (1 - 7\nu/16)/(1 - \nu)^2$ in the approximation proposed by Henderson and Verlet&Levesque [50, 51], dependent only on the volume fraction of the granular gas $\nu = \pi a^2 N/V$. The constant coefficients in (7) and (8) are

$$A = (1 - r^2)/4 + \eta(1 - \eta)/2 \ ,$$

$$B = \eta^2/(2q) \ , \text{ and} \tag{10}$$

$$C = \eta\left(1 - \eta/q\right)/(2q) \ , \tag{11}$$

with the abbreviation $\eta = \eta(r_t) = q(1 + r_t)/(2q + 2)$, as derived and used in [6, 9].

Typical steady-state configurations for translational driving for different r values are shown in Fig. 1a–c, and a snapshot for rotational driving is shown in Fig. 1d.

3.1 Smooth Particles – No Rotation

The time evolution equation of $T = T_{\rm tr}$ was derived for the case of a freely cooling granular gas by means of a pseudo-Liouville operator formalism [6, 9]. We adopt the nomenclature and account for the driving by adding (2) to the mean field (MF) equation for the translational degree of freedom

$$\frac{d}{dt}T(t) = -G_r A T^{3/2} + H_{\rm dr}T^{\delta} \ . \tag{12}$$

For our case of a homogeneous monolayer of smooth spheres, one has $G_r = 8an\sqrt{\pi T_r/m}g(\nu)$, and $A = (1-r^2)/4$, with the number density $n = N/V$, the pair correlation function at contact $g(\nu)$, and the area fraction ν covered by particles [6, 9, 12]. For $\delta = 0$ the driving is homogeneous and independent of the local granular temperature (or velocity), and one can identify our energy input rate with the term $m\xi_0^2$ in [47]. In the case $\delta \neq 0$ the driving is a function of T. Imposing $\frac{d}{dt}T(t) = 0$ one gets from (12) the MF temperature in the steady state

$$T^{\mathrm{mf}} = \left(\frac{H_{\mathrm{dr}}}{G_r A}\right)^{2/(3-2\delta)} , \tag{13}$$

the generalization of the Enskog equilibrium solution for a homogeneously driven granular gas [47]. The scaling exponent of T^{mf} in (13) is 2/3 for $\delta = 0$, while it is 2 for $\delta = 1$. For $\delta \geq 3/2$ the MF theory does not admit a stable equilibrium state. If $\delta > 3/2$ the driving rate grows faster than the dissipation rate. $\delta = 3/2$ is a limiting case for which the equilibrium state is unstable against density fluctuations. A more detailed stability analysis is far from the scope of this paper, since the typical perturbation around the steady state often also relies on a Maxwellian velocity distribution, which we do not find numerically.

The final approach to the steady state can be obtained by linearizing (12) around T^{mf}, what leads to an exponentially fast approach to equilibrium

$$T^{\mathrm{mf}} - T(t + t_0) = T(t_0) \exp\{-[3At_n^{-1} + \delta H_{\mathrm{dr}}(T^{\mathrm{mf}})^{\delta-1}]t\} , \tag{14}$$

where $t_n^{-1} = G_r\sqrt{T^{\mathrm{mf}}}/2$ is the Enskog collision frequency for elastic particles with temperature T^{mf}, and $T(t_0)$ is the initial temperature at time t_0. By inserting (13) in the expression for t_n^{-1}, one can express the characteristic relaxation time $t_{\mathrm{relax}} = [\ldots]^{-1}$ as a function of the model parameters, which reduces for $\delta = 1$ to $t_{\mathrm{relax}}^{-1} = (5/2)H_{\mathrm{dr}}$. Thus, for $\delta = 1$ the characteristic time $t_{\mathrm{relax}} = [3At_n^{-1} + \delta H_{\mathrm{dr}}(T^{\mathrm{mf}})^{\delta-1}]^{-1}$ for the evolution of T towards its equilibrium value does *not* depend on A which contains all the information about the inelasticity. This characteristics is confirmed by numerical simulations (see Fig. 3 below).

3.2 Rough Particles with Translational Driving

Setting to zero the temporal derivatives in (7) and (8), one obtains the steady state properties of the driven system with $J_{\mathrm{dr}} = 0$ and, for the sake of simplicity, $\delta = 0$:

$$T_{\mathrm{rot}}^{\mathrm{mf}} = \left(\frac{H_{\mathrm{dr}}}{G\Gamma_{\mathrm{tr}}}\right)^{2/3} , \quad \text{and} \quad T_{\mathrm{tr}}^{\mathrm{mf}} = T_{\mathrm{rot}}^{\mathrm{mf}}/R , \tag{15}$$

with $\Gamma_{\mathrm{tr}} = \left(C/B^3\right)^{1/2}\left(CA - B^2\right)$, and $R = B/C$.

Starting from this mean field result for both translational and rotational degrees of freedom, an analysis similar to the one in the previous subsection is possible. Since this did not lead to dramatically new insights, we do not discuss it further. Note, however, that the relaxation times for the translational and rotational degrees of freedom can be strongly different from each other [6].

3.3 Rough Particles with Rotational Driving

Setting to zero the temporal derivatives in (7) and (8), one obtains the steady state properties of the rotationally driven system with $H_{\rm dr} = 0$:

$$T_{\rm rot}^{\rm mf} = \left(\frac{J_{\rm dr}}{G\Gamma_{\rm rot}} \right)^{2/3} , \quad \text{and} \quad T_{\rm tr}^{\rm mf} = T_{\rm rot}^{\rm mf}/R , \tag{16}$$

with $\Gamma_{\rm rot} = 2\left(B/A^3\right)^{1/2}\left(CA - B^2\right)$, and $R = A/B$. A more detailed analysis of this quasi steady state mean field solution, the approach to the steady state, the parameter dependencies, and the velocity distributions is to be published elsewhere.

4 Numerical Simulations

Most of our event driven (ED) molecular dynamics simulations, see [6, 9, 12, 13] for details, with the driving specified above, are first equilibrated without driving and with elastic interactions ($r = 1$, and $r_t = -1$), until the velocity distribution is close to a Maxwellian. Then, dissipation and driving are switched on. However, we checked that the steady state does not depend on the initial conditions. The simulations are performed at a volume fraction $\nu = 0.34$ with $N = 1089$ ($f_{\rm dr} = 133\,{\rm s}^{-1}$) or $N = 11025$ ($f_{\rm dr} = 67\,{\rm s}^{-1}$), and different values of r and r_t. (In our simulations with translational driving, we have chosen $a = 10^{-3}\,{\rm m}$ and $H_{\rm dr} = 1.0\,{\rm s}^{-1}$ so that, for example, $G_r = 8\nu v_r/\sqrt{\pi} a g(\nu) = 3.1 \times 10^3\,{\rm s}^{-1}$, $T^{\rm mf} = 0.0358$ and thus $t_n^{-1} = 2.9 \times 10^2\,{\rm s}^{-1}$, if $r = 0.90$ is used.) With these typical values and a homogeneous driving ($\delta = 0$), the model of a driven granular gas is very close to a homogeneous state; no clusters are observed and the velocity distribution is almost Maxwellian. This is in contradiction to the experimental findings [15, 19], and suggests that the correct representation of the driving in those experiments is *not the homogeneous white noise* usually implemented.

4.1 Approach to Steady State

In Figs. 2a–d we compare the solution of the MF equation (13) for a multiplicative driving with $\delta = 1$ to numerical simulations with $N = 11025$, $\nu = 0.34$, and different $r = 0.99$, 0.995, 0.997, 0.998. As one can see, the

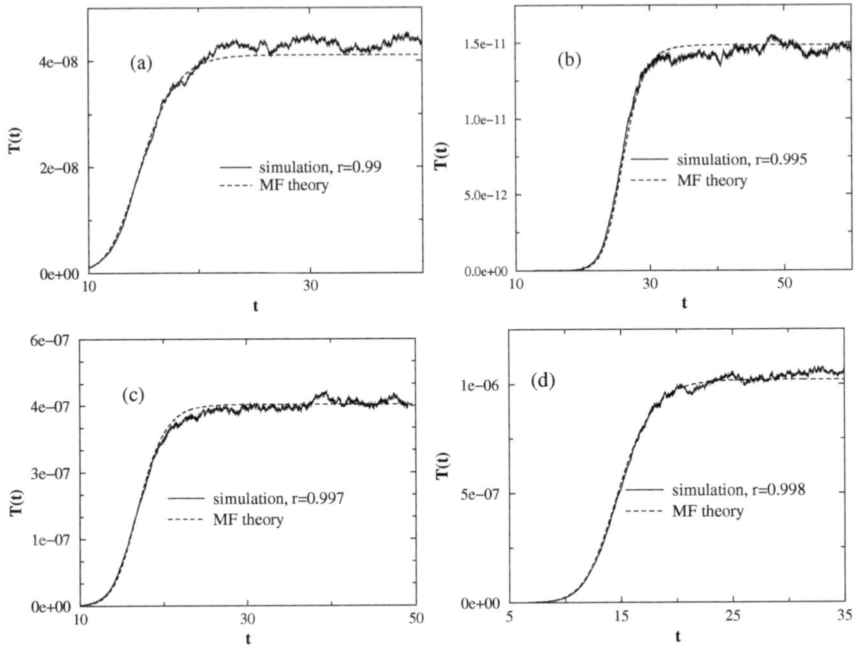

Fig. 2. a–d Comparison between the solution of the MF equation (13) and numerical simulations for $N = 11025$, $\nu = 0.34$, $\delta = 1$ and $r = 0.99$ **a**, $r = 0.995$ **b**, $r = 0.997$ **c** $r = 0.998$ **d**

transient dynamics is quite well reproduced by the MF theory, in the limit of low dissipation (quasi-elastic) considered in the simulations. In the steady state, intermittent behavior is evidenced, however, we do not present a more detailed statistical analysis here.

4.2 Relaxation Time

In Fig. 3 we check another prediction of the MF theory, i.e. the independence of the characteristic time t_{relax} of the near-to-equilibrium relaxation to the steady state, from the parameter $A = (1 - r^2)/4$ which contains inelasticity. In the figure we compare the normalized MF temperature $T(t)/T^{\text{mf}}$ and the normalized numerical temperature $T(t)/T^{\text{eq}}$ computed for $r = 0.99$, 0.995, 0.997, 0.998. Because of noise in the initial conditions and dynamics with respect to the MF steady state, the simulation time t has to be translated by an offset time t_0 (with a different value of t_0 for each simulation) in order to collapse the data. The fact that the numerical data collapse onto the MF curve, supports the MF prediction of the independence of the characteristic time from inelasticity in the case $\delta = 1$.

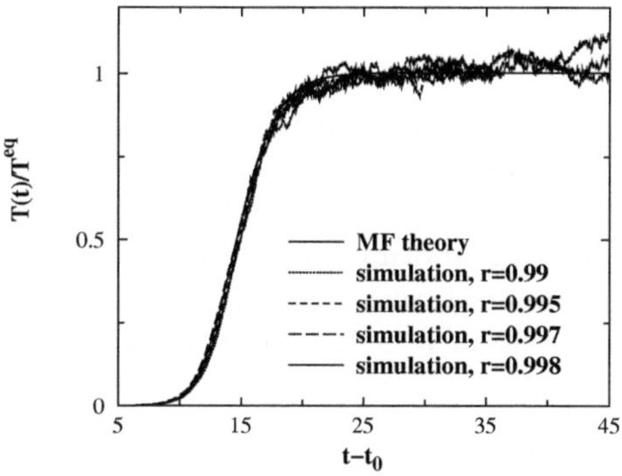

Fig. 3. Comparison between the approach to equilibrium of the normalized MF Temperature $T(t)/T^{\mathrm{mf}}$ and the normalized numerical values $T(t)/T^{\mathrm{eq}}$ for $N = 11025$, $\nu = 0.34$, $\delta = 1$ and $r = 0.99, 0.995, 0.997, 0.998$. The numerical data collapse onto the MF curve, supporting the MF previsions of the independence of the characteristic time t_{relax} from inelasticity in the case $\delta = 1$

4.3 Steady State Temperature

In Fig. 4 we plot the ratio between the numerical results for long times T^{eq} and the theoretical equilibrium temperatures T^{mf} as function of r for different δ. The agreement of the simulations with the MF prediction is optimal for $r \to 1$. For $\delta < 0$, the range of agreement extends to much smaller r values, i.e. to stronger dissipation, as in the case of $\delta = 1$ and even in the case $\delta = 0$. One could naively think that the more negative is δ the more favorized should be the homogeneous state. In fact, this is not true. We have performed simulations with $\delta = -1$, and found that the driving is very singular in the low velocity limit, since an excessive amount of energy is given to slow particles.

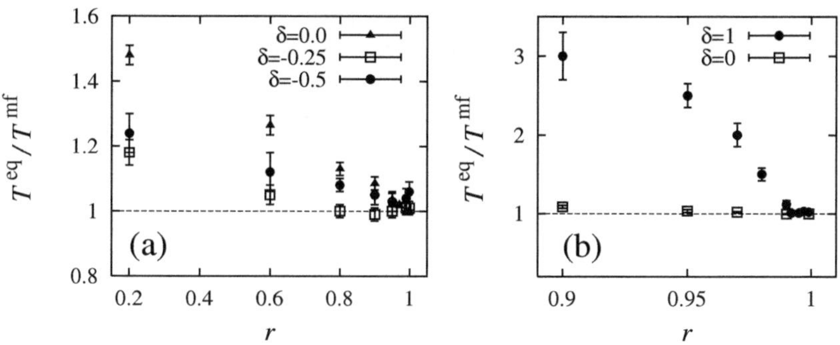

Fig. 4. Rescaled translational temperature $T^{\mathrm{eq}}/T^{\mathrm{mf}}$ plotted against the restitution coefficient r for $N = 11025$ and different values of δ as given in the insets. Note the different axis scaling in **a** and **b**

4.4 System Size Dependence

In Fig. 5, the ratio between equilibrium temperature from simulations and MF temperature from the theory, $T^{\rm eq}/T^{\rm mf}$, is plotted against r for two system sizes $N = 11025$ and 5476. The deviation from unity, i.e. the discrepancy between theoretical prediction and numerical simulation of the model, is somewhat larger for the smaller system. For both system sizes, the quality of the theory is perfect in the limit $r \to 1$, within the statistical fluctuations.

4.5 Steady State Clustering

In Fig. 1a–c snapshots of the system's steady state were shown for $r = 0.999$ (a) $r = 0.97$ (b) and $r = 0.6$ (c), for the more interesting case $\delta = 1$, rather than for the classical $\delta = 0$. Different regimes are observed: A homogeneous state exists for very weak dissipation ($r = 0.999$, Fig. 1a, whereas dense, persistent clusters with a crystalline structure, domain boundaries, and vacancies are found for higher dissipation ($r = 0.97$). The region between these clusters appears rather homogeneous and dilute (gas-like), very similar to the structures observed in experiments [15, 19]. For higher dissipation ($r = 0.6$) the clusters appear less symmetric, are smaller, and evolve more dynamically.

Note that the clustering in the case $\delta = 1$ is qualitatively different from the case of homogeneous driving, and it appears already for quite high values of r (see Fig. 1). Homogeneous driving, in fact, leads to transient clusters, i.e. they appear and disappear continuously, while in the experiments [15, 19] and in the case of multiplicative driving, the individual clusters are in equilibrium with a gas phase and are stable for rather long times. Simulations with negative δ, (we used $\delta = -0.5$ and $\delta = -0.25$) give a behavior qualitatively similar to the case of homogeneous driving ($\delta = 0$). For $\delta = 1$, the particles inside the cluster, with rather small relative velocities, are much less agitated than the particles in the surrounding gas phase, so that a cluster is stable.

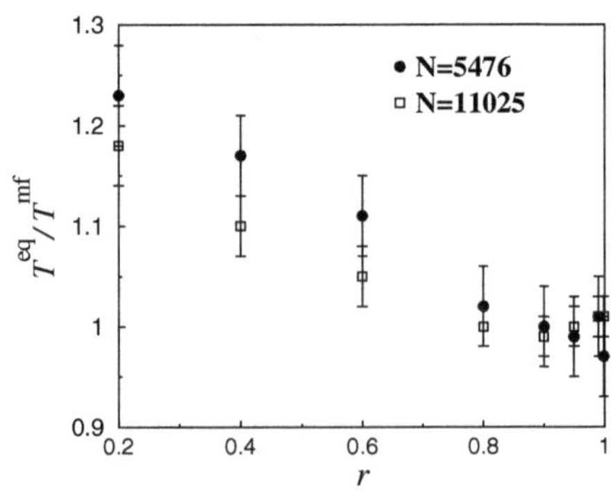

Fig. 5. Rescaled translational temperature $T^{\rm eq}/T^{\rm mf}$ plotted against the restitution coefficient r for two values $N = 11025$, 5476, and $\delta = -0.25$

For $\delta \leq 0$, the particles in the cluster are driven comparatively strong, what leads to less stable, dynamic clusters.

5 Simulations with Rotation

In Fig. 6 we present the stationary (steady-state) values of T_{rot}, normalized by the MF value $T_{\text{rot}}^{\text{mf}}(r = 0)$, and of the ratios $R = T_{\text{rot}}/T_{\text{tr}}$, as obtained from numerical simulations and theory for a system of $N = 11025$ particles, with volume fraction $\nu = 0.34$, $r_t = 1$, and r ranging from 0.99 to 10^{-4}. Surprisingly, the agreement with the MF prediction is very good, even *for the lowest value* $r = 10^{-4}$ of the normal restitution, which corresponds to very strong dissipation, where the deviation from MF theory is of the order of only 10%.

To give an example, if the system is driven on the translational degrees of freedom, the stationary temperatures show deviations of $30 - 40\%$ from

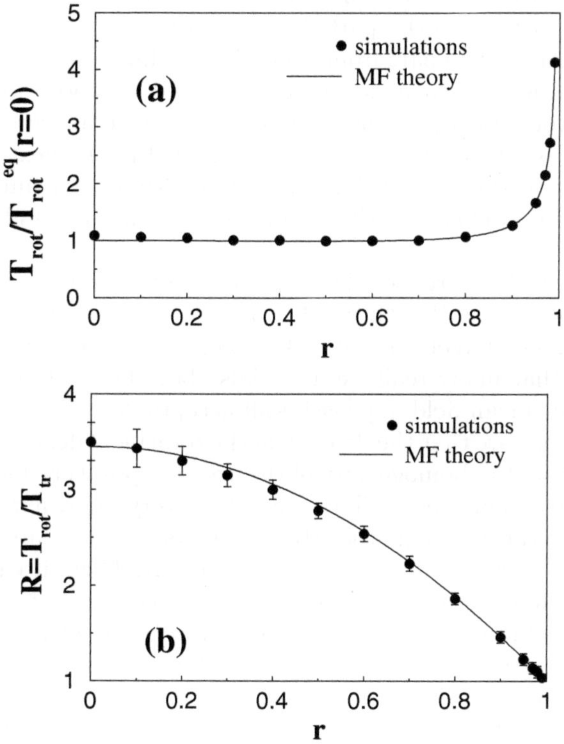

Fig. 6. Simulation (*points*) and theory (*lines*) for the parameters $\nu = 0.34$, $N = 11025$, and $r_t = 1$, plotted against r. **a** Stationary rotational temperature T_{rot}, normalized by the MF value $T_{\text{rot}}^{\text{mf}}(r = 0)$ at $r = 0$. **b** Ratio of stationary rotational and translational temperature $R = T_{\text{rot}}/T_{\text{tr}}$

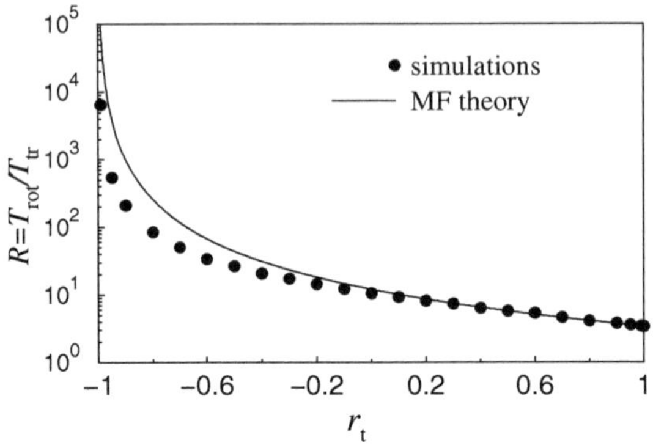

Fig. 7. Simulation (*points*) and theory (*lines*) results for $R = T_{\text{rot}}/T_{\text{tr}}$, with parameters $\nu = 0.34$, $N = 11025$, and $r = 0.1$, plotted against r_t

MF predictions already for $r = 0.6$, see [12] and the previous section. The snapshot in Fig. 1d shows the particle distribution for $r = 0.1$ and appears spatially homogeneous – apart from small density fluctuations not quantified here. Thus, rotational velocities are characterized by good homogenization at low r; however, also the translational velocity distribution shows strong deviations from a Maxwellian as will be quantified in the next section. This deviation is due to the rather high dissipation. Numerical simulations with $r = 0.99$ give a Maxwellian distribution for both rotational and translational velocities.

In order to check the role of the tangential restitution, we show in Fig. 7 the stationary values of R with $r = 0.1$ and $r_t \in [-1, 1]$. While for positive r_t there is still good agreement with MF theory, strong deviations appear as $r_t \to -1$. Note that many realistic materials obey the relation $r_t \approx 0.4$ [52], what renders our mean field approach still acceptable.

Our conclusions are that the driving on the rotational degrees of freedom is able to keep the spatial homogeneity of the system up to very high dissipation rates, for positive values of r_t. This leads to a very good agreement of the stationary temperatures with the MF predictions. There are two possible reasons for this. First, the driving acts on rotations. Then, it cannot favorize collisions, since it does not increase the normal component of the relative velocity of the colliding particles. Second, the increase of rotational energy triggered by the driving leads to a shearing force between particles, which reduces density fluctuations and should destroy the translational velocity correlations - but astonishingly does not. When $r_t \to -1$, i.e. in the smooth limit, the agreement with the MF is lost. To explain this result one has to remember that $1 + r_t$ is a measure for the strength of the coupling. Not enough rotational energy is transferred to the translational degree, so that the randomization on collisions does not take place. Thus, it is not surprising that MF is no more valid in this very singular limit. Snapshots of the particle distribution for $r = 0.1$ and r_t near to -1 (not displayed here) show indeed stronger density fluctuations in the system as reported in Fig. 1d.

6 Analytical Study of the Velocity Distribution

In order to study analytically the shape of the velocity distribution function, we extend the approach used by van Noije et al. [36, 37] in the special case $\delta = 0$ to arbitrary δ. The Enskog-Boltzmann equation for the freely cooling gas of spheres, with the standard collision integral $I(f, f)$, has to be extended by the multiplicative driving term proportional to the particle velocity. The Enskog-Boltzmann equation, see Eq. (26) in [36, 37], corrected by a Fokker-Planck diffusion term, is

$$\frac{\partial}{\partial t} f(\vec{v}_1, t) = g(\nu) I(f, f) + \frac{H_{\rm dr}}{2m} \frac{\partial^2}{\partial \vec{v}_1^2} \left[v_1^{2\delta} f(\vec{v}_1, t) \right] . \tag{17}$$

Considering the stationary limit $\partial f(\vec{v}_1, t)/\partial t = 0$, and introducing a scaled distribution function, $f(\vec{v}) = (n/v_0^2)\, \tilde{f}(c)$, with the dimensionless velocity $c = v/v_0$ and the mean thermal velocity v_0, one obtains a dimensionless evolution equation for the scaled $f(\vec{v})$, as in [36, 37]. Multiplied by c_1^p and integrated over $\vec{c}_1$, this leads to the following set of equations which couple the moment $\langle c^{p-2+2\delta} \rangle$ to the p-th moment μ_p of the collision term:

$$\frac{H_{\rm dr}}{2 v_0^3 g(\nu) n a} p^2 \left\langle c^{p-2+2\delta} \right\rangle = \mu_p . \tag{18}$$

In the special case $\delta = 1$ and $p = 2$, one obtains (for dimensions $d = 2$)

$$\frac{H_{\rm dr}^{1-\delta}}{2 v_0^3 g(\nu) n a} 2 \langle c^2 \rangle = \mu_2 . \tag{19}$$

and observing that $\langle c^2 \rangle = v_0^2$ we obtain the stationary value of the thermal velocity in terms of μ_2.

$$v_0 = H_{\rm dr}/(g(\nu) \mu_2 a n). \tag{20}$$

For $\delta \neq 1$ and $\delta \neq 0$ the MF thermal velocity can be obtained by assuming

$$\langle c^{2\delta} \rangle = [\langle c^2 \rangle]^\delta = v_0^{2\delta}. \tag{21}$$

This step is certainly valid for $\delta = 0$ and $\delta = 1$. For real values of δ, we see *a posteriori* that the step leads to the correct MF theory (see previous section), but we have no clear justification for it. If this assumption, as it seems, turns out to be true, it could probably have interesting implications.

For high c, the rescaled collision term can be approximated by

$$\tilde{I}(\tilde{f}, \tilde{f}) \approx -\beta_1 c_1 \tilde{f}(c_1), \tag{22}$$

with $\beta_1 = \pi^{1/2}$ in 2D and the equation for $f(c)$ becomes

$$-\beta_1 c\tilde{f}(c) + \frac{\mu_2}{2d}\left(\frac{\mathrm{d}^2}{\mathrm{d}c^2} + \frac{d-1}{c}\frac{\mathrm{d}}{\mathrm{d}c}\right)\left[c^{2\delta}\tilde{f}(c)\right] = 0 \ . \tag{23}$$

Inserting a solution of the form $\tilde{f}(c) \propto \exp(-Bc^\alpha)$, we obtain the large c solution $\alpha = \frac{3-2\delta}{2}$ and $B = \frac{2}{3-2\delta}\sqrt{2d\beta_1/\mu_2}$. In particular, for $\delta = 1$ we find $\alpha = 1/2$ what corresponds to the exponent $\alpha = 0.49(3)$ we find in the low r limit (see Fig. 8). Our explanation is that there is a crossover velocity c^* above which the tail with exponent 1/2 appears, which decreases as r decreases. Such a tail is not observed numerically for high r values, since c^* is outside the velocity range we can examine within reasonable averaging time.

6.1 Translational Driving with $\delta = 1$

In the following, we focus on the steady state velocity distribution. Therefore, we performed a large series of simulations for different values of r, with $\nu = 0.34$, $\delta = 1$, and both $N = 1089$ and $N = 11025$. For each value of r we performed about 10^3 simulations for $N = 1089$ and about 200 simulations for $N = 11025$, each with different initial configurations. The distribution of velocity is symmetric and isotropic so that we present only the data for the x component.

For data analysis, a Gaussian with the width obtained from the numerical simulations can be superposed to the velocity distribution function $f(v_x)$ in order to visualize the differences. Furthermore, as a more quantitative approach, a three parameter fit function $f_\alpha(v_x) = f_0 \exp(-B|v_x - \langle v_x\rangle|^\alpha)$ is used to estimate the exponent α of the tail of the distribution. The results are reported in Fig. 8, where a representative velocity distribution is shown in the inset. The main outcome of our simulations is that the velocity distribution is not Gaussian for high inelasticity and that the exponent α in the stretched exponential depends on the normal restitution r. The non-Gaussian behavior is limited to the tail for large r, but seems to extend over the whole range of velocities if r is small enough.

In Fig. 8, the exponent α, for multiplicative driving with $\delta = 1$, varies continuously between 0.49(3) and 1.99(3) and, in particular, for $N = 1089$ and $r = 0.9$ we get $\alpha = 1.05(4)$, while for $N = 11025$ and $r = 0.95$ we observe $\alpha = 1.02(3)$. Such an almost exponential velocity distribution has recently been observed in experiments on a vibrated monolayer of spheres, together with the cluster structure discussed above [15, 19, 20].

6.2 Multiplicative Driving with Non-integer Values of δ

In Fig. 9, for comparison with the results obtained with $\delta = 1$ we show the values of the exponent α, for multiplicative driving with $\delta = 0.5$. The

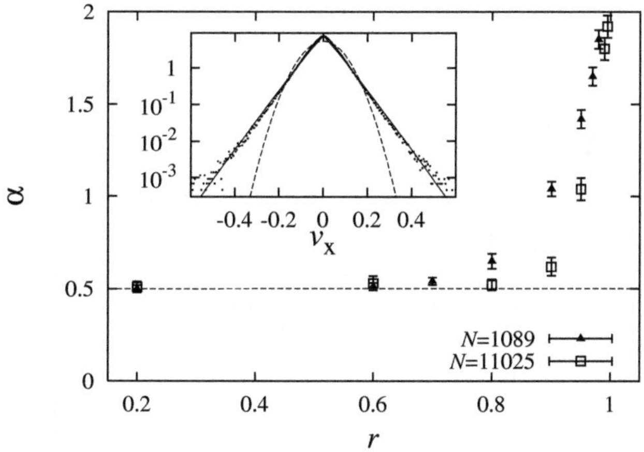

Fig. 8. Plot of the exponent α as a function of r, for $\delta = 1$ and system sizes $N = 1089$ (*solid triangles*) and $N = 11025$ (*open squares*). In the inset, the *dots* give the averaged, normalized velocity distribution from simulations with $N = 1089$ and $r = 0.9$, the *solid line* is a fit according to $f_\alpha(v_x)$ with $f_0 = 7.6$, $B = 18.9$ and $\alpha = 1.05(9)$, and the *dashed line* is a Gaussian with the same standard deviation as the data

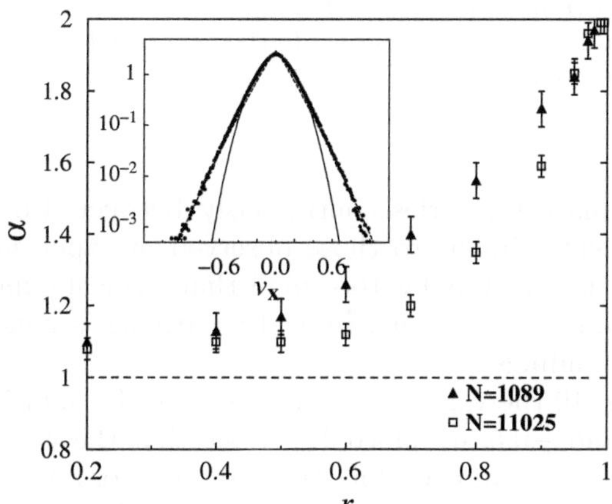

Fig. 9. Plot of the exponent α as a function of r, for $\delta = 0.5$ and system sizes $N = 1089$ (*solid triangles*) and $N = 11025$ (*open squares*). In the inset, the *dots* give the averaged, normalized velocity distribution from simulations with $N = 1089$ and $r = 0.6$, the *solid line* is a fit according to $f_\alpha(v_x)$ with $f_0 = 3.1$, $B = 1.1$ and $\alpha = 1.12(5)$, and the *dashed line* is a Gaussian with the same standard deviation as the data

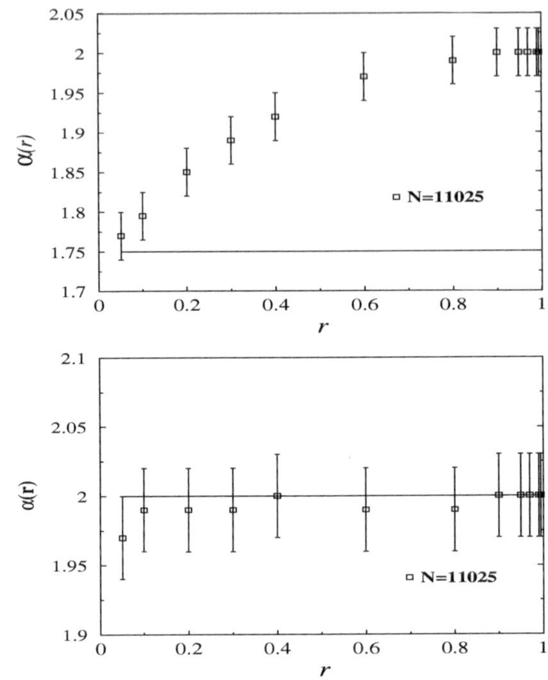

Fig. 10. Plot of the exponent α as a function of r, for system size $N = 11025$, and multiplicative driving with $\delta = -0.25$ **a**, and $\delta = -0.5$ **b**. The *symbols* give the simulation results, the *error-bars* denote the uncertainty in the fit, and the *line* gives the theoretical prediction

exponent α, in this case, varies continuously between 1.08(3) and 1.99(3) and to get exponents similar to those observed in experiments we have to go down to $r = 0.6$, which is far too small than typical experimental values ($r \sim 0.9$). The value $\delta = 1$ for multiplicative driving, seems then to best fit the experimental findings.

Finally, in Fig. 10 the values of the exponent α, for multiplicative driving with $\delta = -0.25$ and -0.5 are plotted. We see that the velocity distribution is very near to a Gaussian even for strong dissipation, as one would expect from the good agreement of mean field theory and simulations even for small r. For $\delta = -0.5$ the distribution is Gaussian for all value of r.

Simulations with $\delta = -0.5$, -0.25, and 0.5 (see Figs. 9 and 10) and small r ($r = 0.05$ for $\delta = -0.5$, -0.25 and $r = 0.2$ for $\delta = 0.5$) give 1.99(3), 1.77(3), and 1.08(3), quite near to the values 2.0, 7/4, and 1, given by the large c analysis. It interesting to note that for $\delta = -0.5$ the theory predicts $\alpha = 2.0$ for all values of r, i.e. some Gaussian distribution. Our numerical simulations support this result. This does not mean, however, that MF is valid everywere. In fact, we observe both for $\delta = -0.5$ and for $\delta = -0.25$ that the velocity distribution has an anomalously large variance for very low r, so for $\delta = -0.5$ the distribution is a Gaussian, in the sense that is it quadratic in the velocity, but it is not a Maxwellian, since it has anomalously large fluctuation, which do not correspond to the MF predictions. This explains the small, but not negligible, deviations from MF the one observes for low r in Fig. 4a. Furthermore, we see from Fig. 4 that a discrepancy between simulations and MF exists even for $\delta = -0.5$. Our explanation is that the driving with $\delta = -0.5$ is such that it preserves the Gaussian shape of the velocity distribution but not its parameters. The velocity fluctuations, i.e. the temperature fluctuations, are anomalously large with respect to the mean.

The behavior at low and intermediate velocity and the effect of spatial inhomogeneities are more difficult to analyze theoretically. In fact, one should solve the Fokker-Planck equation with the full collision integral taking into account the inhomogeneities – maybe an unachievable task. However the following heuristic argument can give an insight in the origin of the non-universal behavior of the velocity distribution at intermediate velocities. Since the driving depends on the velocity, it acts on all the velocity scales, affecting the shape of $f(v)$ in a more complex way than in the case of homogeneous driving. Moreover, when inhomogeneities are present, the driving will depend on the local thermal velocity $v_0(\vec{x})$ *and* on the local density $n(\vec{x})$, while homogeneous driving is *identical everywhere.* Since inhomogeneities are generated by the dissipation, a *feed-back* between driving and dissipation could be at the origin of the non-universal behavior of $f(v)$.

6.3 Rotations and Their Distribution

In Fig. 11, we show the stationary rotational and translational velocity distributions for $r = 0.1$, with the other parameters as above. The rotational velocity distribution is very near to a Maxwellian. A three parameter fit $f(x) = A\exp(-B|(x - \langle x\rangle)/\sigma|^\alpha)$, where $\sigma = \left(\langle(x - \langle x\rangle)^2\rangle\right)^{1/2}$, and x either equals ω or v, is plotted as dashed line in Fig. 11. The parameters $\langle\omega\rangle$, $\langle v\rangle$ and σ are taken from the simulations, and the fit gives $\alpha = 1.92(6)$ for ω, while we obtain $\alpha = 1.41(6)$ for v. This last value is quite near to the value 3/2 obtained theoretically in [36]. The applicability of the approach of [12, 36] to the present case, however, has to be discussed since originally a translationally driven, granular gas of smooth particles was considered.

7 Summary and Conclusions

We presented a study of a granular gas subject to both translational driving *proportional* to a power δ of the local velocity and to rotational driving (independent of the angular velocity). Numerical simulations and a theoretical analysis of this model for positive values of δ reproduce qualitatively some

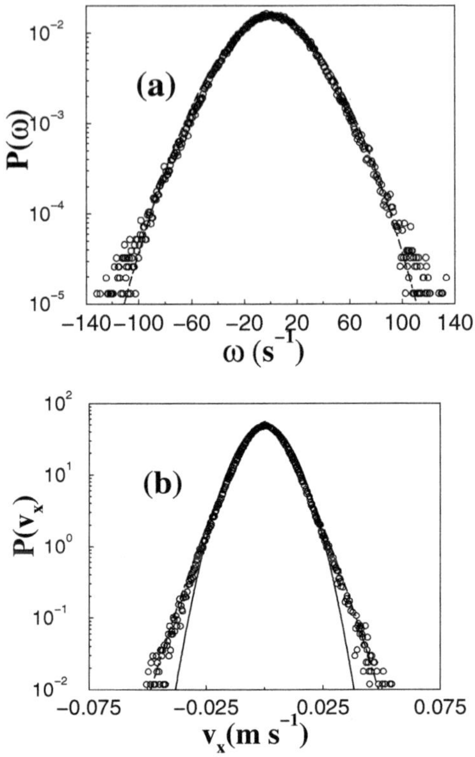

Fig. 11. Steady state rotational **a** and translational (b) velocity distributions for $N = 11025$, $\nu = 0.34$, $r_t = 1.0$ and $r = 0.1$. A power law fit (*dashed line*) gives an exponent $\alpha = 1.92(6)$ for the rotational distribution and $\alpha = 1.41(6)$ for the translational distribution (see text for details). For comparison, a Maxwellian (*solid line*) is plotted in **b**

experimental findings of [15, 19], which could not be accounted for by homogeneous driving ($\delta = 0$). Furthermore, numerical simulations and theory for negative values of δ and for rotational driving show good homogeneization of the system up to very low values of the restitution coefficient, r, i.e. very strong dissipation.

A microscopic justification for the multiplicative driving is still lacking and requires more detailed experiments or three-dimensional simulation [38] studies. However, as suggested in [12], there is a possible experimental check of our ideas. From MF theory the scaling of the equilibrium temperature of the vertically vibrated gas with the area fraction ν is given by $T^{\mathrm{mf}} \propto [\nu g(\nu)]^{-2/(3-2\delta)}$. Experimental measurement of the equilibrium temperature of the vertically vibrated monolayer for different values of ν could allow to estimate the value of δ and to verify the hypothesis of a multiplicative effective driving.

Another result of the present study is that both the driving of the rotational degrees of freedom and a multiplicative driving with negative δ are able to keep the spatial homogeneity of the system up to very high dissipation rates – at least for positive values of r_t, i.e. strong coupling between rotational and translational degrees of freedom. This leads to a very good agreement of the stationary temperatures with the MF predictions.

Given the good agreement between MF prediction and simulations, this indicates that deviations from a Maxwellian are not necessarily related to clustering. Moreover, one can have a good agreement of the second moment (the temperature) of the velocity distribution with MF theory together with a non Maxwellian velocity distribution. This poses a theoretical challenge, since some theories for translationally driven granular gases, relate clustering to fat tails in the velocity distribution. The reason why clusters do not occur in our situation, possibly due to the fact that vortices in their early stage are destroyed by the rotational driving, is another issue for future studies.

A possible setup for an experiment with rotational driving is the following: Each extremely rough granular sphere contains a small magnetic bar (to reduce the effect of dipole-dipole interaction at collision). The plane on which the spheres move should be extremely smooth, in order to avoid energy dissipation. Then, spatially homogeneous magnetic pulses periodically spaced in time can be applied in the horizontal directions. This would be the magnetic analogon of the oscillating plane. If the magnetic field is really spatially homogeneous, the magnetic dipoles of the spheres will receive angular momentum from the field, so only rotations are driven, and this angular momentum will be "randomized" by the collisions, if they are frequent enough. To reach a steady-state, it is necessary to give an initial translational velocity to the particles. We are aware that such an experiment is extremely difficult to realize, but a similar setup seems already operational in Dortmund [M. Markus, private communication]. Another reason to look at rotational driving via magnetic forces is the recent interest in electrostatically driven granular media [53] and in magnetic particles with dipolar interactions [A. Kudrolli, private communication, D. Wolf, private communication].

Acknowledgements

R.C. and H.J.H. acknowledge financial support under the European network project FMRXCT980183; S.L. and H.J.H. acknowledge funding from the Deutsche Forschungsgemeinschaft (DFG). We thank M.A.+Muñoz, E. Trizac, and J.S. Urbach for helpful discussions.

References

1. H. M. Jaeger, S. R. Nagel, and R. P. Behringer. Granular solids, liquids, and gases. *Reviews of Modern Physics*, 68(4):1259–1273, 1996.
2. H. J. Herrmann, J.-P. Hovi, and S. Luding, editors. *Physics of dry granular media - NATO ASI Series E 350*, Dordrecht, 1998. Kluwer Academic Publishers.
3. T. Pöschel and S. Luding, editors. *Granular Gases*, Berlin, 2001. Springer. Lecture Notes in Physics 564.
4. K. R. Mecke and D. Stoyan, editors. *Statistical Physics and Spatial Statistics*, Berlin, 2000. Springer.
5. I. Goldhirsch and G. Zanetti. Clustering instability in dissipative gases. *Phys. Rev. Lett.*, 70(11):1619–1622, 1993.
6. S. Luding, M. Huthmann, S. McNamara, and A. Zippelius. Homogeneous cooling of rough dissipative particles: Theory and simulations. *Phys. Rev. E*, 58:3416–3425, 1998.
7. S. Luding. Structures and non-equilibrium dynamics in granular media. *Comptes Rendus Academie des Science*, 3:153–161, 2002.
8. P. K. Haff. Grain flow as a fluid-mechanical phenomenon. *J. Fluid Mech.*, 134:401–430, 1983.
9. M. Huthmann and A. Zippelius. Dynamics of inelastically colliding rough spheres: Relaxation of translational and rotational energy. *Phys. Rev. E*, 56(6):6275–6278, 1998.
10. E. Ben-Naim and P. Krapivsky. The inelastic maxwell model. In Pöschel and Brilliantov [54], page 65.
11. S. McNamara and W. R. Young. Dynamics of a freely evolving, two-dimensional granular medium. *Phys. Rev. E*, 53(5):5089–5100, 1996.
12. R. Cafiero, S. Luding, and H. J. Herrmann. Two-dimensional granular gas of inelastic spheres with multiplicative driving. *Phys. Rev. Lett.*, 84:6014–6017, 2000.
13. R. Cafiero, S. Luding, and H. J. Herrmann. Rotationally driven gas of inelastic rough spheres. *Europhys. Lett.*, 60(6):854–860, 2002.
14. A. Kudrolli and J. P. Gollub. Studies of cluster formation due to collisions in granular material. In *Powders & Grains 97*, page 535, Rotterdam, 1997. Balkema.
15. J. S. Olafsen and J. S. Urbach. Clustering, order and collapse in a driven granular monolayer. *Phys. Rev. Lett.*, 81:4369, 1998. cond-mat/9807148.
16. W. Losert, D. G. W. Cooper, and J. P. Gollub. Propagating front in an excited granular layer. *Phys. Rev. E*, 59:5855, 1999.
17. W. Losert, D. G. W. Cooper, J. Delour, A. Kudrolli, and J. P. Gollub. Velocity statistics in vibrated granular media. *Chaos*, 9(3):682–690, 1999. cond-mat/9901203.
18. W. Losert, L. Bocquet, T. C. Lubensky, and J. P. Gollub. Particle dynamics in sheared granular matter. *Phys. Rev. Lett.*, 85(7):1428–1431, 2000. cond-mat/0004401.
19. J. S. Olafsen and J. S. Urbach. Velocity distributions and density fluctuations in a 2d granular gas. *Phys. Rev. E*, 60:R2468, 1999.
20. A. Prevost, D. A. Egolf, and J. S. Urbach. Forcing and velocity correlations in a vibrated granular monolayer. *Phys. Rev. Lett.*, 89:084301, 2002.

21. J. T. Jenkins and S. B. Savage. A theory for the rapid flow of identical, smooth, nearly elastic, spherical particles. *J. Fluid Mech.*, 130:187–202, 1983.
22. J. T. Jenkins and M. W. Richman. Kinetic theory for plane shear flows of a dense gas of identical, rough, inelastic, circular disks. *Phys. of Fluids*, 28:3485–3494, 1985.
23. A. Goldshtein and M. Shapiro. Mechanics of collisional motion of granular materials. Part 1. General hydrodynamic equations. *J. Fluid Mech.*, 282:75–114, 1995.
24. S. B. Savage. Analyses of slow high-concentration flows of granular materials. *J. Fluid Mech.*, 377:1–26, 1998.
25. José Maria Montanero and Andrés Santos. Computer simulation of uniformly heated granular fluids. *Granular Matter*, 2(2):53–64, 2000. cond-mat/0002323.
26. I. Goldhirsch and T. P. C. van Noije. Green-Kubo relations for granular fluids. *Phys. Rev. E*, 61:3241–3244, 2000.
27. V. Garzó. Tracer diffusion in granular shear flows. cond-mat/0204268, 2002.
28. A. Barrat and E. Trizac. Lack of energy equipartition in homogeneous heated binary granular mixtures. *Granular Matter*, 4(2):57–63, 2002.
29. I. Pagonabarraga, E. Trizac, T. P. C. van Noije, and M. H. Ernst. Randomly driven granular fluids: Collisional statistics and short scale structure. *Phys. Rev. E*, 65(1):011303, 2002.
30. M. H. Ernst and R. Brito. Asymptotic solutions of the nonlinear Boltzmann equation for dissipative systems. In Pöschel and Brilliantov [54], page 3.
31. T. Pöschel and N. V. Brilliantov. Kinetic integrals in the kinetic theory of dissipative gases. In *Granular Gas Dynamics* [54], page 131.
32. I. Goldhirsch. The homogeneous cooling state revisited. In Pöschel and Brilliantov [54], page 37.
33. Y. L. Duparcmeur, H. J. Herrmann, and J. P. Troadec. Spontaneous formation of vortex in a system of self motorised particles. *J. Phys. I France*, 5:1119–1128, 1995.
34. A. Puglisi, V. Loreto, U. M. B. Marconi, and A. Vulpiani. Clustering and non-gaussian behavior in granular matter. *Phys. Rev. Lett.*, 81:3848, 1998.
35. C. Bizon, M. D. Shattuck, and J. B. Swift. Linear stability analysis of a vertically oscillated granular layer. *Phys. Rev. E*, 60:7210–7216, 1999.
36. T. P. C. van Noije and M. H. Ernst. Velocity distributions in homogeneously cooling and heated granular fluids. *Granular Matter*, 1(2):57–64, 1998.
37. T. P. C. van Noije and M. H. Ernst. Velocity distributions in homogeneous granular fluids: the free and the heated case. cond-mat/9803042, 1998.
38. X. Nie, E. Ben-Naim, and S. Y. Chen. Dynamics of vibrated granular monolayers. *Europhys. Lett.*, 51(6):679–648, 2000. cond-mat/9910371.
39. I. Pagonabarraga and E. Trizac. Kinetics of fragmenting freely evolving granular gases. In Pöschel and Brilliantov [54], page 163.
40. S. McNamara and S. Luding. Energy flows in vibrated granular media. *Phys. Rev. E*, 58:813–822, 1998.
41. A. Goldshtein, A. Alexeev, and M. Shapiro. Shock waves in granular gases. In Pöschel and Brilliantov [54], page 187.
42. A. Puglisi, V. Loreto, U. M. B. Marconi, and A. Vulpiani. A kinetic approach to granular gases. *Phys. Rev. E*, 59:5582–5595, 1999.

43. C. Bizon, M. D. Shattuck, J. B. Swift, and H. L. Swinney. Velocity correlations in driven two-dimensional granular media. In J. Karkheck, editor, *Dynamics: Models and Kinetic Methods for Nonequilibrium Many-Body Systems*, pages 361–371, Dordrecht, 2000. Kluwer Academic Publishers. cond-mat/9904135.
44. T. P. C. van Noije, M. H. Ernst, and R. Brito. Spatial correlations in compressible granular flows. *Phys. Rev. E*, 57:R4891, 1998.
45. T. P. C. van Noije, M. H. Ernst, E. Trizac, and I. Pagonabarraga. Randomly driven granular fluids: Large scale structure. *Phys. Rev. E*, 59:4326–4341, 1999.
46. A. Kudrolli and J. Henry. Non-gaussian velocity distributions in excited granular matter in the absence of clustering. *Phys. Rev. E*, 62(2):R1489–92, 2000. cond-mat/0001233.
47. T. P. C. van Noije, M. H. Ernst, and R. Brito. Ring kinetic theory for an idealized granular gas. *Physica A*, 251:266–283, 1998.
48. S. McNamara and S. Luding. Energy nonequipartition in systems of inelastic, rough spheres. *Phys. Rev. E*, 58:2247–2250, 1998.
49. S. McNamara and E. Falcon. Vibrated granular media as experimentally realizable granular gases. In Pöschel and Brilliantov [54], page 347.
50. D. Henderson. A simple equation of state for hard discs. *Mol. Phys.*, 30:971–972, 1975.
51. L. Verlet and D. Levesque. Integral equations for classical fluids. III. the hard disks system. *Mol. Phys.*, 46(5):969–980, 1982.
52. S. F. Foerster, M. Y. Louge, H. Chang, and K. Allia. Measurements of the collision properties of small spheres. *Phys. Fluids*, 6(3):1108–1115, 1994.
53. I. S. Aranson, B Meerson, P. V. Sasorov, and V. M. Vinokur. Phase separation and coarsening in electrostatically driven granular media. *Phys. Rev. Lett.*, 88(20):204301, 2002.
54. T. Pöschel and N. V. Brilliantov, editors. *Granular Gas Dynamics*, volume 624 of *Lecture Notes in Physics*. Springer, Berlin, Heidelberg, New York, 2003.

Van der Waals-Like Transition in Fluidized Granular Matter

Rodrigo Soto, Médéric Argentina, and Marcel G. Clerc

Departamento de Física, Facultad de Ciencias Físicas y Matemáticas, Universidad de Chile, Casilla 487-3, Santiago, Chile

Abstract. A phase separation of fluidized granular matter is presented. Molecular dynamics simulations of a granular system, in two spatial dimensions, with a vibrating wall and without gravity exhibit appearance, coagulation, and evaporation of bubbles. The instability is produced by the existence of a negative compressibility region, that is caused by the energy dissipation at collisions. The phenomenon is analogous to the spinodal decomposition of the gas-liquid transition of the van der Waals model. A hydrodynamic model gives account for the negative compressibility and predicts a critical point that is in qualitative agreement with the results of the simulations. In the onset of phase separation, we have deduced a macroscopic model that agrees quite well with molecular dynamics simulations.

Granular matter, when fluidized by continuous energy injection exhibits a variety of phenomena that resembles those of molecular fluids: patterns and instabilities appear, Rayleigh-Benard like convection is developed, etc [1–3]. The main difference with molecular fluids is that, at collisions, grains dissipate kinetic energy into the internal degrees of freedom of the grains. Hence energy must be supplied continuously to sustain a fluidized regime. Experimentally, energy is usually injected through vibrating walls or by the gravitational field.

Fluidized granular systems have the tendency to create large gradients in the hydrodynamic fields, and microscopic and macroscopic time and length scales are not enough separated as in elastic fluids [4]. However, when dissipation is not too large, fluidized regimes of granular matter are described successfully using hydrodynamic models. These models are similar to the Navier-Stokes equations for elastic fluids, with the addition of an energy dissipation term.

Here, we describe a new type of instability observed in fluidized granular matter, analogous to the spinodal decomposition of the gas-liquid transition in the van der Waals model [5]. The instability was first predicted using a linear stability analysis of the hydrodynamic equations [6] and later in [7]. An nonlinear analysis, as well as the time evolution of the instability was performed by the authors in [8].

The origin of the instability relies on the fact that for granular media the average granular temperature is a decreasing function of the density for fixed boundary conditions (see later for a detailed explanation). This results in negative compressibility for dissipations larger than a critical value giving rise

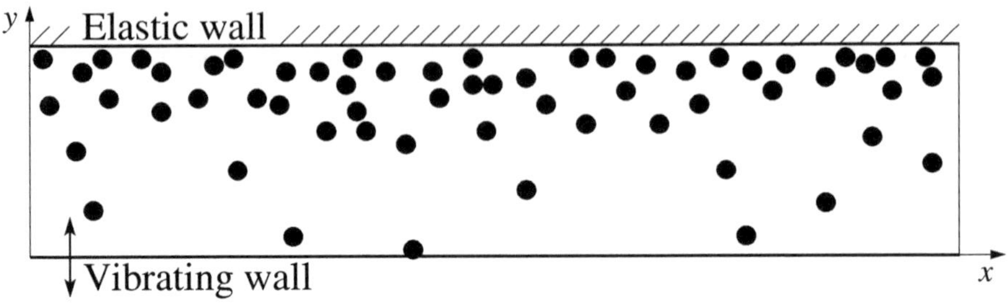

Fig. 1. Schematic representation of the studied system. Grains are placed in an horizontal box. The bottom wall is vibrating while the top one reflects grains elastically. The system is periodic in the x direction

to a spatial instability. We also show that the instability can be understood using a simple hydrodynamic model.

We consider a two-dimensional system of grains on an horizontal surface, with friction ignored, placed in a box with large aspect ratio (see Fig. 1). Henceforth, we will refer to horizontal and vertical directions as the long and short directions, respectively; the system is periodic in the horizontal direction. The top wall reflects grains elastically while the bottom one injects energy into the system by means of a vibrating wall at high frequency ω and small amplitude A. The collisions with the wall are elastic with no friction, thus conserving horizontal momentum. Due to the high frequency, the collisions with the wall are uncorrelated, being modeled in a stochastic way. For simplicity, the vibrating wall is well modeled by a stochastic wall: each time a grain collides with the wall, it is reflected conserving the tangential component of the velocity, but the normal component is sorted from a Maxwellian distribution at a certain temperature, that scales as $T \sim m(A\omega)^2$, m being the mass of the gains [9]. We define the granular temperature (the *temperature* from now on), like in molecular fluids, to be proportional to the kinetic energy per particle in the reference frame of the fluid. We emphasize that both collisions with the walls and between the grains conserve horizontal momentum.

The system is studied using molecular dynamics simulations of the Inelastic Hard Sphere model (IHS) [10–12]. Grains are modeled as smooth rigid disks and the collisions are characterized by a constant normal restitution coefficient α. Grains only have translational degrees of freedom and there is no tangential friction between grains at collisions. The IHS has been widely studied and reproduces well many of the observed phenomena in granular fluids at moderate densities, where rotation is not fundamental (see, for example, [10–12]). Units are chosen such that the diameter σ and mass m of each disk is one. Also, taking the wall temperature as one, energy units are fixed. Under these conditions, the system is completely defined by the total number of grains N, the aspect ratio $\lambda = L_x/L_y \gg 1$, the global number den-

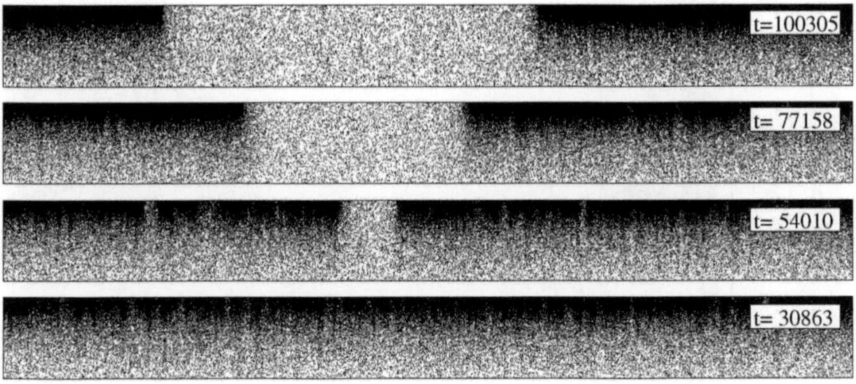

Fig. 2. Snapshots of a system with $N = 153600$, $\lambda = 102.4$, $n_0 = 0.15$, $L_x = 10240$, and $q = 0.02$. The configurations correspond to different simulation times. Each *black dot* represents a simulated grain and the aspect ratio has been distorted to make the system visible. The bubble appears at $t \approx 40000$

sity $n_0 = N/(L_x L_y)$, and the restitution coefficient of the grains $\alpha = (1-2q)$; the elastic limit corresponds to $q = 0$.

At low dissipation, the granular media develops vertical density and temperature profiles induced by the dissipation at collisions and the energy injection mechanism, but both fields are homogeneous in the horizontal direction. The system is hotter and less dense near the injecting wall, and colder and denser by the opposite wall (see Fig. 1). For a detailed study of this state see [13].

For a larger dissipation ($N = 153600$, $\lambda = 102.4$, $n_0 = 0.15$, $L_x = 10240$, and $q = 0.02$) an spatial instability is observed: the system exhibits the coexistence of two fluid phases, characterized by different densities (see Fig. 2). Initially the fluid remains horizontally homogeneous and suddenly a bubble (low density region) appears and grows until it achieves its final size. After that, the system remains stationary with the two phases coexisting.

To characterize in more detail this instability, we analyze the temporal evolution of the vertically averaged density (*coarse grained density*) $\rho(x,t) = L_y^{-1} \int_0^{L_y} n(x,y,t)dy$, where $n(x,y,t)$ is the density field. In Fig. 3 the spatio-temporal evolution of $\rho(x,t)$ is presented. It is clearly seen that the system remains in a homogeneous state for a finite large time until the bubble is nucleated due to density fluctuations. Therefore, this homogeneous state is a metastable one. Afterward, the bubble grows with a nearly constant velocity while two densification fronts propagate away from the bubble at a larger, but also constant, velocity (illustrated in Fig. 3). When the fronts reach the growing bubble —through the periodic boundary conditions— it is pushed inward, leading to damped oscillations in its size. Finally, the system achieves a stationary state with the bubble.

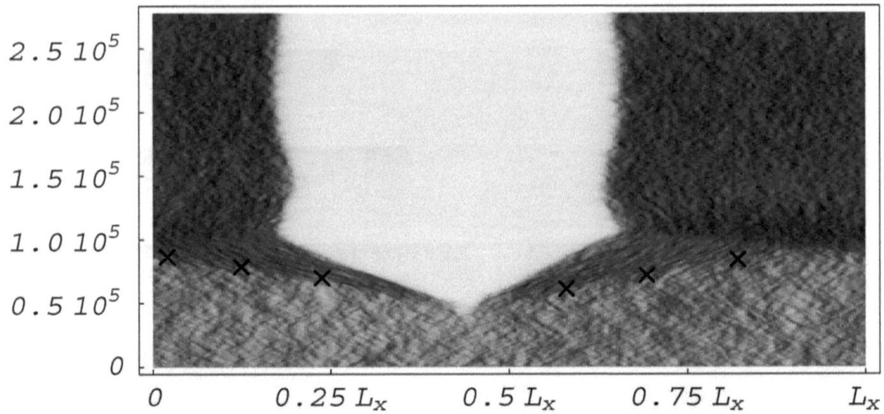

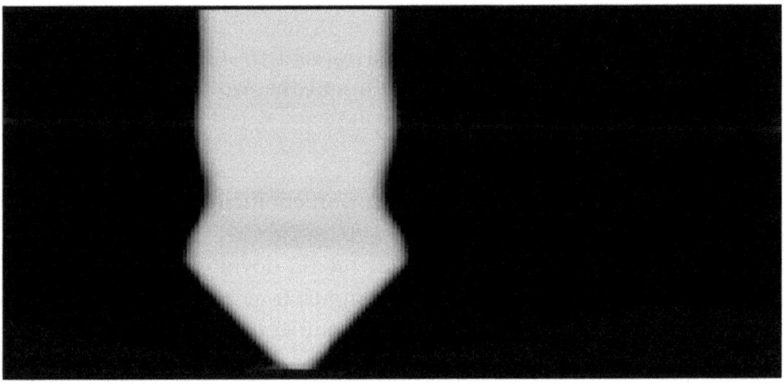

Fig. 3. Spatiotemporal evolution of the coarse grained density $\rho(x,t)$, with time on the vertical axis and increasing upward. The gray scale is proportional to density, with darker regions representing denser regions in the system. The top graph corresponds to the molecular dynamics simulation with the same parameters as in Fig. 2, where the bubble nucleation is triggered by internal noise. In the final state, the vertically averaged density of the bubble is $\rho = 0.025$, while in the dense region $\rho = 0.257$. The densification fronts are marked with *lines of crosses*. The bottom graph is obtained by the simulation of the model defined by (11) with $\varepsilon = -6.6 \times 10^{-4}$ and $\nu = 2$. The system size is 5400 and the total simulation is time $T = 3.5 \times 10^5$. An initial condition (with $u = 1.4 \times 10^{-2}$) that overcomes the nucleation barrier is imposed. The minimum (*light gray*) and maximum (*dark gray*) densities are $u = -2.6 \times 10^{-2}$ and $u = 2.9 \times 10^{-2}$, respectively

The system shows a different behavior, according to the value of the dissipation parameter q. In Fig. 4 the spatiotemporal evolution of a system with a smaller dissipation is shown ($q = 0.01$). In this case four bubbles are created in the fluid with no apparent metastable time. Two of them merge into a single one, and later on, the smallest one disappears or evaporates. Later on

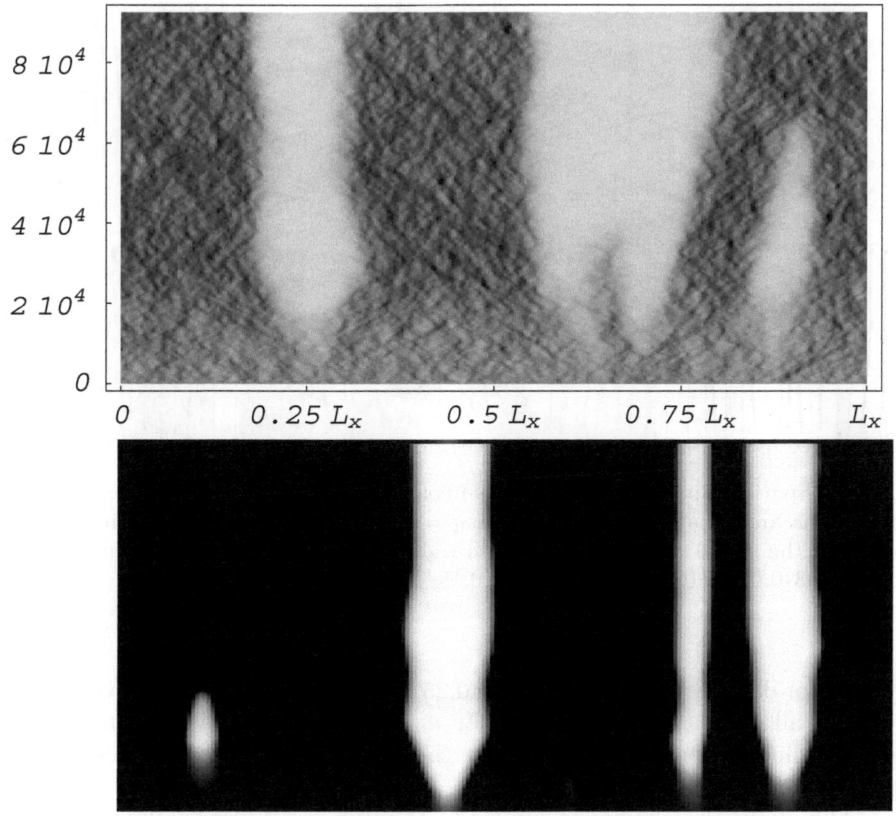

Fig. 4. The same representation as in Fig. 3, but changing the dissipation parameter to $q = 0.01$. In the final simulated state, the vertically averaged density of the bubbles is $\rho = 0.046$, while in the dense regions $\rho = 0.216$. In the bottom graph, we use the same parameters and gray scale as in Fig. 3 and the initial condition is $u = 7.4 \times 10^{-3}$. In this case there is no nucleation barrier and the initial condition used in the solution of the model is homogeneous with small fluctuations that are amplified by the instability

the two remaining bubbles evolve slowly, one of them growing while the other one decreases. Densification fronts, created with the bubbles, are also seen.

We have also made a series of simulations keeping constant the value of q and changing the value of the global density n_0. All the simulations were done with the following parameters $N = 153600$, $q = 0.02$, and $L_y = 100$. The spatio-temporal evolution of the coarse grained density $\rho(x,t)$ is represented in Fig. 5 for different values of n_0. For global densities smaller that $n_0 = 0.03$ the system remains stable; for densities between $n_0 = 0.033$ and $n_0 = 0.15$ the system is unstable or metastable and depending on the case low density regions (bubbles) or large density ones (droplets) are created;

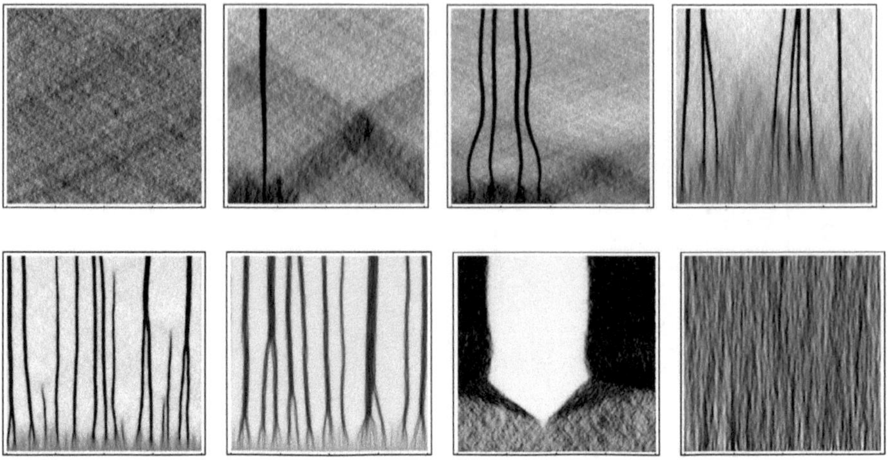

Fig. 5. Spatiotemporal evolution of the coarse grained density $\rho(x,t)$. All the simulations are done with $N = 153600$, $q = 0.02$, and $L_y = 100$. The simulations differ in the global density; from left to right and top to bottom the densities are $n_0 = 0.03, 0.033, 0.035, 0.04, 0.05, 0.08, 0.15$, and 0.30

finally for densities larger than $n_0 = 0.25$ no instabilities were observed. Note that in all cases (specially visible for $n_0 = 0.03$) density waves created by density fluctuations are observed in the background regions. These waves propagate with well defined velocities.

The observed behavior is similar to the van der Waals gas liquid transition, where for temperatures lower that the critical one, there is a density range for which the homogeneous state is unstable. Here, in the granular fluid, the control parameter is the dissipation coefficient q instead of the thermodynamic temperature.

The origin of the instability can be understood as follows. In thermodynamic equilibrium, the pressure is a monotonous increasing function of the density and the temperature [14]. But, in this system, the granular temperature adjusts itself to a stationary profile given by the energy balance between dissipation at collisions and injection at the vibrating wall. As the dissipation rate is proportional to density, the stationary temperature is a decreasing function of ρ. As an outcome, the effective pressure (that is independent of y due to the absence of gravity) can present a region where it decreases with an increasing density. That is, the temperature drop produced by collisions can be large enough to give rise to a decrease in pressure when increasing the global density [15]. Then, there can be a region of negative compressibility ($\partial p_{xx}/\partial n_0 < 0$) in the same sense as the van der Waals loop in classical fluids, that triggers the instability.

The existence of the negative compressibility region can be verified numerically as follows. We have performed molecular dynamics simulations of

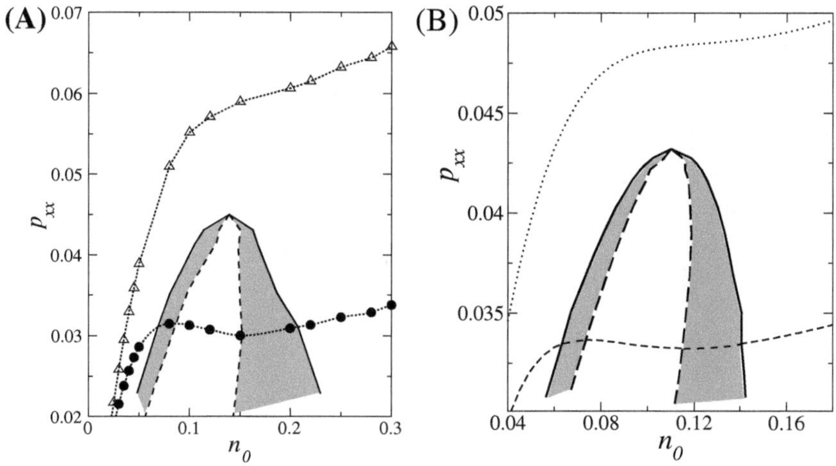

Fig. 6. **A** Phase diagram obtained from molecular dynamics simulations in tall, narrow boxes. The dotted curves correspond to the horizontal component of the pressure tensor p_{xx} for the dissipations $q = 0.0032$ (*open triangles*) and $q = 0.0070$ (*filled circles*), smaller and larger than the critical one, respectively. The second case presents a van der Waals loop. The coexistence curve (*solid line*) and spinodal curve (*dashed line*) are plotted. **B** Phase diagram obtained from the hydrodynamic model. The solid line (*long-dashed*) curve is the coexistence (spinodal) curve. The gray region is the metastability region. The dotted (*short-dashed*) curve correspond to the hydrodynamic pressure for the dissipation $q = 0.007$ ($q = 0.0130$)

the IHS system in tall boxes ($L_x \ll L_y$) to prevent the spatial instability to develop; when the horizontal system size is small, there is a large surface energy cost in creating a bubble, and the system remains stable. We compute the horizontal momentum flux (x–x component of the pressure tensor, p_{xx}), averaged over the vertical direction for different values of the global density n_0 and the dissipation coefficient q. For constant q, the pressure p_{xx} exhibits the appearance of van der Waals loops, with a critical point located at $q \approx 0.0047$ and $n_0 \approx 0.15$ (see Fig. 6). For larger values of q there is a region with negative compressibility that is bounded by the spinodal curve. Also, we can compute the coexistence curve using Maxwell's construction. The region between the coexistence and spinodal curves define the metastability region, where bubbles or droplets can nucleate thanks to density fluctuations. Inside the spinodal curve, the system is mechanically unstable.

In the previous paragraph, we used concepts of equilibrium thermodynamics like surface energy and Maxwell's construction. Below we show that the system dynamics is described by a free energy (Landau type) allowing the use of this concepts.

The existence of the negative compressibility region can be studied theoretically using a hydrodynamic description of granular fluids [16]. In [17] the negative compressibility region was also observed, but due to the use of constitutive relations of dilute granular gases does there was no saturation.

Here we investigate the presence of the negative compressibility region with a different perspective. As in [16] we consider simple hydrodynamic equations that mimic the Navier-Stokes ones with a dissipation term in the energy equation, modeling the energy dissipation at collisions. For sufficiently small dissipation ($q \ll 1$), one has [18–21]

$$\begin{aligned} \frac{\partial n}{\partial t} &+ \nabla \cdot (n\vec{v}) = 0, \\ n\left(\frac{\partial \vec{v}}{\partial t} + (\vec{v}\cdot\nabla)\vec{v}\right) &= -\nabla \cdot I\!P, \\ n\left(\frac{\partial T}{\partial t} + (\vec{v}\cdot\nabla)T\right) &= -\nabla\cdot\vec{J} - I\!P : \nabla\vec{v} - \omega, \end{aligned} \tag{1}$$

with the usual constitutive equations for the pressure tensor (hydrostatic pressure and Newton's law) and the Fourier's law for the heat, and the new constitutive equation for the energy dissipation rate flux,

$$\begin{aligned} I\!P_{i,j} =& p(n,T)\delta_{i,j} - \eta(n,T)\left(\frac{\partial v_i}{\partial x_j} + \frac{\partial v_j}{\partial x_i}\right) - (\xi(n,T) - \eta(n,T))\nabla\cdot\vec{v}\delta_{i,j} \\ \vec{J} =& -k(n,T)\nabla T \\ \omega =& \omega_0(n,T)n^2T^{3/2}. \end{aligned} \tag{2}$$

where p is the hydrostatic pressure, η is the shear viscosity, ξ is the bulk viscosity, k is the thermal conductivity, and ω_0 is a factor that accounts for the energy dissipation at collisions.

The transport coefficients in the limit of small dissipation are the same as for elastic spheres in 2D [22], except for the one associated to the energy dissipation, that is calculated directly using the Enskog theory [23]. All the transport coefficients are expressed in terms of the pair correlation function of elastic disks at contact, χ [23].

$$\chi(n) = \frac{1 - 7\pi n/64}{(1 - \pi n/4)^2} \tag{3}$$

$$p(n,T) = nT\frac{(1 + \pi^2 n^2/128)}{(1 - \pi n/4)^2} \tag{4}$$

$$\eta(n,T) = \frac{1}{2\chi}\left(1 + \pi n\chi/2 + 0.87(\pi n\chi/2)^2\right)\sqrt{\frac{T}{\pi}} \tag{5}$$

$$\xi(n,T) = \frac{1.246}{2\chi}(\pi n\chi/2)^2\sqrt{\frac{T}{\pi}} \tag{6}$$

$$k(n,T) = \frac{2}{\chi}\left(1 + 3\pi n\chi/4 + 0.87(\pi n\chi/2)^2\right)\sqrt{\frac{T}{\pi}} \tag{7}$$

$$\omega_0 = 4\sqrt{\pi}q\chi(n) \tag{8}$$

The use of hydrodynamic equations for describing granular fluids is controversial because hydrodynamic theory is based on the local tendency of fluids toward equilibrium, condition that is not necessarily satisfied in granular fluids. However, the use of kinetic theory give a support for this kind of description at least in the dilute and quasielastic case (see for example [18, 24]). In this article we will not put into question the validity of the hydrodynamic model, and we will use it to obtain qualitative and quantitative predictions.

In the model (1), density and horizontal momentum satisfy conservation equations. We consider that the the velocity field is small enough to neglect nonlinear terms in $\vec{v}$. Density and horizontal momentum can be vertically averaged giving $\rho(x,t)$ and $j(x,t)$, that obey the equations

$$\partial_t \rho\,(x,t) = -\partial_x j\,(x,t)\,, \tag{9a}$$

$$\partial_t j\,(x,t) = -\partial_x \Phi, \tag{9b}$$

where Φ is the vertical average of the x–x component of the pressure tensor.

Equations (9) show that the conserved fields ρ and j have an evolution given by a time scale proportional to the value of the horizontal gradients. When the horizontal gradients are much smaller than the vertical ones, ρ and j behave as the only slow variables and govern the system dynamics. The other fields (vertical momentum, temperature, and Φ) have fast dynamics and behave as slave variables of ρ and j, that is their values follow the evolution of the slow fields.

The value of Φ can be computed using the equations for the temperature and vertical momentum. The vertical momentum relaxes fast to its equilibrium value, zero, due to the boundary conditions. Then, the pressure is independent of y ($\Phi = p$). Energy is not conserved implying that temperature is also a fast variable and satisfies the equation $\nabla \cdot \vec{J} + \omega = 0$. The equations of T and Φ can be solved numerically with the boundary conditions $T = 1$ at the bottom and $\vec{J} = 0$ at the top, and the integral condition $\int_0^{L_y} n(x,y,t)dy = L_y\,\rho(x,t)$.

For fixed q, the pressure exhibits, as proposed before, van der Waals loops as a function of $n_0 = \rho(x,t)$ (see Fig. 6). The negative compressibility region has a critical point at $q \approx 0.0086$ and $n_0 \approx 0.11$.

The results on the hydrodynamic model are in qualitative agreement with the molecular dynamics simulations in tall boxes. The difference is originated mainly in the boundary condition for T. Indeed, even though particles are forced to come out of the bottom wall with velocities characteristic of a temperature equal to 1, the granular temperature at the boundary is less than one because, due to dissipation, the particles that arrive at the wall come with a smaller kinetic energy. This is a well known effect in dilute gases (Knudsen layer effects [25]), that become also important in granular fluids.

Having verified that the mechanism that originates the instability is a negative compressibility in the effective pressure, we can describe the dynamics of the dominant mode near the bifurcation point (critical point in the van der Waals language). The critical point is defined by the condition that both $\partial\Phi/\partial\rho$ and $\partial^2\Phi/\partial\rho^2$ vanish. Using symmetry and scaling arguments it can be argued that the averaged pressure Φ close to the critical point can be written as (see [8])

$$\Phi \approx \Phi_o + \frac{\partial\Phi}{\partial\rho}\bar{\rho} + \frac{\partial^2\Phi}{\partial^2\rho}\frac{\bar{\rho}^2}{2} + \frac{\partial^3\Phi}{\partial^3\rho}\frac{\bar{\rho}^3}{6} + \frac{\partial\Phi}{\partial j^2}j^2 + \frac{\partial\Phi}{\partial\rho_{xx}}\bar{\rho}_{xx} + \frac{\partial\Phi}{\partial j_x}j_x, \tag{10}$$

where $\bar{\rho} = \rho - n_0$.

For sake of simplicity, we define now $u = \rho - \rho_M$, where ρ_M is the density at the Maxwell point (i.e. where $\partial^2\Phi/\partial^2\rho\big|_{\rho_M} = 0$). Scaling u and x, (9) are approximated at the dominant order by the *Van der Waals normal form*

$$\begin{aligned}\partial_{tt}u &= \partial_{xx}\left(\varepsilon u + u^3 - \partial_{xx}u + \nu\partial_t u\right), \\ &= \partial_{xx}\frac{\delta\mathcal{F}}{\delta u} + \nu\partial_{xxt}u,\end{aligned} \tag{11}$$

where $\varepsilon = \partial\Phi/\partial\rho\big|_{\rho_M}$ is the control parameter and $\nu\partial_{xxt}u$ is a diffusion term. The variables scale as $u \sim \varepsilon^{1/2}$, $j \sim \varepsilon$, $\partial_x \sim \varepsilon^{1/2}$, $\partial_t \sim \varepsilon$ and $\nu \sim o(1)$. Note that, the convection term in the momentum flux ($\frac{\partial\Phi}{\partial j^2}j^2 \sim \varepsilon^2$) has been neglected, in comparison with the dominant part (order $\varepsilon^{3/2}$). The sign of $\partial\Phi/\partial\rho_{xx}\big|_{\rho_M}$ has been chosen to be negative, in order to saturate the linear instability and to impose the existence of a global minimum for the free energy

$$\mathcal{F} = \int dx \left\{ \varepsilon\frac{u^2}{2} + \frac{u^4}{4} + \frac{(\partial_x u)^2}{2} \right\}. \tag{12}$$

Equation (11), that describes the time evolution of the system close to the critical point, implies that the system evolves in a irreversible manner toward an stationary state characterized by the minimization of the $\mathcal{F}$. It can be directly checked form (11) that $(\mathcal{K} + \mathcal{F})$ decreases monotonically, where $\mathcal{K} = \int dx(\partial_t\Lambda)^2$ and Λ is the primitive of u. Then, the final state is stationary and with the minimum value of $\mathcal{F}$. Also, if a small noise was added to the equation (to describe fluctuations) it can be shown that the stationary probability distribution of u is of the form $P(u) \sim e^{-\mathcal{F}[u]/\eta}$, where η is the noise intensity [26–28]. These properties make appropriate to call $\mathcal{F}$ a *free energy* or a *Lyapunov functional.*

The study of instabilities in non-linear physics have shown that the existence of a non-equilibrium free energy, allows to use some concepts of equilibrium thermodynamics in non-equilibrium instabilities. In particular, Maxwell's construction can be used to determine the coexisting densities. Also, the structure of $\mathcal{F}$ indicates that large density gradients (thus interfaces) are energetically suppressed.

We emphasize that the Landau free energy $\mathcal{F}$ has the classical phase diagram, with coexistence and spinodal curves [29]; as those observed in binary fluids, binary alloys, and $^3He-^4He$ mixtures, to mention a few examples (see [5] and references cited therein).

For negative ε, the homogeneous solution ($u = 0$) undergoes a spatial instability, characterized by the appearance of equally distributed bubbles. Later on, closest bubbles merge into a bigger one, as a consequence of a coalescence process. Subsequently, bubble dynamics is led by the interaction mediated by waves (see Figs. 4 and 5). The above dynamics is a consequence of the system tendency to minimize its free energy (see [30] for a complementary discussion). With periodic boundary conditions, the global minimum is a unique bubble, as the solution shown in Fig. 3. Further work in this coarsening process is in progress.

In the metastability region (see Fig. 6), i.e. the region bounded by the spinodal and coexistence curves, the homogeneous state is stable. Nevertheless, a finite fluctuation that overcomes the nucleation barrier gives rise to a bubble along with two *state-waves (densification waves)*, which propagate away from the bubble (see Fig. 3). These fronts are a result of mass conservation: when the bubble is created the excess mass must be ejected. Due to the periodic boundary conditions, later on, the waves collide with the bubble producing its oscillation. Afterward, due to viscosity, the oscillations are damped and the final state is a single bubble at rest.

It has been shown in a full linear stability analysis of the hydrodynamic equations that for aspect ratios $\lambda = L_x/L_y$ smaller than a critical value, the instability is suppressed [6, 16, 31]. The fact that the system becomes stable for small enough values of λ is also captured in the van der Waals equation. A linear stability analysis of 11 around the homogeneous state shows that the perturbations with wave-vectors larger than $k^c = \sqrt{-\varepsilon - 3u_0^2}$ are stable, where u_0 is the reference scaled density. Then, if the horizontal system size is smaller than $L_x^c = 2\pi/k^c$, the system is always stable. This is an effect of the surface energy already mentioned. We have used this result to compute the van der Waals loops in thin columns.

The numerical solutions of (11) (see Figs. 3 and 4) in the unstable and metastable regions show good qualitative agreement with the molecular dynamic simulations. That is, the model captures the main features of the system dynamics close to the instability (density waves, fronts, metastability, and instability).

The model (11) implies also that for global densities smaller than the Maxwell point density ($u < 0$), dense regions (instead of low density regions) can nucleate. This prediction is in agreement with the simulations presented in Fig. 5, where one or many dense regions are nucleated for $n_0 < 0.05$. The dynamics of $\rho(x, t)$ is, however, not completely well described by the model

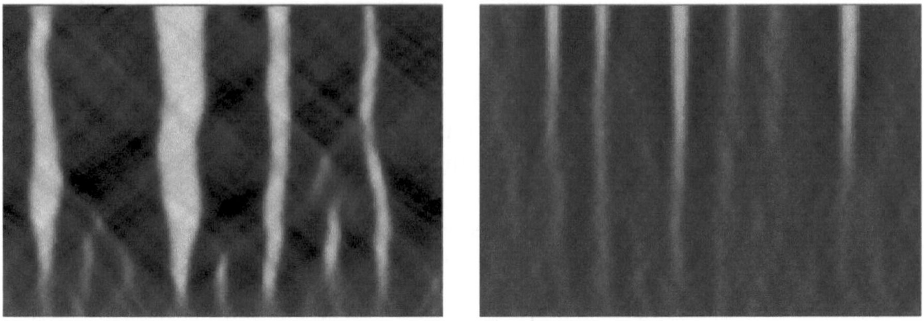

Fig. 7. Spatiotemporal evolution of the coarse grained density in the unstable region, modeled by (16). **Left:** $\varepsilon = -0.51$, $\nu = 0.5$, $\eta_0 = 0.5$, and $\mu = 0.0$. **Right:** the same parameters but with moderate friction connected, $\mu = 0.3$

because these simulations are done with parameters not close to the critical point. The case of $n_0 = 0.03$ is special, because in this case a long wavelength density modulation is created before the droplet appears. This phenomenon is not captured by the model.

In experiments there is always a small amount of friction with the air or the walls. In this case momentum is not exactly conserved and the model must be modified. The momentum density satisfies the following equation

$$\partial_t j = -\partial_x \Phi - \mu j, \tag{13}$$

where μ is the friction coefficient. Inserting (13) into (9a) and doing the appropriate scalings we obtain the modified van der Waals equation

$$\partial_{tt} u + \mu \partial_t u = \partial_{xx} \left(\varepsilon u + u^3 - \partial_{xx} u + \nu \partial_t u \right) . \tag{14}$$

If μ is large, the above equation reduces adiabatically to the usual Cahn-Hilliard equation [29] that differs from the van der Waals equation in that it does not present waves and the dynamics is much slower (see Appendix). Note that, usual spinodal decomposition in binary mixtures, for example, obeys the Cahn-Hilliard and not the van der Waals equation. Therefore, in experiments, the detection of density and shock waves will depend on the intensity of the friction. In Fig. 7 the time evolution of the spinodal decomposition process with and without friction is presented. In both cases noise is added (see Appendix) to trigger nucleation. In the case without friction waves are observed and the coarsening evolution is faster, while in the case with friction waves are almost suppressed. Also, the coarsening process is much slower in this case.

As a final test of the model, we have made MD simulations of the system in the elongated box setup (extended system). For two global densities ($n_0 = 0.08$ and $n_0 = 0.15$) and a number of different inelasticities, the systems were simulated until a final steady state is achieved with one or no bubble inside. When the final configuration contained one bubble (or droplet) the vertical-averaged density profile $\rho(x)$ was extracted. In the Ginzburg-Landau theory [32] it is shown that the density profile that minimizes the Landau free energy (12) is a single bubble were the kink profile is given a by an hyperbolic tangent. Then, we fit $\rho(x)$ to

$$\rho(x) = \rho_1 + \frac{\rho_2 - \rho_1}{2} \left[\tanh\left(\frac{x - x_A}{\Delta_x} \right) - \tanh\left(\frac{x - x_B}{\Delta_x} \right) \right] \tag{15}$$

where ρ_1 and ρ_2 are the homogeneous densities of the two phases, Δ_x is the interface width, and x_A and x_B are the position of the kinks (bubble borders).

In each case we computed also the vertical average of the horizontal component of the pressure tensor p_{xx}. Even in the presence of bubble, it is horizontally homogeneous, as the system is is mechanical equilibrium.

The homogeneous densities ρ_1 and ρ_2 should correspond to the coexistence densities of the van der Waals theory. In Fig. 8 we present these densities as a function of the homogeneous pressure p_{xx}. We note that the coexisting densities in both series of simulations (made with different global densities) coincide. Also the final pressure depends only on the inelasticity q, but not on the global density. These results confirm that the phenomenon is a van der Waals separation.

The coexisting densities are compared in Fig. 8 with the predictions obtained using Maxwell's construction in the simulation on tall columns. The comparison is fairly good indicating that the phenomenological Landau free energy (12), from where we made the Maxwell construction, is qualitatively correct even far from the critical point.

In summary, we have studied the phase separation in fluidized granular matter. Molecular dynamics simulations of a grain system, in two dimensions, with a vibrating wall and no gravity exhibit appearance, coagulation and disappearance of low density regions (bubbles). Rarefaction and densification density waves lead the bubble dynamics. The mechanism of phase separation is triggered by a negative compressibility which, in turns, is a result of the fact that for granular media in closed geometries the granular temperature is a decreasing function of the density.

A simple hydrodynamic model is able to predict the existence of van der Waals loops in the effective pressure. The predicted critical point is in qualitative agreement with the results of molecular dynamic simulations.

Close to the transition, the system is described by the van der Waals normal form (11). This model describes quite well the molecular dynamics simulations. The phase separation is the analog to the spinodal decomposition of the gas-liquid transition of the van der Waals model, but the transient evolution of the system is led by state-waves.

The van der Waals instability should be observed independently of the peculiarities of the energy injection mechanism or of its stochastic modelization, as long as the particle collisions with the wall conserve horizontal momentum.

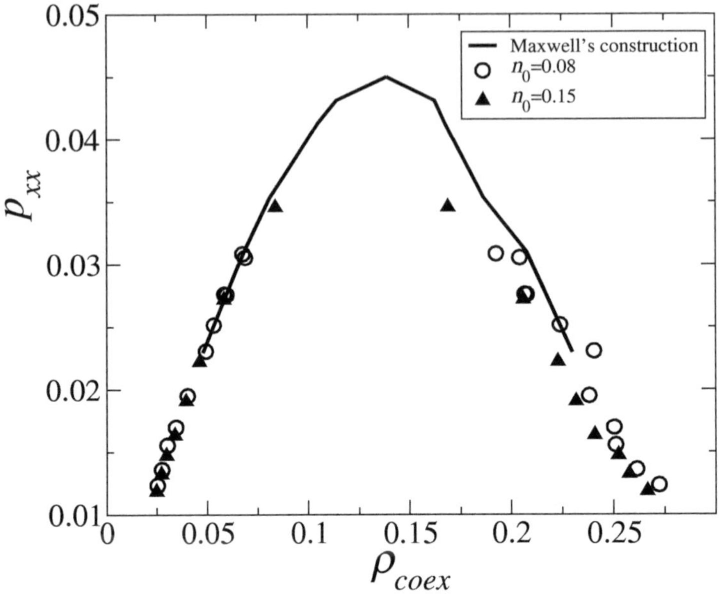

Fig. 8. Coexistence densities ρ_{coex} as a function of the $x - x$ component of the pressure tensor. Symbols are results obtained in MD simulations of extended systems for two global densities: $n_0 = 0.08$ and $n_0 = 0.15$. Different points correspond to different inelasticities. The continuous line is the prediction obtained from the Maxwell construction extracted from the simulations in tall columns

It should also be observed in the presence of a small gravitational field either by inclining slightly the table or by placing the system vertically with characteristic injected energy much larger than the gravitational potential energy $(m(A\omega)^2 \gg mgL_y)$. In both possible generalizations, the general theoretical framework continues to be valid, and preliminary numerical simulations confirm it (in preparation).

Acknowledgements

The authors would like to thank P. Cordero, D. Risso, and E. Tirapegui for fruitful discussions. The simulation software developed at INLN, France, has been used for the simulations of model (11). M.G.C. and R.S. thank the support of *Programa de inserción de científicos chilenos* of Fundación Andes and the support of FONDAP grant 11980002. M.A. and R.S. greatly thank the support of the FONDECYT projects 30000017 and 1010416, respectively.

Appendix: Density Fluctuations

Density fluctuations are a important aspect of the described phenomenon. They are responsible for the nucleation of bubbles in the metastable region.

Also, in the homogeneous phases, they seem to propagate with well defined velocities (see Figs. 3-5). The dynamics of these fluctuations can be understood using the van der Waals model. A study of the system dynamics close to the instability threshold, where noise is amplified was done in [31].

In fact, density fluctuations are created by internal noise, and they obey the linearized van der Waals equation, with the addition of a noise term. In the spirit of Landau and Lifshitz's theory [33], the noise term must be added to the constitutive relation (10). If n_0 is the homogeneous density, the density fluctuations $n(x,t) = \rho(x,t) - n_0$ and the momentum density $j(x,t)$ are described by

$$\begin{aligned} \partial_t n &= -\partial_x j \\ \partial_t j &= -\partial_x \left(cn - \partial_{xx} n - \nu \partial_x j \right) - \mu j - \partial_x \eta \end{aligned} \tag{16}$$

where $\eta(x,t)$ is a noise term and $c = \varepsilon + 2n_0^2$.

The above equations can be solved in Fourier space $n(x,t) = n_k(t)e^{ikx}$, $j(x,t) = j_k(t)e^{ikx}$, in terms of the noise. The statistical properties of the noise are unknown, especially because the reference system is out of equilibrium and fluctuation-disipation theorems cannot be applied. However, close to the critical point the only slow variables are the order parameters, all the others having fast dynamics and short correlation times. Then, close to the critical point, it can be safely assumed that the noise is white and delta correlated in space. That is, we assume that $\langle \eta_k(t)\eta_{k'}(t')\rangle = \eta_0 \delta(t-t')\delta(k+k')$.

The statistical properties of the density fluctuations are obtained computing the intermediate scattering function

$$F(k,t) = \lim_{t' \to \infty} \langle n_k(t+t') n_{-k}(t') \rangle \tag{17}$$

where the average is over realizations. With the assumption of white noise it reduces to

$$F(k,t) = \eta_0 k^2 \int_0^\infty ds \left[e^{A_k(t+s)} \right]_{12} \left[e^{A_{-k}s} \right]_{12} \tag{18}$$

where the matrix A_k is

$$A_k = \begin{pmatrix} 0 & -ik \\ -ikc - ik^3 & -\mu - \nu k^2 \end{pmatrix} \tag{19}$$

and the subscript 12 means that this component of the matrix must be taken.

The instantaneous density fluctuations are described by

$$F(k) = F(k,0) = \frac{\eta_0 k^2}{2(c+k^2)(\mu + \nu k^2)} \tag{20}$$

that has a maximum at $k^* = (c\mu/\nu)^{1/4}$. Note that if there is no friction ($\mu = 0$) long wavelength fluctuations are the predominant ones.

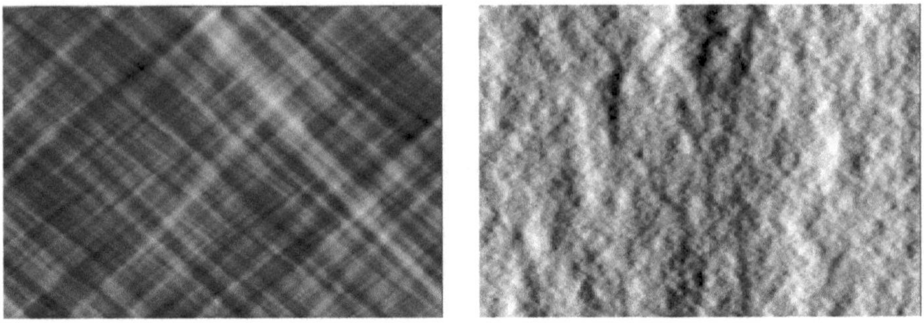

Fig. 9. Spatiotemporal evolution of the density fluctuations in the stable region, modeled by (16). **Left:** $\varepsilon = 0.47$, $\nu = 0.1$, $\eta_0 = 0.5$, and $\mu = 0.0$. **Right:** the same parameters but with friction connected, $\mu = 0.52$

The time evolution of the density fluctuations are given by the eigenvalues of A_k, that can be real or complex conjugate. In the real case, the evolution is purely diffusive, while in the complex case waves propagate. For a given wave-vector k, waves propagate if

$$4(ck^2 + k^4) - (\mu + k^2\nu)^2 > 0, \tag{21}$$

condition that is fulfilled in general when the dissipation mechanisms (μ and ν) are small. The critical condition for the suppression of all waves is that

$$\nu = \frac{c^2 + \mu^2}{c\mu} \tag{22}$$

when all fluctuations diffuse. Note that in the Cahn-Hilliard limit ($\mu \to \infty$) the evolution, then, is simply diffusive. In Fig. 9 numerical solutions of (16) are presented. It is seen that when friction is connected ($\mu \neq 0$), waves are suppressed and the evolution is diffusive.

References

1. P.B. Umbanhowar, F. Melo, and H.L. Swinney, Nature **382**, 793 (1996)
2. R. Ramírez, D. Risso, and P. Cordero, Phys. Rev. Lett. **85**, 1230 (2000)
3. Y. Forterre and O. Pouliquen, Phys. Rev. Lett. **86**, 5886 (2001)
4. M-L. Tan and I. Goldhirsch, Phys. Rev. Lett. **81**, 3022 (1998); L. Kadanoff, Rev. Mod. Phys. **71**, 435 (1999). B.J. Glasser and I. Goldhirsch, Phys. Fluids **13**, 407 (2001)
5. J.D. Gunton, M. San Miguel, and P.S. Sanhi, in *Phase Transitions and Critical Phenomena, volume 8*, edited by D. Domb and J.L. Lebowitz (Academic Press, London, 1983), pp. 267–466
6. E. Livne, B. Meerson, and P.V. Sasorov, Phys. Rev. E **65**, 021302 (2002)
7. J. J. Brey, F. Moreno, and R. García-Rojo, M. J. Ruiz-Montero, Phys. Rev. E **65**, 061302 (2002)

8. M. Argentina, M. Clerc, and R. Soto, Phys. Rev. Lett. **89**, 044301 (2002)
9. S. Warr and J.M. Huntley, Phys. Rev. E **52**, 5596 (1995)
10. I. Goldhirsch and G. Zanetti, Phys. Rev. Lett. **70**, 1619 (1993)
11. P. Zamankhan, A. Mazouchi, and P. Sarkomaa, Appl. Phys. Lett. **71**, 3790 (1997)
12. D.C. Rapaport, Physica A **249**, 232 (1998).
13. E. L. Grossman, T. Zhou, and E. Ben-Naim, Phys. Rev. E **55**, 4200 (1997)
14. L.D. Landau and E.M. Lifshitz, *Statistical Physics* (Pergamon Press, New York, 1969)
15. This mechanism is similar to the one giving rise to the clustering phenomena [10]. Nevertheless, is this later case the granular temperature is not an slave variable
16. E. Khain and B. Meerson, Phys. Rev. E **66**, 021306 (2002)
17. J.J. Brey, M.J. Ruiz-Montero, F. Moreno, and R. García-Rojo, Phys. Rev. E **65**, 011305 (2002).
18. J.T. Jenkins and M.W. Richman, Arch. Rat. Mech. **87**, 355 (1985)
19. J.J. Brey, F. Moreno, and J.W. Dufty, Phys. Rev. E **54**, 445 (1996)
20. C. Bizon, M.D. Shattuck, J.B. Swift, and H.L. Swinney, Phys. Rev. E **60**, 4340 (1999)
21. R. Soto, M. Mareschal, and D. Risso, Phys. Rev. Lett. **83**, 5003 (1999)
22. D. Gass, J. Chem. Phys **54**, 1898 (1971)
23. S. Chapman and T.G. Cowling, *The Mathematical Theory of Non-Uniform Gases*, 3rd ed. (Cambridge University Press, New York, 1970)
24. N.V. Brilliantov and T. Pöschel, *Kinetic Theory of Granular Gases*, (Oxford University Press, 2004)
25. C. Cercignani, *Theory and Application of the Boltzmann Equation* (Elsevier, Amsterdam, 1975)
26. M.S. Green, J. Chem. Phys. **20**, 1281 (1952)
27. H. Grabert, R. Graham, and M. S. Green, Phys. Rev. A **21**, 2136 (1980)
28. M. C. Cross and P. C. Hohenberg, Rev. Mod. Phys. **65**, 851 (1993)
29. J.W. Cahn and J.E. Hilliard, J. Chem. Phys. **28**, 258 (1958)
30. E. Livne, B. Meerson, and P.V. Sasorov, Phys. Rev. E **66**, 050301(R) (2002)
31. B. Meerson, T. Pöschel, P. V. Sasorov, and T. Schwager,`cond-mat/0208286`, (2002)
32. K. Kawasaki , Prog. Theor. Phys. **79**, 161 (1984); Prog. Theor. Phys. **80**, 123 (1984)
33. L.D. Landau and E.M. Lifshitz, *Fluid Mechanics*, Pergamon Press, Oxford (1984)

Birth and Sudden Death of a Granular Cluster

Ko van der Weele, Devaraj van der Meer, and Detlef Lohse

Department of Applied Physics and J.M. Burgers Centre for Fluid Dynamics, University of Twente, P.O. Box 217, 7500 AE Enschede, The Netherlands

Abstract. Granular material is vibro-fluidized in N connected compartments. For sufficiently strong shaking the particles are statistically uniformly distributed over the compartments, but if the shaking intensity is lowered this uniform distribution gives way to a clustered state. The clustering transition is experimentally shown to be of 2nd order for $N = 2$ compartments and of 1st order for $N \geq 3$. In particular, the latter is hysteretic, involves long-lived transient states, and exhibits a striking lack of time reversibility. In the strong shaking regime, a cluster breaks down very abruptly and in its further decay shows anomalous diffusion, with the length scale going as $t^{1/3}$ rather than the standard $t^{1/2}$. We focus upon the self-similar nature of this process. The observed phenomena are all accounted for within a flux model.

1 Introduction

Granular matter is everywhere, yet our understanding of it is still far from complete. At rest, it resembles a solid. When it is flowing, it resembles a fluid. And when it is shaken it resembles a gas. Nevertheless, the dissipative nature of the particle interactions makes it fundamentally different from any standard solid, liquid, or gas, and for this reason it is often called the "fourth state of matter" [1–3]. A better understanding is essential not only for physicists, but also for the many industries that handle granular materials (mining, food production, pharmaceutical industry, construction works, chemical reactors, etc.). It has been estimated [1] that no less than 40% of the capacity of these industries is wasted due to problems that have to do with the transportation and sorting of granular matter. In the gas-phase, which is found for instance in gas-fluidized beds and shakers [4], the main problem is the tendency of the particles to spontaneously form clusters.

This clustering effect can be traced back to the fact that the collisions between the granules are inelastic [5]. Some energy is dissipated in every collision, which means that a relatively dense region (where the particles collide more often than elsewhere) will dissipate more energy, and thus become even denser, resulting in a cluster of slow particles. Vice versa, relatively dilute regions will become more dilute. The few particles in these regions are very rapid ones.

A striking illustration of the clustering phenomenon is provided by the so-called Maxwell Demon experiment [6, 7], depicted in Fig. 1a. It consists of a box divided into two compartments by a wall of height h, with a few hun-

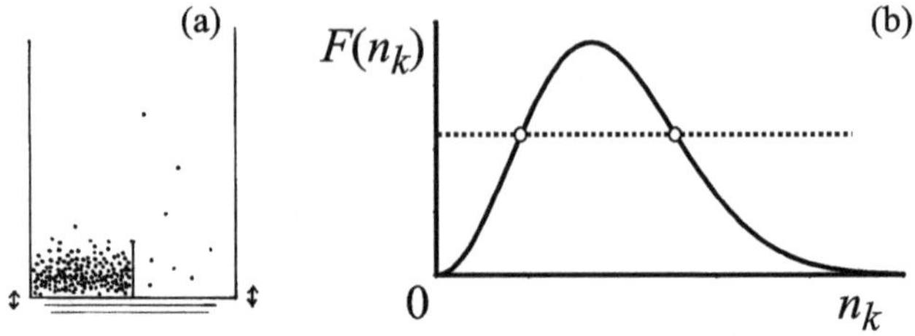

Fig. 1. a Clustering in the Maxwell-demon experiment. Below a certain critical shaking level the particles cluster into one compartment, due to the non-elasticity of the collisions. In terms of the granular temperature (i.e., the mean kinetic energy of the particles) the cluster is cold. The few, much faster particles in the other compartment are warm. This separation in a cold and a hot compartment is reminiscent of what Maxwell's Demon was supposed to accomplish [8]. **b** The flux function $F(n_k)$, i.e., the particle flux from compartment k as a function of the fraction n_k within the compartment. As a result of the inelasticity of the collisions between the particles, F does not grow anymore beyond a certain value of n_k, but decreases instead. The horizontal dashed line indicates that the flux from a relatively empty compartment (left intersection point) can be equal to the flux from a well-filled compartment (right intersection point): this is exactly what happens in the clustered state [7]

dred small beads in each compartment. The beads are brought in a gaseous state by shaking the system vertically, with frequency f and amplitude a. If the shaking is vigorous enough, the dissipation from the inelastic collisions is overpowered by the energy input into the system, and the beads divide themselves uniformly over the two compartments, as in any ordinary molecular gas. But if the driving is lowered below a certain level, the beads cluster in one of the two compartments. One ends up with a "cold" compartment containing a lot of beads (moving rather sluggishly, hardly able to jump over the wall anymore) and a "hot" compartment containing only a few (much more lively) beads. Cold and hot here refer to the so-called granular temperature, i.e., the mean kinetic energy of the particles in each compartment.

In this paper we analyze this experiment quantitatively in terms of a dynamical flux model, and we generalize it to an arbitrary number of connected compartments. The flux model is introduced, and tested against experimental data in Sect. 2. The extension to N compartments is made in Sect. 3. Already for $N = 3$ it turns out that this is more than just an upscaled version of the 2-box system: the physics changes qualitatively. Whereas the clustering for $N = 2$ takes place via a continuous (2nd order) phase transition, for $N = 3$ we find a discontinuous (1st order) phase transition, including a strong hysteretic effect. In Sect. 4 two special phenomena associated with this transition for $N = 3$ are discussed. Section 5 is about the continuum limit for $N \to \infty$. We conclude (in Sect. 6) with a brief overview of further extensions and applications.

2 Flux Model

The clustering can be described quantitatively by means of a flux model [7]. The central idea behind it is that in equilibrium the average particle flux from left to right is equal to the flux from right to left. Hence, the asymmetric equilibrium of the clustered situation can only be explained if the flux F from one compartment to the other does not monotonously increase with the number of particles within that compartment. Rather, it is a function that attains a maximum (at a certain particle number) and thereafter decreases again, as sketched in Fig. 1b. So the flux from a nearly empty box can balance the flux from a well-filled box, and that is exactly what is the case in the clustered situation.

This one-humped form of F can be understood as follows. If there are no particles in the box, it can give nothing to its neighbors: so the flux starts out from $F(0) = 0$. Initially, the flux increases with growing particle number, just as in an ordinary gas. Beyond a certain filling level, however, the increasingly frequent collisions make the particles in the compartment so slow that the flux starts to decrease. Eventually F goes to zero again.

In agreement with this, and based on the kinetic theory for dilute granular gases, Eggers proposed the following analytic approximation for the flux (rewritten here in a form suited for an arbitrary number of N connected compartments) [7, 9]:

$$F(n_k) = An_k^2 e^{-N^2 B n_k^2}. \tag{1}$$

This is indeed a function of the form sketched in Fig. 1b, with a maximum at $n_k = 1/(N\sqrt{B})$. Here n_k is the fraction of particles in the k-th box, normalized to $\sum_k n_k = 1$.

The factors A and B depend on the particle properties (their radius r, and the restitution coefficient e of the collisions between particles), the experimental setup (e.g. the height h of the wall), and the driving parameters a and f. The factor A determines the absolute rate of the flux, and can simply be incorporated in the time scale. The dimensionless factor B is more important: it governs the phase transition towards the clustering state. It is given by

$$B \propto \frac{gh}{(af)^2}(1 - e^2)^2 \left(\frac{r^2 P_{tot}}{\Omega N}\right)^2, \tag{2}$$

with P_{tot} the total number of particles, and Ω the ground area of each box. For a given granular material (r and e fixed), B can be raised either by increasing the value of h or by decreasing the value of af, i.e., by reducing the driving. Note that for elastically colliding particles ($e = 1$) the factor B is equal to zero, which means that in that case the flux function F is monotonously increasing, and clustering is impossible.

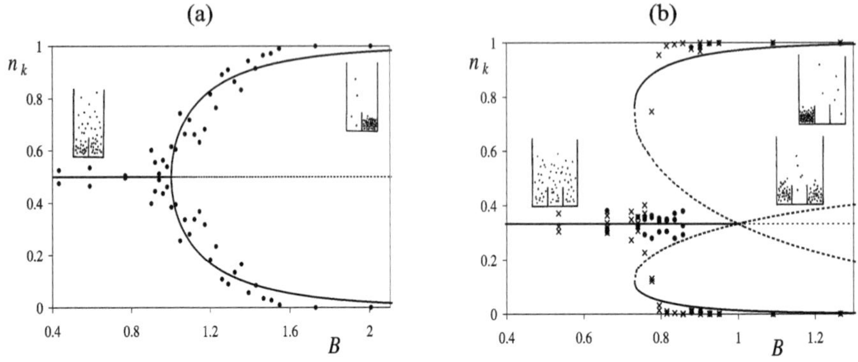

Fig. 2. **a** Bifurcation diagram for the 2-box system ($k = 1, 2$). The solid and dashed lines correspond to the flux model, and the dots are experimental measurements. The transition to the clustering state is seen to take place via a pitchfork bifurcation. **b** Bifurcation diagram for the 3-box system ($k = 1, 2, 3$). The *solid and dashed lines* correspond to the flux model, and the *dots and stars* are experimental measurements (*dots* for measurements that were started from the uniform distribution $\{1/3, 1/3, 1/3\}$, *stars* for those that were started from a single peaked distribution). Note the hysteresis in the clustering transition

Let us see how the flux model works out for the 2-box system. The dynamics is given by the balance equation

$$\frac{dn_1}{dt} = -F(n_1) + F(n_2) = -F(n_1) + F(1 - n_1). \tag{3}$$

In equilibrium one has $dn_1/dt = 0$. For $B < 1$ there exists only one equilibrium, namely the stable uniform $\{1/2, 1/2\}$ distribution. For $B > 1$ it becomes unstable and gives way to an asymmetric equilibrium. In other words, there is a symmetry breaking bifurcation at $B_{bif} = 1$. In Fig. 2a we have drawn the corresponding bifurcation diagram [7, 9].

For the experimental verification of this diagram we put 600 glass beads ($r = 1.25$ mm, $e = 0.97$) in a perspex tube of inner radius 27.5 mm, divided into two equal compartments by a wall of height $h = 23.0$ mm. The tube was mounted on a shaker with an adjustable frequency and amplitude, so that B could be varied. In the present experiment this was done by means of the frequency f, at a fixed value of $a = 6.5$ mm. The measurements are included as solid dots in Fig. 2a. Clearly, theory and experiment agree on the fact that the clustering transition takes place via a pitchfork bifurcation.

3 More Than Two Compartments: Hysteresis

The flux function was invented for 2 compartments, but we want to go beyond that and apply it to an arbitrary number of compartments. The dynamics for the general N-box system, assuming only nearest neighbor interactions, is given by [9, 10]:

$$\frac{dn_k}{dt} = F(n_{k-1}) - 2F(n_k) + F(n_{k+1}) \qquad \text{for } k = 1, ..., N. \tag{4}$$

The first three terms on the right hand side denote the flux out of box k into its two neighbors (hence the factor 2), and the flux into the box from its neighbors [12].

The case with $N = 3$ is already considerably richer than the original two-compartment system [9]. Again, we find a uniform distribution $\{1/3, 1/3, 1/3\}$ at high driving levels (small B) and a single-cluster distribution at milder driving levels (large B), but in contrast to the $N = 2$ case the transition between them is hysteretic, see Fig. 2b. That is, the value of B at which the transition occurs when one goes from vigorous to mild driving is different from the B-value when one goes in the opposite direction. The experimental measurements were done with 600 glass beads ($r = 1.25$ mm, $e = 0.97$) in a cyclic 3-compartment tube mounted on a shaker.

The flux model confirms these experimental findings. The uniform distribution $\{1/3, 1/3, 1/3\}$ turns unstable at $B_{bif} = 1$. Already at $B_{sn} = 0.73$, however, a stable clustered solution comes into existence by means of a saddle-node bifurcation. The basin of attraction of this clustered state grows as B increases, while its unstable counterpart closes in upon $\{1/3, 1/3, 1/3\}$ and finally destabilizes this uniform solution when it coincides with it at $B_{bif} = 1$. (A more detailed analysis is given in refs. [9, 10]). Thus, if one gradually increases the value of B from zero upwards (i.e, if one reduces the driving), the transition from the uniform to the clustered state takes place at $B_{bif} = 1$. If the value of B is then gradually turned down again, the reverse transition occurs at the lower value $B_{sn} = 0.73$ (hysteresis). Physically, this is easily understood: the particles in the cluster keep each other slow, dissipating a lot of energy, and it requires an extra effort in terms of driving power (i.e., a lower value of B) to overcome this.

The unstable counterpart of the clustered state is not only instrumental in destabilizing the uniform solution, but it plays an important role also beyond $B = 1$. If we start out from the (unstable) uniform distribution, the system can go to the clustered state either directly, or - and this is actually more likely - via this counterpart state, which forms a transient situation in which two boxes compete for dominance. This approach towards the 1-cluster state can be interpreted as a coarsening process. We will come back to it in the next section.

4 Coarsening and Sudden Death

What is true for 3 compartments, holds also for $N > 3$ [10], so it is not the 3-box system that is exceptional but the 2-box system. As an example,

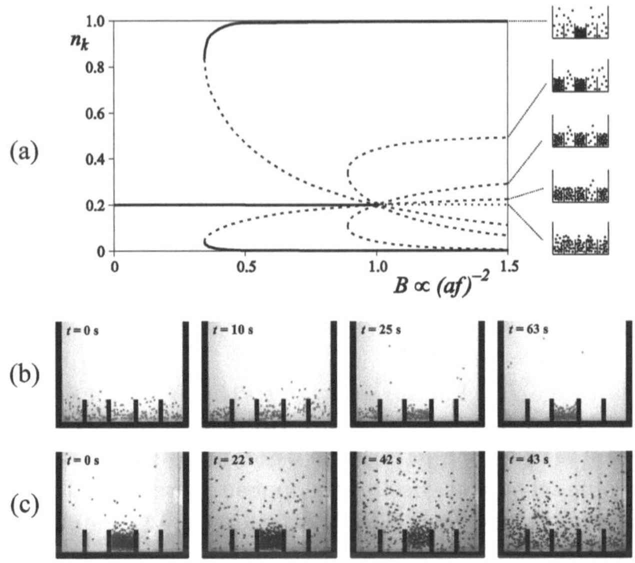

Fig. 3. **a** Bifurcation diagram for the 5-compartment system. The *solid lines* correspond to stable equilibrium states, the *dashed lines* to unstable ones. **b** Four snapshots from a clustering experiment. Note that the particles do not immediately cluster into one compartment, but first go through a transient 2-cluster state. **c** Declustering at higher shaking strength. The cluster is clearly present until $t = 42$ s, then suddenly collapses, leaving no trace one second later

in Fig. 3a the bifurcation diagram for $N = 5$ compartments is given. The hysteresis is even more pronounced than in the case for $N = 3$ (the first saddle-node bifurcation already takes place at $B_{sn} = 0.34$) and there are also more transient states.

In Fig. 3b we show four stages in the clustering process for the 5-box system, starting out from the uniform distribution, at B slightly larger than 1. The particles do not cluster immediately into one compartment, but form a transient state first, in which two of the boxes (here the leftmost and the center compartment) are in competition. This transient state lives for roughly half a minute, until finally one of the boxes wins. The system always ends up in a one-cluster state, since this is the only truly stable equilibrium (for $B > 1$), but it does not necessarily have to be in the middle box. In fact, the end boxes have a slightly higher probability to draw the cluster towards them, since the extra sidewalls of these boxes form an additional source of energy dissipation.

It is also interesting to have a look at the reverse process (*declustering*), which occurs for $B < B_{sn}$. In industrial situations such as conveyor belts and sorting machines, declustering is often much more desirable than clustering. It is also very interesting from a purely physical point of view, as Fig. 3c (at $B \approx 0.96 B_{sn}$) shows.

The cluster persists for a remarkably long time. After 42 seconds of shaking it is still clearly visible, and then suddenly (within one second) it vanishes without leaving a trace. At $t = 43$ s it has given way to the uniform distribution. We call this phenomenon "sudden death" [13]. Obviously, declustering is by no means the same as clustering in reverse time order. This lack of reversibility is yet another consequence of the inelastic collisions. It is the rule in dissipative systems, and as such does not come as a surprise, yet the intensity of the effect is striking.

The observed declustering behavior can be understood from the bifurcation diagram. It occurs for B-values just below B_{sn}; i.e., just to the left of $B = 0.73$ for $N = 3$ (Fig. 2b), or $B = 0.34$ for $N = 5$ (Fig. 3a). Initially, all particles are in the middle box; i.e., $n_k = \{0, 1, 0\}$ or $n_k = \{0, 0, 1, 0, 0\}$ respectively. When we start to shake, the material slowly leaks out of the cluster into the surrounding compartments, and the system seems to be heading for a dynamical equilibrium (one well-filled compartment surrounded by nearly empty compartments), just as if the corresponding lines in the bifurcation diagram were extrapolated to the left of B_{sn}. Only when this imaginary line has been passed (and hence the corresponding equilibrium has proven to be false) the system shoots towards the uniform distribution. The time from the start of the experiment until this sudden collapse, the lifetime τ of the cluster, can be calculated from the flux model. The result turns out to agree very well with the experimentally measured lifetime, for all $N \geq 3$. A full account of the sudden death phenomenon is given in [11].

To get an idea how the cluster lifetime is calculated, consider the case of $N = 3$ compartments. Let the initial cluster be positioned in box 2. The material spills symmetrically out of the cluster, so $n_1 = n_3 = \frac{1}{2}(1 - n_2)$, and thus the equation of motion for the cluster fraction is:

$$\frac{dn_2}{dt} = F(n_1) - 2F(n_2) + F(n_3) = 2\left(F\left(\frac{1}{2}(1 - n_2)\right) - F(n_2)\right), \tag{5}$$

and this is readily integrated to give:

$$\tau = \int_{n_{thr}}^{1} \frac{dn_2}{2\left(F(n_2) - F(\frac{1}{2}(1 - n_2))\right)}. \tag{6}$$

Here n_{thr} is the threshold value where the system passes through the imaginary equilibrium line, but its precise value is not too critical. In Fig. 4a the above theoretical expression for τ is compared with the experimental data for the 3-box system, for various values of B, and the agreement is seen to be excellent. Naturally, at $B = B_{sn}$ the lifetime becomes infinite, since the cluster becomes stable there.

In Fig. 4b we present lifetime data for further values of N, based on experiments and numerical simulations. The first thing to note is that the vertical axis (τ) spans no less than 8 orders of magnitude. As explained above, τ goes to infinity at $BN^2 = B_{sn}N^2$, i.e., at the first saddle-node bifurcation

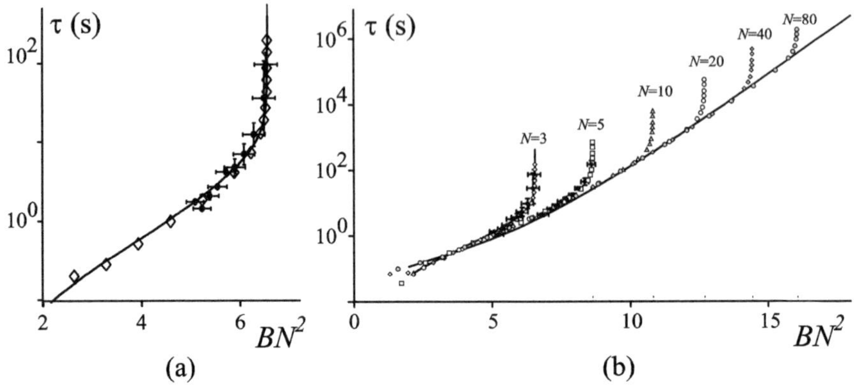

Fig. 4. a The lifetime τ of a particle cluster in a 3-box system, as a function of the shaking parameter. The *dots with the error bars* represent experimental data (based on 15 repetitions of the experiment), the *diamonds* are numerical data from the flux model, and the *solid curve* is the theoretical value from (6). At $B = B_{sn} = 0.73$ (or equivalently, $BN^2 = 6.57$) the cluster becomes stable and hence its lifetime diverges. **b** Lifetime data for various values of N, determined from experiments with $N = 3, 5$ boxes (*dots with error bars*) and from numerical evaluation of the flux model with $N = 3, 5, 10, 20, 40, 80$ boxes (*empty symbols*). The *solid curves* are analytical solutions for $N = 3$ (see also Fig. a) and for the envelope curve, which goes roughly as $\exp[(3/2)BN^2]$ (cf. (7))

for each N. Away from these divergencies, the lifetimes are seen to lie on a universal envelope curve, independent of the value of N. This reflects the fact that for these values of BN^2 the sudden death occurs *before* the particles leaking out of the cluster have had time to fill the outermost boxes of the system to any significant level. Therefore, the behavior does not depend on the value of N. (Or stated otherwise: the system does not feel its finite size during the cluster's lifetime and hence the number of boxes can be taken to be infinite, which brings us to the continuum description of the next section). The form of the envelope curve can be calculated from the continuum version of the flux model and is given by [11]:

$$\tau = K \left[\int_{n_{thr}}^{1} \frac{e^{(3/4)\widetilde{B}n^2} dn}{n^{3/2}(1 - \widetilde{B}n^2 e^{-\widetilde{B}n^2})} \right]^2, \qquad \text{with } \widetilde{B} = BN^2. \tag{7}$$

The only free parameter in this expression is the constant K, which sets the absolute scale of τ.

All the above effects (hysteresis, coarsening, sudden death) become more pronounced with growing number of compartments. For example, in Fig. 5a the bifurcation diagram for $N = 80$ is depicted. The hysteresis is seen to reach down almost all the way to $B = 0$, and there is a whole forest of transient states, in which the system can easily get stuck. In order to see this happen, we perform a simulation based on the flux model including a

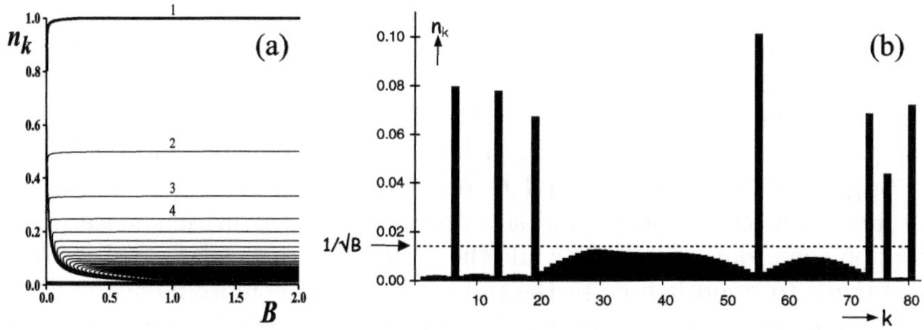

Fig. 5. a Bifurcation diagram for the 80-compartment system. The number 1 indicates the single-cluster state (the only clustered state that is mathematically stable); the other numbers indicate transient multi-cluster states. **b** A snapshot from a flux model simulation (including a statistical noise term) at $B = 0.90$. The depicted situation, a transient state with 7 clusters, has arisen from an initial condition in which the 80 boxes were filled more or less uniformly

noise term. We take $B = 0.9$ and start out from a situation in which the granular material is distributed more or less evenly over the boxes. The basin of attraction of the uniform equilibrium is quite narrow at $B = 0.9$, so even a small statistical fluctuation is sufficient to throw the system out of it, and from that moment on the system is heading towards the 1-cluster situation. However, as we know, it does not go there in one stroke but via transient multi-cluster states. In Fig. 5b we show the situation after a few minutes of shaking, when the system happens to be in a transient state consisting of 7 clusters. Next to the clusters there are depletion zones, and in between we see a characteristic hilly pattern.

This 7-cluster transient state may be compared with the two-cluster transient state for the 5-box system in Fig. 3b. And just like there, the clusters will compete each other until eventually we end up in the 1-cluster state, which after all is the only clustered state that is truly stable. But whereas in the 5-box system this coarsening process takes roughly a minute, in the present case it may take years. The competition between clusters is extremely slowed down when the distance between them becomes larger. In practice, the 80-box experiment will be turned off long before the 1-cluster state is reached.

5 The Limit for $N \to \infty$ Compartments: Anti-diffusion

Finally, we consider the limit of an infinite number of boxes. In this case the fraction $n_k(t)$ is replaced by its continuum version $n(x,t)$, and the equation of motion takes the following form [11]:

$$\frac{\partial n(x,t)}{\partial t} = (\Delta x)^2 \frac{\partial^2 F\left(n(x,t)\right)}{\partial x^2} = C(n)\left(\frac{\partial n(x,t)}{\partial x}\right)^2 + D(n)\frac{\partial^2 n(x,t)}{\partial x^2} \quad (8)$$

Here $C(n) = (\Delta x)^2 \partial^2 F/\partial n^2$ and $D(n) = (\Delta x)^2 \partial F/\partial n$, with Δx the (infinitesimal) distance between boxes. The conservation condition now reads $\int n(x,t)dx = 1$.

This partial differential equation looks very similar to the Kardar-Parisi-Zhang equation [13], which is used to describe the roughening of interfaces. In the original KPZ equation C and D are positive constants, and in that case the first term on the right hand side (the nonlinear term) has a steepening effect on the profile of $n(x,t)$, while the second term (which represents diffusion) tends to smoothen it. In the present case this is more intricate: C and D are functions of $n(x,t)$ and can become negative, which means that both terms can either have a steepening or a smoothening effect. For instance, the diffusion coefficient D will be negative on the decreasing right part of the plot in Fig. 1b (i.e., for $n > 1/\sqrt{\widetilde{B}}$, with $\widetilde{B}$ taking the place of $N^2 B$ in this limit). We then get anti-diffusion, local accumulation of granular material, which indeed is the continuum analog to the clustering.

Also on the increasing left part of $F(n)$, where D is positive, the diffusion shows an uncommon feature. The width of a profile in this regime grows as $t^{1/3}$, and its height decreases as $t^{-1/3}$, instead of $t^{1/2}$ and $t^{-1/2}$ as in ordinary diffusion [13]. This anomalous exponent 1/3 comes out straightforwardly from an analysis of the vigorous-shaking limit $\widetilde{B} = 0$, for which (8) reduces to

$$\frac{\partial n(x,t)}{\partial t} = 2\widetilde{A}\left(\frac{\partial n(x,t)}{\partial x}\right)^2 + 2\widetilde{A}n\frac{\partial^2 n(x,t)}{\partial x^2}, \tag{9}$$

with $\widetilde{A} = A(\Delta x)^2$. The diffusion coefficient ($D = 2\widetilde{A}n$) is invariably positive in this case, so a given profile always spreads out, just as one might expect for very strong shaking. Moreover, in simulations it is observed that the profile diffuses without losing its basic form, indicating that the process is a self-similar one. Therefore, given that $n(x,t)$ has the dimension length^{-1} (as follows directly from the conservation condition) and $\widetilde{A}$ has the dimension length^3/time (as can be seen from (9)), we try a solution of the following form [11, 14]:

$$n(x,t) = \frac{1}{(\widetilde{A}t)^{1/3}}G(\eta), \quad \text{with } \eta = \frac{x}{(\widetilde{A}t)^{1/3}}. \tag{10}$$

It is here that the coefficient 1/3 comes in. The function G does not depend on x or t separately, but only on the combined dimensionless variable η. Inserting the ansatz (10) into (9), the partial differential equation for $n(x,t)$ is transformed into an ordinary differential equation for $G(\eta)$:

$$G + \eta\frac{\partial G}{\partial \eta} + 6\left(\frac{\partial G}{\partial \eta}\right)^2 + 6G\frac{\partial^2 G}{\partial \eta^2} = 0. \tag{11}$$

It is non-linear, but nevertheless not very difficult to solve: $G(\eta) = G_0 - (1/12)\eta^2$. This inverted parabola represents, in one single curve, the whole evolution of the diffusing profile.

6 Extensions and Applications

There are many interesting generalizations and applications. Presently we are working on a system in which the compartments are not on one level, but form a staircase, resembling an industrial conveyor belt. The central topic here is the competition between clustering and the natural tendency of the particles to stream downwards. The experiment becomes even more interesting when the linear array of boxes is extended to a 2-dimensional grid of compartments. In that case, the clusters are no absolute barriers to the flux anymore (they can be circumvented) and this gives rise to beautiful front propagation patterns.

We also study mixtures consisting of particles of different sizes [15]. This is done with an eye to practical applications, where granular material is rarely mono-disperse. One of the main themes here is the competition between clustering and the tendency of the large and small particles to segregate.

A third application is the traffic jam problem. There exists a strong analogy between granular and traffic flow. In the past decade this has already led to several successful traffic strategies, especially in situations where the cars could still be treated as individual particles [16]. On a much larger scale, e.g. the Dutch network of highways, it is no longer feasible to consider the cars individually and here a coarse-grained approach may help. To this end, we divide the highway in imaginary cells (1 to 5 km per cell) and concentrate on the *density* ρ_k of cars in these cells. The traffic flow from one cell to the next is given by a flux function $F(\rho_k)$. Actually, this function has already been measured a number of times in real-life situations (among traffic analysts it is known as the "fundamental diagram" [17]), and it shows a marked resemblance with the flux function in Fig. 1b. At low densities the cars flow freely from cell to cell and $F(\rho_k)$ is an increasing function of ρ_k, whereas at high densities they hinder each other and $F(\rho_k)$ decreases. As we have seen in the present paper, clustering (traffic jam formation) is adequately described by this type of flux model. It should be possible, given the traffic situation at a certain moment, to predict in which cells traffic jams are likely to be formed within the next hour - and perhaps take measures to prevent them.

In conclusion, the Maxwell Demon experiment is a rich dynamical system full of surprising effects. As for many problems in granular dynamics, the questions are interesting from a fundamental point of view, and at the same time important for practical applications. And the answers are obtained by an approach in which experiment, theory, and numerical simulations go hand in hand.

References

1. H.M. Jaeger, S.R. Nagel, R.P. Behringer: Granular solids, liquids, and gases, Rev. Mod. Phys. **68**, 1259 (1996)
2. L.P. Kadanoff: Built upon sand: theoretical ideas inspired by granular flows, Rev. Mod. Phys. **71**, 435 (1999)
3. P.G. de Gennes: Granular matter: a tentative view, Rev. Mod. Phys. **71**, S374 (1999)
4. J.A.M. Kuipers, W.P.M. van Swaaij: Computational Fluid Dynamics applied to Chemical Reaction Engineering, Adv. Chem. Engineering **24**, 227 (1998)
5. I. Goldhirsch, G. Zanetti: Clustering instability in dissipative gases, Phys. Rev. Lett. **70**, 1619 (1993); A. Kudrolli, M. Wolpert, J.P. Gollub: Cluster formation due to collisions in granular material, Phys. Rev. Lett. **78**, 1383 (1997)
6. H.J. Schlichting, V. Nordmeier: Strukturen im Sand, Math. Naturwiss. Unterr. **49**, 323 (1996)
7. J. Eggers: Sand as Maxwell's demon, Phys. Rev. Lett. **83**, 5322 (1999)
8. A.S. Leff, A.F. Rex: *Maxwell's Demon: Entropy, Information, Computing* (Adam Hilger, Bristol, 1990)
9. K. van der Weele, D. van der Meer, M. Versluis, D. Lohse: Hysteretic clustering in granular gas, Europhys. Lett. **53**, 328 (2001)
10. D. van der Meer, K. van der Weele, D. Lohse: Bifurcation diagram for compartmentalized granular gases, Phys. Rev. E **63**, 061304 (2001)
11. D. van der Meer, K. van der Weele, D. Lohse: Sudden Collapse of a Granular Cluster, Phys. Rev. Lett. **88**, 174302 (2002)
12. Equation 4 is written down for a cyclic array of boxes, for which box N and 1 are (connected) neighbors. If the array is non-cyclic, the equation for the end box 1 reads $dn_1/dt = -F(n_1) + F(n_2)$, and analogously for the end box N. The differences between the two cases are small [10]
13. M. Kardar, G. Parisi, Y.C. Zhang: Dynamic scaling of growing interfaces, Phys. Rev. Lett. **56**, 889 (1986)
14. G.J. Barenblatt: *Scaling, self-similarity, and intermediate asymptotics* (Cambridge Univ. Press, Cambridge, 1996)
15. R. Mikkelsen, D. van der Meer, K. van der Weele, D. Lohse: Competitive Clustering in a Bidisperse Granular Gas, Phys. Rev. Lett. **89**, 214301 (2002)
16. M. Schreckenberg, D.E. Wolf (eds.): *Traffic and Granular Flow '97* (Springer, Singapore, 1998)
17. D. Helbing: Traffic and related self-driven many-particle systems, Rev. Mod. Phys. **73**, 1067 (2001)

Vibrated Granular Media as Experimentally Realizable Granular Gases

Sean McNamara[1,3] and Eric Falcon[2]

[1] Centre Européen de Calcul Atomique et Moléculaire
[2] Laboratoire de Physique, École Normale Supérieure de Lyon, 46, allée d'Italie, 69007 Lyon, France
[3] current address: Institute for Computer Applications 1, Universität Stuttgart, 70569 Stuttgart, Germany

Abstract. We report numerical simulations of strongly vibrated granular materials designed to mimic recent experiments performed both in presence [1–3] or absence [4] of gravity. We show that a model with impact velocity dependent restitution coefficient is necessary to bring the simulations into agreement with experiments. We measure the scaling exponents of the granular temperature, collision frequency, impulse and pressure with the vibrating piston velocity. As the system changes from a homogeneous gas state at low density to a clustered state at high density, these exponents are all found to decrease continuously with the particle number. In absence of gravity, a loose cluster appears near the upper wall, opposite the piston, and acts as a buffer for fastest particles leading to unexpected non-extensive scaling exponents ; whereas in presence of gravity, the cluster bounces as a single inelastic body. All these results differ significantly from classical inelastic hard sphere kinetic theory and previous simulations, both based on a constant restitution coefficient.

1 Introduction

Granular gases [5] often seem to be the exclusive domain of theory and simulation. For example, the "freely cooling" granular gas [5] is impossible to experimentally realize: even if an experimentalist somehow managed to implement periodic boundaries in all directions, he would still have to contend with energy and time scales that vary over many orders of magnitude. Another frequently studied example of granular gases is the vibrated granular medium. This case is much closer to experiments, because it is relatively simple to put particles in a box and shake them, or drive particles with a vibrating piston. Indeed, there are many experimental [6–9], numerical [9–13], and theoretical [6, 14–16] studies of this system. However, numerous questions remain about the link between experiments on one hand, and theory and simulations on the other. Most numerical and theoretical studies were not intended to be compared with experiments. Therefore, they have parameter values far from the experimental ones, and none of them predict even the most basic features of the experimental results. For example, numerical and theoretical studies often assume that the vibration frequency is very high so that the shaking wall can be replaced by a thermal boundary. The vibrated granular gas thus attains a steady state. But very high frequencies with an amplitude sufficient to fluidize the granular materials are difficult to attain experimentally; the vibration amplitude in experiments is often a significant fraction of the size of the container. In experiments, the pressure and granular temperature all vary strongly over the period of one vibration cycle. It has therefore been impossible to compare the experimental and numerical results in a meaningful way.

In this paper, we bridge the gap between experiments and numerics by presenting simulations of strongly vibrated granular materials designed to mimic recent experiments [1–4]. We present the first simulations which resemble the experiments for a large range of parameters. We show that two parameters are especially important for the agreement between experiment and simulation. First of all, an impact velocity dependent restitution coefficient is necessary to bring the simulations into agreement with experiment. Most previous studies consider only constant restitution coefficient [9–13]. Secondly, it is important to explicitly consider the number of particles N. Studying only one value of N or comparing results obtained at different N can lead to interpretive difficulties.

We measure the scaling exponents of the granular temperature, collision frequency, impulse and pressure with the vibrating piston velocity, and find results that differ significantly from classical inelastic hard sphere kinetic theory and previous simulations. Finally, we show that the system undergoes a smooth transition from a homogeneous gas state at low density to a clustered state at high density. Depending on whether gravity is present or not, the transition manifests itself in different ways: due to inelastic collisions (through the restitution coefficient properties), a nearly motionless loose cluster is observed in microgravity with non-extensive scaling properties, whereas a bouncing cluster appears under gravity.

2 Description of the Simulations

2.1 The Variable Coefficient of Restitution

The greatest difference between our simulations and the previous numerical studies of vibrated granular media [9–13] is that we use a restitution coefficient that depends on impact velocity. The restitution coefficient r is the ratio between the relative normal velocities before and after impact. In all previous simulations of strongly vibrated granular media, the coefficient of restitution is considered to be constant and lower than 1. However, since a century, it has been shown from impact experiments that r is a function of the impact velocity v [17–23]. Indeed, for metallic particles, when v is large ($v \gtrsim 5$ m/s [20]), the colliding particles deform fully plastically and $r \propto v^{-1/4}$ [17–21]. When $v \lesssim 0.1$ m/s [20], the deformations are elastic with mainly viscoelastic dissipation, and $1 - r \propto v^{1/5}$ [21–26]. Such velocity-dependent restitution coefficient models have recently shown to be important in numerical [27–34]

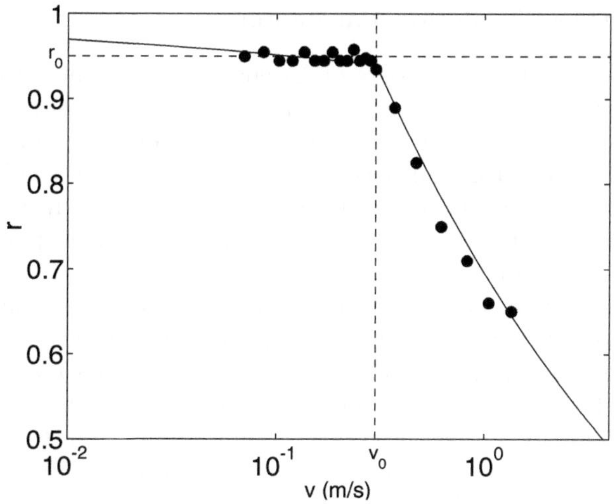

Fig. 1. The restitution coefficient r as a function of impact velocity v, as given in (1) (*solid line*). The dashed lines show $v_0 = 0.3$ m/s and $r_0 = 0.95$. Experimental points (•) for steel spheres were extracted from Fig. 1 of [39]

and experimental [23, 35–38] studies. Applications include: granular fluidlike properties (convection [28], surface waves [29]), collective collisional processes (energy transmission [30], absence of collapse [23, 31]) and planetary rings [32–38]. But surprisingly, such model has not yet been tested numerically for strongly vibrated granular media.

In this paper, we use a velocity dependent restitution coefficient $r(v)$ and join the two regimes of dissipation (viscoelastic and plastic) together as simply as possible, assuming that

$$r(v) = \begin{cases} 1-(1-r_0)\left(\frac{v}{v_0}\right)^{1/5}, & v \le v_0, \\ r_0\left(\frac{v}{v_0}\right)^{-1/4}, & v \ge v_0, \end{cases} \tag{1}$$

where $v_0 = 0.3$ m/s is chosen, throughout the paper, to be the yielding velocity for stainless steel particles [20, 39] for which r_0 is close to 0.95 [39]. Note that $v_0 \sim 1/\sqrt{\rho}$ where ρ is the density of the sphere [20]. We display in Fig. 1 the velocity dependent restitution coefficient of (1), with $r_0 = 0.95$ and $v_0 = 0.3$ m/s, that agrees well with experimental results on steel spheres from [39]. As also already noted by [20], the impact velocity to cause yield in metal surfaces is indeed relatively small. For metal, it mainly comes from the low yield stress value ($Y \sim 10^9$ N/m^2) with repect to the elastic the Young modulus ($E \sim 10^{11}$ N/m^2). Most impacts between metallic bodies thus involve some plastic deformation.

2.2 The Other Simulation Parameters

The numerical simulation consists of an ensemble of identical hard disks of mass $m \approx 3\times10^{-5}$ kg excited vertically by a piston in a two-dimensional box. Simulations are done both in the presence and absence of uniform gravity g. Collisions are assumed instantaneous and thus only binary collisions occur. For simplicity, we neglect the rotational degree of freedom. Collisions with the wall are treated in the same way as collisions between particles, except the wall has infinite mass.

Motivated by recent 3D experiments on stainless steel spheres, 2 mm in diameter, fluidized by a vibrating piston [1–3], we choose the simulation parameters to match the experimental ones: in the simulations, the vibrated piston at the bottom of the box has amplitude $A = 25$ mm and frequencies 5 Hz $\leq f \leq$ 50 Hz. The piston is nearly sinusoidally vibrated with a waveform made by joining two parabolas together, leading to a maximum piston velocity given by $V = 4Af$. The particles are disks $d = 2$ mm in diameter with stainless steel collision properties through v_0 and r_0 (see Fig. 1). The box has width $L = 20$ cm and horizontal periodic boundary conditions. Since our simulations are two dimensional, we consider the simulation geometrically equivalent to the experiment when their number of layers of particles, $n = Nd/L$, are equal. Hence in the simulation, a layer of particles, $n = 1$, corresponds to 100 particles. We checked that n is an appropriate way to measure the number of particles by also running simulations at $L = 10$ cm and $L = 40$ cm. None of this paper's results depend significantly on L, except in Sect. 4.2, where this point is discussed. As in the experiments, the height h of the box depends on the number of particles in order to have a constant difference $h - h_0 = 15$ mm, where h_0 is the height of the bed of particles at rest. Heights are defined from the piston at its highest position.

3 Comparison of Simulation and Experiment

3.1 The Importance of the Variable Coefficent of Restitution

We examine first the dependence of the pressure on the number of particle layers for maximum velocity of the piston $1 \lesssim V \lesssim 5$ m/s ($V = 4Af$). The time averaged pressure at the upper wall is displayed in Fig. 2 as a function of n for various f: from the experiments of Falcon et al. [1–3] (see Fig. 2a), from our simulations with velocity dependent restitution coefficient $r = r(v)$ proposed in (1) (see Fig. 2b), and with constant restitution coefficient $r = 0.95$, often used to describe steel particles (see Fig. 2c), and finally with an unrealistic constant restitution coefficient $r = 0.7$ (see Fig. 2d). Simulations with $r = r(v)$ give results in agreement with the experiments: At constant external driving, the pressure in both Figs. 2a and 2b passes through a maximum for a critical value of n roughly corresponding to one particle layer. For $n < 1$, most particles are in vertical ballistic motion between the piston and the lid.

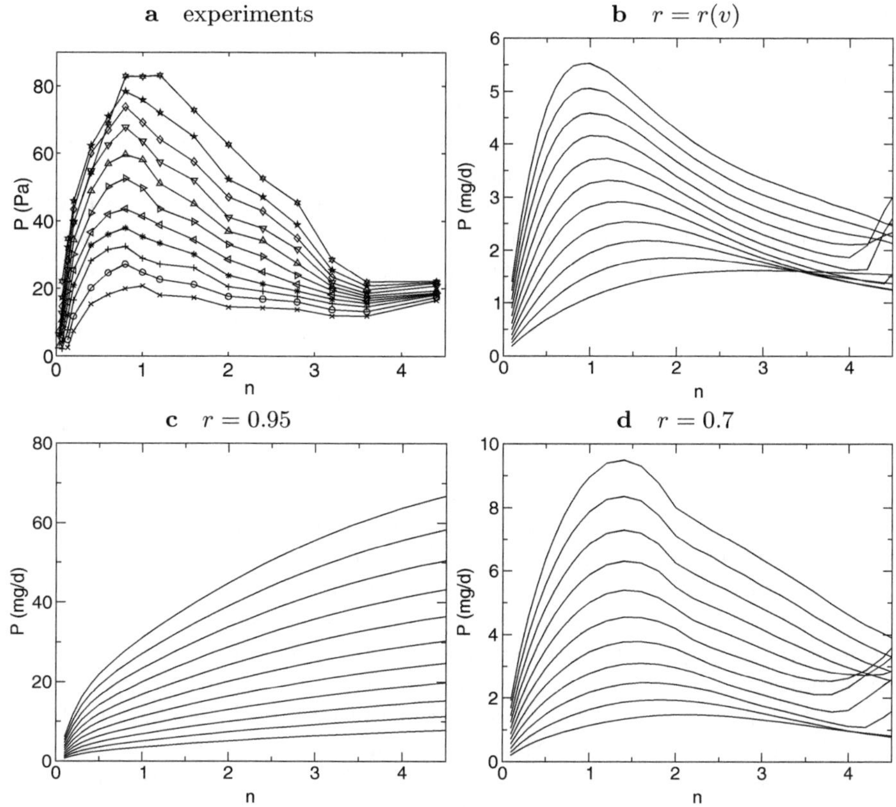

Fig. 2. Time averaged pressure P on the top of the cell as a function of particle layer, n, for various vibration frequencies, f : **a** Experimental results from [1–3] for stainless steel beads 2 mm in diameter, with $A = 25$ mm, $10 \leq f \leq 20Hz$ with a 1 Hz step (from lower to upper) and $h - h_0 = 5$ mm. **b** Numerical simulation where the coefficient of restitution is given by (1), **c** Numerical simulation with a coefficient of restitution is 0.95, independent of impact velocity, **d** Numerical simulation with a coefficient of restitution is 0.7, independent of impact velocity. The simulations **b–d** are 2D with gravity, done for 2 mm disks, with $A = 25$ mm, $10 \leq f \leq 30$ Hz with a 2 Hz step (from lower to upper) and $h - h_0 = 15$ mm. In the simulations, the two-dimensional pressure is given in units of $mg/d \approx 0.15$ Pa $\cdot$ m

Thus, the mean pressure increases roughly proportionally to n. When n is increased such that $n > 1$, interparticle collisions become more frequent. The energy dissipation is increased and thus the pressure decreases. As we show below in Sect. 4.1, this maximum pressure is not due to gravity because it also appears in simulations with $g = 0$ and $r = r(v)$. Turning our attention to Fig. 2c, we see that setting $r = 0.95$ independently of impact velocity gives pressure qualitatively different from experiments. The difference between Figs. 2b and 2c can be understood by considering a high velocity collision (e.g. $v = 1$ m/s). In Fig. 2b, this collision has a restitution coefficient of $r = r(1 \text{ m/s}) \approx 0.7$ (see Fig. 1), whereas in Fig. 2c, r is fixed at 0.95 for all collisions. This means that for equal collision frequencies, dissipation is much stronger for $r = r(v)$ than for $r = 0.95$, because the high velocity collisions dominate the dissipation. Stronger dissipation leads to lower granular temperatures and thus to lower pressures.

We can check this interpretation by changing the constant restitution coefficient to $r = 0.7$ and then comparing it to $r = r(v)$. In these two cases, the high velocity collisions will have roughly the same restitution coefficient. We indeed observed a pressure that decreases for large n for constant $r = 0.7$ (see Fig. 2d). Therefore, surprisingly, constant $r = 0.7$ reproduces more precisely the experimental pressure measurements than constant $r = 0.95$, even though $r = 0.95$ or $r = 0.9$ is often given as the restitution coefficient of steel. Consulting Fig. 1, one can see that $r(v) = 0.7$ is well inside the plastic regime. Therefore, our results suggest that plasticity is much more important than viscoelasticity in vibrated granular media. Although constant $r = 0.7$ gives good agreement with the experiments in Fig. 2, this will not be true in all situations. If the inter-particle collisions were much gentler, the effective coefficient of restitution would rise, and another choice of constant r would be needed. Furthermore, the pressure is just one property of the system. If we look at other property, we see that $r = 0.7$ does not give good results. An example is shown in Fig. 3, where we show three snapshots from three different simulations with $n = 3$ in absence of gravity. When $r = r(v)$, the particles are concentrated in the upper half of the chamber, but they are evenly spread in the horizontal direction (see Fig. 3a). The system is hotter and less dense near the vibrating wall, and colder and denser by the opposite wall. Such a loose cluster has been already observed in microgravity during parabolic flight experiments [40], where grains are agitated with a piston. In contrast, when $r = 0.95$, the particles are uniformly distributed throughout the domain (see Fig. 3b). This occurs because the particle move much faster than the piston, and thus can fill the space left by the descending piston. Finally, when $r = 0.7$, the majority of the particles are confined to a tight cluster, pressed against the upper wall, coexisting with low density region (see Fig. 3c). This instability has been already been reported numerically [41, 42], although for much different parameters (constant restitution coefficient $r = 0.96$, thermal walls, no gravity, and large n). However, nothing like this was ever seen experimentally. Therefore, if one is seeking information about particle positions, $r = 0.7$ gives incorrect results even though it gives acceptable results for the pressure. We conclude, therefore, that the only way to successfully describe all the properties in all situations is to use a velocity dependent restitution coefficient model.

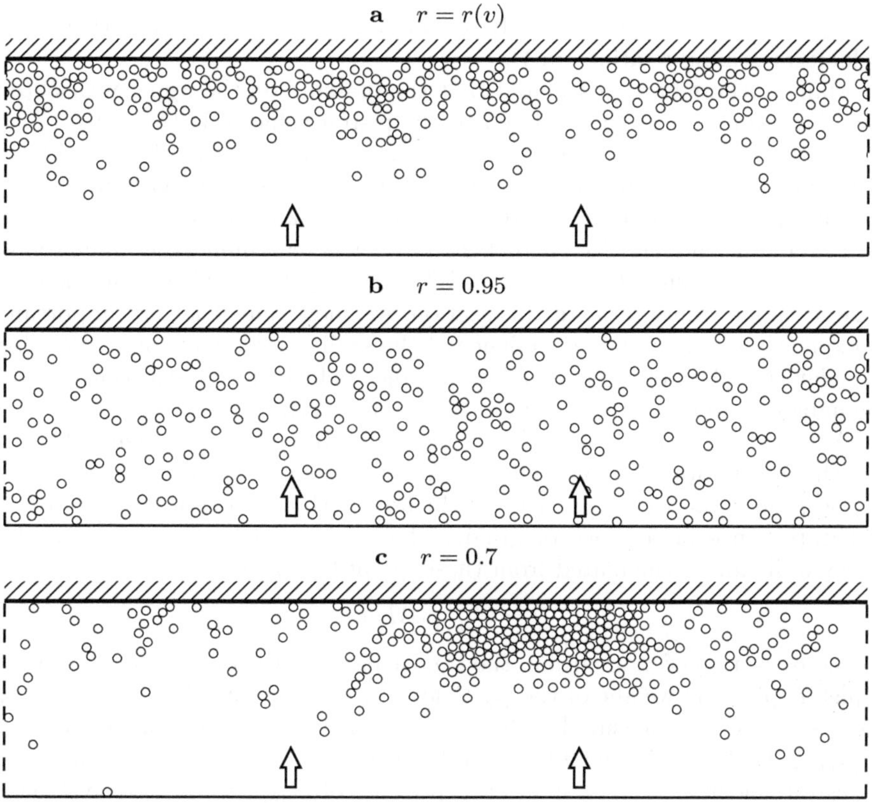

Fig. 3. Snapshots from the simulations with $n = 3$, gravity $g \neq 0$, driving frequency $f = 30$ Hz and $h - h_0 = 15$ mm. The upper wall is stationary, and the lower wall is the piston, which is at its lowest position in all three snapshots. The horizontal boundaries are periodic (indicated by dashed lines). Gravity points downwards. **a** $r = r(v)$, as given in (1), **b** constant $r = 0.95$, and **c** constant $r = 0.7$. In **c** we see a tight cluster which was not observed in the experiments

3.2 The Importance of the Particle Number

Observations suggest that the pressure (or granular temperature) of strongly vibrated granular medium is controlled by the piston vibration velocity V. It is important to point out that V is not the only way to characterize the vibration. One could also use the maximum piston acceleration Γ. When Γ is close to g, it controls the behavior of the system, i.e., adjusting A and f while keeping Γ constant does not change the system's behavior much. But in the simulations presented here, $\Gamma \gg g$, and the system's behavior is controlled by V. This can be checked by multiplying the frequency by 10 while dividing A by 10, thus keeping V the same (while Γ increases by an order of magnitude).

Doing so changes the pressure only by about 20%. It is reasonable, therefore, to look for scaling relations in V.

Many authors have postulated that the pressure on the upper wall P (or granular temperature T) is related to the piston velocity V through $P, T \propto V^\theta$. However, it is not clear what the correct "scaling exponent" θ should be. This question has been addressed several times in the past, without a clear resolution of the question [13–16]. For example, kinetic theory [6, 15] and hydrodynamic models [16] predict $T \propto V^2$ whereas numerical simulations [9–12] or experiments [6–9] give $T \propto V^\theta$, with $1 \leq \theta \leq 2$. These studies were done at single values of n. In this section, we show that it is very important to explicitly consider the dependence of the scaling exponents on n. We also consider the effect of gravity and a variable coefficient of restitution. Doing so enables us to explain and unify all previous works.

At the upper wall, we measured numerically the collision frequency, C_{upper} and the mean impulsion per collision, ΔI for various frequencies of the vibrating wall and numbers of particles in the box, with $r = r(v)$ or with $r = 0.95$, in the presence or absence of gravity. The time averaged pressure on the upper wall can be calculated from these quantities using

$$P = C_{\mathrm{upper}} \Delta I / L \; . \tag{2}$$

(By conservation of momentum, the time averaged pressure on the lower wall is just P plus the weight of the particles Nmg.) The total kinetic energy of the system is also measured to have access to the granular temperature, T. C_{upper}, ΔI, P, and T are all found to fit with power laws in V^θ for our range of piston velocities. Fig. 4 shows θ exponents of C_{upper}, ΔI, P, and T as a function of n. When $g = 0$ and r is constant (see Fig. 4a), we have $P \sim V^2$, $\Delta I \sim V$ and $C_{\mathrm{upper}} \sim V$ for all n. We call these relations the classical kinetic theory scaling. This scaling can be established by simple dimensional analysis when the vibration velocity V provides the only time scale in the system. This is the case for $g = 0$ and r independent of velocity. However, in the experiments, two additional time scales are provided, one by gravity and another by velocity dependent restitution coefficient. Numerical simulations can separate the effects of these two new time scales on the scaling exponents θ. This is done in Fig. 4b [where $g = 0$ but $r = r(v)$] and Fig. 4c [where r is constant but $g \neq 0$]. In both figures, all the exponents become functions of n. However, the time scale linked to $r = r(v)$ leads to much more dramatic departure from the classical scaling. After considering the two time scale separated, let us consider the case corresponding to most experiments, where both gravity and $r = r(v)$ are present (see Fig. 4d). The similarity between this figure and Fig. 4b confirms that the velocity dependent restitution coefficient has a more important effect than the gravity. Furthermore, only the variation of the restitution coefficient on the particle velocity explains the experiment performed in low-gravity [4]. This experiment gives a $V^{3/2}$ pressure scaling (•-mark on Fig. 4b) for $n = 1$ and a motionless clustered state for $n > 2$. Only the simulation with $r = r(v)$ can reproduce these results

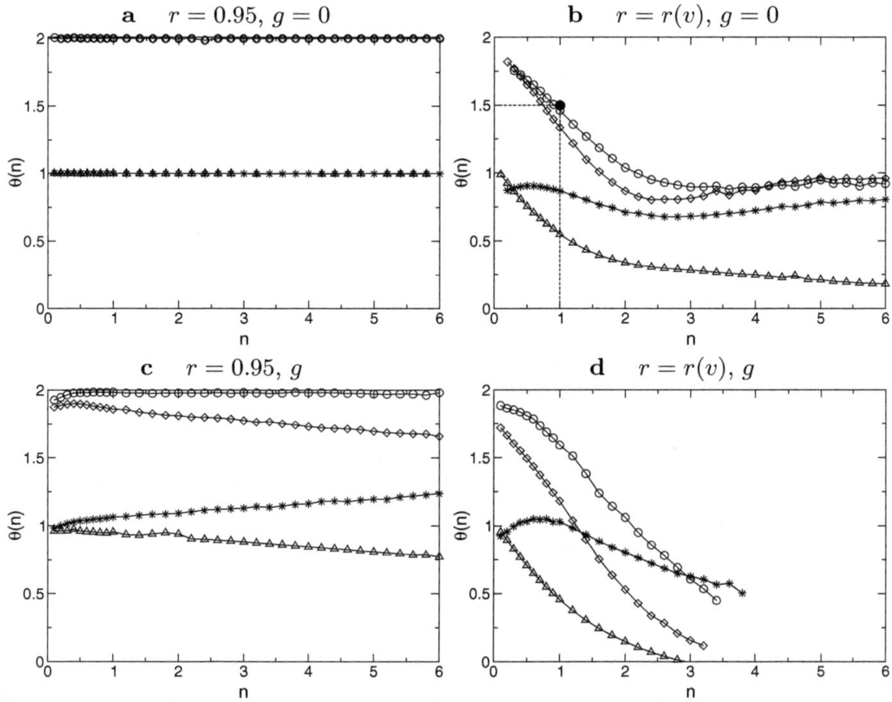

Fig. 4. The exponents θ as a function of n which give the scaling of the granular temperature T ($\Diamond$), collision frequency C_{upper} ($*$), mean impulsions ΔI ($\triangle$) and pressure P ($\circ$). All these quantities are proportional to $V^{\theta(n)}$. Without gravity: **a** for $r = 0.95$ and **b** for $r = r(v)$. With gravity: **c** for $r = 0.95$ and **d** for $r = r(v)$. The exponents are obtained by fixing n and performing eleven simulations, varying f from 10 Hz to 30 Hz. Then $\log X$ (where X is the quantity being considered) is plotted against $\log V$. The resulting curve is always nearly a straight line (except for $n > 3$ in **d** – see text), and the exponent is calculated from a least squares fit. The pressure scaling point ($\bullet$) on **b** is from experiment [4] performed in low-gravity. See Fig. 3a (resp. Fig. 3b) for typical snapshots corresponding to $n = 3$, $g \neq 0$ and $r = r(v)$ (resp. $r = 0.95$)

(see Fig. 4b) whereas constant r simulations leads to the classical scaling ($P \propto V^2$, see Fig. 4a) and only a gaseous state for all n shown in the figure.

As shown in Fig. 4, it is thus very important to explicitly consider the dependence of θ on n. In all cases, except the unrealistic case of Fig. 4a, θ depends on n. To our knowledge, the only experiment [1–3] to systematically investigate this effect shows that $T \propto V^{\theta(n)}$, with θ continuously varying from $\theta = 2$ when $n \to 0$, as expected from kinetic theory, to $\theta \simeq 0$ for large n due to the clustering instability. These experiments [1–3] performed under gravity (shown in Fig. 5) are well reproduced by the simulations of Fig. 4d. In

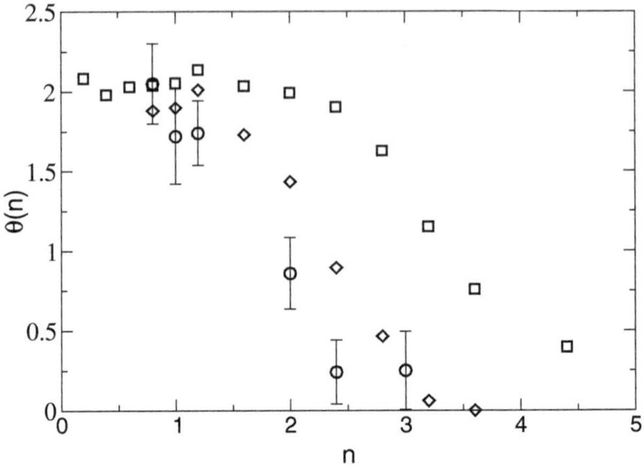

Fig. 5. Experimental data performed under gravity from [1–3]: The exponents $\theta(n)$ of time averaged pressure (□) (see Fig. 2a), and kinetic energy extracted from density profile (○) or volume expansion (◇) measurements. These data should be compared with the simulations of Fig. 4d

both cases, the observed pressure and granular temperature scaling exponents strongly decrease with n.

We finish this section by noting two curious facts about Fig. 4. First of all, in Fig. 4b [$g = 0$ and $r = r(v)$], $\theta \approx 1$ for the pressure and temperature when $n > 2$. This is the sign of new robust scaling regime where $P, T \propto V$, which is the subject of Sect. 4.1. Secondly, in Fig. 4d [$g \neq 0$ and $r = r(v)$], the scaling exponents are not shown for $n \geq 3$, because the dependence of P, T C_{upper} and ΔI on V is no longer a simple power law. (More precisely, we do not plot a point on Fig. 4 when $|\log X_{\text{observed}} - \log X_{\text{fitted}}| \geq 0.25$ for any of the eleven simulations used to calculate the exponent – see caption.) The power law breaks down because there is a resonance between the time of flight of the cluster under gravity and the vibration period. This is the subject of Sect. 4.2.

4 Effects of Clustering ($n > 2$ or 3)

4.1 Clustering without Gravity

In this section, we examine the situation concerning realistic particles [$r = r(v)$] in microgravity ($g = 0$). At high enough density, a loose cluster is formed that resembles Fig. 3a (even though $g \neq 0$ in Fig. 3a), and we noted that $P \propto V$ for $n > 2$ (see Fig. 4b). We would like to know if there is some simple physical explanation for this aparently robust scaling relation. To investigate this question, we present in Fig. 6 the dependence of P on V, n and $h - h_0$. Fig. 6a shows the same quantities as Fig. 2, except that the scale of the x-axis

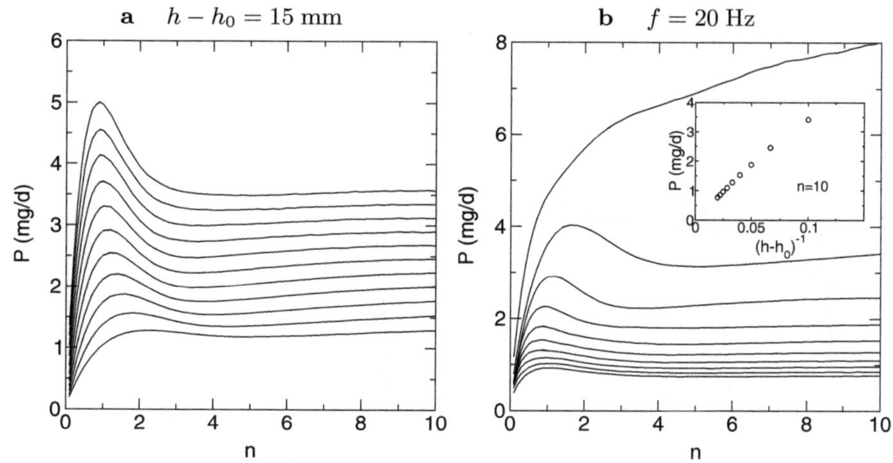

Fig. 6. Time averaged pressure P on the top of the cell as a function of particle layer, n: **a** for various vibration frequencies: $f = 10$ Hz (lowest curve) to $f = 30$ Hz (uppermost curve) with steps of 2 Hz; **b** for various heights: $h - h_0 = 5$ mm (uppermost curve) to $h - h_0 = 50$ mm (lowest curve) with steps of 5 mm. Inset: P vs. $(h - h_0)^{-1}$ for $n = 10$ and 10 cm $\leq h - h_0 \leq 50$ cm. Simulations were done with $g = 0$ and $r = r(v)$

has changed: instead of $0 < n < 4.5$, we show $0 < n < 10$. Note that Fig. 6a concerns the same situation as in Fig. 2b, except here the gravity has been "turned off". Thus it is not surprising that for $n < 3$, the two figures are almost the same: after a rapid increase of the pressure for small n, there is a maximum pressure near $n \approx 1$. But for $n > 3$, the figures are different: in Fig. 2b, the pressure decreases as particles are added, but in Fig. 6a the pressure is nearly independent of n. Therefore, when $g = 0$, the pressure is not changed by adding more particles, as long as there are already enough particles present. Considering now the dependence of P on V, we note that the evenly spaced curves in Fig. 6a suggest that $P \propto V$. We have confirmed this proportionality by calculating the pressure scaling exponent and checking that it remains close to 1. (The exponent for $n < 6$ has already been shown in Fig. 4b.) In Fig. 6b, we examine how changing $h - h_0$ affects the pressure. We see that the pressure is independent of n for $n > 3$ and $h - h_0 \geq 10$, with the pressure decreasing as the height increases. Examining the dependence of the pressure on $h - h_0$ shows that $P \propto (h - h_0)^{-1}$. (For example, see the inset in Fig. 6b.) Thus, the pressure obeys the simple non-extensive relation

$$P \propto \frac{N^0 V^1}{h - h_0} \tag{3}$$

Recall that this relation concerns high enough number of realistic particles in microgravity. Therefore, this pressure scaling of (3) may be observable in

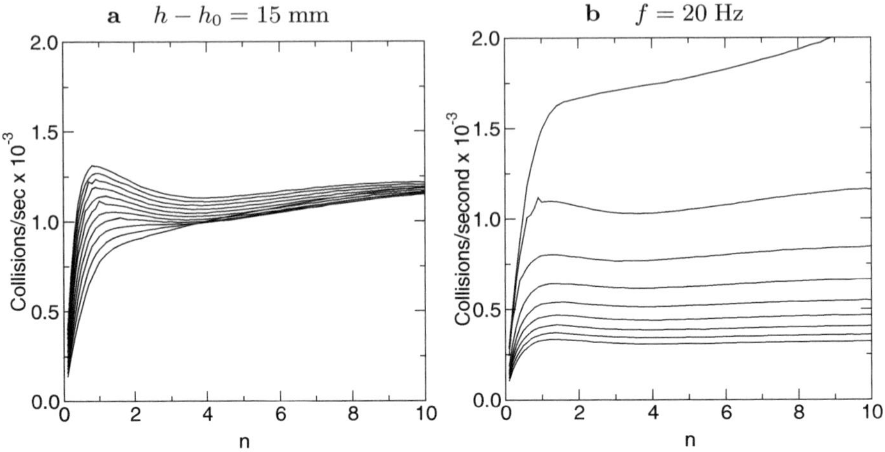

Fig. 7. The particle-piston collision rate C_{lower} as a function of particle layer, n: **a** for various vibration frequencies: $f = 10$ Hz (*lowest curve*) to $f = 30$ Hz (*uppermost curve*) with steps of 2 Hz; **b** for various heights: $h - h_0 = 10$ mm (*uppermost curve*) to $h - h_0 = 50$ mm (*lowest curve*) with steps of 5 mm. Simulations are done with $g = 0$ and $r = r(v)$

microgravity experiments. One complication is that in microgravity experiments, one does not usually agitate the grains with a piston (except very recently [40]) – it is much simpler to shake a box of fixed size [4].

In Fig. 3a, we observe that the majority of particles remain in a loose cluster pushed against the stationary wall, opposite the piston. Only those particles that "evaporate" from the cluster are struck by the piston. The evaporation rate can be estimated from the rate of collisions between the piston and the particles C_{lower}. This collision rate has a very curious behavior, as shown in Fig. 7. As can be seen from Fig. 7a, C_{lower} is roughly independent of n and V when $n > 3$. This behavior hold for other values of $h - h_0$, as can be seen from Fig. 7b. Thus, at high enough density, the collision frequency on the vibrating wall is found to be

$$C_{\text{lower}} \propto \frac{N^0 V^0}{h - h_0}. \tag{4}$$

We now present a simple model and a physical interpretation of the unexpected scaling of (3) for such a loose cluster. We suppose that the cluster against the stationary wall has a granular temperature independent of both the number of particles N and the vibration velocity V. The granular temperature is independent of N and V because the impact velocity dependent restitution coefficient acts as a kind of "granular thermostat". As one can see from (1) or Fig. 1, when the typical particle velocity v is greater than v_0, $r(v)$ is small, and energy is dissipated rapidly. On the other hand, when $v < v_0$, $r(v)$ is nearly unity and energy dissipation is slow. Thus, in a weakly

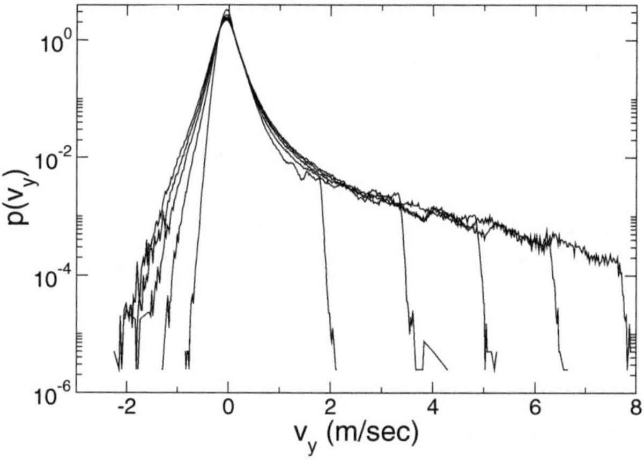

Fig. 8. The probability density function of the vertical velocities at high density, for $n = 5$, $r = r(v)$, and $g = 0$. The driving frequencies are $f = 10$ Hz (narrowest distribution) to $f = 50$ Hz (broadest distribution) with steps of 10 Hz

forced cluster, like the one found at the stationary wall, $T \sim v^2$ will be near v_0^2, independent of the strength of the forcing and the size of the cluster.

Since the granular temperature of the cluster is independent of N and V, so is the flux F_{evap} of grains "evaporating" from the cluster and moving towards the piston. If all the grains that evaporate from the cluster reach the piston, then we have $C_{\text{lower}} = F_{\text{evap}} \propto N^0 V^0$. In their collision with the piston, the grains acquire an upwards velocity proportional to V. Thus the flux of momentum entering the system at the piston is $VC_{\text{lower}} \propto N^0 V^1$. Since momentum is conserved, the flux of momentum leaving the system through the stationary wall must have the same value. But the time averaged pressure P is just the time averaged momentum flux divided by the area of the upper wall. Thus our model predicts $P \propto N^0 V^1$, in agreement with (3).

This theory also explains why P and C_{lower} are inversely proportional to $h - h_0$. When a particle evaporates from the cluster, it must travel a certain distance before it encouters the piston. This distance increases with $h - h_0$ and thus the particle's travel time also increases. During its voyage, the evaporated particle could be struck by another particle that has just encountered the piston. If this happens, both particles are scattered back into the cluster. Thus the evaporated particle never reaches the piston. If we assume that the probability of an evaporated particle being scattered back into the cluster is independent of time, the number of particles reaching the piston is inversely proportional to $h - h_0$.

This explanation can be checked by looking at the probability density of vertical (y) velocities parallel to the direction of vibration. In Fig. 8, we show these density functions. Note the assymmetry of the distibution's positive and negative wings. The edge of the positive wing increases linearly with V

whereas the negative wing extends much more slowly. The reason for this is that the positive wing is populated by particles which have just been struck by the piston. Their velocity is thus proportional to V. On the other hand, the negative wing is due to fluctuations in the cluster. The negative wing thus expands slowly because of the "thermostat" effect described above.

We conclude by emphasizing that this effect can only be seen with the velocity dependent restitution coefficient. If one uses a constant restitution coefficient, one obtains $P \propto V^2$, no matter the value of r.

4.2 Clustering with Gravity

In this section, we consider the situation that applies to most of the experiments, i.e. realistic particles [$r = r(v)$] under gravity. We will show that the relatively simple situation described in the previous section is complicated by the presence of gravity. Under gravity, the loose cluster will not remain near the upper wall (unless f is large), but will fall downwards towards the piston. At certain values of f, $h - h_0$ and large enough n, a resonance occurs between the driving frequency and the time of flight of the particles, leading to a cluster that bounces back and forth between the piston and the wall. As f increases, however, we recover the situation found in the previous section.

In Fig. 9, we show two pictures from the experiments of [1–3]. At low densities, the particles move like atoms in a gas (Fig. 9a). At higher densities, a cluster forms that bounces up and down on the piston (Fig. 9b). This cluster

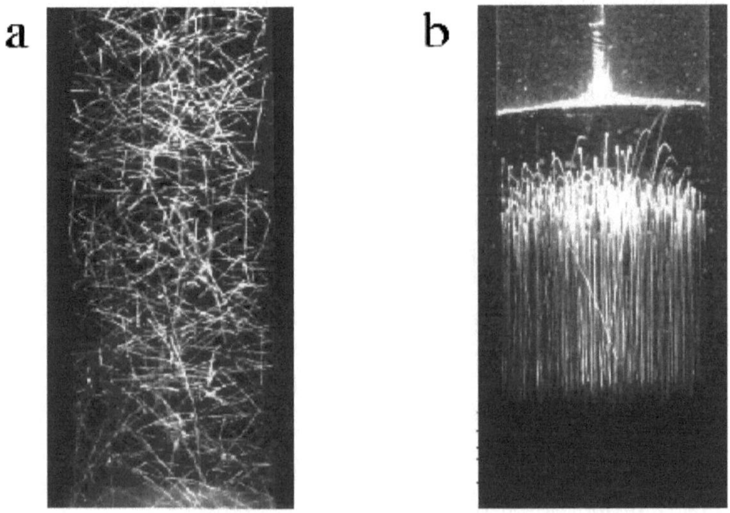

Fig. 9. Photographs of particle trajectories from the experiments of [1–3], showing the cluster at high densities. In both pictures f = 20 Hz and A = 40 mm. The number of layers of particles is **a** $n = 1$ **b** $n = 4$. The driving piston is at the bottom (not visible), the inner diameter of the tube being 52 mm

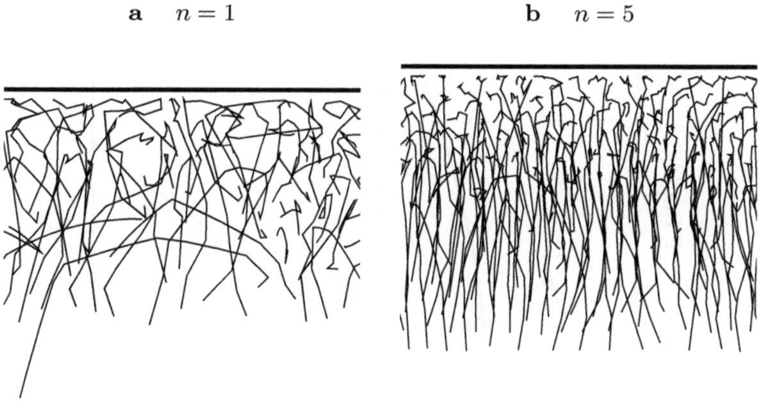

Fig. 10. Particle trajectories from the simulations with $f = 20$ Hz, $A = 25$ mm, $r = r(v)$, and $h - h_0 = 15$ mm. The number of layers of particles is **a** $n = 1$ and **b** $n = 5$. The containers have different sizes, because the size of the container is increased as more particles are added, as was done in the experiments

was observed only for certain frequencies, densities and container heights. It was not possible to establish a criterion for the existence of the cluster based on the experimental data. Both of these behaviors are reproduced in the simulations (Fig. 10). Therefore, the simulations should give insight into the origin of the bouncing cluster.

We first examine the relation between P and V in a case where the bouncing cluster is observed. One does not obtain the simple scaling in (3) for the pressure. Instead, one obtains a much more complicated relation between P and V. Indeed, as we show in Fig. 11, the pressure displays two maxima as a function of V. These maxima correspond to the bouncing cluster shown in Figs 9b and 10b. The bouncing cluster transports momentum very effectively between the piston and the wall. When the bouncing cluster disappears at $n < 3$, the maxima also disappear. These maxima make it impossible to describe the relation between the pressure and the forcing by a simple power law. This is the reason why there are no points on Fig. 4d for $n > 3$.

Unlike the other results of this paper, Fig. 11 changes significantly when L and N are changed while holding n fixed. When L and N are both divided by two, the first maximum moves to $V \approx 1$ m/sec and the second maximum disappears. When L and N are doubled, the second maximum moves to near $V \approx 1.4$ m/sec. Therefore, the results of this section should not be considered as a systematic investigation of the resonance, but rather a descriptive introduction. The resonance is sensitive to L because it can involve horizontal collective motions of the particles, not just vertical ones. Indeed, for $L = 40$ cm, one can observe waves similar to those discussed in [29].

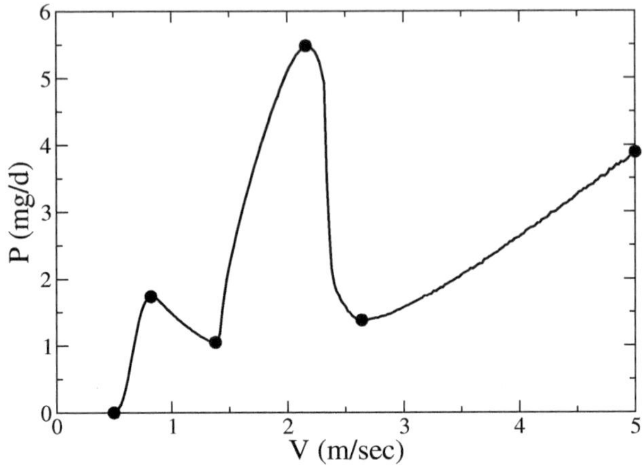

Fig. 11. Pressure as a function of the piston velocity, $V = 4Af$, for $n = 6$, $r = r(v)$, $h - h_0 = 15$ mm under gravity. Simulations where done for $f = 5$ Hz to $f = 50$ Hz with steps of 0.2 Hz. The dots mark the particular frequencies shown in Fig. 12

To investigate the origins of the bouncing cluster, we show in Fig. 12 the position of the piston and the approximate location of the cluster as a function of time for various driving frequencies. The frequencies are chosen to correspond to the maxima and minima of the pressure, and are indicated by the large points in Fig. 11. At very low frequencies (Fig. 12a), the pressure vanishes because no particles hit the top of the container. The cluster bounces on the vibrating plate. As the piston velocity increases, the cluster flies higher and higher until it strikes the top. The pressure reaches a first maximum near $V = 0.82$ m/s (Fig. 11) where there is a first resonance between the time of flight of the cluster and the vibration period. These two times are such that the cluster lands on the piston just when the piston is at its maximum upwards velocity (Fig. 12b). Note that a large number of particles descend below the piston's maximum height. Then as the piston rises, it sweeps up all the particles, increasing the density of the cluster. On the other hand, as the cluster falls from the stationary wall, it slowly expands.

As the piston velocity increases further, the pressure decreases, because the cluster lands on the piston when it is near its maximum height (Fig. 12c). Then the pressure rises to a new maximum near $V = 4Af = 2.2$ m/s (Fig. 11) because there is a second resonance: the cluster's flight time is equal to two piston cycles (Fig. 12d). Note that the frequencies of these two resonances ($f = 8.2$ Hz and $f = 21.6$ Hz) are not related by a simple factor of 2. The reason for this is that the time of flight of the cluster depends in a complicated way on the piston velocity. Comparing Figs. 12b and 12d, one can see that the velocity of the cluster when it leaves the piston is close to V, but the time it takes the cluster to fall back from the upper wall onto the vibrating is roughly independent of V. Note also that the collisions between the cluster

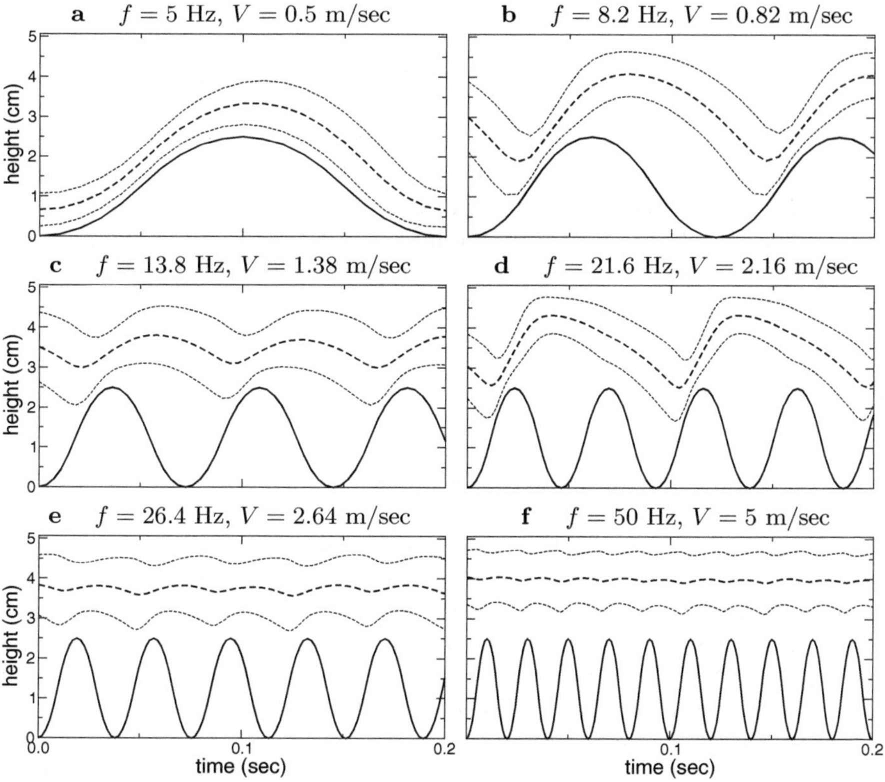

Fig. 12. The motion of the grains under gravity at various vibration frequencies during 0.2 seconds, for $n = 6$, $h - h_0 = 25$ mm and $r = r(v)$. In all plots, the vertical axis is height, with the lower boundary at the piston's lowest point, and the upper boundary being the upper wall of the container. Solid line shows the position of the piston. The *heavy dotted line* shows the height $y_{\rm cm}$ of the center of mass. The *upper and lower thin dotted lines* show $y_{\rm cm} + \sigma_{\rm cm}$ and $y_{\rm cm} - \sigma_{\rm cm}$ respectively, where $\sigma_{\rm cm}$ is the standard deviation of the particle heights at a given instant. The majority of the particles are thus contained between the two *thin dotted lines*

and the piston or wall takes a finite time, as can be seen by comparing the times of the minima of $y_{\rm cm} + \sigma_{\rm cm}$, $y_{\rm cm}$, and $y_{\rm cm} - \sigma_{\rm cm}$ in Fig. 12b. All of these effects exclude a simple resonance condition for the pressure maxima. The resonant frequencies depend on g and $h - h_0$. The resonant frequencies are proportional to $g^{1/2}$ but decrease for as $h - h_0$ is increased. The resonant frequencies are not sensitive to changing n, as long as n is sufficiently large.

At higher driving frequencies (Fig. 12e and Fig. 12f), the center of mass of the particles changes very little over one cycle, and there are no more resonances. In fact, setting $g = 0$ changes Fig. 12f very little. The reason for this is that average particle kinetic energy is much larger than gh, so gravity is irrelevant. Higher order resonances, (where the cluster's flight time equals three or more vibration periods) are not observed. These resonances are probably not observed because there is not enough time for the particles to fall below the piston's highest point. Thus, the piston does not gather up all the particles into a single dense cluster, but rather acts to disperse the cluster. For example, imagine that the driving frequency is suddenly doubled in Fig. 12d just at the moment when the cluster starts its descent from the wall. When the cluster arrives at the piston, perhaps a half of the particles will be swept up by one vibration cycle, and a half in the next cycle. This is how energetic forcing breaks up clusters.

5 Conclusions

In Sect. 3, we brought simulations as close as possible to the experiments. To bring simulations into agreement with the experiments, it is essential to use a velocity dependent coefficient of restitution. It is especially important to take into account plastic deformations that cause the restitution coefficient to decrease rapidly with increasing impact velocity. Indeed, the restitution coefficient for strongly vibrated steel spheres is very far from the constant values of $r = 0.95$ or $r = 0.9$ that are often cited in simulations as typical for steel spheres. We also noted that it is very important to take into account the number of particle layers n. The dependence of the pressure P on the piston velocity V changes with n. It is not accurate to speak of “a” scaling exponent for the pressure in terms of V: this exponent depends continuously on n, and does not exist at high density ($n > 3$) under gravity, due to the clustering instability.

In Sect. 4, we investigated this question of clustering. When there is no gravity, a loose cluster forms against the stationary wall opposite the vibrating piston. Due to the velocity dependent restitution coefficient, this cluster acts as a buffer for fastest particles, and leads to a simple non-extensive scaling $P \propto N^0 V$. Unlike the case with gravity, the scaling holds for a wide range of parameters. We discussed the origin the scaling, and remarked that it should be observable in microgravity experiments. We also studied the effect of clustering in the presence of gravity. In this case, the clusters give a very complicated behavior, because there are interactions between the period of vibration and the flight time of the cluster. Finally, if the vibration frequency is strong enough, the cluster can be broken up, because the cluster-piston collision time becomes longer than vibration period.

Acknowledgments

We thank Stéphan Fauve for fruitful discussions. The authors gratefully acknowledge the hospitality of the ENS-Lyon physics department which made this collaboration possible.

References

1. E. Falcon, S. Fauve, and C. Laroche. Experimental Study of Granular Gas Fluidized by Vibrations. In Pöschel and Luding [5], page 244
2. E. Falcon, S. Fauve, and C. Laroche. Cluster formation, pressure and density measurements in a granular medium fluidized by vibrations. *Eur. Phys. J. B.*, 9:183, 1999
3. E. Falcon, C. Laroche, and S. Fauve. Experimental determination of a state equation for dissipative granular gases. *J. Chim. Phys.*, 96:1111, 1999
4. E. Falcon, R. Wunenburger, P. Evesque, S. Fauve, C. Chabot, Y. Garrabos, and D. Beysens. Cluster formation in a granular medium fluidized by vibrations in low gravity. *Phys. Rev. Lett.*, 83:440, 1999
5. T. Pöschel and S. Luding, editors. *Granular Gases*, volume 564 of *Lectures Notes in Physics*. Springer-Verlag, Berlin, 2001
6. S. Warr, J.M. Huntley, and G.T.H. Jacques. Fluidization of a two-dimensional granular system: Experimental study and scaling behavior. *Phys. Rev. E*, 52:5583, 1995
7. X. Yang and D. Candela. Potential energy in a three-dimensional vibrated granular medium measured by NMR imaging. *Phys. Rev. Lett.*, 85:298, 2000
8. R.D. Wildman, J.M. Huntley, and D.J. Parker. Granular temperature profiles in three-dimensional vibrofluidized granular beds. *Phys. Rev. E*, 63:061311, 2001
9. S. Luding, E. Clément, A. Blumen, J. Rajchenbach, and J. Duran. Studies of columns of beads under external vibrations. *Phys. Rev. E*, 49:1634, 1994
10. S. Luding, H.J. Herrmann, and A. Blumen. Scaling behavior of 2-dimensional arrays of beads under external vibrations. *Phys. Rev. E*, 50:3100, 1994
11. S. Luding. Granular materials under vibration: Simulations of rotating spheres. *Phys. Rev. E*, 52:4442, 1995
12. H.J. Herrmann and S. Luding. Modeling granular media with the computer. *Continuum Mech. and Thermodyn.*, 10:189, 1998
13. S. McNamara and S. Luding. Energy flows in vibrated granular media. *Phys. Rev. E*, 58:813, 1998
14. J.M. Huntley. Scaling laws for a two-dimensional vibro-fluidized granular material. *Phys. Rev. E*, 58:5168, 1998
15. V. Kumaran. Temperature of a granular material "fluidized" by external vibrations. *Phys. Rev. E*, 57:5660, 1998
16. J. Lee. Scaling behavior of granular particles in a vibrating box. *Physica A*, 219:305, 1995
17. C.V. Raman. The photographic study of impact at minimal velocities. *Phys. Rev.*, 12:442, 1918
18. D. Tabor. A simple theory of static and dynamic hardness. *Proc. Roy. Soc. A*, 192:247, 1948
19. W. Goldsmith. *Impact: The theory and physical behavior of colliding solids.* Edward Arnold, London, 1960
20. K.L. Johnson. *Contact Mechanics.* Cambridge Univ. Press, Cambridge, 1985
21. L. Labous, A.D. Rosato, and R. N. Dave. Measurement of sphere collision properties. *Phys. Rev. E*, 56:5717, 1997
22. G. Kuwabara and K. Kono. Restitution coefficient in a collision between two spheres. *Jap. J. of Appl. Phys.*, 26:1230, 1987

23. E. Falcon, C. Laroche, S. Fauve, and C. Coste. Behavior of one inelastic ball bouncing repeatedly off the ground. *Eur. Phys. J. B*, 3:45, 1998. See also references therein
24. J.-M. Hertzsch, F. Spahn, and N.V. Brilliantov. On low-velocity collisions of viscoelastic particles. *J. Phys. II (Paris)*, 5:1725, 1995
25. J. Schäfer, S. Dippel, and D.E. Wolf. Force schemes in simulations of granular materials. *J. Phys. I (Paris)*, 6:5, 1996
26. S. Luding, E. Clément, A. Blumen, J. Rajchenbach, and J. Duran. Anomalous energy dissipation in molecular dynamics simulations of grains: The "detachment effect". *Phys. Rev. E*, 50:4113, 1994
27. N.V. Brilliantov and T. Pöschel. Velocity distribution in granular gases of viscoelastic particles. *Phys. Rev. E*, 61:5573, 2000
28. C. Salueña, T. Pöschel, and S.E. Esipov. Dissipative properties of vibrated granular materials. *Phys. Rev. E*, 59:4422, 1999
29. C. Bizon, M.D. Shattuck, J.B. Swift, W.D. McCormick, and H.L. Swinney. Patterns in 3d vertically oscillated granular layers: Simulation and experiment. *Phys. Rev. Lett.*, 80:57, 1998
30. T. Pöschel and N.V. Brilliantov. Extremal collision sequences of particles on a line. *Phys. Rev. E*, 63:021505, 2001
31. D. Goldman, M.D. Shattuck, C. Bizon, W.D. McCormick, J.B. Swift, and H.L. Swinney. Absence of inelastic collapse in a realistic three ball model. *Phys. Rev. E*, 57:4831, 1998
32. H. Salo, J. Lukkari, and J. Hänninen. Velocity dependent coefficient of restitution and the evolution of collisional systems. *Earth, Moon, and Planets*, 43:33, 1988
33. F. Spahn, U. Schwarz, and J. Kurths. Clustering of granular assemblies with temperature dependent restitution under Keplerian differential rotation. *Phys. Rev. Lett.*, 78:1596, 1997
34. H. Salo. Numerical simulations of the collisional dynamics of planetary rings. In Pöschel and Luding [5], page 330
35. F.G. Bridges, A. Hatzes, and D.N.C. Lin. Structure, stability and evolution of Saturn's rings. *Nature*, 309:333, 1984
36. A. Hatzes, F.G. Bridges, and D.N.C. Lin. Collisional properties of ice spheres at low impact velocities. *Mont. Not. R. Astr. Soc.*, 231:1091, 1988
37. K.D. Supulver, F.G. Bridges, and D.N.C. Lin. The coefficient of restitution of ice particles in glancing collisions: Experimental results for unfrosted surfaces. *Icarus*, 113:188, 1995
38. M. Higa, M. Arakawa, and N. Maeno. Measurements of restitution coefficients of ice at low temperatures. *Planet. Space Sci.*, 44:917, 1996
39. J.M. Lifshitz and H. Kolsky. Some experiments on anelastic rebound. *J. Mech. Phys. Solids*, 12:35, 1964
40. E. Falcon, S. Fauve, P. Evesque, F. Palencia, C. Chabot, Y. Garrabos, and D. Beysens. Collision statistics in a dilute granular gas fluidized by vibration in low gravity, 2003. Unpublished
41. M. Argentina, M.G. Clerc, and R. Soto. van der Waals-like transition in fluidized granular matter. *Phys. Rev. Lett.*, 89:044301, 2002
42. R. Soto, M. Argentina, and M.G. Clerc. Van der Waals-like transition in fluidized granular matter. In T. Pöschel and N.V. Brilliantov, editors, *Granular Gas Dynamics*, volume 624 of *Lecture Notes in Physics*, page 317, Berlin, Heidelberg, New York, 2003. Springer

Lecture Notes in Physics

For information about Vols. 1–586
please contact your bookseller or Springer-Verlag
LNP Online archive: springerlink.com

Vol.587: D. Heiss (Ed.), Fundamentals of Quantum Information. Quantum Computation, Communication, Decoherence and All That.

Vol.588: Y. Watanabe, S. Heun, G. Salviati, N. Yamamoto (Eds.), Nanoscale Spectroscopy and Its Applications to Semiconductor Research.

Vol.589: A. W. Guthmann, M. Georganopoulos, A. Marcowith, K. Manolakou (Eds.), Relativistic Flows in Astrophysics.

Vol.590: D. Benest, C. Froeschlé (Eds.), Singularities in Gravitational Systems. Applications to Chaotic Transport in the Solar System.

Vol.591: M. Beyer (Ed.), CP Violation in Particle, Nuclear and Astrophysics.

Vol.592: S. Cotsakis, L. Papantonopoulos (Eds.), Cosmological Crossroads. An Advanced Course in Mathematical, Physical and String Cosmology.

Vol.593: D. Shi, B. Aktaş, L. Pust, F. Mikhailov (Eds.), Nanostructured Magnetic Materials and Their Applications.

Vol.594: S. Odenbach (Ed.),Ferrofluids. Magnetical Controllable Fluids and Their Applications.

Vol.595: C. Berthier, L. P. Lévy, G. Martinez (Eds.), High Magnetic Fields. Applications in Condensed Matter Physics and Spectroscopy.

Vol.596: F. Scheck, H. Upmeier, W. Werner (Eds.), Noncommutative Geometry and the Standard Model of Elememtary Particle Physics.

Vol.597: P. Garbaczewski, R. Olkiewicz (Eds.), Dynamics of Dissipation.

Vol.598: K. Weiler (Ed.), Supernovae and Gamma-Ray Bursters.

Vol.599: J.P. Rozelot (Ed.), The Sun's Surface and Subsurface. Investigating Shape and Irradiance.

Vol.600: K. Mecke, D. Stoyan (Eds.), Morphology of Condensed Matter. Physcis and Geometry of Spatial Complex Systems.

Vol.601: F. Mezei, C. Pappas, T. Gutberlet (Eds.), Neutron Spin Echo Spectroscopy. Basics, Trends and Applications.

Vol.602: T. Dauxois, S. Ruffo, E. Arimondo (Eds.), Dynamics and Thermodynamics of Systems with Long Range Interactions.

Vol.603: C. Noce, A. Vecchione, M. Cuoco, A. Romano (Eds.), Ruthenate and Rutheno-Cuprate Materials. Superconductivity, Magnetism and Quantum Phase.

Vol.604: J. Frauendiener, H. Friedrich (Eds.), The Conformal Structure of Space-Time: Geometry, Analysis, Numerics.

Vol.605: G. Ciccotti, M. Mareschal, P. Nielaba (Eds.), Bridging Time Scales: Molecular Simulations for the Next Decade.

Vol.606: J.-U. Sommer, G. Reiter (Eds.), Polymer Crystallization. Obervations, Concepts and Interpretations.

Vol.607: R. Guzzi (Ed.), Exploring the Atmosphere by Remote Sensing Techniques.

Vol.608: F. Courbin, D. Minniti (Eds.), Gravitational Lensing:An Astrophysical Tool.

Vol.609: T. Henning (Ed.), Astromineralogy.

Vol.610: M. Ristig, K. Gernoth (Eds.), Particle Scattering, X-Ray Diffraction, and Microstructure of Solids and Liquids.

Vol.611: A. Buchleitner, K. Hornberger (Eds.), Coherent Evolution in Noisy Environments.

Vol.612: L. Klein, (Ed.), Energy Conversion and Particle Acceleration in the Solar Corona.

Vol.613: K. Porsezian, V.C. Kuriakose, (Eds.), Optical Solitons. Theoretical and Experimental Challenges.

Vol.614: E. Falgarone, T. Passot (Eds.), Turbulence and Magnetic Fields in Astrophysics.

Vol.615: J. Büchner, C.T. Dum, M. Scholer (Eds.), Space Plasma Simulation.

Vol.616: J. Trampetic, J. Wess (Eds.), Particle Physics in the New Millenium.

Vol.617: L. Fernández-Jambrina, L. M. González-Romero (Eds.), Current Trends in Relativistic Astrophysics, Theoretical, Numerical, Observational

Vol.618: M.D. Esposti, S. Graffi (Eds.), The Mathematical Aspects of Quantum Maps

Vol.619: H.M. Antia, A. Bhatnagar, P. Ulmschneider (Eds.), Lectures on Solar Physics

Vol.620: C. Fiolhais, F. Nogueira, M. Marques (Eds.), A Primer in Density Functional Theory

Vol.621: G. Rangarajan, M. Ding (Eds.), Processes with Long-Range Correlations

Vol.622: F. Benatti, R. Floreanini (Eds.), Irreversible Quantum Dynamics

Vol.623: M. Falcke, D. Malchow (Eds.), Understanding Calcium Dynamics, Experiments and Theory

Vol.624: T. Pöschel (Ed.), Granular Gases

Vol.625: R. Pastor-Satorras, M. Rubi, A. Diaz-Guilera (Eds.), Statistical Mechanics of Complex Networks

Vol.626: G. Contopoulos, N. Voglis (Eds.), Galaxies and Chaos

Vol.627: S.G. Karshenboim, V.B. Smirnov (Eds.), Precision Physics of Simple Atomic Systems

Vol.628: R. Narayanan, D. Schwabe (Eds.), Interfacial Fluid Dynamics and Transport Processes

Vol.629: F. Sylos Labini, A. Gabrielli, L. Pietronero, Statistical Physics for Cosmic Structures

Vol.630: T. Brandes, S. Kettemann (Eds.), Anderson Localization and Its Ramifications

Vol.631: D. J. W. Giulini, C. Kiefer, C. Lämmerzahl (Eds.), Quantum Gravity, From Theory to Experimental Search

Springer Nature Customer Service Center GmbH
Europaplatz 3, 69115 Heidelberg, Germany
Printed by [illegible] GmbH
[illegible], Germany

MIX
Papier aus verantwortungsvollen Quellen
Paper from responsible sources
FSC® C105338

If you have any concerns about our products,
you can contact us on
ProductSafety@springernature.com

In case Publisher is established outside the EU,
the EU authorized representative is:
Springer Nature Customer Service Center GmbH
Europaplatz 3, 69115 Heidelberg, Germany

Printed by Libri Plureos GmbH
in Hamburg, Germany